MEASUREMENTS
IN HEAT TRANSFER

MEASUREMENTS IN HEAT TRANSFER

————————— S E C O N D   E D I T I O N —————————

EDITED BY

Ernst R. G. Eckert
Richard J. Goldstein

both of
School of Mechanical and Aerospace Engineering
University of Minnesota

HEMISPHERE PUBLISHING CORPORATION

Washington   London

McGRAW–HILL BOOK COMPANY

New York   St. Louis   San Francisco   Auckland   Düsseldorf
Johannesburg   Kuala Lumpur   London   Mexico   Montreal
New Delhi   Panama   Paris   São Paulo   Singapore
Sydney   Tokyo   Toronto

This book is a revised version of a study which was originally published in November 1970 as AGARDograph No. 130 by the Advisory Group for Aerospace Research and Development of the North Atlantic Treaty Organization (AGARD/NATO), 7 rue Ancelle, 9200 Neuilly sur Seine, France. The original material is reprinted by permission of AGARD.

MEASUREMENTS IN HEAT TRANSFER

1 2 3 4 5 6 7 8 9 0   K P K P   7 8 3 2 1 0 9 8 7 6

This book was set in Press Roman by Hemisphere Publishing Corporation. The editors were Evelyn Walters Pettit, Martha M. Mahuran, and Editech; the designer was Lilia N. Guerrero; the production supervisor was Rebekah McKinney; and the compositors were Bernie Doenhoefer and Pat Hopper.
The printer and binder was The Kingsport Press, Inc.

Library of Congress Cataloging in Publication Data

Main entry under title:

Measurements in heat transfer.

  (Series in thermal and fluids engineering)
  "Revised version of a study originally published in November 1970 [under title: Measurement techniques in heat transfer] as AGARDograph no. 130."
  Includes indexes.
    1. Heat−Transmission−Measurement.   2. Thermometers and thermometry.   I. Eckert, Ernst R. G., date.
II. Goldstein, Richard J.
QC320.M4   1976   536'.2'0028   75-38514
ISBN 0-07-018926-9

CONTENTS

PREFACE

Textbooks on heat transfer usually concentrate on the analysis of heat transfer processes and either devote only a small amount of space to the discussion of measurement techniques or exclude this subject completely. But because special problems are encountered in heat transfer measurement, knowledge and skill are required to achieve accurate results. This book focuses on methods of, difficulties in, and instruments for measuring heat transfer.

Temperature measurements in a heat transfer situation should not be considered without a check on possible systematic errors caused by conduction, radiation, or, in an unsteady situation, heat capacity effects. Chapter 1 discusses these errors and the means to calculate them. Chapter 2 discusses resistance thermometers, and Chap. 3 describes thermocouples, including their calibration and use for temperature measurements. Heat transfer processes under cryogenic conditions require special techniques for temperature measurements, which are discussed in Chap. 4.

Optical techniques have the advantage of not disturbing the temperature field in which the measurements are to be made. Optical systems based on variations in index of refraction are reviewed in Chap. 5. At high temperatures, spectroscopy can be used as a diagnostic tool; this technique is discussed in Chap. 6. A very useful tool for measurement of the enthalpy in a high temperature gas stream is the enthalpy probe discussed in Chap. 7.

The measurement of heat flux poses special difficulties as illustrated by the techniques described in Chap. 8. The close analogy between heat transfer, on the one hand, and mass transfer, on the other, makes it possible to obtain information on a heat transfer process by measurements in an analogous mass transfer situation. This offers an advantage where mass transfer experiments are simpler to perform or where clearly defined boundary conditions are required. The electrochemical method offers additional advantages: it requires only electrical measurements, and local and instantaneous measurements can be performed. Chapter 9 discusses such analogy measurements. Thermal radiation as a means of heat transfer has increased in importance in recent years because of the trend in engineering systems toward higher temperatures and because it is the only mechanism for heat transfer from vehicles moving through space outside the atmosphere. Techniques used for the investigation of thermal radiation are discussed in Chap. 10.

Knowledge of thermodynamic and transfer properties is required to calculate heat transfer in specific engineering problems as well as to generalize the results of heat transfer measurements with the use of dimensionless parameters. Measurements of such properties are often performed concurrently with heat transfer investigations. Techniques used for such measurements are discussed in Chaps. 11 and 12.

The measurement of fluid velocity is often fundamental to measurements of convective heat transfer. The laser-Doppler method described in Chap. 13 permits velocities to be inferred from the frequency shift of a scattered laser beam. The light beam, unlike a normal velocity probe, does not affect the flow field. Velocity measurements which are usually performed by hot wires at low temperatures require special instruments when they are to be performed at higher temperature levels. Such instruments are discussed in Chap. 14.

The first edition of this book developed from lectures delivered at a special summer course on "Measurement Techniques in Heat Transfer" held at the University of Minnesota. Lectures were presented by the staff of the Thermodynamics and Heat Transfer Division, Mechanical Engineering Department, University of Minnesota, and by outside speakers who are specialists in the subjects of their lectures. Papers based on these lectures were originally published in 1970 with support by the Advisory Group for Aerospace Research and Development. The many inquiries that we received since then encouraged us to prepare this second, revised and augmented edition. Major changes from the earlier edition include the new chapters on resistance thermometers and thermocouples, the completely revised chapter on laser-Doppler anemometry, and updating of most of the other chapters.

We hope that this volume will be useful for those in the various branches of science and engineering who have to perform measurements in heat transfer studies.

1 Error estimates in temperature measurement

E. M. SPARROW
Heat Transfer Laboratory, Department of Mechanical Engineering, University of Minnesota, Minneapolis

It is widely recognized that the output of a sensor such as a thermocouple or thermometer represents an approximation of the temperature at some location in a fluid or a solid. There are a variety of factors which can cause deviations between the probe output and the actual temperature at the point of interest (i.e., in the absence of the probe). First of all, the presence of the probe itself may modify the thermal conditions at the point and in its surroundings, thereby altering the temperature distribution. This happens, for instance, when heat is conducted to or away from a thermocouple junction through the lead wires. A second major factor is that the sensor may communicate with other environments beside the one whose temperature is being measured. For example, a thermocouple whose function is to measure the temperature in a flowing gas may exchange heat by conduction and radiation with the duct walls.

In addition, certain basic characteristics of the energy transfer and energy storage processes tend to favor the occurrence of errors in temperature measurement. One such characteristic is that convective heat exchange cannot take place without a temperature difference (i.e., the heat transfer coefficient is not infinite). Another is that viscous dissipation (aerodynamic heating) occurs in the boundary layer adjacent to a body situated in a high-speed flow. In addition, in transient processes, the heat capacity of the sensor brings about a temperature difference between sensor and fluid.

The task of designing low-error temperature sensors is aggravated by the fact that near-perfect thermal insulators do not exist. This is in contrast to the situation in electrical measurements, where essentially perfect insulators are readily available.

By careful design, it is possible to reduce the measurement errors that result from one or more of the aforementioned causes. For instance, convective heat transfer coefficients can be increased by locally increasing the fluid velocity adjacent to the sensor. Similar desirable effects can be achieved by manipulating the size and the shape of the sensor. Radiative exchange with the surroundings

1

can be diminished by shielding the sensor. Design considerations are treated at length in the published literature, for instance, [1–4].

It will be assumed here that specific information related to probe design can be obtained from the aforementioned references and from the great wealth of papers that deal with this subject. The purpose of this chapter is to review the sources of error in temperature measurement and to discuss analytical models that may be employed in the estimation of such errors.

A seemingly natural goal of analysis would be to furnish more or less precise correction formulas that would be used to adjust the output of the sensor and thereby provide accurate values for the temperature. However, in practice, the heat transfer problems involved with temperature probes are so complex as to defy precise analysis. Even when large electronic computers are employed to facilitate the solution, relatively simple analytical models are still appropriate. At present, a realistic goal of analysis is to furnish estimates of the order of magnitude of the errors that may be expected in temperature measurement. Analysis also provides guidelines for probe design and suggests appropriate correlation parameters for calibration tests.

To illustrate the nature of the heat transfer problems that are encountered in analyzing temperature measurement errors, consider a sensor situated in a flowing gas stream such as that pictured schematically in Fig. 1.1. The sensor is positioned by some sort of support structure or by its own lead wires. The dashed lines in the figure are intended to represent paths for energy transfer. The sensor itself is shown to communicate by various paths to several environments that may, in general, have different temperatures. By conduction through the support and/or its leads, the sensor communicates with the wall temperature at the point of attachment. Radiation provides paths that connect the sensor to one or more

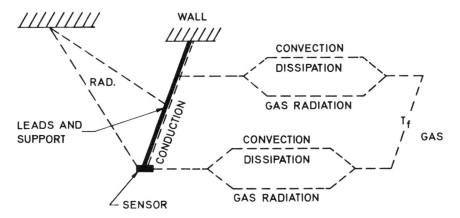

FIG. 1.1 Illustrative heat transfer paths for a temperature sensor in a gas flow.

visible surfaces (i.e., the wall or other structural elements) and to the gas stream (provided that the gas is radiatively participating). In general, neither the visible surfaces nor the gas has temperatures that are spatially uniform. Heat also passes between the sensor and the gas by convection. In the presence of a high-speed flow, the convective transfer is modified by viscous dissipation in the boundary layer.

The support structure and/or the leads also exchange heat by radiation and convection with the walls and with the gas, and such exchanges may have a significant influence on the conduction heat transfer between the sensor and the wall. Furthermore, the sensor itself may be a fairly complex unit with radiation shields, velocity shroud, etc. Even if the sensor is a bare thermocouple junction, it represents a nonelementary geometrical configuration for each one of the transport mechanisms—conduction, convection, and radiation.

The foregoing discussion points up the great difficulty in making precise analytical corrections for temperature measurement errors and reaffirms that the essential role of analysis is to provide order-of-magnitude estimates. This is the spirit that will be adopted in the forthcoming portions of the chapter.

The quality of the error estimates furnished by analysis will depend on the faithfulness of the analytical model to the actual physical situation. The next several sections of the chapter will discuss illustrative analytical models relevant to temperature measurements of solids and of fluids. In most cases the discussion will be concerned with thermocouples; however, the basic ideas are applicable to other types of sensors. Common to the analysis of both solid and fluid temperature measurements is the need for models of heat transfer in lead wires, and this is the next subject to be discussed.

1.1 LEAD WIRE MODELS

Most simply, lead wires may be envisioned as providing a heat conduction path between the temperature sensor and an isothermal zone located at the other extremity of the leads (Fig. 1.2). If the lead wires pass through a fluid, there will be heat transfer from the surface of the leads to the fluid by convection and, perhaps, by radiation. If the fluid is radiatively transmittant, heat is also exchanged by radiation between the leads and the surrounding visible surfaces.

The simplest lead is a single uninsulated wire. In view of the small diameter and relatively high thermal conductivity of such a wire, it is reasonable to neglect variations of temperature in the wire cross section. That is, the temperature may be regarded as a function of the axial coordinate x. The heat transfer in the lead wire may then be analyzed in a manner identical to that used for fins (for instance [5]).

More commonly, the wire may be covered by an annular layer of insulation as pictured in Fig. 1.3. With such a configuration, some alternatives present themselves for modeling the axial and the transverse heat transfer. For the wire itself, it is reasonable to assume that the temperature does not vary in the cross

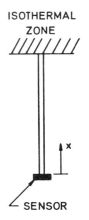

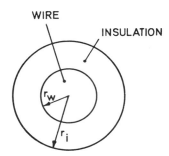

FIG. 1.2 Lead wire schematic.   **FIG. 1.3** Cross section of single insulated wire.

section, so, $T = T(x)$ alone. On the other hand, for the insulation, the temperature varies both radially and axially. An exact solution of such a heat conduction problem, with a convection and/or radiation boundary condition, is quite involved. Furthermore, such a solution probably is not justified because of various other uncertainties (e.g., contact resistance).

A simple model, sufficient for most purposes, is to assume that all of the axial heat flow is confined to the wire, while the heat flow through the insulation is purely radial. Thus, if k_w and A_w, respectively, denote the thermal conductivity and cross-sectional area of the wire and $T(x)$ is its local temperature, then the axial heat flow may be represented as

$$Q_x = -k_w A_w \frac{dT}{dx} \tag{1.1}$$

On the other hand, the radial heat flow per unit length dQ_r/dx is given by

$$\frac{dQ_r}{dx} = \frac{2\pi k_i (T - T_{\text{surf}})}{\ln (r_i/r_w)} \tag{1.2}$$

in which k_i is the thermal conductivity of the insulation and T_{surf} is the temperature on the outer surface of the insulation.

Under steady state conditions, the dQ_r given by Eq. (1.2) is equal to the convective and/or radiative heat transfer at the surface; for instance, for convection

$$dQ_r = h(2\pi r_i dx)(T_{\text{surf}} - T_f) \tag{1.3}$$

where T_f is the temperature of the fluid. By bringing together Eqs. (1.2) and (1.3), one finds

$$\frac{dQ_r}{dx} = \frac{T - T_f}{R} \tag{1.4}$$

$$R = \frac{1}{h2\pi r_i} + \frac{\ln (r_i/r_w)}{2\pi k_i} \tag{1.5}$$

The symbol R denotes the thermal resistance, while $T - T_f$ is the temperature difference between the wire and the fluid.

By employing Eqs. (1.1), (1.4), and (1.5), the insulated wire problem can be solved like a fin problem. It may be noted, in passing, that Eq. (1.1) may somewhat underestimate the axial conduction. An alternative representation, which overestimates Q_x, is obtained by replacing $k_w A_w$ by the quantity $\widetilde{kA}$, which is defined as

$$\widetilde{kA} = k_w A_w + k_i A_i \tag{1.6}$$

In thermocouple practice, a common lead configuration involves two wires and appropriate insulation as shown schematically in Fig. 1.4. The wires have identical radii r_w, but different thermal conductivities k_{w1} and k_{w2}. Under the assumption that the wires have the same temperature at a given x, the axial conduction of heat may be represented as

$$Q_x = - \widetilde{kA} \frac{dT}{dx} \tag{1.7}$$

in which

$$\widetilde{kA} = (k_{w1} + k_{w2})A_w \tag{1.8}$$

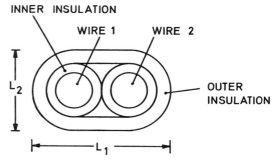

FIG. 1.4 Double wire configuration.

wherein axial conduction through the insulation is neglected. Adding terms $k_i A_i$ for the various layers of insulation to Eq. (1.8) overestimates the axial conduction.

To model the transverse conduction from the wires to the outer surface of the insulation, consider an equivalent composite cylinder consisting of a central wire and an annular layer of insulation. Let the outer and inner radii, r_2 and r_1 respectively, be defined as

$$r_2 = \frac{L_1 + L_2}{4} \qquad r_1 = \sqrt{2} r_w \tag{1.9}$$

The radial heat flux through the insulation is given by Eq. (1.2), with r_i and r_w respectively replaced by r_2 and r_1. For convective heat exchange at the surface of the insulation, the radial heat flow dQ_r continues to be represented by Eqs. (1.4) and (1.5), provided that the just-mentioned substitutions are made.

The foregoing computational models are representative of those that are employed to characterize the axial and radial heat conduction in lead wires.

1.2 TEMPERATURE MEASUREMENTS
IN SOLIDS

Attention will now be given to computational models for estimating errors in the measurement of steady state temperatures of solids. Three illustrative situations will be discussed: (1) surface temperature of a relatively massive solid, (2) imbedded thermocouple in a relatively massive solid, and (3) measurement of the temperature of a thin plate. Errors in the measurement of time-dependent temperatures in solids will be considered in Sec. 1.2.4.

1.2.1 Surface Temperature of a Relatively
Massive Solid

The physical situation is pictured schematically in Fig. 1.5. A thermocouple is affixed to the surface of a relatively massive solid, well bonded to the surface so that the thermal contact resistance is negligible. Alternatively, the thermocouple may be seated in a shallow hole drilled into the surface. The thermocouple lead wires pass through a fluid whose temperature is T_f. As will now be discussed, this type of thermocouple arrangement can alter the temperature at the surface location where the measurement is being made, thereby giving rise to erroneous information.

If the fluid temperature is lower than that of the solid, the thermocouple conducts heat away from the surface. In most instances, the heat transported away from the surface per unit time and area by the thermocouple is substantially greater than the corresponding convective heat loss from the surface to

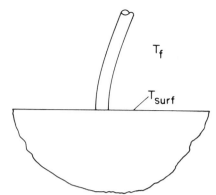

T_f

T_{surf}

FIG. 1.5 Surface temperature measurement of a massive solid.

the fluid. Consequently, temperature gradients are set up in the solid so as to cause heat to flow toward the thermocouple junction. In particular, consistent with such a heat flow, the temperature of the junction must be depressed. Therefore, the thermocouple reads low. If the fluid temperature exceeds that of the solid, then the thermocouple functions as an efficient pipeline through which heat flows into the solid, thereby causing a local increase in the temperature of the surface. Correspondingly, the thermocouple reads high.

The temperature measurement errors associated with the presence of the surface-mounted thermocouple have been analyzed in [6], and the results of that analysis are presented here. Let T_{surf} denote the true surface temperature (i.e., in the absence of the thermocouple) and T_{tc} the temperature sensed by the thermocouple. Suppose that the thermocouple wires have an effective circular cross section of radius r_1. If there is one wire whose radius is r_w, then $r_1 = r_w$. When there are two wires, each of radius r_w, then r_1 is given by Eq. (1.9). In the analysis, it is assumed that the temperature on the circular area of contact between the thermocouple junction and the surface is uniform and equal to T_{tc}. If the thermocouple wires are insulated, then the equivalent outer radius of the insulation is denoted by r_2 (see Eq. [1.9] and related text).

The conductivity area product kA for axial conduction in the thermocouple wires and the radial thermal resistance R are evaluated as outlined in the prior section of this chapter (Sec. 1.1). The length of the thermocouple leads is denoted by L. The solid has a thermal conductivity k_s, and the convective heat transfer coefficient between the surface of the solid and the fluid is h_s. In general, h_s will be different from the thermocouple-to-fluid convection coefficient h that is used in evaluating the radial resistance R.

The results will be presented by employing three dimensionless groups. The first is the dimensionless temperature error

$$\frac{T_{surf} - T_{tc}}{T_{surf} - T_f} \tag{1.10}$$

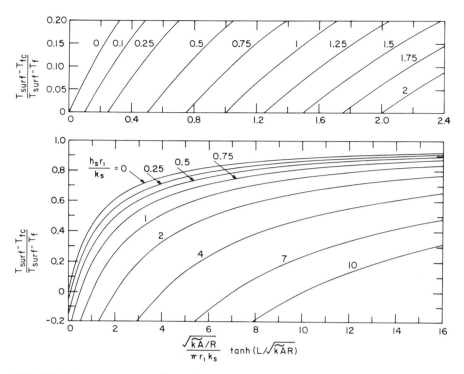

FIG. 1.6 Measurement errors for surface-mounted thermocouple.

in which $T_{\text{surf}} - T_{\text{tc}}$ is the measurement error and $T_{\text{surf}} - T_f$ is the characteristic temperature difference of the problem. The second dimensionless group is

$$\frac{\sqrt{\widetilde{kA}/R}\,\tanh\,(\widetilde{kA}R)^{-\frac{1}{2}}L}{\pi r_1 k_s} \tag{1.11}$$

in which the numerator comes from fin theory [5, 7]. In most applications, the lead length L is sufficiently large so that the tanh is equal to one. Finally, the third group is the Biot number of the solid

$$\frac{h_s r_1}{k_s} \tag{1.12}$$

The results are presented in Fig. 1.6, with the dimensionless temperature error on the ordinate and the group expressed by Eq. (1.11) on the abscissa. The curves are parameterized by the Biot number. The figure is subdivided into two graphs. The upper graph is for smaller values of the abscissa group, whereas the lower graph is for larger values of the abscissa group.

The magnitude of the abscissa group is a measure of the conductance of the thermocouple compared to the conductance of the solid. The figure shows that for a fixed value of the Biot number $h_s r_1 / k_s$ the measurement error is accentuated when the thermocouple has a relatively high conductance compared to that of the solid. On the other hand, the measurement error is of lesser importance when the conductance of the solid is high compared with that of the thermocouple. Further inspection of the figure indicates that for a fixed value of the abscissa group, smaller temperature errors are encountered for larger values of the Biot number.

It is also seen from the figure that the dimensionless temperature error may be negative under certain conditions. This will occur in the unlikely situation in which the thermocouple functions as an insulator rather than as a conductor.

1.2.2 Imbedded Thermocouple in a Relatively Massive Solid

A schematic diagram of the problem being studied is shown in Fig. 1.7. A thermocouple is situated in a hole drilled into the surface; bonding is provided by some sort of adhesive (indicated as blackened in the figure), for example, epoxy. Upon emerging from the solid, the thermocouple leads pass through a fluid whose temperature is T_f. The leads are long enough so that they ultimately take on the fluid temperature T_f.

The thermocouple serves as a channel through which heat can flow into (or out of) the solid. Such heat flow gives rise to two sources of temperature measurement error. First, there is a raising (or lowering) of the temperature of the solid. Second, owing to the presence of the adhesive, there will be an additional temperature difference between the solid and the thermocouple junction. If the solid is a metal, then the second of these mechanisms is dominant. That is, as a first approximation, the temperature of the solid is assumed to be

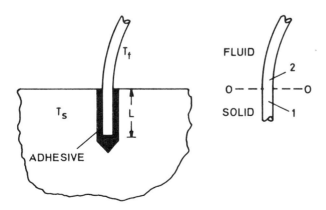

FIG. 1.7 Imbedded thermocouple in a massive solid.

uninfluenced by the presence of the thermocouple. This is the case that will be considered here.

The analytical model and the resulting temperature error expression are due to Moffat [8]. He envisions the thermocouple leads to be made up of two sections: (1) the segment that is imbedded in the solid and (2) the segment that passes through the fluid. This is shown schematically at the right of Fig. 1.7. The dashed line 0-0 represents the location of the surface. Moffat begins by separately solving the heat conduction problems for Regions I and II (see Fig. 1.7). These solutions involve the temperature T_{00} at the interface between the two regions. Then, the condition of heat flux continuity at 0-0 is imposed and, by this, T_{00} is determined. Once T_{00} is known the temperature error at the junction is readily calculated.

In Region I the thermocouple leads behave like a fin situated in a uniform temperature environment (i.e., the solid at temperature T_s). The adhesive gives rise to a thermal resistance

$$\frac{\ln (r_3/r_2)}{2\pi k_a} \tag{1.13}$$

in which k_a is the thermal conductivity of the adhesive, while r_2 and r_3 are, respectively, the outer radii of the thermocouple leads and of the adhesive. Then, the overall resistance R_I for radial heat flow dQ_r/dx from the thermocouple wire (or wires) to the solid is:

$$R_I = \frac{\ln (r_2/r_1)}{2\pi k_i} + \frac{\ln (r_3/r_2)}{2\pi k_a} \tag{1.14}$$

where r_2 and r_1 may represent equivalent radii when the thermocouple is composed of two wires and several layers of insulation (see discussion in Sec. 1.1).

If T_{00} is the temperature at Plane 0-0, fin theory gives the following expression for the rate of heat flow across Plane 0-0 into Region I.

$$Q_I = \sqrt{\frac{\widetilde{kA}}{R_I}} \; (T_{00} - T_s) \; \tanh \; (\widetilde{k}AR_I)^{-\frac{1}{2}}L \tag{1.15}$$

The quantity $\widetilde{kA}$ is the conductivity area product for axial conduction through the thermocouple wires, and L is the length of the leads in Region I. Equation (1.15) neglects the axial heat transfer at the junction itself, but this can be included by employing an alternate expression from fin theory.

The heat transfer passing across Plane 0-0 from Region II is expressed by an equation similar to Eq. (1.15), but with appropriate changes in nomenclature:

$R_I \rightarrow R_{II}$ and $(T_{00} - T_s) \rightarrow (T_f - T_{00})$. In addition, the lead length in Region II will generally be sufficiently large so that $\tanh \approx 1$.

$$Q_{II} = \sqrt{\frac{\widetilde{kA}}{R_{II}}} \, (T_f - T_{00}) \qquad (1.16)$$

The expressions for Q_I and Q_{II} must be equal, from which it follows

$$T_{00} - T_s = \frac{T_f - T_s}{1 + \sqrt{R_{II}/R_I} \tanh (\widetilde{kA}R_I)^{-\frac{1}{2}}L} \qquad (1.17)$$

Now, returning to Region I and once again using fin theory, the temperature T_{tc} at the thermocouple junction can be expressed in terms of T_{00} as follows:

$$\frac{T_{tc} - T_s}{T_{00} - T_s} = \frac{1}{\cosh (\widetilde{kA}R_I)^{-\frac{1}{2}}L} \qquad (1.18)$$

Upon combining Eqs. (1.17) and (1.18), one obtains an equation for the temperature error $T_{tc} - T_s$,

$$\frac{T_{tc} - T_s}{T_f - T_s} = \frac{1}{\cosh (\widetilde{kA}R_I)^{-\frac{1}{2}}L} \left[\frac{1}{1 + \sqrt{R_{II}/R_I} \tanh (\widetilde{kA}R_I)^{-\frac{1}{2}}L} \right] \qquad (1.19)$$

For a given value of $T_f - T_s$, the temperature error is accentuated by small imbedding depths L, large values of $\widetilde{kA}$, large values of the thermal resistance R_I, and small values of the thermal resistance R_{II}. These qualitative rules are physically reasonable.

1.2.3 Measurement of Temperature of a Thin Plate

Consider next a thin plate (thickness δ) situated between two flows, respectively having temperatures T_{f1} and T_{f2} (Fig. 1.8). The corresponding convective heat transfer coefficients are h_1 and h_2. In the absence of the thermocouple, and assuming that the variation of the temperature across its thickness is negligible, the equilibrium temperature T^* of the plate is

$$T^* = \frac{h_1 T_{f1} + h_2 T_{f2}}{h_1 + h_2} \qquad (1.20)$$

The effect of the thermocouple is to channel heat into (or out of) the plate, with the consequence that the temperature of the thermocouple junction and the

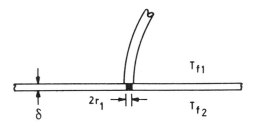

FIG. 1.8 Measurement of temperature of a thin plate.

adjacent portion of the plate is increased (or decreased). The corresponding temperature error has been calculated in [7] (pp. 173–179). If T_{tc} is the temperature indicated by the thermocouple and T^* is the plate temperature in the absence of the thermocouple, the error $T_{tc} - T^*$ can be expressed as

$$\frac{T_{tc} - T^*}{T_{f1} - T^*} = \left\{ 1 + 2\pi k_s \epsilon r_1 \left[\frac{K_1(\epsilon r_1)}{K_0(\epsilon r_1)} \right] \frac{\sqrt{R/\widetilde{kA}}}{\tanh (\widetilde{kA}R)^{-\frac{1}{2}}L} \right\}^{-1} \quad (1.21)$$

in which

$$\epsilon = \sqrt{\frac{h_1 + h_2}{k_s \delta}} \quad (1.21a)$$

The symbols K_1 and K_0 denote modified Bessel functions of the second kind. Numerical values of these quantities are listed in standard tables.

In deriving Eq. (1.21) it was assumed that only the actual thermocouple wire (or wires) is imbedded within the plate, that is, the insulation has been removed. For a single wire of radius r_w, $r_1 = r_w$. If there are two wires, each of radius r_w, then $r_1 = \sqrt{2} \, r_w$. The quantity $\widetilde{kA}$ is the conductivity area product for axial conduction in the thermocouple leads, while R is the thermal resistance for radial heat flow in the leads. Both of these parameters have already been discussed in detail. L is the length of the thermocouple leads and k_s is the thermal conductivity of the solid.

The relative temperature error—left-hand side of Eq. (1.21)—is plotted in Fig. 8.5 of [7]. The figure shows that the error is accentuated at small values of both ϵr_1 and $k_s \delta \sqrt{(R/\widetilde{kA})}/\tanh (\widetilde{kA}R)^{-\frac{1}{2}}L$.

1.2.4 Errors in Measurement of Unsteady Temperatures

The estimation of errors in the measurement of unsteady temperatures in solids is a much more difficult task than is the estimation of steady state temperature errors. A survey of published literature indicates that even a simple physical model leads to a formidable mathematical problem. The heart of the difficulty is that one must deal with partial differential equations even in cases where the spatial dependence is one dimensional. Furthermore, owing to the thermal interaction between the sensor (e.g., thermocouple) and the solid, coupled pairs of partial differential equations are encountered. An analytical solution of such mathematical systems is possible only for the simplest conditions and, even then, nonelementary mathematical methods are necessary. A published analysis for a semi-infinite body, subjected to a step change in its internal temperature distribution, and fitted with a surface-mounted thermocouple, typifies the mathematical task [9].

Numerical solutions, carried out with the aid of a digital computer, appear to be more or less mandatory for time-dependent situations. Representative numerical solutions for temperature errors associated with imbedded thermocouples are reported in [10-12].

It may be noted that the estimation of measurement errors for transient problems is by no means a casual undertaking, regardless of whether analytical or numerical solutions are contemplated. Each problem must be modeled more or less individually in order to justify the substantial solution effort that is required.

1.3 TEMPERATURE MEASUREMENTS IN FLUIDS

Consideration is now given to estimates of errors in temperature measurements in fluids, both under steady state and transient operating conditions. For the present, radiative exchange will be set aside. The role of radiation will be discussed in Sec. 1.4.

1.3.1 Steady State Temperature Measurements

The usual computational model employed to estimate the error in steady state fluid temperature measurements envisions the thermocouple or other sensors to behave, in essence, like a fin. Although there are many variants depending on the specific application, the general nature of the problem can be illustrated by a diagram such as Fig. 1.9. The thermocouple leads (and/or support) are attached to a wall whose temperature is T_w. The fluid temperature is T_f. Owing to heat conduction in the leads, the temperature at the thermocouple junction will be between T_w and T_f, provided that the viscous dissipation is negligible (low Mach number).

For high-speed gas flows, viscous dissipation plays an important role, so that, in the absence of heat conduction in the leads, the thermocouple would read the recovery temperature T_r defined by

$$T_r = T_f\left(1 + \Re\,\frac{k-1}{2}\,\mathrm{M}^2\right) \tag{1.22}$$

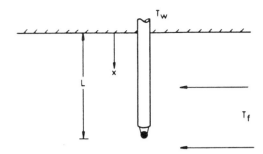

FIG. 1.9 Schematic of fluid temperature measurement.

where $\mathfrak{R}$ is the recovery factor (a number on the order of unity), k is the ratio of specific heats c_p/c_v, and M is the Mach number. In such a flow, the temperature sensed by the thermocouple junction lies somewhere between T_w and T_r.

The error analysis typically ignores possible variations of the heat transfer coefficient, recovery factor, and fluid temperature along the length of the leads. By applying fin theory and taking account of the viscous dissipation effects, the temperature at the thermocouple junction can be expressed as

$$\frac{T_{tc} - T_r}{T_w - T_r} = \frac{1}{\cosh{(\widetilde{kA}R)^{-\frac{1}{2}}L}} \tag{1.23}$$

in which the quantities $\widetilde{kA}$ and R have already been discussed in prior sections of this chapter. Equation (1.23) also ignores axial conduction at the thermocouple junction itself, but this is probably a lesser cause for uncertainty than is the irregular geometry of the junction and the doubtful accuracy of the heat transfer coefficient in the junction region. Equation (1.23) can be employed for estimating the temperature error.

In the case of a nondissipating flow (that is, $T_r \approx T_f$), it is evident that the temperature error can be minimized by making $L/(\widetilde{kA}R)^{\frac{1}{2}}$ as large as possible. There are, of course, certain natural limitations in such an undertaking. The wires must have a certain minimum cross section for strength; the insulation cannot be so thick as to provide a flow blockage; and the length is limited by a variety of considerations. One approach to increasing the length is to position the thermocouple leads axially in part of their passage through the fluid.

In a high-speed gas flow, dissipative effects more or less preclude the possibility of designing for equality of T_{tc} and T_f. Instead, one may attempt to design so as to minimize $(T_{tc} - T_r)$ and then to apply Eq. (1.22) to determine T_f from T_r. To perform the latter calculation, it is necessary to know the recovery factor $\mathfrak{R}$. Moffat [1] (p. 555) has surveyed available experimental information and gives

$$\mathfrak{R} = 0.68 \pm 0.07 \quad \text{wires normal to flow} \tag{1.24a}$$

$$\mathcal{R} = 0.86 \pm 0.09 \qquad \text{wires parallel to flow} \tag{1.24b}$$

These results are for gases whose Prandtl numbers are approximately 0.7.

In principle, the just-discussed error estimation could be improved by taking account of factors such as variations of fluid temperature, velocity, and heat transfer coefficient along the length of the leads. However, in practice, this information is usually not known or is, at best, uncertain. As a consequence, such refinements are rarely incorporated into the error estimation calculations.

Radiation effects are frequently accounted, as will be discussed in Sec. 1.4.

1.3.2 Unsteady Temperature Measurements

The traditional model for estimating errors in the measurement of unsteady fluid temperatures assumes negligible spatial temperature variations within the sensor and neglects lead wire conduction to or from the sensor. Thus, if M, c, and A_s are the mass, specific heat, and surface area of the sensor, the basic energy balance is[*]

$$Mc \frac{dT_{tc}}{dt} = hA_s(T_f - T_{tc}) \tag{1.25}$$

The temperature error $T_f - T_{tc}$ at any time is then equal to $(Mc/hA_s)(dT_{tc}/dt)$ and so, in principle, the error can be estimated from measured values of the temperature derivative dT_{tc}/dt.

For given time variations $T_f(t)$ in the fluid temperature, the error in the thermocouple output can be calculated by integrating Eq. (1.25). Some representative cases will now be discussed. Consider first a very rapid change (i.e., step change) in fluid temperature from T_{f0} to T_f. If the initial temperature of the sensor is also T_{f0}, then its response to the step change in fluid temperature is given by

$$\frac{T_{tc} - T_f}{T_{f0} - T_f} = e^{-(hA_s/Mc)t} \tag{1.26}$$

It is customary to define the time constant τ of the sensor such that

$$\frac{T_{tc} - T_f}{T_{f0} - T_f} = 0.5$$

Thus

$$\tau = 0.693 \frac{Mc}{hA_s} \tag{1.27}$$

[*]Viscous dissipation effects will not be explicitly included here.

For instance, for a cylindrical copper sensor of radius 0.020 in. and with $h = 25$ Btu/hr ft^2 °F, $\tau \approx 6$ sec.

Although the time constant provides a convenient index of the response characteristics of a sensor, it may be necessary to take t considerably larger than τ to achieve the desired accuracy in the measurement of T_f. Thus, if it is desired that $(T_{tc} - T_f)/(T_{f0} - T_f) = 0.10$, then $t \approx 3.3\tau$.

A second interesting case of potential practical interest is that in which the fluid temperature increases linearly with time, that is

$$T_f - T_{f0} = \beta t \qquad (1.28)$$

The thermocouple response to this imposed time variation is obtained by solving Eq. (1.25), which gives, for times subsequent to an initial transient period

$$T_{tc} - T_f = \beta \frac{Mc}{hA_s} \qquad (1.29)$$

Thus, there will be a constant temperature error equal to $\beta(Mc/hA_s)$. Alternatively, the quantity $\beta(Mc/hA_s)$ may be interpreted as the amount by which the thermocouple temperature lags behind the fluid temperature. Note also that $dT_{tc}/dt = \beta$, so that the temperature-time output of the thermocouple can, in principle, be employed to facilitate the estimation of the temperature error. For a given variation in fluid temperature, the error is diminished when M and c are as small as possible, and when h and A_s are as large as possible.

Periodic timewise variations of the fluid temperature are also of practical interest. Since any periodic function can be expanded in a Fourier series, it is natural to consider

$$T_f = \bar{T}_f + \gamma \sin \omega t \qquad (1.30)$$

By seeking a steady periodic solution of Eq. (1.25), one finds

$$T_{tc} - T_f = \frac{-\gamma}{\sqrt{1 + (hA_s/Mc\omega)^2}} \sin\left(\omega t + \tan^{-1} \frac{hA_s}{Mc\omega}\right) \qquad (1.31)$$

Evidently, only when $hA_s/Mc\omega$ is large will the measurement error $T_{tc} - T_f$ be small. On the other hand, when $hA_s/Mc\omega$ is small, the sensor records the mean value of the fluid temperature, that is, $T_{tc} \approx \bar{T}_f$.

The foregoing treatment of unsteady temperature measurements is based on the postulates that spatial temperature variations within the sensor are negligible and that heat conduction to or from the sensor (through lead wires or supports) is very small. The validity of these assumptions will now be examined.

Consider first the question of spatial temperature uniformity. Suppose, for concreteness, that the sensor is a cylinder of radius r having thermal conductivity k and thermal diffusivity α. Initially, the sensor and the fluid are at the same uniform temperature T_{f0}. Then, at $t = 0$, the fluid temperature is step changed to T_f and maintained uniform thereafter. If the possibility of radial temperature variations is admitted, the heat transfer process is governed by a partial differential equation in which the temperature is a function of the radial position and of time. This equation is to be solved subject to a convective boundary condition at the surface of the cylinder and to the above-mentioned thermal conditions at $t = 0$.

A solution of this mathematical system, given by Jakob [13] (pp. 275–278), has been adapted to the problem of the temperature sensor by Clark [14]. To characterize the extent of the spatial nonuniformity, he employs the quantity

$$\frac{T_{\mathbb{C}} - T_{\text{surf}}}{T_{f0} - T_f} \tag{1.32}$$

where $T_{\mathbb{C}}$ and T_{surf} are, respectively, the temperatures at the centerline and the surface of the cylinder. A plot of this temperature ratio, taken from [14], is given in Fig. 1.10. For values of $\alpha t/r^2 > 0.4$, the curves can be extended by evaluating the equation

$$\frac{T_{\mathbb{C}} - T_{\text{surf}}}{T_{f0} - T_f} = \frac{2}{\chi_1} \frac{J_1(\chi_1)[1 - J_0(\chi_1)]}{J_0^2(\chi_1) + J_1^2(\chi_1)} e^{-\chi_1^2 (\alpha t/r^2)} \tag{1.33}$$

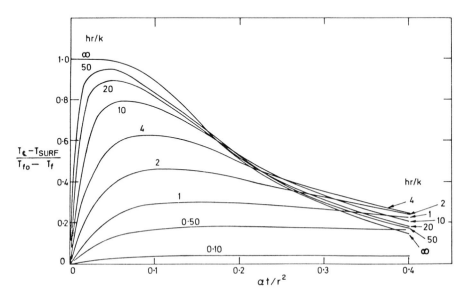

FIG. 1.10 Transient temperature difference between centerline and surface for a cylinder.

where J_0 and J_1 are Bessel functions and the χ_1 are constants, dependent on the values of hr/k, which are listed in Jakob's Table 13-9* [13].

With information such as Fig. 1.10 at hand, a judgment can be made about the correctness of neglecting spatial variations. In particular, if the magnitude of $(T_{\mathcal{C}} - T_{\text{surf}})/(T_{f0} - T_f)$ is much less than the magnitude of $(T_{\text{tc}} - T_f)/(T_{f0} - T_f)$ given by Eq. (1.26), then the neglect of spatial variations is justified. For instance, if the sensor is a copper wire of radius 0.020 in. and $h = 25$ Btu/hr ft^2 °F, the aforementioned criterion is met.

Although the just-concluded examination provides information for the specific case of a step change in fluid temperature, the same approach can be employed for the ramp and sinusoidal variations that were considered earlier.

Having considered spatial temperature variations within the sensor, attention may now be directed to the matter of conduction to or from the sensor, through leads or supporting structure. To examine this point, a simple physical situation will be studied: a thermocouple such as that shown in Fig. 1.9. Initially, the thermocouple, the fluid, and the wall are all at a common temperature T_{f0}. At $t = 0$, the fluid temperature is step changed to T_f, while the wall temperature is maintained at its initial value. The temperature is assumed to be a function of axial position along the leads and of time (no spatial variations in the wire cross section).

A solution for the variation of the temperature at the thermocouple junction with time is [15]

$$\frac{T_{\text{tc}} - T_f}{T_{f0} - T_0} = \psi + (1 - \psi) \exp -\left[\frac{hA_s}{Mc(1 - \psi)} t \right] \tag{1.34}$$

where ψ is the steady state temperature solution, an expression for which is given by the right-hand side of Eq. (1.23). By comparing Eq. (1.34), which includes conduction effects, with Eq. (1.26), which neglects conduction effects, it is seen that the omission of conduction is justified when $\psi \ll 1$, that is, when

$$\cosh (\widetilde{kA}R)^{-\frac{1}{2}}L \gg 1 \tag{1.35}$$

A more quantitative estimate of the influence of conduction can be obtained by numerically evaluating Eq. (1.34).

1.4 THE ROLE OF RADIATION IN ERROR ESTIMATES

Up to this point, only heat transfer by conduction and convection have been included in the computational models employed for estimating temperature measurement errors. Now, consideration will be given to the contribution of

---

*In the table, bs is equivalent to hr/k.

thermal radiation. Although radiative transfer is more commonly encountered in situations in which the sensor and/or lead wires pass through a gas, there are also technically interesting applications involving partially transparent solids. It is widely recognized that, qualitatively speaking, the role of radiation will be small at moderate temperature levels and will become more important as the temperatures increase.

To provide some approximate quantitative guidelines, it is useful to estimate the magnitudes of the radiation heat transfer coefficient. Consider an enclosure whose walls have arbitary radiation properties. If the walls are isothermal (temperature* T_w), the thermal radiation within such an enclosure corresponds to black body radiation at temperature T_w. The radiation heat loss per unit time and area of a small body situated within the enclosure is

$$q_{rad} = \epsilon\sigma(T^4 - T_w^4) \tag{1.36}$$

in which ϵ and T denote the emittance and temperature of the small body. Equation (1.36) is based on the assumption that the small body has gray radiation characteristics.

An alternate form of Eq. (1.36) can be written in terms of the radiation heat transfer coefficient h_{rad}:

$$q_{rad} = h_{rad}(T - T_w) \tag{1.37}$$

where

$$h_{rad} = \frac{\epsilon\sigma(T^4 - T_w^4)}{T - T_w} \tag{1.38}$$

If the temperature difference $T - T_w$ is substantially smaller than T_w, then h_{rad} becomes

$$h_{rad} \approx 4\epsilon\sigma T_w^3 \tag{1.39}$$

At any T_w, the largest value of h_{rad} is achieved when $\epsilon = 1$. At $T_w = 500°R$, $h_{rad} = 0.85$ Btu/hr ft^2 °F. This value of h_{rad} is comparable to heat transfer coefficients for natural convection in gases. It is, however, much lower than the heat transfer coefficients for most forced convection gas flows. Therefore, at this temperature level, radiation may be neglected except where natural convection processes are taking place.

The radiation coefficient h_{rad} increases markedly with increasing T_w. For instance, when $T_w = 1000°R$, $h_{rad} \approx 6.8$ Btu/hr ft^2 °F. However, this is still well below forced convection coefficients for airflow across fine wires. When

*In computing radiation fluxes, all temperatures must be absolute temperatures.

$T_w = 2000°\text{R}$, $h_{\text{rad}} \approx 55$ Btu/hr ft^2 °F, which is definitely competitive with coefficients for forced convection.

Radiation arriving at the sensor and/or its leads may originate at solid surfaces (e.g., duct walls) or in the adjacent gas. Many common gases such as air, nitrogen, oxygen, etc., are virtually radiatively nonparticipating except at very high temperature levels. Others such as water vapor and carbon dioxide partici- pate at intermediate temperature levels. Combustion gases and gas flows contain- ing solid particles also may be radiatively active.

With respect to temperature measurement in gases, radiant energy arriving at the sensor and/or leads from solid surfaces tends to increase the error (if the temperature of such surfaces differs from the gas temperatures), while gas radiation tends to decrease the error. The role of these processes may be reversed in the case of surface-mounted or imbedded thermocouples for temperature measurement of solids.

In cases where radiation is expected to cause a significant error in gas temperature measurements, it is common to shield the sensor so that its view of the error-inducing radiating solid surface is obstructed. In some temperature sensors, several such shields are employed.

Simple computational models for estimating temperature measurement errors in the presence of thermal radiation will now be discussed. For this purpose, consider an unshielded thermocouple immersed in a gas flow as illustrated in Fig. 1.9. Suppose first that the gas is radiatively nonparticipating.

If the duct walls are isothermal at temperature T_w, then the thermocouple can be assumed to behave like a small body in a large isothermal enclosure, and the local radiative loss per unit area is given by Eq. (1.36) or (1.37). To facilitate the analysis, it is convenient to employ the radiation heat transfer coefficient and to neglect its variation along the length of the leads.

At any location x (see Fig. 1.9), the surface heat loss from the leads by convection and radiation is given by

$$h2\pi r_2 (T_{\text{surf}} - T_f) + h_{\text{rad}}2\pi r_2 (T_{\text{surf}} - T_w) \tag{1.40}$$

where r_2 is the effective outer radius of the insulation and T_{surf} is the surface temperature of the insulation (T_{surf} is a function of x). It is convenient to rephrase the surface heat loss in terms of an effective heat transfer coefficient $\tilde{h}$ and effective environment temperature $\tilde{T}$. To this end, the surface heat loss may be represented as

$$\tilde{h}2\pi r_2 (T_{\text{surf}} - \tilde{T}) \tag{1.41}$$

where

$$\tilde{h} = h + h_{\text{rad}} \tag{1.41a}$$

$$\widetilde{T} = \frac{hT_f + h_{\mathrm{rad}}T_w}{h + h_{\mathrm{rad}}} \tag{1.41b}$$

In terms of these variables, the radial heat transfer from the thermocouple wires, dQ_r/dx of Eq. (1.4), can be re-expressed as

$$\frac{dQ_r}{dx} = \frac{T - \widetilde{T}}{R} \tag{1.42}$$

in which

$$R = \frac{1}{\widetilde{h}2\pi r_2} + \frac{\ln (r_2/r_1)}{2\pi k_i} \tag{1.43}$$

The foregoing formulation facilitates the direct application of fin theory, which yields the following expression for the temperature at the thermocouple junction.

$$\frac{T_{\mathrm{tc}} - \widetilde{T}}{T_w - \widetilde{T}} = \frac{1}{\cosh (\widetilde{k}AR)^{-\frac{1}{2}}L} \tag{1.44}$$

from which the temperature error can be evaluated. If the conduction heat transfer at the thermocouple junction is negligible, i.e., if $\cosh(\widetilde{k}AR)^{-\frac{1}{2}}L \gg 1$, then

$$T_{\mathrm{tc}} = \widetilde{T} = \frac{hT_f + h_{\mathrm{rad}}T_w}{h + h_{\mathrm{rad}}} \tag{1.45}$$

The application of the foregoing error estimates requires that the numerical values of h_{rad} be known. In this connection, it is recommended that some reasonable initial guess be made for h_{rad}, and that this guess be refined using Eq. (1.38) in conjunction with the solution for the temperature distribution along the leads, that is

$$\frac{T(x) - \widetilde{T}}{T_w - \widetilde{T}} = \frac{\cosh (\widetilde{k}AR)^{-\frac{1}{2}}(L - x)}{\cosh (\widetilde{k}AR)^{-\frac{1}{2}}L} \tag{1.46}$$

The h_{rad} values thus evaluated from Eq. (1.38) are functions of x and must, therefore, be averaged before being introduced into Eqs. (1.41a) and (1.41b).

The accounting of gas radiation introduces further uncertainties into the analysis because the radiative properties of gas bodies are not known to high accuracy. It is believed sufficient to treat the radiative interchange between the thermocouple and the gas by employing a radiation heat transfer coefficient similar to that used for radiation between the duct surface and the thermo-couple.

NOMENCLATURE

A cross-sectional area
A_s surface area
c specific heat
h convective heat transfer coefficient
h_{rad} radiative heat transfer coefficient
$\tilde{h}$ convective-radiative coefficient, Eq. (1.41a)
k thermal conductivity
$\widetilde{kA}$ conductivity-area product for axial conduction
L length
M   Mach number
M mass of sensor
Q heat transfer rate
Q_r radial heat transfer rate
Q_x axial heat transfer rate
q heat transfer rate per unit area
R thermal resistance based on dQ_r/dx, Eq. (1.4)
$\mathcal{R}$ recovery factor
r radius
r_1 outer radius of an equivalent single thermocouple wire
r_2 outer radius of an equivalent annulus of insulation
T temperature
$\tilde{T}$ effective environment temperature, Eq. (1.41)
t time
x axial coordinate
α thermal diffusivity
ϵ emittance
σ Stefan-Boltzmann constant
τ time constant

Subscripts

f fluid
$f0$ initial fluid temperature
i insulation
r recovery
s solid
surf   surface
tc   thermocouple
w wire

w   wall

00   interface

$\mathcal{C}_L$ centerline

REFERENCES

1. Dahl, A. I. (ed.), "Temperature, Its Measurement and Control in Science and Technology," vol. 3, pt. 2, sec. 4, Reinhold, New York, 1962.
2. Baker, H. D., Ryder, E. A., and Baker, N. H., "Temperature Measurement in Engineering," vol. 1, Wiley, New York, 1953.
3. Baker, H. D., Ryder, E. A., and Baker, N. H., "Temperature Measurement in Engineering," vol. 2, Wiley, New York, 1961.
4. Dean, R. C., Jr., (ed.), "Aerodynamic Measurements," chap. 2, MIT, Cambridge, Mass., 1953.
5. Eckert, E. R. G., and Drake, R. M., Jr., "Analysis of Heat and Mass Transfer," chap. 3, McGraw-Hill, New York, 1972.
6. Hennecke, D. K., and Sparrow, E. M., Local Heat Sink on a Convectively Cooled Surface—Application to Temperature Measurement Error, *Int. J. Heat Mass Transfer*, vol. 13, pp. 287-304, 1970.
7. Schneider, P. J., "Conduction Heat Transfer," Addison-Wesley, Reading, Mass., 1955.
8. Moffat, R. J., Temperature Measurement in Solids: Errors Due to Thermal Resistance between the Thermocouple and the Specimen, personal communication, February, 1968.
9. Henning, C. D., and Parker, R., Transient Response of an Intrinsic Thermocouple, *J. Heat Transfer*, vol. C89, pp. 146-154, 1967.
10. Beck, J. V., and Hurwicz, H., Effect of Thermocouple Cavity on Heat Sink Temperature, *J. Heat Transfer*, vol. C82, pp. 27-36, 1960.
11. Beck, J. V., Thermocouple Temperature Disturbances in Low Conductivity Materials, *J. Heat Transfer*, vol. C84, pp. 124-132, 1962.
12. Pfahl, R. C., Jr., and Dropkin, D., Thermocouple Temperature Perturbations in Low Conductivity Materials, paper 66-WA/HT-8, American Society of Mechanical Engineers, New York, 1966.
13. Jakob, M., "Heat Transfer," vol. 1, Wiley, New York, 1949.
14. Clark, J. A., Transient Temperature Distribution within Thermal Sensing Elements, personal communication, March 1967.
15. Scadron, M. D., and Warshawsky, I., Experimental Determination of Time Constants and Nusselt Numbers for Bare-Wire Thermocouples in High-Velocity Air Streams and Analytic Approximation of Conduction and Radiation Errors, *NACA Tech. Note* 2599, January 1952.

2 Platinum resistance thermometry

JOHN L. RIDDLE, GEORGE T. FURUKAWA,
and HARMON H. PLUMB
*Institute for Basic Standards, National Bureau of Standards
Washington, D.C.*

2.1 BACKGROUND AND BASIC CONCEPTS

The development of the platinum resistance thermometer resulted in an internationally acceptable practical temperature scale defined to give values of temperatures close to those on the thermodynamic scale. The birth of platinum resistance thermometers as useful precision instruments occurred in 1887 when H. L. Callendar [1] reported that platinum resistance thermometers exhibited the prerequisite stability and reproducibility if they were properly constructed and treated with sufficient care. In the next 4 decades the platinum resistance thermometers gained such wide acceptance that their use was proposed and adopted by the Comité International des Poids et Measures in 1927 in defining values on a practical scale of temperature, the first International Temperature Scale [2]. Since that time improvements in the purity of platinum and other materials of construction, in thermometer design, and in calibration techniques have yielded improvements in the precision, accuracy, and range of temperatures that can be measured with platinum resistance thermometers. The international scale has been redefined both to take advantage of these improvements and to bring the scale more nearly into agreement with the thermodynamic scale.

2.1.1 Thermodynamic Temperature Scale

Temperature scales based on functions that can be derived from the first and second laws of thermodynamics, and hence that give values of temperature

This chapter is adapted from John L. Riddle, George T. Furukawa, and Harmon H. Plumb, Platinum Resistance Thermometry, *Natl. Bur. Std. Mon.* 126, April 1973.

The authors gratefully acknowledge the critical review and many suggestions given by Martin L. Reilly which helped to improve the original monograph. We also acknowledge the help of William J. Hall who wrote the computer program for reducing the SPRT calibration data between 13.81 and 90.188K and William R. Bigge who made the measurements for some of the illustrations.

consistent with the entire system of logic (thermodynamics) derivable from these laws, are said to be thermodynamic scales. The present thermodynamic scale was established in 1954 by assigning the value 273.16 kelvins to the triple point of water [3]. On the Kelvin thermodynamic scale, the values of temperature and the designation of the appropriate units and symbols have been adopted [4] as follows: The unit of measurement on this thermodynamic scale is the kelvin indicated by the symbol K. The same name and symbol are also used to denote an interval of temperature. The designation of an interval of temperature by "degrees Celsius" is also permitted. The symbol of the value of temperature on the Kelvin thermodynamic scale is T. The Celsius thermodynamic scale is defined by the equation: $t = T - 273.15K$. The zero of the Celsius scale is 0.01K below the triple point of water. (Although the value of the temperature of the ice point is close to and can be taken for all practical purposes to be $0°C$, the value of the ice point is no longer defined as $0°C$.) The unit of a Celsius thermodynamic temperature is the degree Celsius, symbol $°C$, which is equal to the unit of temperature on the Kelvin scale. The symbols of the values of temperature, t and T, contain the unit of temperature $°C$ and K, respectively.

2.1.2 Practical Temperature Scales

Accurate measurements of values of temperature on the thermodynamic scale are fraught with experimental difficulties, yet all values of temperature should be referable to the Kelvin thermodynamic scale. "Practical" temperature scales are intended to give values of temperature that are comparatively easily reproduced and, therefore, utilitarian, but most are based on functions that are not related to the first and second laws of thermodynamics. In contrast, vapor pressure scales, e.g., that of ^4He [5], are based on the laws of thermodynamics as is also the radiation pyrometry scale. Because many physical laws are based on thermo-dynamic temperatures, the values of temperature on any practical scale should be as close to the thermodynamic scale as is possible or the differences between a practical scale and the thermodynamic scale should be appropriately documented so that conversions from the practical scale to the thermodynamic temperature scale are possible.

Many practical scales have been used in the last 50 years, but the only scales employing platinum resistance thermometers that have had widespread use in the United States were those defined and sanctioned by either the International Committee of Weights and Measures or the National Bureau of Standards. Successive scales that were defined by the International Committee of Weights and Measures are the International Temperature Scale of 1927 (ITS-27) [2], the International Temperature Scale of 1948 (ITS-48) [6], and the International Practical Temperature Scale of 1948 with the text revision of 1960 (IPTS-48) [7]. The differences in the definition of these scales primarily arose from steps to improve the reproducibility of the scale and the values of temperature between -182.97 and $630.5°C$ remained essentially unchanged—the changes in the values of temperature were less than the uncertainty of the values of the scale at the time. These differences are summarized in the Appendix to this chapter.

Because the ITS-27 [2] extended only 7°C below the normal boiling point of oxygen (− 182.962°C), the National Bureau of Standards was motivated to develop a practical scale from the oxygen point down to approximately 11K [8]. This scale has been variously referred to as the NBS provisional scale, the Hoge and Brickwedde scale, and most recently as the NBS–39 scale; directly related to the scale is the NBS-55 scale ($T_{NBS-55} = T_{NBS-39} − 0.01K$) [9]. These NBS scales have had widespread use but, together with the previously mentioned international scales, have now been supplanted by the International Practical Temperature Scale of 1968 (IPTS-68) [4] (see the Appendix at the end of this book for the complete text of IPTS-68).

The text of the IPTS-68 introduced the first major changes in the international practical scale since 1927. Changes were made to extend the range of the scale down to 13.81K and to enhance its reproducibility as well as to improve its agreement with the thermodynamic scale. Temperatures on the IPTS-68 may be expressed in either kelvins or degrees Celsius. The symbols and units of the international practical scale are like those described earlier in this section for use with the thermodynamic scale. The symbol for the value of temperature and for the unit of temperature on the International Practical Kelvin Scale are T_{68} and K (Int. 1968), respectively. For the International Practical Celsius Scale they are correspondingly, t_{68} and °C (Int. 1968). The relation between T_{68} and t_{68} is

$$t_{68} = T_{68} − 273.15K \qquad (2.1)$$

The subscripts or parenthetical parts of the designation need not be used if it is certain that no confusion will result from their omission.

In the text of the IPTS-1968, platinum resistance thermometers are defined as standard interpolation instruments for realizing the scale from − 259.34 to 630.74°C (or from 13.81 to 903.89K). The resistance values (or ratios of resistance values, $R(t)/R(0)$, where $R(t)$ is the resistance at temperature t and $R(0)$ the resistance at 0°C) of any particular standard platinum resistance thermometer are related to the values of temperature with specified formulae. The constants of the formulae are determined from the resistance values of the thermometer at stated defining fixed points and, usually, from the derivatives of the formula specified for the temperature range immediately above. Hereafter, platinum resistance thermometers that meet the following specifications, based primarily on the text of the IPTS-68, will be referred to as *standard platinum resistance thermometers* and abbreviated as SPRT:

- The platinum resistor shall be very pure annealed platinum supported in such a manner that the resistor remains as stain free as practical.
- The value of $R(100)/R(0)$ shall not be less than 1.3925.

- The stability of the thermometer shall be such that, when the thermometer is subjected to thermal cycling similar to that encountered in the normal process of calibrating it, the value of $R(0)$ does not change by more than $4 \times 10^{-6} R(0)$.
- The platinum resistor shall be constructed as a four-lead element and both the resistor and its leads shall be insulated in such a manner that the measured resistance of the platinum resistor is not affected by the insulator more than about $4 \times 10^{-7} R(0)$ at the temperatures of calibration or use.
- The four-lead resistor shall be hermetically sealed in a protective sheath. (An SPRT that has an $R(0)$ value of about 25.5 Ω will be referred to as a 25-Ω SPRT.)

The text of the scale assigns an exact numerical value of temperature to each of the defining fixed points. The defining points and their assigned values are listed in Table 2.1. The realization of the fixed points and the use of interpolation formula are discussed in detail in Secs. 2.6 and 2.5, respectively.

The IPTS-68 is defined by the text of the scale. However, even if perfect experimental work were possible, the values of temperature would not be unique because of a lack of uniqueness in the definition of the scale itself. For example, the text of the scale specifies the minimum quality of the platinum to be used in the SPRT; yet different samples of suitable platinum may not give exactly the same value of a temperature between defining fixed points. The different values of temperature obtained would, however, be in accordance with IPTS-68 because the samples of platinum are within the definition of the scale. Another source of difference is the variation in the realization of the fixed points where small differences in the purity of the fixed-point samples could easily occur. This spread of temperatures (the "legal" spread of the scale) is the result of the looseness of the definition of the scale; it is small compared to the usual errors of experimental measurement, but does exist and is a practical necessity if the physical properties of real materials are to be used in realizing the scale.

2.2 PLATINUM RESISTANCE THERMOMETER CONSTRUCTION

This discussion is primarily directed toward thermometers that are suitable as defining standards on the IPTS-68; however, many of the techniques described are applicable to any resistance thermometer. To be suitable as an SPRT as described in the text of the scale (Appendix to this book) the resistor must be made of platinum of sufficient purity that the finished thermometer will have a value of $R(100)/R(0)$ not less than 1.3925 or α, defined as $[R(100) - R(0)]/100 R(0)$, not less than 0.003925. This requirement provides a scale that is more closely bounded and is not unreasonable since platinum wire of sufficient purity to yield the above ratio is now produced in several countries. The typical SPRT has an

TABLE 2.1 Defining fixed points of the IPTS–68[a]

	Assigned value of International Practical Temperature		
Equilibrium state	T_{68} (K)	t_{68} (°C)	W^*
Equilibrium between the solid, liquid, and vapor phases of equilibrium hydrogen (triple point of equilibrium hydrogen)	13.81	− 259.34	0.0014 1207[b]
Equilibrium between the liquid and vapor phases of equilibrium hydrogen at a pressure of 33 330.6 N/m² (25/76 standard atmosphere)	17.042	− 256.108	0.0025 3445[b]
Equilibrium between the liquid and vapor phases of equilibrium hydrogen (boiling point of equilibrium hydrogen)	20.28	− 252.87	0.0044 8517
Equilibrium between the liquid and vapor phases of neon (boiling point of neon)	27.102	− 246.048	0.0122 1272
Equilibrium between the solid, liquid, and vapor phases of oxygen (triple point of oxygen)	54.361	− 218.789	0.0919 7253[b]
Equilibrium between the liquid and vapor phases of oxygen (boiling point of oxygen)	90.188	− 182.962	0.2437 9911[b]
Equilibrium between the solid, liquid, and vapor phases of water (triple point of water)[c]	273.16	0.01	
Equilibrium between the liquid and vapor phases of water (boiling point of water)[c, d]	373.15	100	1.3925 9668
Equilibrium between the solid and liquid phases of zinc (freezing point of zinc)	692.73	419.58	2.5684 8557

[a]Except for the triple points and one equilibrium hydrogen point (17.042K) the assigned values of temperature are for equilibrium states at a pressure $p_0 = 1$ standard atmosphere (101 325 N/m²).

[b]This value is slightly different from that given in the text of the IPTS–68 (see Appendix A).

[c]The water used should have the isotopic composition of ocean water (see Sec. III.4, Appendix A).

[d]The equilibrium state between the solid and liquid phases of tin (freezing point of tin) has the assigned value of $t_{68} = 231.9681°C$ and may be used as an alternative to the boiling point of water. At this value of temperature $W^* = 1.8925\ 7109$.[b] The value of $t' = 231.929163°C$.

ice-point resistance of about 25.5 Ω and its resistor is wound from about 61 cm of 0.075-mm wire. The wire is obtained "hard drawn," as it is somewhat easier to handle in this condition, but it is annealed after the resistor is formed. Although wires between 0.013-mm and 0.13-mm diameter are commonly used in industrial platinum thermometers, experience shows that finer wires tend to have lower values of $R(100)/R(0)$ (implying the presence of more impurities or strains).

The insulation material that supports the resistor and leads must not contaminate the platinum during the annealing of the assembled thermometer nor when subjected for extended periods of time to temperatures to which the thermometer is normally exposed. The insulation resistance between the leads must be greater than 5×10^9 Ω at 500°C if the error introduced by insulation leakage in the leads of a 25-Ω thermometer is to be less than the equivalent of 1 $\mu\Omega$. For SPRT's the most commonly used insulation is mica. The primary difficulties in the use of mica are the evolution of water vapor at high temperatures and the presence of the iron oxide impurity which, if reduced, leaves free iron that will contaminate the platinum. The most usual mica is muscovite [$H_2KAl_3(SiO_4)_3$], sometimes called *India* or *ruby mica*; it contains about 5% water of crystallization. The dehydration starts at approximately 540°C [10] and not only causes mechanical deterioration of the mica but supplies free water vapor that reduces the insulation resistance. Phlogopite mica [H_2KMg_3-$Al(SiO_4)_3$] or "amber" mica is used for thermometers that are expected to function up to 630°C. Although the room temperature resistivity of phlogopite mica is slightly lower than that of muscovite, it does not begin to dehydrate until well above 700°C [10] .

Some designs employ very high-purity alumina or synthetic sapphire insulation (see Fig. 2.1) [11-14]. Fused silica has also been used [12]. Although the properties of these materials are superior to mica at high temperatures, the materials are more difficult to fabricate and, additionally, they must be of very high purity because migration of impurities and the consequent contamination occurs much more easily at high temperatures. In cutting either mica or alumina great care must be taken to avoid or eliminate traces of metal that originate from either the cutting tool or from the metal clamps that are used to support the insulation during machining. A "carbide-tipped" tool should be used for the cutting process. Unfortunately, attempts to remove the metal chemically from the mica result in contaminating the mica. Synthetic sapphire, however, can be chemically cleaned after machining.

The configuration of the resistor is inevitably the result of compromise between conflicting requirements. The resistor must be free to expand and contract without constraint from its support. This characteristic is the so-called "stain-free" construction. If the platinum were not free to expand, the resistance of the platinum would not only be a function of temperature but would also relate to the strain that results from the differential expansion of the platinum and its support. Seven methods of approach toward achieving strain-free construction are illustrated in Fig. 2.2 [15, 16, 18-21]. Because of the lack of adequate mechanical support, the wire in each of these designs may be strained by acceleration, e.g., shock or vibration. The thermal contact of the resistor with the protecting envelope or sheath is primarily through gas which, even if the gas is mostly helium, is obviously poor compared to the thermal contact that is possible through many solid materials. This poor thermal contact increases the self-heating effect and the response time of the thermometer. The designs shown in Fig. 2.2a, b, and g suffer less in these respects than the others. On the other

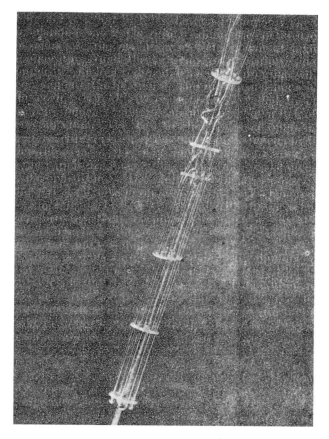

FIG. 2.1 Synthetic sapphire form for thermometer. Synthetic sapphire disks are employed to keep the platinum wires separated in a "bird-cage"-type SPRT. A centrally located, heavier-gage platinum wire maintains the spacings between the disks. (From Evans and Burns, 1962, Ref. [13].)

hand, for calorimetric work the instrument of the lowest heat capacity is often preferred.

The sensing elements of all SPRT's have four leads (see Fig. 2.2). The four leads define the resistor precisely by permitting measurements that eliminate the effect of the resistance of the leads. The resistor winding is usually "noninductive," often bifilar, but occasionally other configurations that tend to minimize inductance are used. This serves to reduce the pickup of stray fields and usually improves the performance of the thermometer in ac circuits. (If the resistor is to be measured using ac, the electrical time constant, i.e., reactive component, should be minimized.)

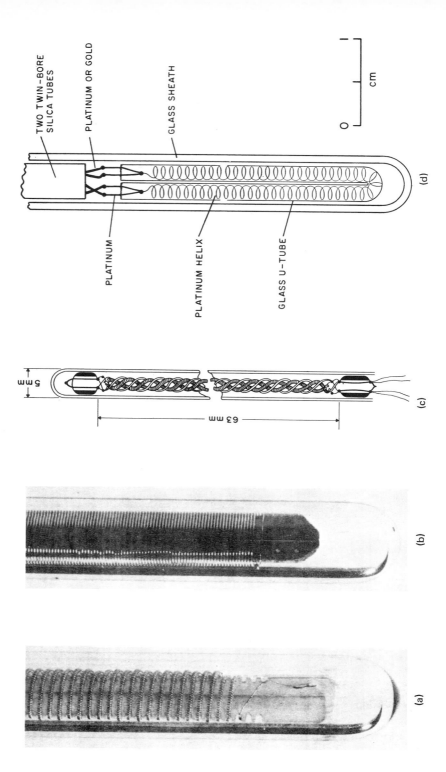

TWO TWIN-BORE SILICA TUBES

PLATINUM OR GOLD

GLASS SHEATH

PLATINUM

PLATINUM HELIX

GLASS U-TUBE

cm

0

(d)

5 mm

63 mm

(c)

(b)

(a)

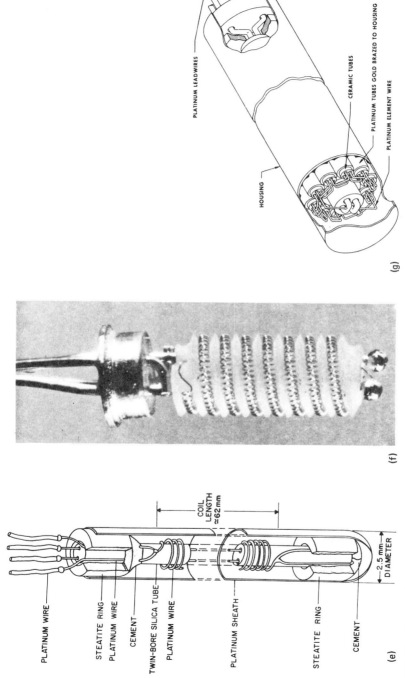

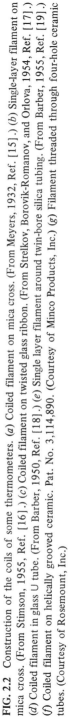

FIG. 2.2 Construction of the coils of some thermometers. (*a*) Coiled filament on mica cross. (From Meyers, 1932, Ref. [15].) (*b*) Single-layer filament on mica cross. (From Stimson, 1955, Ref. [16].) (*c*) Coiled filament on twisted glass ribbon. (From Strelkov, Borovik-Romanov, and Orlova, 1954, Ref. [17].) (*d*) Coiled filament in glass U tube. (From Barber, 1950, Ref. [18].) (*e*) Single layer filament around twin-bore silica tubing. (From Barber, 1955, Ref. [19].) (*f*) Coiled filament on helically grooved ceramic. Pat. No. 3,114,890. (Courtesy of Minco Products, Inc.) (*g*) Filament threaded through four-hole ceramic tubes. (Courtesy of Rosemount, Inc.)

33

Because the junction of the leads is electrically a part of the measured resistor, the leads extending immediately from the resistor must also be of high-purity platinum; the lengths of these leads are often as short as 8 mm. Either gold or platinum wire is employed in continuing these leads within the thermometer. Gold does not seem to contaminate the platinum and is easily worked. Measurement of the resistor may be facilitated if the four leads are made of the same material with the length and diameter the same so that the leads have about equal resistances at any temperature within the temperature range of the thermometer. This statement is also applicable to the leads that are external to the protecting envelope. Figure 2.3 shows the arrangement of thermometer leads near the head of three SPRT's.

The hermetic seal through the soft glass envelope at the thermometer head is frequently made using short lengths of tungsten wire, to the ends of which platinum lead wires are welded. The external platinum leads are soft soldered to copper leads that are mechanically secured to the head. For dc measurements, satisfactory external copper leads can be made from commercially available cable.

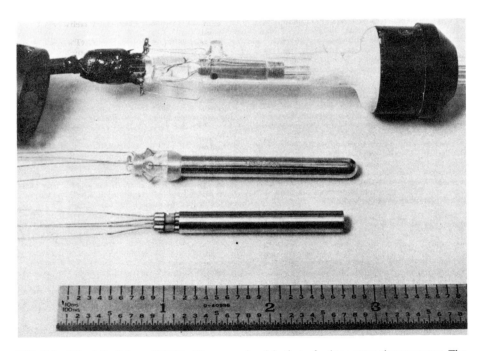

FIG. 2.3 "Heads" of three SPRT's. The upper head is that of a long stem thermometer. The internal leads are brought out through hermetic seals and are connected to external copper leads at the left. In the center is a capsule-type SPRT with leads brought out through glass seals. Capsule body is platinum. The bottom capsule-type SPRT shows the thermometer leads brought out through individual metal-ceramic-metal-type seals. Capsule body is stainless steel.

The cable consists of 12 wires (No. 26 AWG solid bare copper wire) each insulated with double silk windings; additionally, another double silk wrapping encloses each group of three insulated wires. A double silk braid and a tight, waterproof, polyvinyl coating cover the entire four groups. After the appropriate end insulation is removed, the ends of the three wires in each group are twisted together to form a lead to which a lug is soldered. Stranded leads which do not have individually insulated strands should not be used, as the breakage of a single strand may cause "noise" in the resistance thermometer circuits which is difficult to locate and eliminate. The leads external to small capsule-type thermometers are usually solid copper wires, although wires of other materials such as manganin are sometimes used; the leads to the thermometer must be placed in good thermal contact with the system to be measured by the capsule thermometer.

2.3 USING THE THERMOMETER

2.3.1 Mechanical Treatment of SPRT

An SPRT is a mechanically delicate instrument. As discussed in the section on thermometer construction, the platinum wire cannot be rigidly supported and at the same time be free to expand and contract with temperature changes. Shock, vibration, or any other form of acceleration may cause the wire to bend between and around its supports, thus producing strains that change its temperature-resistance characteristics. Strains in the platinum resistor normally will increase the resistance and decrease the value of α. If a tap of a thermometer with glass sheath is hard enough to be audible, but still not cause breakage, the action will typically increase the triple-point resistance of a 25 Ω SPRT by an amount between 1 and 100 $\mu\Omega$. (A change of 100 $\mu\Omega$ is equivalent to 0.001°C in temperature.) Thermometers that have received repeated shocks of this kind through rough handling over a 1-year period have increased in resistance at the triple point of water by an amount equivalent to 0.1°C. Similar changes may be caused by using the thermometer in a bath that transmits vibrations to the thermometer or by shipping the thermometer not suitably packed.

It is preferable to "hand carry" the thermometer to maintain the integrity of its calibration. If the thermometer must be shipped, it should first be placed in a rigid and moderately massive container that has been lined with material which softly conforms to the thermometer and protects it from mechanical shocks or vibrations. This container should then be packed in an appreciably larger box with generous room on all sides for soft packing material that will absorb or attenuate the shocks that might occur during shipment. Two reasonable package configurations are shown in Figs. 2.4 and 2.5.

In arranging storage for the thermometer in the laboratory, a container should be used that minimizes or eliminates the possibility of "bumping" the thermometer. This is a most reasonable precaution when one considers the amount of handling that many SPRT's receive during the course of many measurements. This precaution may be even more pertinent for SPRT's that are routinely used as a standard for calibrating other thermometers. Figure 2.6 shows a storage arrangement employed at NBS.

Care should also be exercised in protecting the thermometer from cumulative shocks that might be received during insertion into apparatus. Figure 2.19 (Sec. 2.6.1) shows the simple provisions made at NBS for reducing mechanical shock when thermometers are inserted into triple-point of water cells. The polyethylene plastic tube at the top of the cell guides the SPRT into the thermometer well without "bumping," and the tapered entrance of the aluminum sleeve near the bottom of the well then guides the SPRT into the sleeve and onto the soft polyurethane sponge at the bottom.

2.3.2 Thermal Treatment of SPRT

With exception of specific instances, SPRT's, as they are generally used, are not greatly susceptible to damage from thermal shocks. In the case of capsule-type thermometers, the metal-to-glass seals may be broken by rapid cooling, e.g., when a capsule thermometer that has been at ambient temperature is quickly immersed

FIG. 2.4 A method for packaging an SPRT for shipment. The metal case contains an SPRT snugly nested in polyurethane foam. The metal case in turn is protected during shipment by a tightly fitted polyurethane foam-lined box.

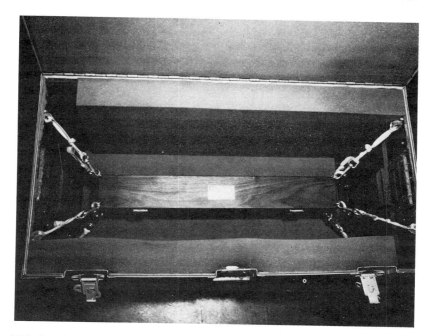

FIG. 2.5 Special container for SPRT shipment. The wooden case contains an SPRT snugly nested in polyurethane foam. Each of the eight corners of this wooden case is attached elastically to the large metal container. Blocks of polyurethane foam give additional protection in the event of "overtravel." (Courtesy of Rockwell International under contract to NASA-JC.)

in liquid nitrogen. Sudden exposure of capsule thermometers to temperatures that are much higher than ambient seldom occurs since they are rarely used above 100°C. Another specific instance of thermal shock is subjecting long-stem SPRT's to temperatures that are well above 600°C and subsequently quenching them. This treatment can mechanically damage the thermometer but, even if no visible damage has been done, the calibration of the thermometer may have been altered by the freezing in of defects that occurred during the quenching [22].

The upper temperature limit of an SPRT is restricted by the softening point of the material of the protecting sheath, the temperature at which the thermometer was annealed before calibration, the evolution of water and other contaminants from the sheath and insulators, and grain growth in the platinum wire. The concomitant changes are a function of both time and temperature and most are predictable in only a qualitative way.

Thermometer sheaths of borosilicate glass soften noticeably at temperatures above 500°C, e.g., their use at 515°C (1000°F) is limited to a few hours unless they are specially supported to prevent deformation. Fused silica sheaths should be used for measurements above 500°C. Platinum grain growth has been observed in SPRT's that have been used over a period of several hundred hours at 420°C.

FIG. 2.6   A storage for SPRT's. The thermometers are stored
vertically by slipping them into the wire braid sleeves. The
cover to the box has been removed only to show the arrangement of the wire braid sleeves.

Figure 2.7, made with an electron microscope, shows in detail a short section of
platinum wire from a thermometer that was subjected to such a treatment.
Grains which are one- to two-wire diameters long and as large as the wire in
diameter may be clearly seen. The growth of large grains results in a thermometer that is more susceptible to calibration changes caused by mechanical shock;
consequently, the thermometer may be considered unstable. The evolution of
water within the thermometer sheath at temperatures below 500°C has been
observed in a few thermometers but seems to be avoidable if sufficient baking
and evacuation are furnished during the fabrication of the thermometer. The
presence of water may become more conspicuous when the thermometer is
cooled, e.g., in a triple-point cell. The presence of water, either within the
thermometer or on the external insulated leads, will be evidenced during electrical measurements by the kick of the galvanometer when the current through

the thermometer is reversed. The kick stems from the polarization of water on the insulation. The decay time (the time for repolarization of water on the insulation) of the galvanometer pulse is usually greater than a minute when moisture exists in the thermometer but is only of a few seconds' duration with moisture on silk insulated leads, indicating that the quantity and physical state of moisture and the effect of moisture on the insulation inside the thermometer are different from those on the external silk insulation.

Annealing may occur if the platinum wire was not properly annealed during the manufacturing process or (more likely) if it has been strained by mechanical shock since it was last well annealed. If the strains are sufficiently severe, noticeable annealing will occur at temperatures as low as 100°C. The annealing process does have somewhat beneficial aspects because the thermometer tends to return toward the metallurgical state that it presumably was in during its previous calibration. (At NBS all long-stem SPRT's are annealed between 470 and 480°C for 4 hours, removed from the furnace while still hot, and allowed to cool in air at ambient temperature prior to their calibration.) The comparison of the resistance of the thermometer at a fixed point before and after annealing will show a shift downward in resistance and, usually, a shift upward in $R(t)/R(0)$ (see Sec. 2.5).

In the measurements at temperatures above 500°C the difficulties discussed above are more likely to occur or be accelerated. More care must be taken in baking and evacuating the thermometer after assembly. For work near the upper temperature limit (630°C), the SPRT's should have phlogopite mica or fused

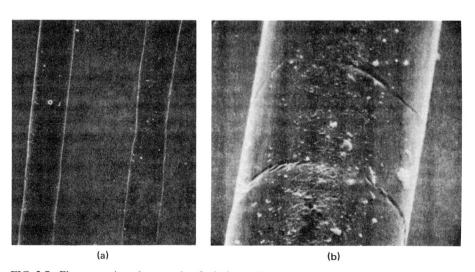

(a)   (b)

FIG. 2.7 Electron microphotograph of platinum thermometer wire subjected to a total of several hundred hours at 420°C. Magnification: Photograph (a) 172X, (b) 850X. The photographs show: (1) grain growth, (2) migration. No attempt has been made to date to identify the white "flecks."

silica coil forms and a fused silica sheath. Also, care should be exercised to avoid the previously mentioned effect of freezing in of high temperature defects when cooling the thermometer.

2.3.3 Thermometer Immersion

A thermometer is sufficiently immersed in the bath when there is no heat flow between the sensor and its environment through the thermometer leads or sheath that extend from the region of sensor. Regrettably, considering the ease with which the adequacy of immersion can be checked, insufficient immersion of a thermometer is still a very common error. The test simply involves equilibrium temperature measurements at two depths of immersion, while the bath is maintained at very nearly constant temperature. If, after taking into account any change in the temperature of the bath, the temperatures at the different immersions do not agree, an immersion problem exists. Experimentalists are frequently led astray by the erroneous idea that the required immersion is strongly temperature dependent. Figure 2.8 shows the difference between the bath temperature and the temperature indicated by the thermometer as a function of immersion for two different thermometers. The immersion characteristics are more clearly seen in Fig. 2.9 where the same data are given in a semilog plot. A linear relationship between the immersion and the logarithm of the temperature difference is to be expected in simple cases, and in practice this is a very useful approximation in the usual region of interest, namely, where the temperature difference is small. Figure 2.9 shows that for thermometer G in an ice bath, the temperature difference is reduced by a factor of 10 for each 3.3 cm of immersion. For thermometer M, the temperature difference is reduced by a factor of 10 for each 1.4 cm of immersion. If the error due to immersion is to be less than 0.025 mK, the temperature difference must be reduced by six orders of magnitude and, therefore, both thermometers must be immersed in the ice bath by six times the respective amounts cited above, or a total of 19.8 cm and 8.4 cm, respectively. Even if the temperature difference between the ambient and the bath were only 2.5 degrees the thermometer immersion could be decreased only slightly because, for the same accuracy, a reduction of the temperature difference by five factors of 10 is still needed.

The similarity of the radial conductance of heat to or from the thermometer in the bath is important. Figure 2.10 shows the immersion characteristics of thermometers G and M in a triple-point-of-water cell rather than an ice bath. The immersion characteristics have changed for both thermometers, but particularly for thermometer M. This is caused primarily by the increase in thermal resistance radially from the thermometer to the ice-water inferface of the triple-point cell: The sheath of water that surrounds the thermometer within the triple-point cell well and the glass wall of the well contribute to the increased thermal resistance (see also Sec. 2.3.4). The immersion problem would be even greater if the space between the thermometer well and the thermometer were filled with air instead

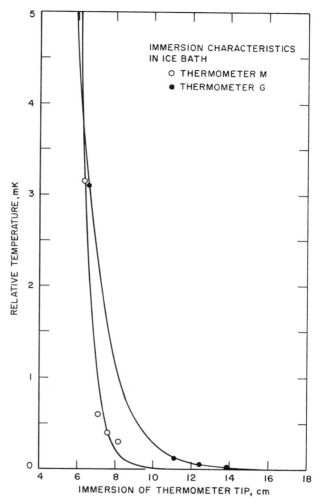

FIG. 2.8 Immersion characteristics in an ice bath of two long-stem SPRT's with different sheath materials and internal construction. The plot shows the relative temperature as a function of the depth of thermometer immersion.

of water. Typical immersion data for thermometers in tin and zinc freezing point cells are given in the discussion of that apparatus (see Sec. 2.6).

2.3.4 Heating Effects in SPRT

The measurement of resistance necessarily involves passing a current through the resistor. The resultant heating that occurs in the resistor and its leads raises their

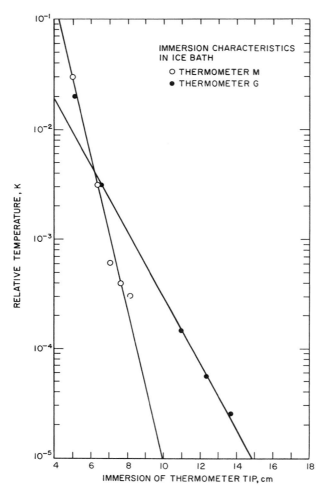

FIG. 2.9 Immersion characteristics in an ice bath of two long-stem SPRT's with different sheath materials and internal construction. The data of Fig. 2.8 have been replotted to show the linear relationship between the logarithm of the relative temperature and the depth of thermometer immersion.

temperature above that of their surroundings until the resistor element attains a temperature sufficiently higher than the surroundings to dissipate the power developed. A typical steady state profile of the radial temperature distribution caused by a current of 1 mA flowing in a 25-Ω SPRT is shown in Figs. 2.11 and 2.12. In Fig. 2.11, the *internal* heating effect of the thermometer, i.e., between the platinum resistor and the outside wall of the protecting sheath, at a given environmental temperature is a function only of the thermometer construction

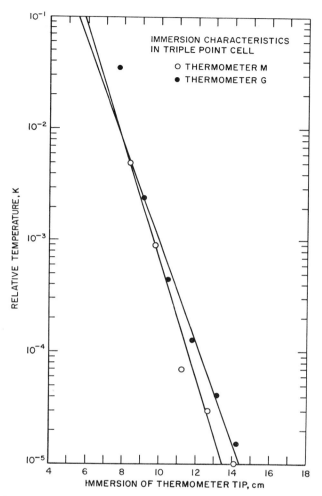

FIG. 2.10 Immersion characteristics in the triple-point-of-water cell of two long-stem SPRT's with different sheath materials and internal construction. The plot shows the relationship between the logarithm of relative temperature and the depth of thermometer immersion. Comparison with Fig. 2.9 shows that the immersion characteristics of the thermometers tend to be poorer in the triple-point cell than those in the ice bath. This change in the observed immersion characteristics is caused primarily by the higher resistance to radial heat flow in the triple-point cell.

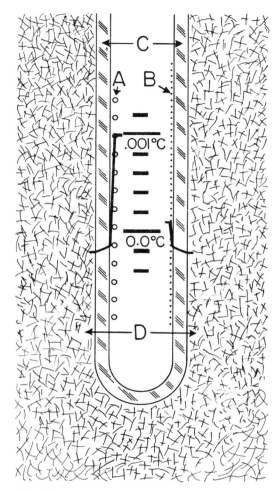

FIG. 2.11 Platinum resistance thermometers at 33-cm immersion in an ice bath. Temperature profile from the middle of thermometer coil out to the ice bath with 1 mA current. A. Platinum coils of coiled filament thermometer; only coils of one side are indicated. B. Platinum coils of single layer helix thermometer; only turns of one side are indicated. C. Borosilicate glass thermometer envelope. D. Finely divided ice and water.

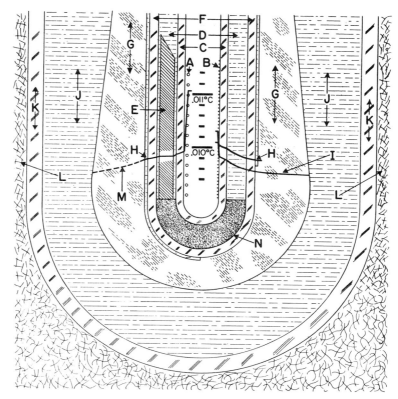

FIG. 2.12 Platinum resistance thermometers at 33-cm immersion in a water triple-point cell. Temperature profile from the middle of thermometer coil out to the ice-water interface with 1 mA current. A. Platinum coils of coiled filament thermometer; only coils of one side are indicated. B. Platinum coils of single layer helix thermometer; only turns of one side are indicated. C. Borosilicate glass thermometer envelope. D. Water from ice bath. E. Aluminum bushing (length not to scale). F. Borosilicate glass thermometer well. G. Ice mantle. H. Inner melt. I. No inner melting; temperature profile relative to the temperature of outer ice-water interface. J. Water in cell. K. Cell well (borosilicate glass). L. Outside ice-water bath. M. Temperature profile of the ice mantle. N. Polyurethane sponge.

and the current and is, therefore, the same during both calibration and use. This assumes that the thermometer resistor does not move within its protecting sheath. At the ice point, the internal heating effect may be measured by direct immersion in the ice bath and is typically between 0.3 mK/(mA)^2 and 1.2 mK/(mA)^2 for a 25-Ω SPRT. If the thermometer is used with the same current that was used during its calibration, the same internal heating effect occurs and no error is introduced in the measurement.

As Fig. 2.12 indicates, there is also an *external* heating effect, an extension

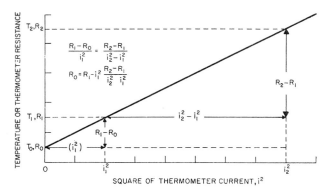

FIG. 2.13 Plot of SPRT resistance (temperature) vs. the square of the thermometer current. The plot illustrates how the value of "zero current" thermometer resistance may be obtained graphically or algebraically from measurements at two currents. R_0 = resistance of SPRT at "zero current." R_1 and R_2 = resistances determined at currents i_1 and i_2, respectively.

of the heating effect beyond the outside of the thermometer sheath because the generated joule heat must flow to some external heat sink. The total heating effect, i.e., the combined internal and external heating effects, can easily be determined by calculating the resistance that would be measured at zero current; this can be performed either algebraically or graphically as shown in Fig. 2.13. When it is desirable to determine the internal heating effect only, the experimental conditions must be such that the external heating effect is negligible. This condition can be closely approximated by directly immersing the thermometer into an ice bath wherein the solid ice particles are in contact with the thermometer sheath, or into a metal (tin or zinc) freezing point apparatus in which the metal freezes directly on the thermometer. (The metal must be remelted before complete solidification or the thermometer will be crushed.) The recommended measurement and calculation procedures are identical to those previously described for determining the total heating effect (see Fig. 2.13). For the most precise work, all resistance measurements should be made at two currents and use made of the value of resistance calculated for zero current. Reiterating what was said earlier, an error due to the heating effect is introduced if the thermometer is not calibrated and used at the same current or is not in good thermal contact with its surroundings.

The external heating effect may be reduced by making the thermometer well a relatively close fit to the thermometer and placing a material of high thermal conductance in the annular space between the thermometer and the well. The material used to fill the space must not undergo an exothermic (or endothermic) reaction at the temperatures involved because such a reaction would additionally change the temperature of the thermometer from that of the surroundings.

Examples of this difficulty that have been experienced in the NBS calibration laboratory are the slow decomposition of light mineral oil (at 122°C) and oxidation of a steel bushing (at 444°C). Difficulties associated with this type of reaction can be detected by comparing the derived values of resistance at zero current that have been obtained from measurements in which (a) the questionable material was used and (b) the material was not present, or a better substitute was employed.

Another source of heat flux to and away from the platinum coil, and consequently a possible source of error, is radiation. If the sensor can "see" a surface that is appreciably hotter or colder, the power gained or lost by the resistor will result in its temperature being changed. In the triple-point-of-water measurements, radiation from lights in the room which is incident upon the top of the ice bath or triple-point cell can easily produce an error of 0.0001K (see Sec. 2.6 on triple-point cell). The water triple-point cell should, therefore, be immersed in an ice bath in which no extraneous radiation from sources above room temperature can reach the sensor of the SPRT. At a higher temperature (630°C) an error as large as 0.01°C can occur if a clear fused silica thermometer sheath is employed because "light piping" takes place. In this process, radiation is conducted toward ambient temperatures within the wall of the thermometer sheath, being confined there by total internal reflections. Any technique which would eliminate these reflections would, of course, eliminate this source of error. The error can be substantially reduced by coating the exterior wall with graphite paint as is depicted in Fig. 2.14 or by roughening (by sandblasting) the external wall.

2.4 RESISTANCE MEASUREMENTS

In this section, the salient features of instrumentation used to measure the resistance of SPRT's in the NBS calibration laboratory will be described so that the laboratory's general electrical measurement procedures may be understood. The discussion will not give details about other instruments that are available for

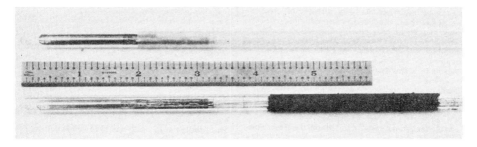

FIG. 2.14 Two methods for reducing radiation "piping" in fused silica thermometer sheaths. The sheath of the thermometer at the top was given a matte finish by "sandblasting." "Aqua dag" is shown painted on the sheath of the SPRT at the bottom.

measuring electrical resistance nor will special design features of particular instruments be described, nor will different measurement techniques be compared. Some of the instruments will be named and literature will be cited to direct an interested reader to more detailed information.

The most suitable methods for and the limitations on achieving both accurate and reproducible resistance measurements depend on several factors. Johnson noise [23], which is inherent in any resistor (caused by the random movement of electrons within a conductor), is given by

$$e = 7.43 \times 10^{-12} \sqrt{TR(\Delta f)} \qquad (2.2)$$

where e is the thermal agitation voltage in volts, Δf is the effective bandwidth in hertz, R is the resistance of the conductor in ohms, and T is the temperature of the resistor in kelvins. The constant 7.43×10^{-12} comes from $2\sqrt{k}$, where k is the Boltzmann constant. However, this frequently mentioned limitation is seldom the predominant source of uncertainty in resistance measurements with SPRT's. For example, using a 25-Ω thermometer at room temperature with an observation (averaging) time of 1 sec, this noise is of the order of 0.6 nV—slightly less than the signal (1.0 nV) that results from a 1-$\mu\Omega$ unbalance when 1 mA is flowing through the thermometer. Spurious emf's, variations in contact resistance, mechanical or electrical disturbance of the detector system, and variations in thermometer lead unbalance are sources of "noise," one or more of which contribute much more to the electrical measurement uncertainty than the Johnson noise in the vast majority of measurement systems.

The signal level at the null detector caused by the resistance unbalance can be raised by simply increasing the current through the thermometer. This may at times be a satisfactory solution; limitations are encountered, however, because increasing the current rapidly increases the uncertainty due to the self-heating of SPRT (see Sec. 2.3) and may even introduce significant error due to the self-heating of reference resistors in the bridge or potentiometer. The increased power dissipated in the thermometer may produce additional difficulties if the thermometer is used in a calorimeter system with a small heat capacity.

In addition to the uncertainty of null detection in a reasonable period of time, there is also the problem of referencing the unknown resistor to a single resistance standard. A large number of arrangements have been proposed and used for this purpose. They may be broadly classified as either bridge or potentiometer circuits that employ either alternating or direct currents with either resistors or inductive dividers to establish the ratio of the unknown resistor to a standard resistor. (A standard resistor is defined, hereafter, as a stable resistor or a combination of stable resistors of known value.)

2.4.1 Mueller Bridge

At the National Bureau of Standards the traditional instrument used with SPRT's is the Mueller bridge [24, 25]. There have been several modifications [26-28] of the bridge since its first appearance, but this discussion will only cover its principal features. A simple form of the bridge is shown schematically in Fig. 2.15. It is basically the classical equal-ratio-arm Wheatstone bridge with provisions for interchanging (commutating) the leads of a four-lead resistor in such a way that the average of two balances is independent of the resistances of the leads. Referring to Fig. 2.15 (normal), when no current flows through the null detector, the voltages e_1 and e_2 are equal, and the bridge is said to be balanced. The equation of balance is

$$R_{D_1} + R_C = R_X + R_T \qquad (2.3)$$

where R_{D_1} is the resistance of the variable decade balancing resistor,
   R_C is the resistance of a lead from the bridge to the thermometer sensing element,
   R_X is the resistance of the thermometer sensing element, and
   R_T is a lead resistance similar to R_C.
After commutating the leads (marked C, T, c, and t) to the positions shown in Fig. 2.15 (reverse) the equation of balance is

$$R_{D_2} + R_T = R_X + R_C \qquad (2.4)$$

where R_{D_2} is the resistance of the adjustable decade balancing resistor that is required for the second balance of the bridge.

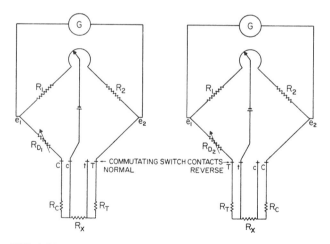

FIG. 2.15 Schematic drawing of a Mueller bridge circuit in normal and reverse thermometer connections. G is the null detector (galvanometer). R_1 and R_2 represent the ratio arm resistors. R_D represents the adjustable resistor. R_C and R_T are the resistances of the "potential" leads of the SPRT. R_X is the resistance of the SPRT.

The addition of Eqs. (2.3) and (2.4) yields the value of R_X in terms of R_{D_1} and R_{D_2}

$$R_{D_1} + R_{D_2} = 2R_X \tag{2.5}$$

or

$$R_X = \frac{R_{D_1} + D_{D_2}}{2} \tag{2.6}$$

An assumption made in the above equations is that R_T and R_C are constant during the time required for making the two balances.

Actually R_T and R_C need not be constant. If Eqs. (2.3) and (2.4) are rewritten as

$$R_{D_1} + R_{C_1} = R_X + R_{T_1} \tag{2.7}$$

and

$$R_{D_2} + R_{T_2} = R_X + R_{C_2} \tag{2.8}$$

and then added, there results

$$R_{D_1} + R_{D_2} + (R_{C_1} - R_{T_1}) - (R_{C_2} - R_{T_2}) = 2R_X \tag{2.9}$$

or

$$R_X = \frac{R_{D_1} + R_{D_2}}{2} + \frac{(R_{C_1} - R_{T_1}) - (R_{C_2} - R_{T_2})}{2} \tag{2.10}$$

Thus, a sufficient condition for the measured resistance to be independent of the lead resistances is that the *difference* in the resistance of the two potential leads is constant during the period of observation. Equation (2.10) demonstrates that experimental emphasis could be placed on insuring that the leads be of equal length and cross section and that the temperature gradients between the leads be constant or only changing slowly, rather than the more difficult option of maintaining the temperature of the leads constant.

Successful operation of the Mueller bridge is dependent upon the reproducibility and self-consistency (linearity) of the adjustable resistor indicated in Fig. 2.15 as R_D. The methods of accomplishing this include thermostating the resistors and employing special circuitry. The circuitry has been designed to reduce the uncertainties that are associated with the variations in the contact resistances of the "dry" switches. For the 1-Ω and 0.1-Ω step decade resistors, the switch contact resistances are placed in series with the bridge ratio arms (R_1 and R_2), which are usually from 500 to 3000 Ω, so that the effect of the possible variations in the contact resistances (about 0.0005 Ω) can usually be neglected. For the measurement of the higher thermometer resistances this arrangement introduces uncertainties which may be significant. For a 25-Ω SPRT the uncertainty of 0.0005 Ω in 3000 Ω or 1 part in 6×10^6 in the ratio arm corresponds to only 1 $\mu\Omega$ or 0.01 mK when measuring 6 Ω (near $-183°C$) but increases to 10 $\mu\Omega$ or 0.1 mK when measuring 60 Ω (near 350°C). The decades with steps of 0.01 Ω or less are the Waidner-Wolff shunted decades [25] which reduce the effect of contact resistance in the switch by a factor of 250 or more. The switches for the 10-Ω step decade and the commutator are directly in series with the resistors in the adjustable arm of the bridge and the thermometer resistor; they have, therefore, mercury-wetted contacts. The mercury-wetted switches in the Mueller bridges employed at the National Bureau of Standards have an uncertainty of less than 2 $\mu\Omega$ when well maintained. All of the mercury contacts are normally cleaned every day before use. The mercury is removed by vacuum using a small polyethylene tip at the end of a vacuum line with a mercury trap. Fresh mercury is placed on each amalgamated contact. If the entire surface of the contact is not wetted by the new mercury, the surface is scrubbed (without removing the mercury) with the flat end of a solid copper rod about 1/4 in. in diameter until the entire surface becomes wetted. The mercury is again removed and replaced with clean mercury. The switch is then re-assembled and operated several times after which it is reopened to remove any mercury that has splashed onto the surrounding surfaces. The switch is then finally assembled for use. Switches with sliding contacts are exercised every day before use by revolving them 10 or 20 times; this is particularly important for the 1-, 0.1-, and 0.01-Ω decade switches. The sliding switches are cleaned occasionally with a lint-free cloth, either dry or moistened with benzene or Varsol. (Carbon tetrachloriae is not recommended as it frequently contains impurities which will result in corrosion.) After cleaning, the contacts are lubri-cated with a light coating of pure petrolatum.

Bridge Ratio Arms

If the two ratio resistors, R_1 and R_2 in Fig. 2.15, change so as to become unequal, this may be compensated by adjusting the "tap" on the slide-wire resistor joining R_1 and R_2. Because this tap is in the battery arm of the bridge, the variations of its contact resistance are unimportant to the determination of thermometer resistances. The ratio arm resistors are adjusted to be equal by varying this tap on the slide wire between R_1 and R_2 until the interchanging of the ratio arms does not change the bridge balance. The accuracy of the ratio is limited by the uniformity of the resistance of the copper leads and switch contacts which are in series with R_1 and R_2 (those connecting the 1-Ω and 0.1-Ω decades) and, of course, by the sensitivity of the null detector.

Some versions of the bridge have incorporated into the commutator addi-tional switch contacts which reverse the ratio arm resistors simultaneously with each commutation of the thermometer leads; the switch contacts of the 1-Ω and 0.1-Ω decades and their leads to the commutator switch are, however, not reversed. The effect of their variations must still be considered in the determina-tion of the bridge resistance. Referring again to Fig. 2.15, but now with the ratio arms R_1 and R_2 interchanged in the reverse bridge connection, the equations of bridge balance with the normal and reverse connections are, respectively

$$\frac{R_{D_1} + R_C}{R_T + R_X} = \frac{R_1}{R_2} = \frac{1 + \epsilon}{1} \tag{2.11}$$

and
$$\frac{R_{D_2} + R_T}{R_C + R_X} = \frac{R_2}{R_1} = \frac{1}{1 + \epsilon} \tag{2.12}$$

where $\epsilon = (R_1 - R_2)/R_2$. After combining Eqs. (2.11) and (2.12) and eliminating R_T and R_C

$$R_X = \frac{R_{D_1} + R_{D_2}(1 + \epsilon)}{2 + \epsilon} \tag{2.13}$$

By adding and subtracting $(R_{D_1} + R_{D_2})/2$ from the right-hand side of Eq. (2.13) and combining, the relation

$$R_x = \frac{(R_{D_1} + R_{D_2})}{2} + \frac{(R_{D_1} - R_{D_2})\epsilon}{2(2 + \epsilon)} \tag{2.14}$$

is obtained. Equation (2.14) shows that if the bridge is operated to yield $R_{D_1} = R_{D_2}$, then $R_X = R_{D_1} = R_{D_2}$, without regard to the lack of equality of the ratio-arm resistors. In practice, because the thermometer leads (T and C) are usually made nearly equal, R_{D_1} and R_{D_2} will be only slightly different, typically much less than 0.002 Ω. Also, the equality of the ratio-arm resistors can be adjusted to better than 1 ppm with ease. Thus, the second term on the right of Eq. (2.14) is completely negligible in normally conducted Mueller bridge measurements which utilize simultaneous commutation of the SPRT leads and the ratio-arm resistors.

Bridge Current Reversal

The indicated balance of any dc bridge is dependent upon the iR voltages across the elements of the circuit and the spurious emf's. The equations of bridge balance involve only resistance values; therefore, the effects of spurious emf's must be eliminated. This could be done by first observing the "indication" of the null detector with no bridge current, then balancing the bridge with current to the same indication. The indication of the null detector with no current includes the effect of any spurious emf's; when the bridge is balanced to the same indication with current in the bridge, the iR voltages are balanced if the effect of the spurious emf's remains unchanged. However, one should recognize that a change in the magnitude of the thermometer current will change the thermometer temperature (due to self-heating) and that enough time must elapse before reading to allow the thermometer to attain a steady thermal state. During this time the effect of spurious emf's may change significantly. This problem can be simply surmounted by reversing the battery current with a snap action toggle switch, so that an essentially continuous heating power is retained in the thermometer (hence, a uniform self-heating effect). In the process, the galvanometer or null detector sensitivity is effectively doubled. The rate at which the current reversals must be made is dependent on the rate of change of the spurious emf's.

2.4.2 AC Bridge

If the current is reversed sufficiently rapidly, it is usually said that the bridge is an "ac" bridge. Bridges operating at 400 Hz have been built at the NBS based on a design by Cutkosky [29]. These bridges were designed for use with SPRT's and include special provisions for use with thermometers having values of $R(0)$ as low as 0.025 Ω. The bridge utilizes an inductive ratio divider that eliminates the necessity of calibrating the bridge because the initial uncertainty of the divider is about 2 parts in 10^8 and appears to be stable. Additionally, the bridge requires only one manual resistance balance, since the phase angle balance is automatic, and incorporates a built-in phase-sensitive null detector with which 1 $\mu\Omega$ in 25 Ω can be easily resolved. Small deviations from balance can be recorded continuously; the accuracy of recording these deviations is limited primarily by the resolution and linearity of the recorder. A small (usually less than 10 $\mu\Omega$) error may be introduced in measuring a 25-Ω SPRT unless coaxial leads are used between the bridge and the thermometer head. (The heads of SPRT's have been modified to contain two BNC coaxial connectors. The two leads from one end of the SPRT coil were connected to the center "female" contacts of the BNC receptacles and the two leads from the other end of the SPRT were connected to the outer shells or the shield contacts.) For precision measurements the length of the pair of coaxial leads should not be greater than 15 m to limit the dielectric losses of the shunt capacitance. Preliminary measurements on 25-Ω SPRT's indicate that, if the leads do not affect the measurements, the accuracy of the measured value in ohms of a thermometer element is limited by the accuracy to which the reference standard resistor is known. However, in the accurate determination of the resistance ratio, $R(t)/R(0)$, the stability, rather than the accuracy, of the reference standard is the important requirement. Further work is in progress at NBS to determine the dc-ac transfer characteristics of SPRT's.

2.4.3 Potentiometric Methods

The Mueller bridge and the Smith bridge methods [30–32] are suitable for a four-terminal determination of thermometer resistance if the lead resistances are relatively stable. When the lead resistances are variable, e.g., in measurements at low temperatures where the cryostat temperature is varying and the thermometer resistance is small, potentiometric methods become better suited to the measurements.

The potentiometric method of resistance measurement depends in principle upon the determination of the ratio of iR voltages developed across the SPRT and across a resistor of known resistance that is connected in series. The ratio is determined by comparisons with iR voltages that are developed in a separate resistance network and usually a separate current supply.

One of the major drawbacks of the potentiometric method has been the requirement of exceptionally high stability of current in both the potentiometer and in the SPRT circuits during the measurement period. However, the recent availability of highly stable current supplies and their continued improvement have made the potentiometric methods more popular.

Three circuits for the more commonly used potentiometric methods are illustrated in Fig. 2.16. To eliminate the effect of spurious emf's in the measurements employing circuits (a) and (b) the currents i_1 and i_2 are reversed. Four balances are necessary in each method, two with the current in one direction and two with the current reversed. Four balances are also necessary employing circuit

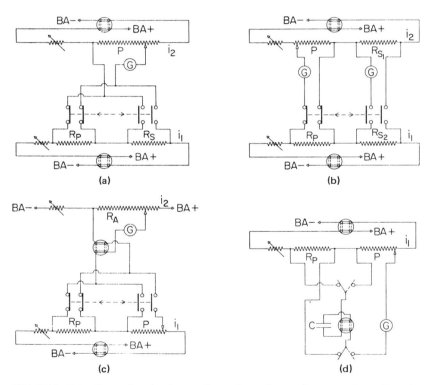

FIG. 2.16 Potentiometric circuits used to determine resistance values. R_p, the resistance to be determined. R_s, standard resistor of precisely known value. P, potentiometer or precisely known adjustable resistor. R_A, adjustable resistor. C, stable capacitor of low dielectric absorption, low absorption time constant, and high leakage resistance (polystyrene). G, null detector. i_1, independent stable current in the unknown resistance circuit. i_2, independent stable current in the potentiometer or in the voltage transfer instrument circuit.

(c) by reversal of the current i_1 and at the connection between the voltages to be compared. The readings for the normal and reverse connections are averaged. By using high-quality reversing switches along with current sources of high resistance (to make the relative effect of small changes in resistance associated with the switching negligible compared to the total resistance of the circuit), the current in the circuit can be made more stable and the switch in series with the detector need not be disturbed during the current reversal process. Nevertheless, a shunt is needed across the detector to protect it because of asynchronous contact operation during the reversal. The discussion of the potentiometric methods to follow assumes that the required balances are made with current reversal.

Using circuit (a) of Fig. 2.16, the potentiometer P is successively balanced against the voltages $i_1 R_p$ and $i_1 R_s$ in terms of $i_2 P_p$ and $i_2 P_s$, respectively, where R_p and R_s are the resistances of SPRT and standard resistor, respectively, and P_p and P_s are in the potentiometer resistance units. (P_p indicates the potentiometer resistance when its iR voltage is balanced against the SPRT; P_s indicates the potentiometer resistance when balanced against a standard resistor.) Thus, with $i_1 R_p = i_2 P_p$ and $i_1 R_s = i_2 P_s$, the SPRT resistance is given by $R_p = P_p/P_s(R_s)$.

There is a wide variety of potentiometers that employ essentially the circuit shown in Fig. 2.16a. The potentiometer iR voltages in some designs [33–36] are developed with different currents in certain decades; in others, the iR voltages are adjusted by varying, by means of the potentiometer switches, the current through a fixed resistor. These designs keep the resistance of the detector circuit constant. In the "double potentiometer," which is designed to make two consecutive voltage balances more conveniently, there are duplicate sets of switches but only a single resistance network to develop iR voltages. A requirement of the potentiometer, whatever the design, is that the "resistance units" or the "iR voltage units" that are developed in the instrument for voltage balance be linear.

Two standard resistors are used in circuit (b). Current i_2 is adjusted until $i_2 R_{s_1} = i_1 R_{s_2}$; then P is adjusted until $i_2 P_p = i_1 R_p$. The SPRT resistance is given by

$$R_p = \frac{R_{s_2}}{R_{s_1}} P_p \tag{2.15}$$

In circuit (c), either current i_2 or resistance R_A is adjusted until $i_2 R_A = i_1 R_p$; then P is adjusted until $i_2 R_A = i_1 P_p$. The thermometer resistance is given directly by $R_p = P_p$.

The "isolating potential comparator" method described by Dauphinee [37] is an adaptation of circuit (c) where the voltage $i_1 R_p$ is set up as $i_2 R_A$ and measured as $i_1 P_p$. The method is shown in circuit (d). The voltage $i_1 R_p$ appears across a high-quality capacitor C and is compared with the voltage $i_1 P_p$, P being adjusted until $i_1 R_p = i_1 P_p$. The break-before-make, double-pole chopper switches the capacitor alternately across R_p and then across P between 20 to 80 times a second. Extraneous voltages are canceled by reversing the current and the capacitor connections and averaging the second reading with the first.

The linearity of the potentiometer can be calibrated by comparing successive steps of one decade agains the total (X) of the next lower decade.

2.5 CALCULATION OF TEMPERATURES
FROM THE CALIBRATION DATA
AND OBSERVED RESISTANCES

The SPRT is the standard interpolation instrument between the defining fixed points in the range 13.81K to 630.74°C. (See Table 2.1 and Appendix A.) The "constants" of the interpolation formulae that relate the resistance of a particular SPRT to the value of its temperature on the IPTS-68 are obtained by resistance measurements at the defining fixed points. At NBS the "long-stem" SPRT's are usually calibrated for application above 90K; the capsule-type SPRT's are calibrated for use between 13K and 250°C or occasionally up to 400°C. The equipment and procedures employed at NBS to achieve the fixed points are described in the next section. The resistance-measuring instruments employed at NBS and other instruments that can be used for resistance measurements of SPRT have been described earlier in Sec. 2.4. This section deals with the methods in use at NBS to obtain the constants of the interpolation formulae from the calibration measurements of an SPRT at the fixed points. Methods of calculating temperatures from the observed resistances, when the constants of the interpolation formulae are known, are also described. At NBS all evaluations of equations and calculations pertaining to the SPRT are performed on a high-speed electronic digital computer (UNIVAC 1108).

2.5.1 Temperatures from 0 to 630.74°C

From 0 to 630.74°C the temperatures on the IPTS-68 are defined by

$$t_{68} = t' + M(t') \tag{2.16}$$

where
$$M(t') = 0.045 \left(\frac{t'}{100}\right)\left(\frac{t'}{100} - 1\right)\left(\frac{t'}{419.58} - 1\right)\left(\frac{t'}{630.74} - 1\right) \tag{2.17}$$

and
$$t' = \frac{1}{\alpha}\left[\frac{R(t')}{R(0)} - 1\right] + \delta\left(\frac{t'}{100}\right)\left(\frac{t'}{100} - 1\right) \tag{2.18}$$

In the official English version of the text of the scale (see Appendix to this book) the unit °C appears in the defining equations, e.g.,

$$t_{68} = t' + 0.045 \left(\frac{t'}{100°C}\right)\left(\frac{t'}{100°C} - 1\right)\left(\frac{t'}{419.58°C} - 1\right)\left(\frac{t'}{630.74°C} - 1\right)°C$$

$$(2.19)$$

In this discussion the equations will be simplified by omitting the units. Similarly $R(t°C)$ and $R(0°C)$ will be simplified to $R(t)$ and $R(0)$. In addition, t_{68} will, henceforth, be abbreviated to t. The constants $R(0)$, α, and δ are determined from calibration resistance measurements of the SPRT at the triple point of water (TP), the steam point or the tin point, and the zinc point. (The tin point is now employed at NBS.) The constants may be more conveniently obtained from Eq. (2.16) and from the relation, equivalent to Eq. (2.18), given by

$$W(t') = \frac{R(t')}{R(0)} = 1 + At' + Bt'^2 \tag{2.20}$$

where $R(t')$ is the observed resistance at the temperature t' and $R(0)$ is the resistance at $0°C$. The constants A and B are related to α and δ by

$$A = \alpha\left(\frac{1 + \delta}{100}\right) \tag{2.21}$$

$$B = -10^{-4}\alpha\delta \tag{2.22}$$

also

$$\alpha = A + 100B \tag{2.23}$$

$$\delta = -\frac{10^4 B}{A + 100B} \tag{2.24}$$

Equations (2.18) and (2.20) appear to be of the same form as the earlier formulations of the International Temperature Scale [2, 6, 7]. But the value of temperature t on the IPTS-68 is not equal to t', and the value of t' is not the value of temperature on the IPTS-48 because the definitions of the two scales are different.

The value t', obtained from Eq. (2.18) or (2.20), may be considered to be a first approximation to t, the value of temperature on the IPTS-68. Equation (2.17) gives the adjustment to be made to the value t' to yield t. This adjustment was included in the definition of the IPTS-68 with the intention of bringing the scale into closer agreement with the thermodynamic scale. Therefore, at any given hotness (except certain fixed points) the values of t' and t are different, but *they both represent the same hotness*; accordingly, the resistances $R(t')$ and $R(t)$ are equal. The observed SPRT resistance will, hereafter, be indicated by $R(t)$.

At the NBS, the SPRT that is received for calibration is first annealed between 470 and 480°C for about 4 hours and then allowed to cool in air at the ambient conditions. The calibration measurements at the fixed points are made in the following sequence: TP, zinc point, TP, tin point, and TP. If any observed resistance is questioned, the measurement is repeated. Any additional tin- or zinc-point measurements will usually be bracketed before and after by TP measurements in order to check any change in the TP resistance that may occur. Whenever the TP resistance of a SPRT changes by more than 75 $\mu\Omega$ (depending upon the design) in the course of calibration the complete sequence of calibration, including the annealing, is repeated.

A measuring current of 1 mA through the resistor of the thermometer is normally used. The sequence of measurements with the Mueller bridge at each of these fixed points is, unless additional measurements are found necessary, NRRN, where N = normal and R = reverse connections of the bridge. The readings are corrected for the "bridge calibration" and "bridge zero" and averaged (see Sec. 2.4 on Mueller bridge resistance measurements).

For the tin- and zinc-point calibration measurements, the "equilibrium temperatures" at the external surface of the SPRT near the resistance coil are computed and assigned to the observed resistance values. The experimental equilibrium temperature value is slightly different from the IPTS–68 value of the fixed-point temperature; the value of the equilibrium temperature at the outer surface of the thermometer sheath is obtained by correcting the IPTS–68 fixed-point temperature value for the hydrostatic head of the liquid metal column (also, for any significant deviation from 1 atm gas pressure over the fixed-point cell) and for the *external* heating effect of the SPRT in the particular cell. In the case of the measurements in a triple-point-of-water cell, the assigned value of the temperature is adjusted for the temperature effect of the hydrostatic head of the water column and external heating effect of the SPRT.

The value of $R(TP)$ that is measured after the $R(tin)$ measurement is employed to obtain $W(tin)$; in a similar manner, the value of $R(TP)$ measured after the $R(zinc)$ measurement is employed to obtain $W(zinc)$. The value of $R(0)$ is calculated from $R(TP)$ according to Eq. (2.20) by assigning $B = 0$ for the small temperature interval of the evaluation. The value of A employed is 3.98485 × 10^{-3}°C^{-1}, which is an average value for SPRT's calibrated at NBS in the past few years. The uncertainty in the adjustment of the value from $R(TP)$ to $R(0)$ in this manner is about $\pm 1 \times 10^{-8}$ $R(0)$ or ± 0.25 $\mu\Omega$ for a 25-Ω SPRT and is, therefore, negligible. The thermometer constants A and B are obtained by simultaneous solution of Eq. (2.20) from the values of $W(tin)$, $t'(tin)$, and $W(zinc)$, $t'(zinc)$. The constants α and δ are obtained from A and B according to Eqs. (2.23) and (2.24). The printed results of the computer analysis of calibration data include, in addition to tables of $W(t)$ or $R(t)$ vs. t, the observed resistances (adjusted for bridge calibration and birdge zero) at the three fixed points, and the values of $R(0)$, α, δ, A, and B.

Several possibilities are available to the user for computing temperatures and for checking the calibration of his SPRT. First, in Eq. (2.18) $R(0)$, α, δ may be taken as invariant and t' calculated from the measured $R(t)$. Second, in Eq. (2.18) only α and δ may be taken as invariant, requiring then the determination of $R(t)/R(0)$. The first method is equivalent to using a table of R vs. t; the second method is equivalent to using a table of $R(t)/R(0)$ vs. t. Equation (2.18) may be rewritten as

$$t' = 100 \, \frac{R(t) - R(0)}{R(100) - R(0)} + \delta \left(\frac{t'}{100}\right)\left(\frac{t'}{100} - 1\right) \tag{2.25}$$

In Eq. (2.25), as a third method, $R(100) - R(0)$, α, and δ may be taken as constant, requiring the determination of $R(t) - R(0)$ to calculate t'. The fourth possibility seen with Eq. (2.25) is to take only δ as constant which requires the determination of $[R(t) - R(0)]/[R(100) - R(0)]$. The ultimate extension of this procedure of reducing the SPRT constants is not to assume that anything has remained constant and to recalibrate the thermometer every time it is used. In 80 years of platinum thermometry each of the above assumptions or its equivalent has been made by investigators.

From the standpoint of accuracy the worst of these assumptions is the first, which requires only the measurement of $R(t)$. This choice may be dictated, however, by convenience or economy. If values of $R(0)$ can be obtained, the choice is not as clear between using $R(t)/R(0)$ or using $R(t) - R(0)$ to calculate the value of temperature. Experience with a particular thermometer may help one make the choice. Lacking this, one may be guided by the experience of the NBS and the National Research Council of Canada which indicates that below 500°C the second assumption, namely, that α and δ are constant and using the value of $R(t)/R(0)$ to calculate a value of temperature, is the most satisfactory. This assumption has the distinctive practical advantage of being dimensionless (i.e., being independent of the size of the resistance unit). If request is made at the time the SPRT is submitted to NBS for calibration, tables based on any one of the assumptions discussed in this paragraph, in a form suitable for linear interpolation, can be provided from the calibration data.

When many observations of $R(t)$, such as in specific-heat measurements, are to be converted to t and a high-speed electronic computer is available, Eqs. (2.16), (2.17), and (2.18) or (2.25) may be more convenient than tables. Iteration methods are best employed in the solution of either Eq. (2.18) or (2.25). For example, an approximate t' can first be obtained neglecting the term containing δ. Improved approximations are obtained by substituting successive solutions into the complete equation, stopping when t' does not change more than desired limits.

If it is desired to generate a table of values of $W(t)$ vs. values of t, entries for integral values of t at 1-degree intervals seem quite useful. The error contributed by using simple linear interpolation within the 1-degree intervals is less than 10^{-4}°C in the range from 0 to 630°C. The calculation of the table is most easily performed employing Eqs. (2.16) and (2.20). The only difficulty is that Eq. (2.20) gives W as a function of t' rather than t, hence if W is to be tabulated as a function of t the value of t' corresponding to the value of t must be first obtained and used in Eq. (2.20). By rearranging Eq. (2.16),

$$t' = t - M(t') \tag{2.26}$$

Below $472°C$ the error introduced by replacing Eq. (2.26) with $t' = t - M(t)$ is less than $1.2 \times 10^{-4}°C$. The error in the value of t' to be used may be reduced below $10^{-5}°C$ by a single iteration of Eq. (2.26). A very efficient equation for calculating the nth entry for tables with intervals of less than 10 degrees is

$$t'_n = t_n - M\{t_n - M[t_n - M(t'_{n-1})]\} \tag{2.27}$$

Equation (2.27) does not require iteration and it may be shown that it introduces an error of only $[dM(t')/dt']^2$ times the tabular interval.

2.5.2 Temperatures Below 0°C

From $0°C$ down to $-259.34°C$, the temperatures on the IPTS-68 are defined by

$$W(t) = W^*(t) + \Delta W(t) \tag{2.28}$$

where $W(t) = R(t)/R(0)$ is the ratio of the observed resistance of the SPRT at temperature t to that at $0°C$ and $W^*(t)$ is a reference value of the resistance ratio given by the function

$$t = \sum_{i=1}^{20} A_i [\ln W^*(t)]^i \tag{2.29}$$

In this chapter the symbol $W^*(t)$ will replace the symbol $W_{CCT-68}(T_{68})$ used in the official text of the IPTS-68 to represent the reference (standard) values of W. The values of $W^*(t)$ represent a fictitious but not "ideal" platinum resistance thermometer. The constants A_i of the reference function are listed in Appendix A. The deviations, $\Delta W(t)$ of Eq. (2.28), are expressed by simple polynomial equations of specified form; the constants of these equations are determined from measurements of $W(t)$ at the appropriate defining fixed points and the corresponding values of $W^*(t)$. The intermediate temperatures are determined from the corresponding measured values of $W(t)$, the appropriate polynomial equation for $\Delta W(t)$, and Eqs. (2.28) and (2.29). (See Appendix at the end of the book, which lists the four temperature ranges, the associated polynomial equations, and the fixed points defined by the IPTS-68 in the range -259.34 to $0°C$.) The text of the scale, given in that Appendix, states the deviation equations for use below the oxygen point as functions of T (in kelvins). This form of these equations will also be used in this chapter because of their greater simplicity. It follows that Eq. (2.28) becomes

$$W(T) = W^*(T) + \Delta W(T) \tag{2.28a}$$

and the reference function Eq. (2.29) becomes

$$T = 273.15 + \sum_{i=1}^{20} A_i [\ln W^*(T)]^i \tag{2.29a}$$

Procedures for obtaining the calibration constants for the polynomial equations will be described later in this section.

The calculation of $W^*(t)$ must be carried out with a large number of digits. At NBS approximately 17 decimal digits are used.

The following procedure may be employed to convert the measured values of $W(t)$ to t. From Eq. (2.28) is obtained

$$W^*(t) = W(t) - \Delta W(t) \tag{2.28b}$$

When Eqs. (2.29) and (2.28b) are combined,

$$t = \sum_{i=1}^{20} A_i \ln [W(t) - \Delta W(t)]^i \tag{2.30}$$

To solve for t, an approximate value of t is first obtained by neglecting $\Delta W(t)$ and evaluating Eq. (2.30) with the measured $W(t)$ only. A closely spaced table of $W^*(t)$ could be employed for this evaluation. The approximate value of t is used with the appropriate polynomial equation to obtain an approximate value of $\Delta W(t)$ which is then employed with $W(t)$ in Eq. (2.30)—or table of $W^*(t)$—to obtain t. The process is repeated until the value of t no longer changes more than the desired limits.

Considering the complexity of the reference function Eq. (2.29), a high-speed electronic computer or a table compiled by such a computer at small enough temperature intervals (0.1 to 1°C, depending upon the value of t) to permit linear interpolation is a necessity.

Although a table of $W^*(t)$ could be employed to obtain t from the measured value of $W(t)$ in accordance with the procedure outlined above, a direct interpolation in a table of t versus $W(t)$ (measured) is obviously more convenient. Such a table can be generated from the relation given by Eq. (2.28). Values of $\Delta W(t)$ can be calculated at appropriate values of t, using the polynomial equations obtained from calibration, and combined with the corresponding values of $W^*(t)$ to obtain $W(t)$.

$-182.962°C$ (90.188K) to 0°C (273.15K)

The polynomial deviation function specified by the text of the IPTS-68 between $-182.962°C$ (90.188K) and $0°C$ (273.15K) is

$$\Delta W(t) = W(t) - W^*(t) = A_4 t + C_4 t^3 (t - 100) \qquad (2.31)$$

The constants A_4 and C_4 in Eq. (2.31) are determined by calibration measurements at the steam point, the triple point of water, and the boiling point of oxygen, or alternatively, at the zinc point, the tin point, the triple point of water, and the boiling point of oxygen. Calibration measurements at the oxygen point and again at the TP are added to the sequence of measurements described earlier (Sec. 2.5.1) for the range 0 to 630.74°C. The value of $W(t)$ at the oxygen point is obtained from $R(\text{oxygen})$ and $R(0)$. The value of α–see Eq. (2.18)–is required from the measurements above $0°C$ in order to evaluate the constant A_4 of Eq. (2.31). From the definition of α,

$$\alpha^* = \frac{W^*(100) - 1}{100} \qquad (2.32)$$

and

$$\alpha = \frac{W(100) - 1}{100} \qquad (2.33)$$

hence

$$\Delta W(100) = W(100) - W^*(100) = 100(\alpha - \alpha^*) \qquad (2.34)$$

The $\alpha^*(= 0.0039259668°C^{-1})$ is the value used in forming the $W^*(t)$ table [4]. Substituting Eq. (2.34) in Eq. (2.31) with $t = 100°C$, there is obtained

$$A_4 = \alpha - \alpha^* = \frac{\Delta W(100)}{100} \qquad (2.35)$$

The value of C_4 is determined by measurement of $W(t)$ at the boiling point of oxygen. The deviation $\Delta W(t) = W(t) - W^*(t)$ at the oxygen point and the value of A_4 from Eq. (2.34) gives the constant C_4

$$C_4 = \frac{\Delta W(t) - A_4 t}{t^3 (t - 100)} \qquad (2.36)$$

where t is the oxygen normal boiling point temperature in °C.

The first and second derivative of the IPTS-68 reference function, Eq. (2.29), have the same value at $0°C$ as those of the function $W(t)$ obtained from Eqs. (2.16), (2.17), and (2.18) with the thermometer constants $\alpha = 3.9259668 \times$

$10^{-3}\,^{\circ}\text{C}^{-1}$ and $\delta = 1.496334^{\circ}\text{C}$. On the other hand, these derivatives of the temperature-resistance function for real SPRT's are not in general continuous through 0°C. The discontinuities are, however, very small and may be neglected.

13.81 to 90.188K

From 13.81 to 90.188K the SPRT's that are received at the NBS are not calibrated at the defining fixed points established by phase equilibrium in accordance with the text of the IPTS–68, but they are calibrated by a procedure that is equivalent to the IPTS–68. The SPRT's are calibrated at NBS by inter-comparison with SPRT standards that maintain the NBS-IPTS–68 (see Sec. 2.6 for details). The temperatures of the intercomparison calibration measurements have been selected so that the measurements, which are made at temperatures very close to those of the defining fixed points, are supplemented by measure-ments suitably placed between these points. Usually 16 measurements are made between 12 and 90K. The data are analyzed by a method of least squares and values of resistances corresponding to the defining fixed-point temperatures are evaluated. The values of $W(T)$ are then computed for the defining fixed-point temperatures: 13.81, 17.042, 20.28, 27.102, 54.361, and 90.188K, and applied to Eq. (2.28) or (2.28a) to obtain the deviation $\Delta W(T)$ at each of these temperatures. The discussion to follow describes procedures for obtaining the constants of the specified polynomial deviation functions.

Below the oxygen point the deviation $\Delta W(T)$ of the value of $W(T)$ from the reference function $W^{*}(T)$ is defined with a succession of polynomial equations, each covering a relatively short range. Each of these deviation equations is described by a three- or four-term power series in T (where T is the value of temperature on the International Practical Kelvin Scale: $T = t + 273.15\text{K}$) and is a smooth downward extension of the deviation curve found for the temperature range immediately above.

54.361 to 90.188K In the range from the oxygen normal boiling point (90.188K) down to the oxygen triple point (54.361K) the deviation function is

$$\Delta W(T) = A_3 + B_3 T + C_3 T^2 \qquad (2.37)$$

The three constants A_3, B_3, and C_3 are determined from the values of $\Delta W(T)$ at the two oxygen fixed points and the value of the first derivative of the deviation equation (2.31) specified for above the oxygen point

$$\left[\frac{d\Delta W(t)}{dt}\right]_{0_2} = A_4 - 300 C_4 t^2 + 4 C_4 t^3 \qquad (2.38)$$

where the t in Eq. (2.38) is the value of the oxygen normal boiling point temperature $(-182.962^{\circ}\text{C})$. At the oxygen point the first derivative of the deviation function Eq. (2.37) extending below the oxygen boiling point is set equal to the first derivative of the deviation function, Eqs. (2.31) and (2.38), above the oxygen normal boiling point, i.e.,

$$\frac{d\Delta W(T)}{dT} = B_3 + 2C_3 T = \left[\frac{d\Delta W(t)}{dt}\right]_{0_2} \tag{2.39}$$

at $t = -182.962°C$ and $T = 90.188K$. Thus, the deviation functions and their first derivatives are forced to be equal at the point of joining.

20.28 to 54.361K The deviation function defined by Eq. (2.37) is joined smoothly at the oxygen triple point ($T = 54.361K$) in a completely analogous manner as described in Sec. 2.5.2 under "54.361 to 90.188K," with the deviation function between 20.28 and 54.361K given by

$$\Delta W(T) = A_2 + B_2 T + C_2 T^2 + D_2 T^3 \tag{2.40}$$

The constants are determined from the values of $\Delta W(T)$ at the normal boiling points of equilibrium hydrogen (20.28K) and neon (27.102K) and the triple point of oxygen (54.361K) and by equating its derivative and the derivative of Eq. (2.37) at the point of joining (54.361K).

13.81 to 20.28K A fourth deviation function

$$\Delta W(T) = A_1 + B_1 T + C_1 T^2 + D_1 T^3 \tag{2.41}$$

covers the range 13.81 to 20.28K. The constants are determined from the values of $\Delta W(T)$ at the triple point (13.81K) and the normal boiling point (20.28K) of equilibrium hydrogen and at the temperature of 17.042K (boiling point of equilibrium hydrogen at 25/76 atm) and by equating the derivative of Eq. (2.41) and the derivative of Eq. (2.40) at the point of joining (20.28K).

Calibration Tables between 13.81 and 273.15K (0°C)

At NBS, the coefficients of the deviation functions obtained through the procedure outlined above allow the values of $\Delta W(T)$ to be calculated over the temperature range specified for each of the deviation functions. Values of $W(T)$ for the thermometer being calibrated are then given by

$$W(T) = W^*(T) + \Delta W(T) \tag{2.28a}$$

Values are calculated to provide tables of $W(T)$ or $R(T)$ vs. T at equal temperature intervals sufficiently small to permit linear interpolation. In the region 13 to 90K the tabulations are usually given at 0.1K intervals and from 90 to 273K at 1K intervals.

2.5.3 Errors of Temperature Determinations

The usefulness of a measured value of temperature depends strongly upon both the amount of its uncertainty and the knowledge of the amount of its uncertainty. Reduction of the errors contributing to the uncertainty is primarily limited by the ingenuity of the experimentalist; there is, however, an inherent source of uncertainty or spread of temperature values on any practical scale which is predicated upon the use of real materials, even if "perfectly" calibrated thermometers are used and no error is introduced in the measurements. Unfortunately only scattered data are available from which one could infer the degree to this ambiguity or possible spread of values of the defined IPTS-68. McLaren [38] did not find a distinctive difference among the measured temperature values at 321 and 231°C of 11 thermometers with α ranging from $0.003921°C^{-1}$ to $0.003926°C^{-1}$. (Five of the eleven thermometers met the requirements of IPTS-68, i.e., α equal to or greater than $0.0039250°C^{-1}$.) The sensitivity of McLaren's test was about ± 0.5 mK; his results indicate a spread of temperature values which are an order of magnitude smaller than the spread reported earlier for seven thermometers of much less pure platinum (α from $0.003909°C^{-1}$ to $0.003925°C^{-1}$) by Hoge and Brickwedde [39]. These latter thermometers, while acceptable on the ITS-27 which was in use at the time of their work, were not made of sufficiently pure platinum to meet the requirement of the IPTS-68 that α equal or exceed $0.0039250°C^{-1}$. The thermometers that were investigated by Hoge and Brickwedde showed a spread of 7 mK between -190 and $0°C$, and a spread of 1.3 mK between 0 and 100°C. The six thermometers measured between 100 and 444°C had a maximum spread of 12 mK. Data which would indicate the spread of interpolated values of temperature on the IPTS-68 below 0°C due to the variations in presently acceptable platinum wire is sparse and inadequate [40-42]. On the basis of existing data, Bedford and Ma [43] estimated that the IPTS-68 is reproducible to ± 3 mK between 14 and 20K, ± 1 mK between 20 and 54K, ± 2 mK between 54 and 90K, and ± 5 mK between 90 and 273K. It should be pointed out that, in addition to unaccounted-for variations in the platinum wire, a spread in the temperature values may also arise from incomplete definitions of the materials that define fixed points (e.g., their isotopic composition).

The remainder of this section will deal with errors that arise both from errors of measurement by the user and from errors of calibration of the thermometer.

Many of the same kinds of error sources plague both the calibrator and the user. The errors may be divided into two categories; first, a temperature error, i.e., a difference between the actual thermometer temperature and the temperature of the point of interest or reference temperature, and second, a resistance error, i.e., a difference between the measured resistance and the resistance of the sensor. For example, temperature error occurs because of an unknown difference between the temperature of the thermometer and the equilibrium temperature realized in a fixed-point apparatus used for calibration. (The calibrator is also plagued, of course, by any unknown difference between the equilibrium temperature realized in his fixed-point apparatus and the equilibrium temperature defined by IPTS–68.) Temperature error also occurs because of unknown temperature differences between the thermometer and the object of interest in the user's apparatus. The experimenter must decide where to place the thermometer to minimize the magnitude of this error. Several possible sources of temperature error are discussed in Sec. 2.3. In considering the errors of thermometer resistance measurements one must not only examine the reproducibility and calibration of his measurement equipment but also his measurement technique. Sections 2.4, 2.5, and 2.3 indicate possible sources of resistance or resistance ratio error.

The following analysis describes the error in the calculated value of temperature that results from errors made in calibrating the SPRT at the fixed points. Any additional imprecision and inaccuracy introduced by the user are assumed to be best known by him.

At any value of temperature t the total differential of t for the SPRT is

$$dt = \sum_i \left(\frac{\partial t}{\partial t_i}\right) dt_i + \sum_i \left(\frac{\partial t}{\partial W_i}\right) dW_i \qquad (2.42)$$

where t_i is the temperature attributed to the calibration fixed point and W_i is the value of the resistance ratio $R(t_i)/R(0)$ attributed to that temperature. The analysis may be simplified by converting any error in the value of temperature attributed to the calibrating fixed point to an equivalent error in W at that fixed point; then dt_i of Eq. (2.42) vanishes and Eq. (2.42) reduces to

$$dt = \sum_i \left(\frac{\partial t}{\partial W_i}\right) dW_i \qquad (2.43)$$

where dW_i now includes the contribution from errors in the values of temperature as well as the errors in the values of $R(t)/R(0)$ at the calibration points. By utilizing Eq. (2.43) the error in the value of temperature t on the IPTS–68 caused by an error in W_i corresponding to a positive unit error in t_i was evaluated as a function of temperature. Figure 2.17 shows the errors in the values of temperature that would result from the calibration errors at each of the fixed points. For each curve the calibration error is taken to have occurred at only one fixed point with no calibration error at the other fixed points.

The measurement at the triple point of water should be a part of platinum resistance thermometry work and $W(t)$, i.e., $R(t)/R(0)$, should be used in calculating temperatures (see Sec. 2.5.1). The value of $W(t)$ is particularly sensitive to the errors of $R(0)$ in applications at high temperatures where any error of $R(0)$ becomes amplified. Figure 2.17 shows the error curve resulting from an error in $R(0)$ corresponding to a positive unit error in temperature.

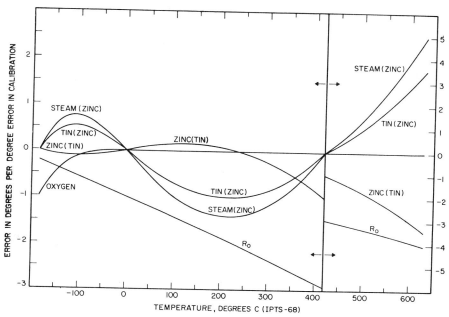

FIG. 2.17 The error at various temperatures propagated from errors made in the calibration of a platinum resistance thermometer. The curves show the error (i.e., departure from the IPTS–68) in the values of temperatures caused by a unit positive (hotter) error in realizing the temperature (hotness) of one given calibration point. The calibration at the triple point of water is assumed to have been made without error. The fixed point at which the error was made is indicated on the curve. The error curve depends not only upon the particular fixed point at which the error occurred, but also upon which other fixed points above $0°C$ were employed in the calibration. The calibration at the other fixed point above $0°C$, indicated in parentheses, is assumed to have been performed without error. A calibration error at the oxygen normal boiling point does not introduce an error in the measured values of temperature above $0°C$. The first derivatives of the zinc(tin) curve and the tin(zinc) curve are not continuous through $0°C$. The curve marked R_0 shows the error that would be introduced if the experimenter makes a unit positive error in realizing the temperature (hotness) of $0°C$ and then calculates the value of a temperature from the value of $R(t)/R(0°C)$.

The estimated uncertainties of the calibration measurements obtained at NBS are 0.002 or 0.003K at the oxygen normal boiling point, 0.0002K at the triple point of water, 0.001K at the tin point, and 0.001K at the zinc point. These uncertainties are estimates; hopefully, in the future, all of the desired documentation of these estimates will become available. The experimenter could estimate his overall error by statistically summing his measurement error with the calibration errors that can be obtained from the curves on Fig. 2.17 and the above estimate of calibration uncertainties.

2.6 CALIBRATION

The calibration of an SPRT on the IPTS–68 involves the measurement of the resistance when the thermometer is at the temperature of the prescribed defining fixed points. This section deals with the equipment, the preparation of fixed-point cells, and the procedures employed at NBS to realize the temperatures of the prescribed defining fixed points for calibrating SPRT's.

In practice, thermometer calibrations are not made at exactly the temperatures of prescribed equilibrium conditions for the fixed points. For small known departures from the prescribed conditions a temperature correction can satisfactorily be made, e.g., such a departure may exist because of the hydrostatic head at the level of the thermometer resistor in calibrations that employ triple-point or freezing-point cells. (The effect of pressure deviations is given in the text of IPTS–68, end-of-book Appendix.) Gross departures from the prescribed conditions usually require extensive efforts to establish that the adjusted temperature and resulting calibration are sufficiently close to the IPTS–68.

The calibration apparatus at the NBS has been designed to be used with the great majority of SPRT's. Capsule-type SPRT's, however, are mounted in special stainless steel holders (Fig. 2.18) before they are calibrated in fixed-point equipment with deep thermometer wells.

2.6.1 Triple Point of Water

The triple point of water (0.01°C) is the most useful and important of the defining fixed points for calibrating SPRT's. The virtues of regular thermometer measurements at the triple point of water are so great that all but the most casual measurements of temperature with an SPRT should include a reference measurement at the triple point. The triple point is realized in a sealed glass cell (Fig. 2.19) containing ice, water, and water vapor. When the cell is in use it may be placed in an ordinary crushed ice-water bath. Figure 2.19 shows the cell and ice-bath system used at NBS.

The cell is first immersed in the ice bath with the mouth of the reentrant thermometer-well above the surrounding ice-water level. The well is thoroughly dried, then filled with crushed Dry Ice* and maintained full for about 20 min by replacing the sublimated Dry Ice. The initial freezing of water within the triple-point cell will occur several degrees below the triple-point temperature because water easily supercools. When the water does start to freeze fine needles of ice crystals (dendrites) are initially formed and protrude from the wall of the well into the liquid. The fine needles quickly cover the well but soon disappear to form a clear coating of ice (on the well) that will grow and become a

---

*Certain commercial materials are identified in this chapter in order to adequately specify the experimental procedure. Such identification does not imply recommendation or endorsement by the National Bureau of Standards or the publishers.

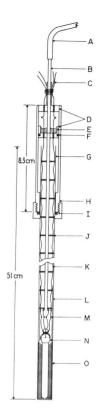

FIG. 2.18 Holder for capsule-type platinum resistance thermometers. Calibration measurements are performed in the holder at the triple point of water, tin point, and the normal boiling point of oxygen. A. Elastomer tubing to helium gas source. B. Thin-wall (0.005 in.) stainless steel tubing for purging the holder with helium gas before the vacuum tubing connector *I* is sealed. C. Leads to measurement equipment (Mueller bridge). D. Sections of mechanical tie-down (brass), soldered to the stainless purge tube, for guiding and fastening the incoming thermometer leads. E. Hard wax for holding and sealing the 0.005-in. gold leads that extend down to the thermometer. F. O ring vacuum seal to brass tube *H*. G. Thin wall stainless steel tube (7/16-in OD) closed at the bottom. H. Brass tube with lead seal at top and vacuum tubing connector *I* at the bottom. I. Vacuum tubing connector for sealing the tube *G*. J. Polytetrafluoroethylene plastic lead spacers. K. Insulated gold leads (four) passing through holes close to the outer diameter of the spacers *J* to attain good tempering. The lead insulation (not shown) is polytetrafluoroethylene tubing cut into a helix and held in tension to eliminate buckling. The four gold leads are welded at the bottom end to short sections of platinum leads. L. Thin polytetrafluoroethylene sheet rolled into a cylinder to insulate the exposed leads (near the connections to the capsule thermometer) against the stainless steel tubing. M. Connections between the short sections of platinum leads of the holder and the platinum leads of the capsule thermometer. N. Capsule thermometer. O. Aluminum sleeve to fit the thermometer and the stainless steel tube. The sleeve reduces the external heating effect of the thermometer.

4-to-8-mm-thick mantle in about 20 min. It is important to keep the well completely full of Dry Ice during this period. (If the Dry Ice level in the well is refilled, the ice mantle is very apt to crack. The desired triple-point temperature may not be achieved if a crack in the ice mantle extends from the well surface into the surrounding liquid water.) In the process of introducing Dry Ice into the well some of the Dry Ice may be deposited around the top of the cell, causing the water within the cell to freeze solidly across the top. The ice at the top of the cell should be melted immediately to avoid breaking the cell glass. Whenever ice is frozen solidly across the top surface of the cell water and a strong bond is formed between the thermometer well and the outer cell wall, any subsequent freezing of water below the surface ice can result in sufficient pressure to rupture the cell. The surface ice can be melted by raising the cell slightly and warming the top of the cell with the hands briefly while gently shaking the top of the cell sideways to "wash" the region with the cell water which facilitates the melting of the layer of ice. After 20 min no additional Dry Ice should be added and the remaining Dry Ice in the well should be allowed to sublime completely. Finally, when the Dry Ice in the well is completely gone, the cell is lowered deeper into

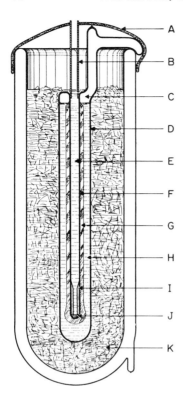

FIG. 2.19 Water triple-point cell. A. Heavy black felt shield against ambient radiation. B. Polyethylene tube for guiding the SPRT into the thermometer well. C. Water vapor. D. Borosilicate glass cell. E. Water from ice bath. F. Thermometer well (precision bore). G. Ice mantle. H. Air-free water. I. Aluminum bushing with internal taper at upper end to guide the SPRT into the close-fitting inner bore. J. Polyurethane sponge. K. Finely divided ice and water.

the ice bath and the well allowed to fill with water. If the cell is raised high enough to see the mantle during the freezing process the magnification of the cylinder of water will give the impression that the cell is or is about to be frozen solid with ice although the coating of ice on the well may still be as little as 1- or 2-mm thick. If the cell is inverted, the true thickness of the ice may be seen. (The cell should not be inverted after the "inner melt," described in the next paragraph, has been made.) An immersion type cooler may be used instead of Dry Ice for freezing the ice mantle [44]. However, care must be taken to avoid admitting the auxiliary heat transfer liquid, e.g., alcohol, into the surrounding ice bath.

A second ice-water interface is formed by melting the ice immediately adjacent to the well surface. This is referred to as the "inner melt." The inner melt is made by inserting a glass tube at ambient temperature into the well for a few seconds. A test for the existence of the ice-water interface over the entire interior surface of the mantle is to give the cell a small rotational impulse and determine whether or not the ice mantle rotates freely around the axis of the thermometer well.

Because of some evidence [45] that the temperature of the triple-point cell is sometimes slightly low (the order of $2 \times 10^{-4}\,^\circ C$) immediately after freezing, the cell should be prepared at least one day prior to its use. The reason for this low initial temperature and the subsequent gradual increase during 1 or 2 days to a steady value is not clearly established but is believed to be connected with structural strains that are produced when the ice is first frozen; presumably the strains are relieved with time as the ice anneals. The magnitude of the lower initial temperature and the rate of increase to a steady temperature value is dependent upon the specific technique that is employed in freezing the cell.

Figure 2.19 shows a triple-point cell immersed in an ice bath. In using the cell, a small soft plastic sponge J is first placed at the bottom of the well to reduce the mechanical shock that the thermometer might otherwise experience when it is lowered. Also, a closely fitting aluminum bushing I about 5-cm long is placed above the sponge (at the bottom of the well) to reduce the external self-heating of the thermometer. In a 1/2-in. well, filled only with water for thermal contact between the SPRT and the well, the external self-heating of a typical 25-Ω SPRT (7.5-mm OD) is about 0.2 mK/(mA)^2. The bushing reduces this heating by a factor of five or more depending upon its fit with the well and the SPRT. (See Sec. 2.3 for the discussion on the self-heating of SPRT.) To eliminate the ambient room radiation (from ceiling lights in particular) a heavy black felt cloth A covers the top of the cell except for a hole through which the thermometer may be inserted. The thermometer is precooled in the ice bath that surrounds the cell before it is inserted into the cell. A polyethylene plastic tube B running from the hole in the felt cloth to the reentrant well provides a guide for inserting the SPRT. Before measurements are made a minimum of 5 min is allowed to elapse (with the thermometer current on and the bridge nearly balanced) for the thermometer to attain thermal equilibrium. If an ice particle is present in the well near the SPRT resistor, an error will occur in the calibration. The water in the well must be free of ice before the SPRT is inserted. Routine measurements made in the calibrating thermometers in triple-point cells have an estimated standard deviation of less than 0.14 mK. Very careful work using two currents and extrapolating to the resistance value for zero current has yielded an estimated standard deviation of less than 0.04 mK.

2.6.2 Metal Freezing Points

Freezing points are advantageous because the effect of a change in pressure on the temperature, the value of dt/dp, is much smaller than that of boiling points. In a metal freezing-point cell, the temperature at the solid-liquid interface depends upon the concentration and kinds of impurities; also, strains in the solid and grain size affect the temperature. The concentrations of impurities (solute) existing at the interface depend upon the amount and kinds of impurities in the sample, the amount of sample frozen, and the rate of freezing. To achieve a true temperature and phase equilibrium the net rate of freezing (or melting) must approach zero. This equilibrium condition may be illustrated by the binary composition vs temperature phase diagram shown in Fig. 2.20 in which the two

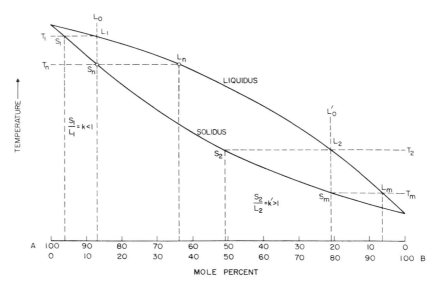

FIG. 2.20 Binary phase diagram of a system that is completely miscible in both solid and liquid phases. The solidus and liquidus curves represent compositions, respectively, of the solid and liquid phases that can coexist in equilibrium. The left side of the diagram shows that the solute concentration in the solid phase is less than that in the liquid phase, i.e., $k < 1$. The right side of the diagram, on the other hand, shows that the solute concentration in the solid phase is greater than that in the liquid phase, i.e., $k > 1$.

constituents are completely miscible in both the liquid and solid phases, e.g., the system Ag-Au or Bi-Sb. The phase diagram of Fig. 2.20 is presented for its simplicity and, since metal samples employed in freezing-point cells are prepared by the zone refining process, the remaining impurities (combined as a single component in the illustration) are expected to form solid solution with the major component. (Most binary metal systems on which data exist do not form a continuous series of solid solutions; various degrees of immiscibility are usually found.) The figure shows the region of the composition and temperature where the solid and liquid phases can coexist. The solidus curve indicates the temperature at which a solid solution of any given composition would begin to melt; the liquidus curve indicates the temperature at which a liquid solution of any given composition would begin to freeze (assuming no supercooling). Alternatively, when both solid and liquid phases are present in equilibrium, the composition of each phase is given by the intersections of the temperature line with the solidus and liquidus curves, respectively. An enlarged section of this phase diagram, showing the effect of a small amount of B in nearly pure A, is shown in Fig. 2.21a. If a completely melted sample of composition L_0 is allowed to cool under equilibrium conditions, no change in composition of the liquid occurs until the temperature reaches the liquidus curve (L_1 at temperature T_1) at which point solid (assuming no supercooling) of the composition S_1, given by the solidus curve at the same temperature, is formed. In the case shown here, the first solid contains a smaller proportion of B than the liquid from which it was formed; as more solid is formed the concentration of B in the liquid increases. When freezing under equilibrium conditions the composition of the solid phase moves from S_1 to S_n; the composition of the liquid phase moves from L_1 to L_n. The last of the liquid solidifies at T_n. Obviously the amounts of A and B in the completely solidified sample must be the same as in the original liquid. S_n must, therefore equal L_0 and L_1 and no segregation occurs. The temperature range of the "freeze" is from T_1 to T_n. Although in most systems the solutes lower the freezing point, there are systems in which solutes raise the freezing point; antimony in tin is an example of the latter [46]. Such a system is illustrated by the extreme right portion of the equilibrium phase diagram where B is the major constituent as in Fig. 2.20. An enlarged section is shown in Fig. 2.21b. The analysis of the phase diagram is similar to that given previously. The most notable difference is that, as the diagram shows, the concentration of the solute is greater in the solid than in the liquid. The ratio (S/L) of solute concentration in the solid (S) to that in the liquid (L) is called the *solute distribution*

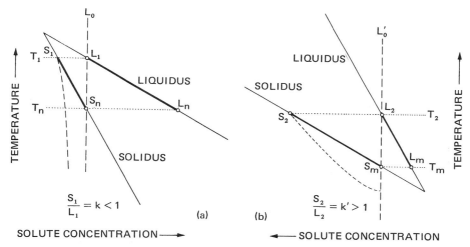

FIG. 2.21 Englargement of a binary phase diagram in the region of high purity of each component. The solute distribution coefficient k is taken to be constant. Figure (a) represents a phase diagram when $k < 1$ and Fig. (b) when $k > 1$. The heavy lines on the liquidus and solidus curves represent compositions of the liquid and solid phase during the equilibrium freezing process. The dashed line beginning at S_1 in (a) or S_2 in (b) and extending downward represents the average composition of the solid phase during a semi-equilibrium freezing process. The composition of the solid at the freezing interface is represented by the solidus curve and that of the liquid phase in equilibrium with the freezing interface is represented by the liquidus curve. The solute distribution coefficient may not be constant over the entire range of fraction frozen for semiequilibrium freeze; therefore, Eq. (2.44) may not be valid when g approaches unity.

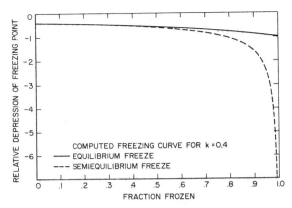

FIG. 2.22 Comparison of computed freezing curves for equilibrium and semiequilibrium freezing processes of a completely miscible (in solid and liquid phases) binary system with constant solute distribution coefficient ($k = 0.4$). The solid line represents the equilibrium freezing process and the dashed line the semiequilibrium freezing process. In the equilibrium freezing process, the temperature, when the last trace of liquid freezes, is shown to be depressed 1 unit for the "sample"; the temperature, when the first solid freezes, is shown to be depressed 0.4 unit. The freezing curve for the semiequilibrium freezing process is shown relative to that of the equilibrium freezing process. The solute distribution coefficient may not be constant over the entire range of fraction frozen for semiequilibrium freeze; therefore, Eq. (2.44) may not be valid when g approaches unity.

coefficient (k) and, as given in Figs. 2.20 and 2.21 is less than unity for solutes that depress the freezing point and is greater than unity for solutes that elevate the freezing point. Except for very dilute solutions, depicted in Fig. 2.21, the solute distribution coefficient depends on concentration.

Phase diagrams are idealized conceptions of systems at equilibrium (often based on relatively little data). For equilibrium freezing, the crystallization process must proceed at such a negligible rate that there is sufficient time for the diffusion of impurities within the solid matrix to achieve uniform distribution throughout the solid, i.e., there is no concentration gradient in the solid. Experimentally this rate of freezing is never realized; however, freezing rates can be achieved which, while large compared to the diffusion rates in the solid matrix, are small compared to the diffusion rate in the liquid. This condition leads to a maximum segregation in the solid but homogeneity in the liquid; this condition will be referred to in this chapter as *semiequilibrium freezing*. The results of semiequilibrium freezing are shown in the phase diagram of Fig. 2.21 and on the cooling curve of Fig. 2.22. The phase diagram still represents the compositions at the solid-liquid interface but the solidus curve no longer is the average composition of the solid phase; the average composition of the solid is given by the dotted line beginning at S_1 in Fig. 2.21a. Compared to an equilibrium freeze, the freezing temperature range for the entire sample is increased due to the increased concentration of impurities at the interface near the end of the freeze. (This assumes no eutectic is formed in the equilibrium freeze.) For the case of semiequilibrium freezing, Pfann [47] * gives an expression equivalent to

$$\frac{S}{L_0} = k(1 - g)^{k-1} \tag{2.44}$$

where L_0 is the overall solute concentration and S is the solute concentration of the freezing interface after the fraction g of the original mass of liquid has frozen. (The equation is not applicable for the entire range of g. The derivation assumes the distribution coefficient k to be constant and the solute diffusion rate in the solid to be zero.) Figure 2.23 gives curves of relative solute composition of

*Pfann [47] refers to semiequilibrium freezing as *normal freezing*.

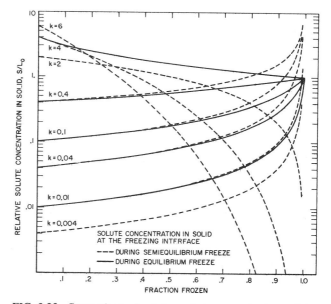

FIG. 2.23 Comparison of relative solute concentrations of the solid phase at the freezing interface during equilibrium and semiequilibrium freezing processes of a completely miscible (in solid and liquid) binary system with constant solute distribution coefficient k. Curves are given for various values of solute distribution coefficient. The solid curve represents an equilibrium freezing process and the dashed curve represents a semiequilibrium freezing process. The solute distribution coefficient may not be constant over the entire range of fraction frozen for semiequilibrium freeze; therefore, Eq. (2.44) may not be valid when g approaches unity.

the freezing interface resulting from this expression for various values of k. Figure 2.22 compares the curve for $k = 0.4$, replotted on a linear scale, with the curve that would be obtained for a corresponding equilibrium freeze. The semiequilibrium freeze is shown to have a broader freezing range.

Sufficiently rapid freezing causes a departure from the semiequilibrium freeze. As the velocity of the advancing solid-liquid interface increases, the rate of solute rejection into the liquid at the interface increases. When the impurity cannot be uniformly distributed throughout the liquid by either diffusion or convective mixing, the concentration of the solute builds up at the interface as shown in Fig. 2.24b. When this occurs, the effective segregation of the solute decreases (i.e., the effective value of k approaches 1). Because of the increase in the solute concentrations at the interface, the temperature of the interface becomes lower. Figure 2.24c shows qualitatively the result of a freeze which is very rapid compared to the impurity diffusion rate in the liquid (and for which there is no other method of homogenization, e.g., stirring).

A notable example of nonequilibrium freezing in metal freezing-point cells is the rapid freezing that occurs after a supercool. When the completely melted sample is cooled through the temperature at which the phase diagram indicates that the solid should first appear (i.e., crosses the liquidus curve) no solid appears. The first solid appears at a somewhat lower temperature (typically 0.02 or 0.06°C lower in zinc, 1 to 25°C lower in tin, water, or antimony). The solid tends to grow most rapidly into the cooler parts of the liquid until the released heat of fusion raises the temperature of the liquid to the equilibrium temperature of the composition at the solid-liquid interface. The temperature to which a liquid supercools is not very reproducible even with the same sample. The amount of supercooling seems to depend on the purity of the sample, the thermal history of the melt, the occurrence of mechanical shock or vibration, and other effects not known or understood. The system may be far from equilibrium during the recovery from the supercool; the solid-liquid interface advances very rapidly and there is very little time for diffusion or convection to homogenize the liquid. As a result, the impurities rejected from the freezing liquid for the case where $k < 1$ become relatively concentrated in the liquid in the region of the interface. When the recovery from the supercool is nominally complete, the high concentration of impurities in the liquid at the interface will be reduced by diffusion and the temperature will rise until the flow of impurities from the freezing liquid to the liquid at the interface is equal to the flow away from the interface through the remaining "bulk" liquid.

Rapid freezing implies a rapid transfer of heat. Authors [48, 49] have suggested that it may be necessary for the temperature at the solid-liquid interface to be significantly below the equilibrium temperature if freezing is to take place rapidly; i.e., the net rate of freezing is zero at the equilibrium temperature and increases only with decreasing solid-liquid interface temperatures. The metal in the two phases at the interface thus constitutes a "proportional" controller, supplying heat on demand and having a temperature "offset"

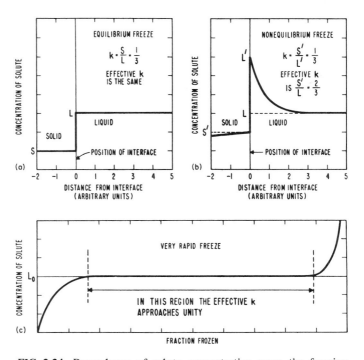

FIG. 2.24 Dependence of solute concentration upon the freezing rate. (*a*) The concentration of solute in the region near the freezing interface during an equilibrium freeze with the solid-liquid interface advancing at a negligible rate. There are no concentration gradients in either liquid or solid phases. (*b*) The solute concentration near the interface during a nonequilibrium freeze with the interface advancing rapidly relative to the rate of diffusion of the solute in the liquid. The solute concentration in the liquid at the interface increases until at steady state the flow of solute from the freezing liquid to the liquid at the interface equals the flow away from the liquid at the interface to the bulk of the liquid. The solute concentration of the frozen solid is shown to be increasing with the advancing interface. (*c*) The solute distribution in the frozen solid approached by a very rapid freeze. The curve shows the initial transient rise of solute concentration of the solid to that corresponding to the solute distribution coefficient of unity.

from the equilibrium value if demand exists. This picture is intuitively appealing and is in qualitative agreement with the theory of reaction kinetics. But, the crucial questions would seem to be what is the rate of energy released per unit of temperature departure from equilibrium and how is the measurement of this separated from other events briefly outlined above that occur in nonequilibrium freezing? The difficulty of this measurement has prevented firm establishment of this model.

2.6.3 Metal Freezing–Point Furnaces

The metal freezing-point cells are used in furnaces of the design shown in Fig. 2.25. At NBS, two furnaces are employed, one for the tin point and the other for the zinc point. Except for the metal sleeves employed at the top section of the furnace core, the two furnaces are the same. The core of the furnace is a stack of three cylindrical coaxial blocks of aluminum (top G, center L, and bottom T) that are thermally insulated from each other by Fiberfrax paper and surrounded by a nichrome wire heater (main heater O) which extends the full length of the three blocks. The heater wire fits closely within each of the two holes of the 51-cm lengths of two-hole, oval alumina tubing. These alumina tubings are very closely spaced around the outer surface of the core (parallel to its axis) and are held tightly against the outside of the blocks with three Inconel "garters." (See the periphery of the furnace core section drawings 1, 2, 3, 4, and 5.) The garters were made of Inconel wire bent into a form resembling a continuing sine-wave about 1-cm in amplitude and wavelength, then rolled to slightly reduce the thickness, and welded into a ring. The top and bottom blocks each have an additional heater (F and U) consisting of nichrome wire that passes through 98-mm lengths of two-hole, round alumina tubings which were selected to fit closely in the 12 holes of each block. To minimize the thermal time lags, the heater assemblies have small clearances between the heaters and the aluminum core blocks. (The leads extending through the core to the heater of the top core block are gold. The leads to the other heaters are heavy nichrome wire.) Extending down through holes that run the length of the three assembled blocks are six thinwall (0.13 mm) stainless steel tubes B nominally 3.2 mm in diameter. The tubes pass through 3.22-mm holes in the end blocks (see a and b of Fig. 2.25). To permit the holes for these tubes to be accurately positioned through the 30.5-cm-long center block, the center block was made in two cylindrical pieces. The end plate Q was attached later to the center core block. Grooves that closely fit the stainless steel tubes were milled on the outside of the inner cylinder which was then fitted tightly inside the second outer cylinder. The two cylinders were "shrunk together," a process that involved precooling the inner cylinder in liquid nitrogen and heating the outer cylinder. The tubes are wells into which thermocouples are placed for controlling the furnace temperature; they also serve as wells for small exploratory resistance thermometers to determine the temperature distribution within the furnace core. The entire core of the furnace, including all three blocks and the heaters, slips into a stainless steel tube A, 11.4 cm in diameter with a 0.76-mm wall. Sheets of mica are wrapped around the main heater near each end to center the core within the tube and the main heater near the ends, and to reduce undesirable heat convection currents. The centering leaves a small annular air space between the tube and the alumina insulators of the main heater in the region of the center block; the thermal contact of the end blocks with the surrounding stainless tube serves as "thermal end guards" and reduces the thermal gradients in the middle section of the tube opposite the center block.

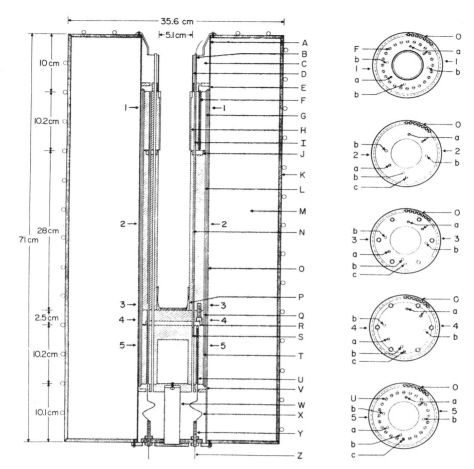

FIG. 2.25 Schematic of furnace body and core. A. Stainless steel tube. 11.4-cm OD × 0.76-mm wall. B. Stainless steel tubes (six); 3.20-mm OD × 0.013-mm wall. C. Insulation, Fiberfrax mats. D. Stainless steel tube. E. Insulation, Fiberfrax mats, and mica sheets. F. Control heaters for top core block. G. Top core block. H. Sleeve (aluminum or Inconel) for metal freezing-point cell. I. Elevation of thermocouple junction (Chromel-P/Alumel), top core block. (See *a*.) J. Insulation, Fiberfrax sheets. K. Brass shell, 4.8-mm thick. L. Center core block. M. Insulation, bulk Fiberfrax. N. Elevation of thermocouple junction (Chromel-P/Alumel), center core block. (See *a*.) O. Main heaters, held on with Inconel "garters." P. "Spider" for centering the freezing-point cell. Q. Center core block end plate. R. Insulation, Fiberfrax sheets. S. Elevation of thermocouple junction (Chromel-P/Alumel) bottom core block. (See *a*.) T. Bottom core block. U. Control heaters for bottom core block. V. Insulation, Fiberfrax mats and mica sheets. W. Stainless steel tube support, 2.5 cm OD × 0.51-mm wall × 11.4 cm long. X. Heater leads. Y. Posts for heater leads. Z. To electric power. a. Wells (B) for control thermocouples (see I, N, and S). b. Wells (B) for testing the temperature profile of furnace core. c. Leads from top core block heaters.

The weight of the core rests on a 0.51-mm wall stainless steel tube W, 11.4 cm long and 2.5 cm in diameter. The space beneath the core is filled with loose Fiberfrax insulation. The outside of the furnace is a 35.6-cm diameter brass tube with end plates and is filled as indicated in Fig. 2.25 with Fiberfrax insulation. A rather widely spaced helical coil of 9.5-mm copper tubing is soldered to the outside brass tube K to permit water cooling; a comparable provision is made for the top end plate. The water cooling is not used when the furnace is at the tin point but is helpful at the zinc point in reducing convection currents around the furnace that cause gradients in the head of the thermometer which, in turn, may result in thermal emf's. The thermal insulation is sufficient so that, even at the zinc point with no water cooling, the outside of the furnace is not "hot to the touch." To achieve further thermal insulation, the furnace core is recessed from the top of the furnace as shown in Fig. 2.25. A major path of heat loss from the top of the core is along the 11.4-cm diameter stainless steel tube that contains the core; to reduce the heat loss the core was recessed and the tube was made relatively longer.

The top core block G of the tin furnace was designed to receive an aluminum sleeve H 10.2 cm long that is bored to fit as closely as possible the glass cell that holds the tin sample. By using removable metal sleeves a reasonable range of glass cell diameters can be accommodated while still achieving good thermal contact with the cell above the crucible where the heat shunts are located. The glass cell is of sufficiently uniform diameter to achieve a suitably close fit. In the design, allowance has been made for the differential expansion of the aluminum and the glass cell. (The NBS glass cells are generally ground on the outside to a uniform diameter.) Because aluminum sleeves bond to the aluminum top core block at the zinc point, split Inconel sleeves are employed in the zinc-point furnace. Also, because of the close tolerances needed for good thermal contact, the zinc-point cells are not permitted to cool to room temperature in the furnace, otherwise the borosilicate glass tube would be crushed owing to the differential contraction between the aluminum top end block and the glass.

The center core block L was bored to provide approximately 1.0-mm clearance for the most common (and largest) size of glass tubes, specifically 51-mm OD; this 1.0-mm annular space provides the thermal insulation that reduces the heat transfer to the crucible and metal sample during freezing or melting experiments. The design also reduces the need for close control or knowledge of the furnace temperature in many operations.

Three Chromel-P/Alumel thermocouples TC enclosed in alumina sheaths control the furnace temperature. Two of the thermocouples (differential) are referenced to the temperature at the middle of the center core block; the measuring junction of one is located approximately 1.2 cm up into the top core block and that of the second is similarly located down into the bottom core block. The third TC is referenced outside the furnace and the measuring junction is located in the middle of the center block.

Each differential TC is connected in series with a stable voltage source (powered by mercury cells) which is adjustable between ± 75 μV (± 1.5°C). The center block TC connects to a reference junction, which is self-compensating for changes in room temperature, and to a voltage source that has (a) a range of 15 mV, (b) a stability of 0.01 percent, and (c) a reproducibility of setting of better than 2 μV. The combined output of each TC and voltage source is amplified by a chopper-type dc amplifier whose output operates a "three-mode" controller. The zero stability of the amplifier is better than 0.5 μV. The output of each controller is an adjustable linear (proportional) function of the input, its rate of change, and its time integral. The signal from the controller operates a gate drive for a full-wave silicon-controlled rectifier which, in turn, controls the power supplied to the heater. The rectifiers do not cause detectable interference with any other equipment (low level null detectors, ac and dc bridges, etc.) in the room. Each aluminum block of the core is grounded by means of a heavy gold wire.

The control of the furnace was made very flexible in that several combinations of manual and automatic controls may be selected for operating the furnace, including the provision for independently setting the temperature offset between the center and top and between the center and the bottom core blocks. A freeze is usually conducted with each of the three heaters under the automatic control of their corresponding TC's. A plot of thermometer resistances (relative to that at the center of the center block) as a function of depth in the core of the furnace, while the furnace and a tin-point cell was at 235°C, is shown in Fig. 2.26. These measurements were made with a small four-lead platinum resistance thermometer, 2.8-mm diameter and 2 cm long, having a nominal ice-point resistance of 50 Ω.

This furnace, which was designed and built at the NBS, is more sophisticated than necessary for calibrating thermometers at the tin and zinc fixed points with reasonable (better than 0.002°C) accuracy. It was designed for special studies of the freezing and melting phenomena of tin and zinc. Tin and zinc freezes for calibrating SPRT's have been performed very successfully in many laboratories with furnaces that contain a single copper block and employed a manually controlled heater. Such calibrations have been performed in furnaces that exhibit core temperature gradients of 1.5°C over the length of the crucible containing the sample [50] and relatively long response times. Realization of the tin and zinc freezing points with reasonable accuracy is principally dependent upon the high purity of freezing-point samples and the use of proper freezing and measurement techniques.

2.6.4 Tin–Point Cell

The realization of either the tin point or the steam point is necessary for the calibration of SPRT's in accordance with the specifications set forth in the text of IPTS–68. The tin point (231.9681°C) has two distinctive advantages; first, the temperature is much closer than the steam point to the midpoint between the

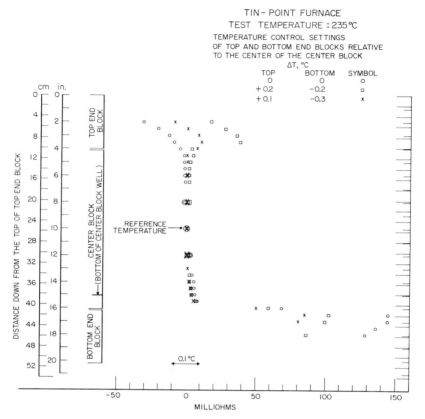

FIG. 2.26 Vertical temperature profile of tin-point furnace at 235°C. The vertical distances are relative to the top of the top core block; the thermometer resistances are relative to that observed at 25.4 cm from the top of the top core block (close to the center of the center core block). The resistance of the thermometer at the reference temperature was about 95 Ω. The resistance thermometer (2.8 mm diameter × 2.0 cm length) sensed the average temperature of a section of well about 2 cm long.

triple point of water and the zinc point and, therefore, tends to produce less average error in the calibration of the thermometer (see Fig. 2.17), second, and of greater practical importance, the solid-liquid equilibrium temperature of tin is 8600 times less sensitive to pressure changes than the liquid-vapor equilibrium temperature of water at 1 atm. For the freezing point of tin, $dt/dp = +4.3 \times 10^{-6}$ °C/torr or $+2.2 \times 10^{-5}$ °C/(cm column of liquid tin). Therefore, knowledge of the pressure within ± 1 torr is adequate for determining the temperature of a tin-point cell.

The tin sample for the freezing-point cell must be of high purity (nominally > 99.999%) and the freezing apparatus must be designed and operated to interpose a solid-liquid interface of tin completely around the resistance element and lower part of the thermometer stem. The immersion of the resistor must be sufficient to prevent sensible heat flow from the resistor up the stem of the thermometer. In addition, the sample holder must be chemically inert and not introduce any impurities that would affect the freezing temperature of tin. At NBS the tin sample is contained in a closed crucible of high-purity graphite. A reentrant well, also of high-purity graphite, is screwed into the lid. The assembly is shown schematically in Fig. 2.27. The cell assembly permits sealing the stem of the thermometer, as well as the crucible with its charge of tin, in helium gas which provides an inert atmosphere for both the crucible and the tin and additionally improves the thermal conductance to the thermometer.

Because of the relatively high thermal impedance of gas, even helium, considerable effort was made to obtain small clearances at the heat shunts to keep the thermal impedance as low as possible. This is practical with components of borosilicate glass and graphite because their thermal coefficients of expansion are similar. The outer borosilicate glass cell was formed from precision bore tubing and its external diameter was ground to fit the furnace sleeve (see H, Fig. 2.25). The two graphite heat shunts G, shown in Fig. 2.27, were fitted closely to

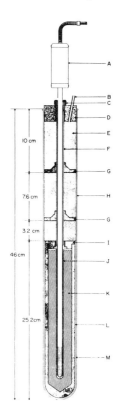

FIG. 2.27 Metal freezing-point cell. A. Platinum resistance thermometer. B. To helium gas supply and pressure gauge. C. Thermometer stem seal with silicone rubber. D. Silicone rubber cap. E. Insulation, washed Fiberfrax. F. Thermometer guide tube, borosilicate glass. G. Heat shunt, graphite. H. Borosilicate glass cell. I. Graphite cap (lid). J. Graphite thermometer well. K. Metal sample. L. Graphite crucible. M. Insulation, Fiberfrax paper.

both the thermometer guide tube F and the glass cell H. The shunts primarily serve to conduct heat to the inner glass tube of the tin-point cell, thereby improving the "immersion characteristics" of the thermometer. (See the end of this section for results of tests on the immersion characteristics of an SPRT in the tin-point cell.) The space between the heat shunts is loosely filled with high purity Fiberfrax (washed) insulation to eliminate convection currents and radiation losses from the top of the crucible and the heat shunts; it also vertically positions the heat shunts.

The equipment arrangement for filling the crucibles is shown in Fig. 2.28. The bell jar (which contains the metal sample and its funnel and crucible) is evacuated to about 0.01 torr by a mechanical pump that is "trapped" with a molecular sieve and the temperature of the enclosed assembly is then raised to a few degrees below the tin point. After 1 hr, the induction heater power is increased so that the tin sample will melt and flow into the crucible within about 15 min. Next, the sample is allowed to cool nearly to room temperature, the crucible is removed from the bell jar, and the crucible is slipped into the special borosilicate glass cell, Fig. 2.29. The glass cell is then purged with helium, lowered into the furnace that is used for realizing the freezing point, and the tin

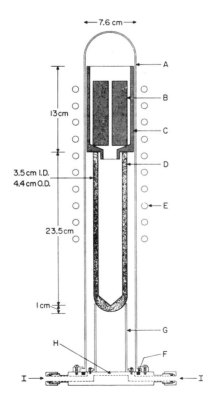

FIG. 2.28 Arrangement for filling the freezing-point cell with metal sample by induction heating. A. Borosilicate glass envelope. B. Metal sample. C. Graphite sample holder and funnel. D. Graphite crucible. E. Induction heater coils. F. O-ring groove. G. Borosilicate glass stand. H. Slot for pumping out and gas purging. I. Connection to vacuum and purified argon supply.

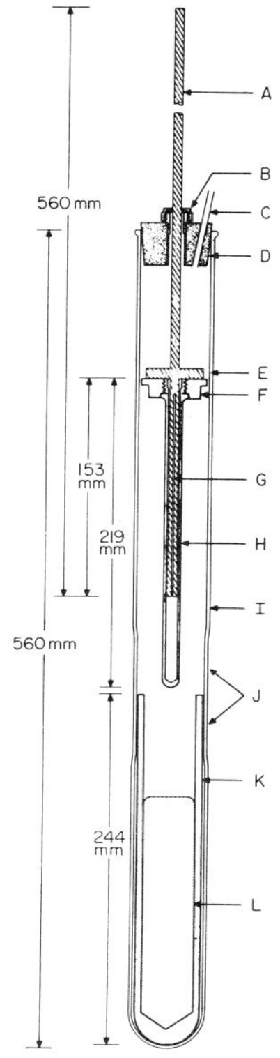

FIG. 2.29 Apparatus and method for installing graphite therometer well and lid in the graphite crucible containing the molten, metal freezing-point sample. A. Stainless steel pusher rod. B. Gas seal with silicone rubber; permits linear motion of the pusher rod A. C. Inlet for purified helium gas that is used in purging and maintaining positive pressure of the gas during assembly. D. Silicone rubber cap. E. Stainless steel flange attached to the pusher rod for pressing against the graphite lid during assembly. F. Graphite lid for the metal sample cell. G. Slit on the pusher rod. The two halves are sprung out to hold up the graphite thermometer well and lid while melting the metal sample. H. Graphite thermometer well. I. Borosilicate glass tube. J. Section of glass tube shrunk to fit the crucible and lid so that lid can be easily guided onto the crucible. K. Graphite crucible. L. Molten metal sample.

reheated to about 250°C. Subsequently, the glass cell is raised out of the furnace and the graphite lid and well assembly are smoothly and quickly pushed into place. To perform the assembly easily the outside diameters of the crucible and lid were made equal and the inside dimensions of the glass tube only slightly (perhaps 0.01 mm) larger. The glass tube served to guide and correctly position the lid to cover the crucible. The lid is closely fitted to the crucible to minimize the possibility of later contamination. Finally, the crucible is again allowed to cool, removed from the special glass cell, and assembled as a freezing-point cell such as is shown in Fig. 2.27.

In the preparation for tin freezing-point measurements, the tin-point cell is placed in the furnace (held about 10°C above the tin point) until complete melting occurs. During the heating process the cell temperature, which is monitored with an SPRT, rises until melting begins, becomes nearly constant until the melting is completed, and then rises again. There is no reason to raise the temperature of the tin more than a few tenths of a degree above the melting point. In a furnace with large temperature gradients, precautions should be taken to make certain that the metal is completely melted. There is no evidence to indicate that heating the tin to several degrees above its melting point in an inert atmosphere is harmful to the sample.

After the melting is completed, the furnace is allowed to cool to the tin point. A thermometer is inserted in the cell well and instrumented so that a temperature range of 25°C below the tin point can be visually monitored within about 0.5°C by a person, standing beside the furnace, holding a tin cell ready to be reinserted. A galvanometer is used for this purpose. When the cell has cooled to the tin point it is removed from the furnace. The cell temperature, after a few seconds, will suddenly decrease very rapidly, perhaps as much as 20 or 25°C. As soon as its temperature *ceases* to drop, the cell is immediately lowered into the furnace again; subsequently, the cell temperature will increase even more rapidly than it previously decreased. If the cell temperature is examined in more detail, the final approach to the tin point (the last few ten-thousandths of a degree) will be seen to take a few minutes (see Fig. 2.30). When the temperature plateau is reached, the SPRT is measured. It may then be removed and a second SPRT inserted. The second thermometer is preheated above the tin point so that its temperature, during insertion into the tin-point cell, slightly exceeds that of the cell. Figure 2.31 shows the results of successive insertions of the same SPRT in a single tin freeze.

An experimental procedure similar to that described above for observing the supercooling of the tin may be used to check the preheating of the SPRT before inserting it into the tin-point cell. The thermometer should be preheated and held in the air over the tin cell while monitoring the decreasing thermometer temperature. As a first approximation, the thermometer can be inserted into the tin-point cell when its indicated temperature is 30°C above the tin point. If subsequent close monitoring indicates that the thermometer temperature is rising, then the thermometer was too cold when it was inserted. Several thermometers may be consecutively calibrated during one freeze if each is properly preheated (see Fig. 2.31). A simple check on the temperature constancy during a freeze is

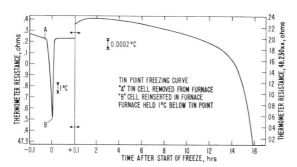

FIG. 2.30 The freezing curve of tin obtained using an ac bridge with the head of the SPRT adapted for co-axial connectors. The furnace was controlled to be 0.9K below the tin point. The mass of tin sample was 1300 g.

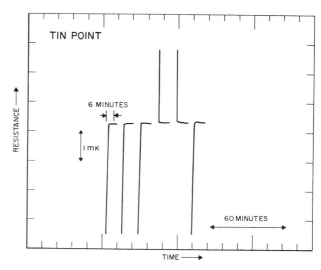

FIG. 2.31 Consecutive measurements of the resistance of a SPRT in the same tin freeze. (An ac bridge was employed in the measurements.) Following each equilibrium resistance observation, the SPRT was completely withdrawn from the tin-point cell, preheated close to the tin point in an auxiliary furnace, and reinserted into the cell. The data show that the preheated SPRT comes to equilibrium in a very short time. In two of the cases shown, the temperature of the SPRT when inserted into the cell was slightly above the tin point. (For comparison when the SPRT was not preheated close to the cell temperature, see Fig. 2.35.)

afforded by employing the same thermometer in both the first and last measurement in a series of SPRT calibrations. At NBS a single SPRT is set aside and only used for this purpose as an experimental control.

The immersion characteristics of an SPRT were tested in one of the tin-point cells that were described earlier. The observed resistances are shown in Fig. 2.32 as a function of the distance from the bottom of the crucible thermometer well. The results show that the immersion of the SPRT in the cell is more than adequate. This figure does not, however, show the temperature gradient expected from the change in pressure with depth.*

In the reduction of the calibration data the temperature at the point of immersion of the SPRT in the tin-point cell is employed. The value of temperature assigned in the text of the IPTS–68 is adjusted for the departure from the

---

*Work done at NBS in the late 1960s indicated that the temperature gradient expected from the change in pressure with depth is distorted or obscured when observations are made from within the graphite thermometer well if the solid liquid interface is not very close to and surrounding the well. Further work in this area is in progress.

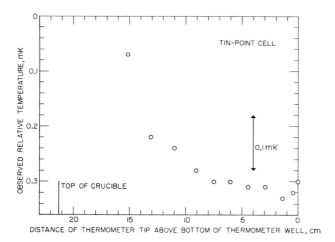

FIG. 2.32 Immersion characteristic of a SPRT in a tin-point cell. The data show that the indications of the SPRT are very nearly the same between about 8 or 10 cm above the bottom and the bottom of the thermometer well.

equilibrium conditions specified for the fixed point; i.e., an adjustment of temperature is made for the effect of the hydrostatic head of liquid tin and for any departure from 1 standard atmosphere of the gas pressure in the cell.

The high reproducibility of the tin point may be seen in the results of tests (made at the NBS) that compare the freezing points of a number of tin samples. The average standard deviation of measurements on a given cell is ± 0.05 mK from freeze to freeze (see Fig. 2.33). All of the measurements are plotted relative to the mean of samples 6C and 6E. The individual measurements (one for each freeze) are shown as short horizontal bars; four freezes are represented for each cell with the exception of 6K on which only two freezes were performed. The values are based on the determinations of $R(t)/R(0)$ that were obtained from extrapolations to zero thermometer current of observations of $R(t)$'s and $R(0)$'s at 1 and 2 mA. The $R(t)$'s were measured 1.5 hr after the initiation of a freeze. The $R(0)$'s were measured immediately after every determination of $R(t)$'s. The samples identified as 5 were reported to be 99.999% pure [51]; the samples identified as 6 were reported to be 99.9999% pure [51]. The sample identified as 6M was accidentally overheated to a temperature above 500°C.

2.6.5 Zinc–Point Cell

The equipment and procedures used at NBS for filling graphite crucibles with zinc are identical to those employed for tin, except that after evacuation at a temperature slightly below the tin point, an atmosphere of purified dry argon is admitted into the bell jar. Because of the relatively high vapor pressure of zinc at the melting point (over 0.1 torr), filling the graphite crucible by melting the sample in vacuum is less practical. The power of the induction heater is raised sufficiently to melt the zinc sample (approximately 1280 g) in about 30 min. In spite of the presence of argon gas, enough zinc vapor diffuses and is deposited on the surface of the bell jar to cause difficulty in observing when the zinc is completely melted. The sample is allowed to cool and the assembly of the zinc-point cell is carried out in the same manner as the tin sample (see Sec. 2.6.4).

To realize the freezing point of zinc with an accuracy of about $\pm\,0.002°\text{C}$ the furnace need not be very complex. However, a zinc sample of high purity

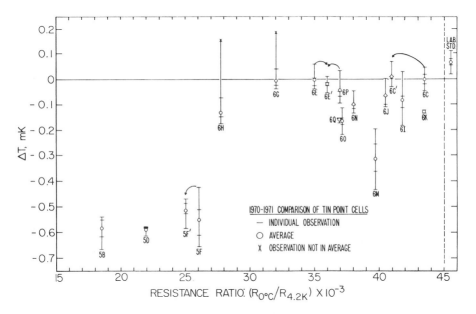

FIG. 2.33 Comparison of tin-point freezing plateau temperatures with resistance ratios: $R(0°\text{C})/R(4.2\text{K})$. The prefixes 5 and 6 indicate samples nominally 99.9999% pure (Lot No. 6637) and samples nominally 99.999% pure (Lot No. 6779), respectively. The alphabetic character identifies the cell. The prime mark identifies a second set of measurements on the same cell. LAB. STD. is a tin-point cell that has been used for over 4 years at the NBS in the calibration of SPRT's. Observations from four separate freezes are shown for each cell. The plateau temperature selected was the reading 1.5 hr after the initiation of the freeze; the temperature corresponded to about 25% frozen. The temperatures of the SPRT during the previous hour differed typically 0.00002 or 0.00003K but never more than 0.0001K from the selected temperatures of any measurements. The observed plateau temperatures are shown relative to the mean value obtained for the tin-point cells 6C and 6E. Cell 6M was accidentally overheated above 500°C before the measurements. Only two measurements were obtained for cell 6K. (Resistance ratio measurements were made by R. L. Powell at the Boulder Laboratories of the National Bureau of Standards.)

and the use of proper freezing-point techniques are important. The experimental procedure for realizing the freezing point of zinc differs appreciably from that employed for tin. The zinc is melted in the furnace while being monitored with an SPRT. The zinc freeze is initiated in the furnace, unlike the tin freeze where the cell is removed from the furnace to initiate the freeze. The furnace temperature is reduced so that the sample temperature, as measured within the well, reaches the zinc-point temperature at a cooling rate of 0.1 to 0.2°C per minute. The zinc supercools below the zinc point by 0.02 to 0.06°C (as measured by the SPRT) before the recalescence occurs. The subsequent rise toward the plateau temperature, while rapid at first, becomes very slow after several minutes (see Fig. 2.34). To hasten this process the induced freeze suggested by McLaren [52] is employed to form a second solid-liquid interface immediately adjacent to the well. (The first solid-liquid interface occurs at the crucible wall.) After the zinc cell has completed the initial rapid freeze following the supercool described above, two cold thermometers are inserted into the well to freeze the zinc around the well; a solid-liquid interface that completely surrounds the well is necessary and a single thermometer may not freeze a complete coating of zinc on the well. The resistance of the second thermometer will rise to a steady value (within 10^{-4} °C) within a few minutes (typically 10 to 15 min) after insertion and will remain nearly constant for a period of time that is determined by the freezing rate of the outer solid zinc layer and the purity of the zinc. (The rate of freezing, of

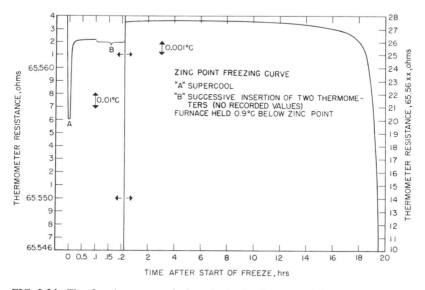

FIG. 2.34 The freezing curve of zinc obtained using an ac bridge. (The head of the SPRT had been adapted for coaxial connectors.) The furnace was controlled at 0.9K below the zinc point. The mass of zinc sample was 1280 g. Two cold SPRT's were inserted in the cell at B to freeze a coating of zinc around the graphite thermometer well.

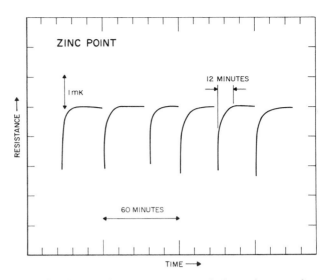

FIG. 2.35 Consecutive measurements of the resistance of a SPRT in the same zinc freeze. (An ac bridge was employed in the measurements.) Following each equilibrium resistance observation, the SPRT was completely withdrawn from the zinc-point cell, exposed to the ambient temperature for 1 min, and reinserted into the cell. In the actual procedure employed in calibration the SPRT is preheated close to the zinc point (see text). The data show that the SPRT comes rapidly to equilibrium even when inserted into the cell relatively cold. (For comparison when the SPRT was preheated, see Fig. 2.31.)

course, depends principally upon the heat loss to the furnace, i.e., the temperature setting of the furnace). Freezes lasting as long as 3 days have been observed at NBS, but 12- to 16-hr freezes are more typical. As with tin, SPRT's may be consecutively calibrated in a single zinc freeze if the thermometers are preheated (see Fig. 2.35). The preheating of the thermometer serves to minimize the growth of the inner freeze on the well and thereby extend the useful life of the freeze. Care must be exercised to avoid melting a hole in the inner freeze or loosening the mantle on the well so that it slides to the bottom of the crucible. To ensure that no melting and no excessive freezing occur, the thermometer is preheated to a temperature slightly below the zinc point before inserting it into the cell. As with tin, an SPRT is set aside at NBS for checking the constancy of the freeze, and it is used for both the first and last measurements in each freeze. The freezing-point measurements are normally not conducted at exactly the specified pressure of 1 std atm and a correction is made for this departure. The pressure at the interface surrounding the thermometer resistor includes both the gas pressure over the sample and the pressure due to the head of liquid zinc. For zinc at the melting point the value of dt/dp is $+ 5.7 \times 10^{-6}$ °C/torr or $+ 2.74 \times 10^{-5}$ °C/(cm column of liquid zinc) [52].

The immersion characteristics of an SPRT were tested in one of the zinc-point cells that were described. The observed resistances are shown in Fig. 2.36 as a function of the distance from the bottom of the crucible thermometer well. The results show that the immersion of the SPRT in the cell is more than adequate.

2.6.6 Oxygen Normal Boiling Point

The oxygen normal boiling point calibration is realized at the NBS by reference to the NBS–55 scale adjusted to correspond to the IPTS–68 (see Sec. 2.6.7). The temperature (hotness) assigned to the oxygen point, now maintained by SPRT standards, is 0.0019K lower [53] than that maintained previously on the NBS–1955 scale [9]. This change resulted from efforts made to achieve uniformity in several national temperature scales; for details see Ref. [53]. Calibrations near the oxygen point are performed by inserting the SPRT's to be calibrated and an SPRT standard in the apparatus shown in Fig. 2.37. In the preparation for calibration the apparatus is evacuated and immersed in liquid nitrogen to the level indicated. There are eight thin-walled Monel wells A extending into the copper block P. Two of these wells are ½ in. in diameter to accommodate capsule-type thermometers in holders. Thermometers are sealed into the wells at the top by a molded band of very soft silicone rubber. The thermometer wells are filled with helium gas to a pressure that is slightly above atmospheric pressure. The helium enhances the thermal contact between the wall of the well and the thermometer and prevents condensible gases from entering the wells. The copper block P and shields F and L are maintained under vacuum. They are

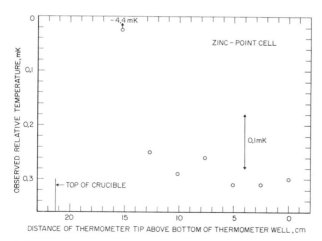

FIG. 2.36 Immersion characteristics of a SPRT in a zinc-point cell. The data show that the indications of the SPRT are very nearly the same between about 8 or 10 cm above the bottom and the bottom of the thermometer well.

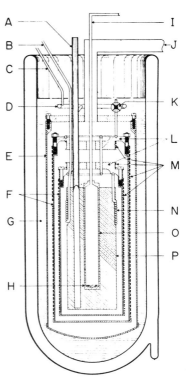

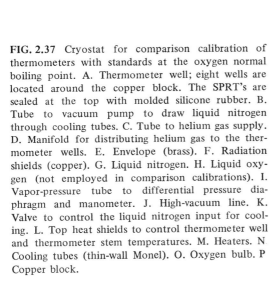

FIG. 2.37 Cryostat for comparison calibration of thermometers with standards at the oxygen normal boiling point. A. Thermometer well; eight wells are located around the copper block. The SPRT's are sealed at the top with molded silicone rubber. B. Tube to vacuum pump to draw liquid nitrogen through cooling tubes. C. Tube to helium gas supply. D. Manifold for distributing helium gas to the thermometer wells. E. Envelope (brass). F. Radiation shields (copper). G. Liquid nitrogen. H. Liquid oxygen (not employed in comparison calibrations). I. Vapor-pressure tube to differential pressure diaphragm and manometer. J. High-vacuum line. K. Valve to control the liquid nitrogen input for cooling. L. Top heat shields to control thermometer well and thermometer stem temperatures. M. Heaters. N Cooling tubes (thin-wall Monel). O. Oxygen bulb. P Copper block.

initially cooled by admitting nitrogen gas and/or liquid through the valve K shown just below the surface of the liquid nitrogen. The cool vapor flows through coils N on the top shields and coils on the massive center block, finally passing through B to a large-capacity vacuum pump. After cooling the block to a temperature near the oxygen point, the valve is closed and the nitrogen gas in the cooling coils is removed; the temperature of the block is brought within 1K of the oxygen point, and the outer shields are controlled at the temperature of the block. Experience shows that the temperature of the block can be maintained more nearly constant by allowing the inner shields to "float" without heating at a temperature near that of the block. (Heaters M on the copper block and the inner shields are used, when desired, to attain higher temperatures.) The outer shields are controlled to be at the temperature of the block by means of one differential thermopile between the top shield and the block, and a second thermopile between the outer top shield and the outer side shield. The signals from these thermopiles go to commercial low level chopper-type dc amplifiers and thence to three-mode controllers; the output signals from the controllers are then raised to power levels that are sufficient to supply the shield heater (less than 1 W). The measurements on the thermometer to be calibrated and the standard thermometer are made simultaneously using two Mueller bridges. The standard deviation of the intercomparison is less than the equivalent of 0.1 mK.

2.6.7 Comparison Calibration between
13.81 and 90.188K

To determine temperatures below the oxygen normal boiling point, the text of IPTS-68 (Appendix A) assigns values to defining fixed points and prescribes the form of interpolation formulae. At the time the IPTS-68 was adopted no national laboratory could satisfactorily realize all the fixed points. The four temperature scales (e.g., NBS-55) upon which the extension of the IPTS-68 below the oxygen point was principally based are each highly reproducible. The text of the IPTS-68, therefore, formally recognized that the use of these national scales, adjusted by the published differences [53], will give a close approximation to the IPTS-68 below 90.188K. At NBS a reference group of capsule-type SPRT's maintains the "national scale" (NBS-55). The NBS version of IPTS-68 (NBS-IPTS-68) is achieved by making reference to the NBS-55 scale and utilizing the published differences between the NBS-55 scale and the IPTS-68 referred to above. Thermometers are calibrated by intercomparison in a nearly isothermal copper block with one of the thermometers of this reference group. A second thermometer from the reference group is also included in the comparison to serve as a check on both the reference thermometer being used and the comparison techniques. These two thermometers will be referred to, henceforth, as the *first* and *second standards*, respectively.

The comparison of capsule-type SPRT's is performed in a cryostat that has been built and operated rather like an adiabatic calorimeter. The apparatus is shown in Fig. 2.38. The copper comparison block V, shown enlarged in Fig. 2.39, has wells for six thermometers, two standards and four thermometers under test. The schematic in Fig. 2.40 shows that the first standard is wired independently of the other five thermometers. The wiring arrangement employed for the five thermometers requires only $2n + 2$ leads for n thermometers, in this case, 12 leads, thus minimizing the number of leads and, therefore, the heat transfer through the leads with the surroundings. The first standard is electrically isolated from the others. This permits the use of two bridges to make simultaneous measurements on the first standard and any one of the thermometers to be calibrated on the second standard. The resistance measurements are made with two Mueller bridges.

To minimize the possible change in the difference of resistance in the potential leads, the leads from the thermometers that pass up through the cryostat are of No. 26 AWG (0.404 mm diameter) manganin wire insulated with a heavy coating of Formvar. The choice represents a compromise among the requirements of low heat conductivity, the resistance seen by the galvanometer, and lead-resistance stability. The leads are "tempered" by being placed in good thermal contact with the series of "thermal tiedowns" N shown in Fig. 2.41. The manganin leads and copper leads for thermocouples and heaters exit through a hard wax seal at C, the top of the cryostat, and then join to heavier copper leads. The external copper leads are thermally insulated to minimize both the temperature change of the leads and any possible gradient change between leads. The resistances of the copper sections of the leads and those of the manganin sections of the leads were separately adjusted for common equality to maintain the lead balance with changing room temperature. The thermometer leads are connected to the bridge through selector switches that employ either mercury-wetted contacts or multiple all-silver contacts in parallel.

The cryostat is constructed so that the central core, including the copper comparison block, is stationary while the remainder of the cryostat assembly may be lowered to expose the block for thermometer installation or removal. The only seals required for sealing the cryostat are at room temperature and consist of simple Viton O rings E.

The cryostat proper (Fig. 2.38) is surrounded with the usual arrangement—an inner Dewar flask M, which contains liquid helium, guarded by an outer Dewar flask (not shown) that contains liquid nitrogen. During the initial cooldown, the inner Dewar flask is filled with liquid nitrogen and the liquid nitrogen vapor pressure is reduced to about 1/3 atm. The vacuum can O of the cryostat is filled with ^{3}He gas to a pressure of about 0.1 torr to facilitate the cooling; under such conditions, the thermometer block cools to about 65K in approximately 12 hr (overnight). The liquid nitrogen remaining in the inner Dewar flask is removed by applying a small overpressure of ^{4}He gas which forces the nitrogen out through a 3.2-mm thin-wall stainless tube (not shown in Fig. 2.38) that extends from the bottom of the Dewar flask to the room. (Precooling the cryostat from about 80K to about 65K by pumping in liquid nitrogen reduces the liquid helium required for cooling by about 2L.) Six to seven liters of liquid helium are usually required to cool the apparatus from 65 to 4.2K and fill the inner helium Dewar flask.

As a part of the calibration procedure the thermometer resistance readings are first made at about 4K, without the use of temperature controls. For the calibrations about 4K, the ^{3}He gas in the vacuum can is removed to reduce the heat transfer and the temperature of the comparison block is controlled by the use of three sets of five-junction Chromel-P/Constantan differential thermopiles that sense temperature differences between the comparison block and the thermal shields. Appropriate automatic controls are employed to power three separate heaters to reduce the indicated temperature differences to zero. When the space in the vacuum can is evacuated by a pressure in the 10^{-6}-torr range, most of the heat transfer to the comparison block from the surroundings is (a) down through the wall of the well F and (b) within the well through the central thin-wall stainless tube and electrical leads. Consequently, two sets of thermopiles and heaters are affixed, one at R on the wall of the well and the other at Q within the well. These two locations are about 15 cm above the comparison block. The third thermopile and heater are mounted on the heavy copper shield which surrounds the lower end of the well where the comparison block rests. The thermopiles are all referenced to the temperature of the comparison block. If adequate temperature control of a system is to be achieved, the thermal lags of both the heaters and the temperature sensors (in this case thermopiles) must

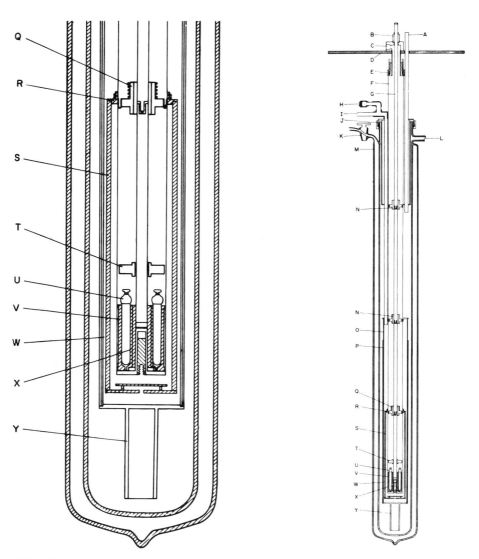

FIG. 2.38 Cryostat for the intercomparison of capsule-type SPRT's. A. Guide for directing transfer tube into liquid helium Dewar. B. Metal bellows to permit differential expansion between the central tube and the supporting thermometer block well. C. Exit for the vacuum line and electrical leads to the thermometers, the thermocouples, and the heater on the copper thermometer block. D. Supporting shelf. E. Demountable O-ring seal to the well around the thermometer block. F. Well around the thermometer block. G. Central tube for supporting the thermometer block. H. Seal for the electrical leads from the vacuum can that surrounds the shield and lower portion of the thermometer block well. I. Line from vacuum can O. J. Demountable O-ring seal to the liquid helium Dewar. K. Glass stopcock to permit reevacuation of liquid helium Dewar. L. Line for pumping the space within the liquid helium Dewar. M. Liquid helium Dewar. N. Thermal tiedown for lead wires. O. Vacuum can that surrounds the lower portion of the thermometer block well and the thermal shield. P. Copper sleeve on vacuum can to maintain uniform temperature when liquid helium level is

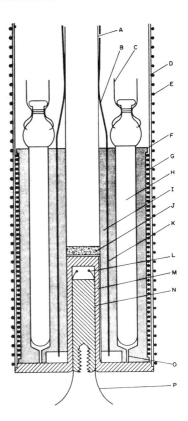

FIG. 2.39 Thermometer comparison block. A. Stainless steel tube to support thermometer comparison block. B. Manganin leads for thermometers and copper leads for block heater. C. Platinum leads of resistance thermometer. D. Stainless steel well for thermometer block. E. Heater distributed along well D. F. Copper thermometer block with wells for six thermometers. G. Heater for thermometer block. H. Platinum resistance thermometer. I. Copper sleeve (attached to central and supporting tube) with longitudinal grooves to serve as thermal tiedowns for manganin leads and thermocouples. J. Brass plug to seal end of central support tube. K. Location of grooves in sleeve I. L. Reference junctions for thermocouples on shield (at W in Fig. 2.38) and thermometer block well ring heater (at R in Fig. 2.38). M. Closely fitted reentrant copper "thumb" (with a groove for venting gas). N. Copper plug with longitudinal grooves to serve as thermal tiedowns for thermocouples. O. Vent. P. Thermocouple leads to shield and thermometer block well ring heater.

be small. In addition, the "immersion error" that is caused by heat conducted along the leads of the sensor should be small. These characteristics were achieved in the cryostat by employing the "thermal tiedowns" shown in Figs. 2.38 and 2.41. (The selection of Chromel-P/Constantan thermocouples was influenced by their low thermal conductivity. There are other materials that possess better thermoelectric characteristics but their relatively high thermal conductivities were considered to be disadvantageous.) The spurious emf of each thermopile is balanced out with a simple emf injector that is adjustable between $\pm 5 \, \mu V$. The

low. Q. Thermal tiedown for leads similar to N with a heater and five-junction Chromel-P/Constantan thermocouple for temperature control. R. Copper ring on the thermometer block well with temperature control components similar to Q. S. Heavy copper shield with temperature control components similar to Q. T. Thermometer lead terminal block of anodized aluminum. U. Capsule-type platinum resistance thermometers thermally attached to copper block with vacuum grease. V. Copper thermometer block with holes for six thermometers. W. Location of thermocouple junctions placed on shield. X. Reentrant "thumb" in the bottom of well. "Thumb" contains reference junctions for thermocouples on R and S. Y. Heavy copper tail on vacuum can to reach liquid helium at a low level.

adjustment for the spurious emf is made at 4.2K after equilibrium has been attained with ^{3}He gas still in the can.

The calibration resistance measurements are made at 16 temperatures; approximately 12.20, 13.81, 15.426, 17.042, 18.661, 20.28, 23.691, 27.102, 33.917, 40.732, 47.547, 54.361, 66.303, 78.245, 84.216, and 90.188K. The values of $W(T)$ at the IPTS–68 defining fixed-point temperatures are obtained by least-square analysis of the comparison data (see Sec. 2.5).

To change from one calibration temperature to the next higher temperature, current is supplied to the heater that is wound directly on the comparison block. Although the control of the shields is adequate to track the temperature of the block when it is rapidly heated at the lower temperatures (below about 40K), the shields lag at higher temperatures when the block is being rapidly heated because of the limitations of the power amplifiers which drive their heaters. The time required for the system to attain thermal equilibrium after the heating of the block is stopped increases with temperature. At 13.8K the thermometer drift will decrease to less than 10 $\mu\Omega$/min in a minute, while at 90K more than 30 min is required to achieve this drift rate. Measurements are considered unsatisfactory if the indicated drift rate is more than 5 $\mu\Omega$/min at temperatures below 20K or more than 10 $\mu\Omega$/min at temperatures above 20K.

In order to achieve these low drift rates rather quickly, particularly at temperatures above 30K, it was found necessary to heat uniformly the lower 20-cm section of the well F when the comparison block was being heated to the next higher calibration temperature. This was accomplished by powering a heater wound around the well E as shown in Fig. 2.39. An adjustable fraction (depending upon the temperature) of the voltage applied to the block heater is applied to the well heater to raise its temperature at the same time and rate as the block temperature.

In the calibration of the capsule-type SPRT's, the resistance measurements are carried out with a continuous current of 1 mA through the thermometer and

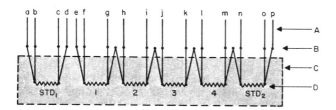

SPRT	Leads used
1	efgh
2	ghij
3	ijkl
4	klmn
STD$_2$	mnop

FIG. 2.40 Schematic drawing of wiring capsule-type platinum resistance thermometers in the copper comparison block of the cryostat for calibration. To avoid excessive heating of the copper block, the bridge connections to the thermometers, wired in series, are made as in the table above. A. Cryostat leads. B. Thermometer connections made among them and with the cryostat leads. C. Schematic of the copper comparison block. D. Platinum resistance thermometers STD$_1$ and STD$_2$ are reference standards; others are to be calibrated.

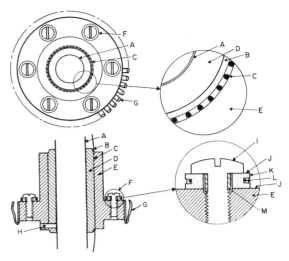

FIG. 2.41 Thermal tiedowns. A. Central support tube (stainless steel) for thermometer block. B. Rounded corner, tangent to the bottom of the grooves for wire. C. Round bottom grooves in cylinder D with wire cemented in place. D. Copper cylinder soldered (tin-lead eutectic) to the central support tube. E. Copper cylinder fitted to D and held in place by pin H. F. Thermocouple tie-down. G. Spring fingers of beryllium copper heavily silver plated and gold "flashed." H. 1.5-mm steel pin to hold cylinder E in position. I. Pan head 2–56 brass screw to clamp thermocouple tiedown assembly to E. J. Insulating washer of 0.005-mm Mylar coated with vacuum grease. K. Copper washer with pretinned groove for thermocouple wire. L. Polyimide-insulated thermocouple wire, placed at the bottom of the pretinned groove and "potted" in eutectic tin-lead solder. M. Epoxy insulation on screw.

with a commutator switching order of NRRNNRRN with the Mueller bridge. This order of measurements is not the optimum one (which is NRRNRNNR), but when a large number of measurements must be made, such as in the calibration of thermometers over this temperature range, the pattern used is less subject to gross errors. At approximately 90.2K, measurements are made at two currents, 1 mA and 2 mA, to determine the self-heating effect of the SPRT. Although infrequent, the glass-to-metal seal of a thermometer may leak; this causes a loss of helium and consequent increase in the self-heating effect of the thermometer. A thermometer having this defect may be unreliable. The treatment of the data resulting from these measurements differs markedly from those obtained above 90.188K (see Sec. 2.5).

REFERENCES

1. Callendar, H. L., On the Practical Measurement of Temperature: Experiments Made at the Cavendish Laboratory, Cambridge, *Phil. Trans. (London)*, vol. 178, pp. 161–230, 1887.
2. Burgess, G. F., The International Temperature Scale, *J. Res. Nat. Bur. Stand. (U.S.)*, RP22, vol. 1, pp. 635–640, 1928.
3. Comité International des Poid et Mesures, *Procès-verbaux des seances de l'année 1954*, vol. 24, pp. 81–82, T20–T23, T44–T47, 1955.
4. The International Practical Temperature Scale of 1968, Adopted by the Comité International des Poids et Mesures, *Metrologia*, vol. 5, no. 2, pp. 35–44, April 1969.
5. Brickwedde, F. G., Van Dijk, H., Durieux, M., Clement, J. R., and Logan, J. K., The 1958 ⁴He Scale of Temperature, *J. Res. Nat. Bur. Stand. (U.S.)*, vol. 64A, no. 1, pp. 1–17, 1960.
6. Stimson, H. F., The International Temperature Scale of 1948, *J. Res. Nat. Bur. Stand. (U.S.)*, RP1962, vol. 42, pp. 209–217, 1948.
7. Stimson, H. F., International Practical Temperature Scale of 1948. Text Revision of 1960, *J. Res. Nat. Bur. Stand. (U.S.)*, vol. 65A, no. 3, pp. 139–145, May–June 1961.
8. Hoge, H. J., and Brickwedde, F. G., Establishment of a Temperature Scale for the Calibration of Thermometers between 14 and 83°K, *J. Res. Nat. Bur. Stand. (U.S.)*, RP1158, vol. 22, pp. 351–373, 1939.
9. Swindells, J. F., National Bureau of Standards Provisional Scale of 1955, in Precision Measurement and Calibration. Temperature, *Nat. Bur. Stand. (U.S.) Spec. Publ. 300*, vol. 2, p. 56, August 1968.
10. Rosebury, F., "Handbook of Electron Tube and Vacuum Techniques," p. 371, Addison-Wesley, Reading, Mass. 1965.
11. Alieva, F. Z., New Platinum Resistance Thermometer for Precision Measurements of High Temperatures, *Meas. Tech. (USSR)*, no. 6, pp. 487–488, June 1964; translation from *Izmeritel. Tekh.*, no. 6, pp. 21–22, June 1964.
12. Curtis, D. J., and Thomas, G. J., Long term Stability and Performance of Platinum Resistance Thermometers for Use to 1063°C, *Metrologia*, vol. 4, no. 4, pp. 184–190, October 1968.
13. Evans, J. P., and Burns, G. W., A Study of Stability of High Temperature Platinum Resistance Thermometers, in American Institute of Physics, "Temperature, Its Measurement and Control in Science and Industry," vol. 3, pt. 1, pp. 313–318, Reinhold, New York, 1962.
14. Toenshoff, D. A., An Evaluation of Some Factors Which May Affect the Stability of Birdcage Thermometers, Rept. of work conducted under contract No. CST383 (National Bureau of Standards) by Engelhard Industries, East Newark, N.J., January 1, 1968.
15. Meyers, C. H., Coiled Filament Resistance Thermometers, *J. Res. Nat. Bur. Stand. (U.S.)*, RP508, vol. 9, pp. 807–817, December 1932.
16. Stimson, H. F., Precision Resistance Thermometry and Fixed Points, in American Institute of Physics, "Temperature, Its Measurement and Control in Science and Industry," vol. 2, pp. 141–168, Reinhold, New York, 1955.
17. Strelkov, P. G., Borovik-Romanov, A. S., and Orlova, M. P., Thermodynamic Studies at Low Temperatures. I. Measurement of Temperatures between 12 and 300°K, *Zh. Fiz. Khim.*, vol. 28, pp. 345–352, 1954.
18. Barber, C. R., Platinum Resistance Thermometers of Small Dimensions, *J. Sci. Instrum.*, vol. 27, pp. 47–49, 1950.
19. Barber, C. R., A Platinum Resistance Thermometer for Use at Low Temperatures, *J. Sci. Instrum.*, vol. 32, pp. 416–417, 1955.
20. Barber, C. R., Resistance Thermometers for Low Temperatures, in K. Mendelssohn (ed.), "Progress in Cryogenics," pp. 147–171 Academic, New York, 1960.

21. Barber, C. R., and Blanke, W. W., A Platinum Resistance Thermometer for Use at High Temperatures, *J. Sci. Instrum.*, vol. 38, pp. 17-19, 1961.

22. Evans, J. P., and Wood, S. D., An Intercomparison of High Temperature Platinum Resistance Thermometers and Standard Thermocouples, *Metrologia*, vol. 7, no. 3, pp. 108-130, July 1971.

23. Johnson, J. B., Thermal Agitation of Electricity in Conductors, *Phys. Rev.*, vol. 32, pp. 97-109, July 1928.

24. Mueller, E. F., Wheatstone Bridges and Some Accessory Apparatus for Resistance Thermometry, *Sci. Papers Nat. Bur. Stand.*, vol. 13, pp. 547-561, plus 1 plate, 1916-1917.

25. Mueller, E. F., Precision Resistance Thermometry, in American Institute of Physics, "Temperature, Its Measurement and Control in Science and Industry," vol. 1, pp. 162-179, Reinhold, New York, 1941.

26. Eppley, M., Modifications of the Resistance Thermometer Bridge (Mueller Bridge) and of the Commutator Selector for Use with It, *Rev. Sci. Instrum.*, vol. 3, pp. 687-711, 1932.

27. Evans, J. P., An Improved Resistance Thermometer Bridge, in American Institute of Physics, "Temperature, Its Measurement and Control in Science and Industry," vol. 3, pt. 1, pp. 285-289, Reinhold, New York, 1962.

28. Furukawa, G. T., Automation Problems in Thermometry and Calorimetry, in Y. Otsubo, H. Kanbe, and S. Seki (eds.), "Netsu, Ondo Sokutei, to Shisanetsu Bunseki, 1969" ("Calorimetry, Thermometry, and Differential Thermal Analysis, 1969"), Kagaku Gijitsu Sha, Tokyo, 1969.

29. Cutkosky, R. D., An A-c Resistance Thermometer Bridge, *J. Res. Nat. Bur. Stand. (U.S.)*, vol. 74C, nos. 1 and 2, pp. 15-18, January-June 1970.

30. Barber, C. R., Gridley, A., and Hall, J. A., A Design for Standard Resistance Coils, *J. Sci. Instrum.*, vol. 29, pp. 65-69, 1952.

31. Barber, C. R., Gridley, A., and Hall, J. A., An Improved Construction of the Smith Bridge, Type 3, *J. Sci. Instrum.*, vol. 32, pp. 213-220, 1955.

32. Gautier, M., A Modification of the Smith Bridge, Type III, *J. Sci. Instrum.*, vol. 30, pp. 381-382, 1953.

33. Schaeffer, W. H., The Six-Dial Thermofree Potentiometer, *Instrum. Contr. Syst.*, vol. 34, no. 2, pp. 283-291, February 1961.

34. White, W. P., Potentiometers for Thermoelectric Measurements Especially in Calorimetry, *J. Amer. Chem. Soc.*, vol. 36, pp. 1868-1885, 1914.

35. White, W. P., Einige neuen Doppelkompensatoren. I. Thermokraftfreie Kontakte, mechanische Anordnungen und Zubehörteile, *Z. Instrumentenk.*, vol. 34, pp. 71-82, 1914.

36. White, W. P., Einige neuen Doppelkompensatoren. II. Die elektrische Einrichtung. *Z. Instrumentenk.*, vol. 34, pp. 107-113, 142-151, 1914.

37. Dauphinee, T. M., Potentiometric Methods of Resistance Measurement, in American Institute of Physics, "Temperature, Its Measurement and Control in Science and Industry," vol. 3, pt. 1, pp. 269-283, Reinhold, New York, 1962.

38. McLaren, E. H., Intercomparison of 11 Resistance Thermometers at the Ice, Steam, Tin, Cadmium, and Zinc Points, *Can. J. Phys.*, vol. 37, pp. 422-432, 1959.

39. Hoge, H. J., and Brickwedde, F. G., Intercomparison of Platinum Resistance Thermometers between − 190 and 445°C, *J. Res. Nat. Bur. Stand., (U.S.)*, RP1454, vol. 28, pp. 217-240, February 1942.

40. Belyansky, L. B., Orlova, M. P., Sharevskaya, D. I., and Astrov, D. N., Investigation of the Resistance-Temperature Properties of Platinum for Resistance Thermometry over the Range from 90 K to 273 K, *Metrologia*, vol. 5, no. 4, pp. 107-111, 1969.

41. Moussa, M. R. M., Van Dijk, H., and Durieux, M., Comparison of Platinum Resistance Thermometers between 63 K and 373.15 K, *Physica*, vol. 40, pp. 33-48, 1968.

42. Sharevskaya, D. I., Orlova, M. P., Belyansky, L. B., and Galoushkina, G. A., Investigation of the Resistance-Temperature Properties of Platinum of Resistance Thermometry over the Range from 14 K to 90 K, *Metrologia*, vol. 5, no. 4, pp. 103–107, 1969.
43. Bedford, R. E., and Ma, C. K., A Note on the Reproducibility of the IPTS-68 below 273.15 K, *Metrologia*, vol. 6, no. 3, pp. 89–94, 1970.
44. Evans, J. P., and Sweger, D. M., Immersion cooler for Freezing Ice Mantles on Triple-Point-of-Water Cells, *Rev. Sci. Instrum.*, vol. 40, no. 2, pp. 376–377, 1969.
45. Berry, R. J., The Temperature-Time Dependence of the Triple-Point of Water, *Can. J. Phys.*, vol. 37, pp. 1230–1248, 1959.
46. Iwase, K., Aoki, N., and Osawa, A., The Equilibrium Diagram of the Tin-Antimony System and the Crystal Structure of the β-Solid Solution, *Sci. Rep. Tohoku Imp. Univ. (Ser. 1)*, vol. 20, pp. 351–368, 1931.
47. Pfann, W. G., "Zone Melting," 2d ed., Wiley, New York, 1966.
48. Saylor, C. P., Calculation of Cryoscopic Data, *Anal. Chim. Acta*, vol. 17, pp. 36–42, 1957.
49. Smit, W. M., Errors Occurring in the Determination of Temperature-Heat Content Curves, *Anal. Chim. Acta*, vol. 17, pp. 23–42, 1957.
50. Unpublished test measurements at NBS with manually controlled copper block furnaces suitable for conducting metal freezing and melting measurements.
51. Michaelis, R. E., Private Communication, Office of Standard Reference Materials (Nat. Bur. Stand.).
52. McLaren, E. H., The Freezing Points of High Purity Metals as Precision Temperature Standards. II. An Investigation of the Freezing Temperatures of Zinc, Cadmium, and Tin, *Can. J. Phys.*, vol. 35, pp. 1086–1106, 1957.
53. Bedford, R. E., Durieux, M., Muijlwijk, R., and Barber, C. R., Relationships between the International Practical Temperature Scale of 1968 and the NBS-1955, NPL-61, PRMI-54, and PSU-54 Temperature Scales in the Range from 13.81 to 90.188 K, *Metrologia*, vol. 5, no. 2, pp. 47–49, April 1969.

APPENDIX: COMPARISON OF ASSIGNED FIXED-POINT VALUES AND INTERPOLATING FORMULAS IN THE TEMPERATURE RANGES DEFINED BY THE PLATINUM RESISTANCE THERMOMETER FOR ITS-27, ITS-48, IPTS-48, AND IPTS-68

TABLE 2.A.1 Comparison of fixed point values

Defining fixed points	ITS-27 (°C)	ITS-48 (°C)	IPTS-48 (°C)	IPTS-68 (°C)	IPTS-68 (K)
TP e-H_2				-259.34	13.81
BP e-H_2, 25/76 atm				-256.108	17.042
NBP e-H_2				-252.87	20.28
NBP Ne				-246.048	27.102
TP O_2				-218.789	54.361
NBP O_2	-182.97	-182.970	-182.97	-182.962	90.188
Ice point	0.000	0			
TP H_2O			0.01	0.01	273.16
NBP H_2O	100.000	100	100	100	373.15
FP Sn				231.9681	505.1181
FP Zn			419.505	419.58	692.73
NBP S	444.60	444.600	444.6		

Some secondary fixed points

	ITS-27 (°C)	ITS-48 (°C)	IPTS-48 (°C)	IPTS-68 (°C)	IPTS-68 (K)
SP CO_2	-78.5	-78.5	-78.5	-78.476	194.674
FP Hg	-38.87	-38.87	-38.87	-38.862	234.288
Ice point			0.000	0	273.15
TP H_2O		0.0100			
FP Sn	231.85	231.9	231.91		
FP Pb	327.3	327.3	327.3	327.502	600.652
FP Zn	419.45	419.5			
NBP S				444.674	717.824
FP Sb	630.5	630.5		630.74	903.89

TP = triple point, BP = boiling point, NBP = normal boiling point, FP = freezing point, SP = sublimation point.

TABLE 2.A.2 Comparison of temperature ranges and interpolating instrument and formulas

	ITS-27	ITS-48	IPTS-48	IPTS-68
$R(100)/R(0)$ of Pt.	$\geqq 1.390$	> 1.3910	$\geqq 1.3920$	$\geqq 1.3925$
Temperature range Interpolating formula	-190 to $0°C$ quartic[a]	-182.970 to $0°C$ quartic[a]	-182.97 to $0°C$ quartic[a]	13.81 to 273.15K reference function[c] plus deviation polynomials[d]
Calibration temperatures	NBP O_2, ice point, NBP H_2O, & NBP S	NBP O_2, ice point, NBP H_2O, & NBP S	NBP O_2, TP H_2O, NBP H_2O, & NBP S or FP Zn	TP e-H_2, BP e-H_2 (25/76 atm), NBP e-H_2 NBP Ne, TP O_2, NBP O_2, TP H_2O, NBP H_2O or FP Sn, & FP Zn.
Temperature range Interpolating formula	0 to 660°C quadratic[b]	0 to 630.5°C quadratic[b]	0 to 630.5°C quadratic[b]	0 to 630.74°C Quadratic plus correction function[e]
Calibration temperatures	ice point, NBP H_2O, & NBP S	ice point, NBP H_2O, & NBP S	TP H_2O, NBP H_2O, & NBP S	TP H_2O, NBP H_2O or FP Sn, & FP Zn.

[a] Quartic interpolation formula: $R(t) = R(0) [1 + At + Bt^2 + C (t - 100) t^3]$.

[b] Quadratic interpolation formula: $R(t) = R(0) (1 + At + Bt^2)$.

[c] Reference function $\quad T_{68} = \sum_{i=0}^{20} A_i [\ln W^*(T)]^i$. (Values of A_i are given in the Appendix to this book.)

[d] Deviation polynomials:

	$\Delta W(T) = W(T) - W^*(T)$
13.81 to 20.28K:	$\Delta W(T) = A_1 + B_1 T + C_1 T^2 + D_1 T^3$
20.28 to 54.361K:	$\Delta W(T) = A_2 + B_2 T + C_2 T^2 + D_2 T^3$
54.361 to 90.188K:	$\Delta W(T) = A_3 + B_3 T + C_3 T^2$
-182.962 to $0°C$:	$\Delta W(t) = A_4 t + C_4 t^3 (t - 100°C)$

where $T = T_{68}$, $t = t_{68}$, $W^* = W_{CCT-68}$

[e] Quadratic interpolation formula plus correction function: $R(t')/R(0) = 1 + At' + Bt'^2$ plus

$$t_{68} = t' + 0.045 \left(\frac{t'}{100°C}\right)\left(\frac{t'}{100°C} - 1\right)\left(\frac{t'}{419.58°C} - 1\right)\left(\frac{t'}{630.74°C} - 1\right)$$

3 Thermocouple thermometry

ROBERT L. POWELL
Consultant; Member ASTM Committee E-20 on Thermometry
Boulder, Colorado

Because of their many advantages, thermocouples have long been used extensively in both scientific research and industrial thermometry. They are inherently simple, consisting usually of two wires, a stable reference junction, and a potentiometer system. More complex thermocouple systems may consist of many separate junctions or thermopiles, but the basic principles and instrumentation remain unchanged. They may be large (to allow mechanical and corrosion protection) or small (to provide rapid response time and small heat capacity). Small wires used in cryogenic systems are very fragile and flexible; the large wires used in furnaces are usually encased and are therefore rigid.

With careful design, thermometer encapsulations can withstand many types of corrosive atmospheres. Thermocouples can be used over a wide range of temperatures, from liquid helium ($-270°C$) to high-temperature furnaces ($2200°C$). However, different alloys are necessary for the extremes in temperatures. Many of the thermocouple combinations give a nearly linear output in a wide range of temperatures. This property leads to calibration and instrumentation methods that are both more simple and more accurate. Unlike the resistance thermometers, thermocouples have no self-heating effect. This is very important in precise calorimetry and in cryogenic research.

The number of potential thermocouple combinations is virtually infinite, but fortunately there has been a large amount of standardization. This standardization of types and thermometric values has led to the widespread availability, at reasonable cost, of largely interchangeable material from several different alloy manufacturers and many different instrumentation companies.

And last, but not least, thermocouple systems are easy to instrument: portable potentiometers, laboratory galvanometers, recording potentiometers, and digital multimeters are all commonly used. The need for a stable reference junction may be negated for most applications by the insertion of a temperature-compensated junction; this is the most common industrial practice.

In 1821 Thomas Johann Seebeck [1] discovered the existence of thermo-electric currents while experimenting on bismuth-copper and bismuth-antimony circuits. He showed that when the junctions of dissimilar metals are heated to different temperatures a net thermoelectric emf is generated. If the wires form a closed circuit, then a thermoelectric current is induced. Within a few years Becquerel had demonstrated that a platinum-palladium couple could be used to measure temperature. About a decade later Jean Peltier [2] discovered an unusual thermal effect when small, externally imposed currents were directed through the junctions of different thermocouple wires. When current flowed across a junction in one direction, the junction was cooled; when the current flow was reversed, the same junction heated. With the aid of the newly de-veloped theories of thermodynamics, William Thomson (later Lord Kelvin) was able to show that the two effects were related. He also derived the fundamental equations used to this day. Theories of thermoelectricity have since become quite refined and complex (though they are unnecessary for using thermocouples in practical applications). A classic for a modern discussion was written by Mac-Donald [3] and a very thorough, but elementary, review has been prepared by Pollack [4] for the American Society for Testing and Materials.

One of the earliest combinations to have widespread usage was copper vs. constantan, now referred to as Type T. Another standard material with a long history is Chromel vs. Alumel (originally developed by Hoskins Manufacturing Company), which is now one of the materials that comply with the standard values of Type K. The most commonly used high-temperature combinations have been platinum vs. platinum-rhodium alloys, Types S, R, and B. The platinum group is referred to as *noble metal combinations*; the others are called *base metal types*. In recent years several new materials have been developed and widely accepted. They include the tungsten-rheniums for very high temperatures, the gold-irons for cryogenic ranges, and Nicrosil vs. Visil for intermediate and moderately high temperatures. These materials will be described in more detail later in this chapter.

Of the many reviews and books written on thermocouples we have found these most useful:

1. On *theory*: MacDonald [3], Pollack [4], Barnard [5], Bridgman [6], and Harman and Honig [7]
2. On *practical applications*: American Society for Testing and Materials [8], Benedick [9], Dike [10], Harman and Honig [7], Heikes and Vine [11], and Kinzie [12]
3. On *standards*: Sparks and Powell [13], Sparks et al. [14], and Powell et al. [15]

3.1 FUNDAMENTAL PRINCIPLES

The fundamental principles necessary for understanding elementary thermo-electric circuits can be expressed in three effects (with the equations that relate them) and three laws that derive from the fundamental equations.

The Seebeck effect can be described with a diagram, Fig. 3.1. If a circuit is formed that consists of two dissimilar conductors A (positive) and B (negative) with junctions at temperatures T, and T_2 (T_2 T_1), then a current will flow through the circuit in the direction indicated. If there is a break in the wire A, then the potential developed across that gap is called the *Seebeck voltage* (other synonyms include *thermoelectric voltage, Seebeck emf*, etc.).

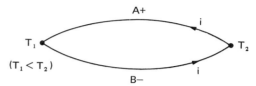

FIG. 3.1 Seebeck effect.

For a small temperature difference, the change in voltage across the leads (the gaps in wire A) is given by

$$dE_s = \alpha_{A,B} \, dT \qquad (3.1)$$

where $X_{A,B}$ is the Seebeck coefficient (or thermoelectric power) for the combination of materials A and B at a temperature T. The total voltage drop for a range of temperatures is given by

$$E_s = \int_{T_1}^{T_2} \alpha_{A,B} \, dT \qquad (3.2)$$

The coefficient $\alpha_{A,B}$ may be obtained by differentiating a functional relation of $E_s(T)$ or it may be obtained by using the relation

$$\alpha_{A,B} = \alpha_{A,R} + \alpha_{R,B} \qquad (3.3)$$

where R indicates a standard reference material.

The Thomson effect may be understood using Fig. 3.2. When a current flows across a junction of dissimilar conductors, heat is absorbed ($T_2 + T$) or liberated ($T_1 - T$) by the junctions. If the electric current flows in the same direction as the Seebeck current, then heat is *absorbed* at the hotter junction (and vice versa). This effect is utilized in thermoelectric heating or refrigeration. The relation is

$$dQ_p = \pi_{A,B} I \, dt \qquad (3.4)$$

where dQ_p is the amount of heat, $\pi_{A,B}$ is the Peltier coefficient, I the electric current, and dt the elapsed time. The Peltier effect is closely related to the Seebeck effect.

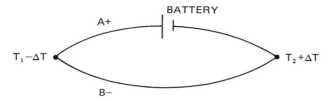

FIG. 3.2 Thomson effect.

The relation of the two was derived by Thomson to be

$$\pi_{A,B} = (\sigma_A - \sigma_B)\, dT \tag{3.5}$$

where σ is the Thomson coefficient for a single conductor defined by

$$E_{T_1, T_2} = \int_{T_1}^{T_2} \sigma_A\, dT \tag{3.6}$$

Verbally the Thomson effect is defined as the change in heat capacity of a single conductor (of unit cross section) when a unit quantity of electric charge flows through it along a temperature gradient of 1K.

The various relations, including the very important Kelvin relations, have been discussed by many authors, including Pollack [4].

The three basic laws of practical thermometry are given in the ASTM STP 470 [8]. They are:

1. "Law of homogeneous metals—a thermoelectric current cannot be sustained in a circuit of a single homogenous material, however varying in cross-section, by the application of heat alone."

   Therefore at least two different materials are required for a thermo-electric circuit. It should be noted that physical or chemical imperfections cause a material to be effectively inhomogeneous.

2. "Law of intermediate metals—the algebraic sum of the thermoelectromotive forces [voltages] in a circuit composed of any number of dissimilar materials is zero if all of the circuit is at a uniform temperature."

   A third inhomogenous material can always be added to a circuit as long as it is in an isothermal region. Because of this law, represented in Fig. 3.3, one consequence is that the method of joining thermocouple wires, e.g., soldering, welding, clamping, mercury contact, etc., does not affect the thermoelectric output if the junction is isothermal. Another consequence is that if the thermoelectric voltages of two materials are known relative to a reference material, their voltages to each other may be determined additively. This is illustrated in Fig. 3.4.

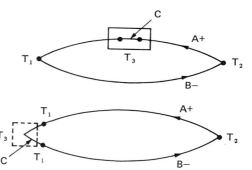

FIG. 3.3 Law of intermediate metals. The thermoelectric voltage is unaffected by insertion of material C.

3. "Law of intermediate temperatures—if two dissimilar homogeneous metals produce a thermal emf of E_1, when the junctions are at temperatures T_1 and T_2, and a thermal emf of E_2, when the junctions are at T_2 and T_3, the emf generated when the junctions are at T_1 and T_3, will be $E_1 + E_2$."

This is illustrated in Fig. 3.5. One result of this law is that thermocouples calibrated for one reference temperature can easily be corrected for another reference temperature. Another convenience resulting from this law is the availability of using extension wires, say in the right hand region of the upper figure, without disturbing the resultant thermoelectric voltage.

3.2 FABRICATION AND MEASUREMENT METHODS

The methods of fabricating and measuring the outputs of thermocouples vary tremendously on the application and temperature. The simplest methods utilize small, lightly insulated wires as in many cryogenic applications; the most complex are the sheathed, ceramic-insulated thermocouple assemblies used in high-temperature nuclear reactors.

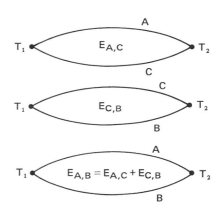

FIG. 3.4 Thermoelectric voltages are simply additive.

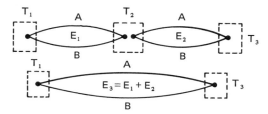

FIG. 3.5 Thermoelectric voltages are additive for intermediate temperatures.

Accurate thermocouple thermometry is possible if care is taken in:

1. material selection,
2. thermocouple testing and calibration,
3. fabrication techniques, insulators, etc., and
4. general experimental design and measurement technique.

The selection of the proper thermocouple for use in a particular situation is based upon many considerations. Probably the most important at high temperatures is compatibility of the environment with the thermocouple material and its insulation. At low temperatures this is also important, but it is generally not a problem. This criterion, satisfied in essentially the same way as at high temperatures, is treated in detail in the ASTM manual [8] and by Caldwell [16, 17]. Insulations commonly available are enamel, polyethylene, polytetrafluorethylene, polyimide, and spun glass. Combinations of glass and one of the others are also frequently available. Probably the most durable, but also most difficult to remove, is polyimide. Wires can also be obtained without insulation for use in special situations.

Many thermocouple materials developed for high temperature usage have too low a sensitivity at low temperatures to be practical; others have reasonable sensitivity, but are erratic because of large inhomogeneities and lot-to-lot variations. Only three common commercial standardized thermocouple combinations have proven themselves for cryogenic use: (1) Type E, KP vs. TN; (2) Type T, TP vs. TN; and (3) Type K, KP vs. KN. Throughout this chapter thermocouples and thermocouple materials are designated by their letter symbols which have been standardized by ANSI, ASTM, and ISA (American National Standards Institute, American Society for Testing Materials, and Instrument Society of America, respectively). The letter types for thermocouples are described, and the main standardized types are discussed in detail in the next section.

Measurement of temperature in the presence of magnetic or radiation fields requires additional care and special techniques. Few temperature sensors retain their normal calibrations under these conditions. Minimum changes to be expected are several percent and may be larger than 100% depending on temperature and field intensity as well as thermocouple material. Materials which have a high sensitivity at low temperature will be strongly affected by magnetic fields. Richards [18] has shown both theoretically and experimentally that thermocouple calibrations are not changed if the segment of wire in the field is isothermal. This, however, is often difficult to achieve. Loscoe and Mette [19] made measurements of the magneto-thermal-voltages and concluded that KN (Alumel) is a poor material to use in a magnetic field at 30°C.

If thermocouples are to be used to measure temperatures in magnetic fields, we recommend avoiding KN (Alumel) at all temperatures and the dilute *Au*-Fe alloys at the lowest temperatures (below 10K) unless in-field calibrations are done. Another alternative would be to design the apparatus so that no temperature gradient exists along the thermocouple wires in the magnetic field, as suggested by Richards [18].

The measurement of temperature in nuclear radiation fields also requires considerable care. Nuclear irradiation of thermocouples can produce changes in calibration because of the physical imperfections introduced into the material and by transmutation of the elemental components of the material. Several detailed studies have been done to determine the magnitudes of these effects. The effect of transmutation has received considerable attention at high temperatures [20-26] and the results are expected to apply at low temperatures as well. The conclusion, based on these studies, is to use nickel base thermocouple materials such as KP (Chromel) and KN (Alumel) because the transmutation effects are relatively small. For considerably more detail on this subject, the reader is referred to Browning [23].

The calibration procedure necessary to make use of a particular thermocouple depends on the application. If temperature differences are needed only to about 10%, one can probably accomplish this by using general calibration tables. These are tables which refer to a general type of thermocouple, not to a particular wire or spool of wire. If one needs 0.01K precision, or better, it will be necessary to do extremely careful in-place calibration against a calibrated thermometer such as a platinum or germanium resistance thermometer.

Standard calibration tables are available for high temperatures [15] and separately for cryogenic temperatures [13, 14]. Standard reference thermocouple tables only represent an approximation of the emf vs. temperature relation for a particular thermocouple. To obtain a more exact calibration, one must measure the emfs of this thermocouple at selected temperatures. The emf difference between a particular thermocouple and the standard table is often small and nearly linear in temperature so that only a few points, sometimes only two, are necessary.

The calibration of a thermocouple is done with a Type 3 probe shown in Fig. 3.6. This probe would normally be used to obtain the thermoelectric emfs between a series of pairs of fixed points. These emfs would then be used in conjunction with a standard reference table and a simple interpolation scheme to form a new and more accurate table for that specific thermocouple. Such a

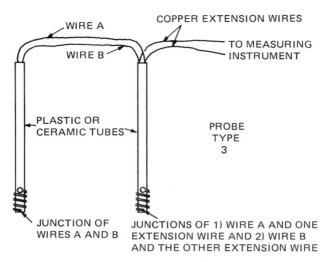

FIG. 3.6 Type 3 probe for thermocouple calibration.

calibration may be accurate to a few millidegrees if the wire has been thoroughly homogenized and is free of physical defects. Before calibrating wire for highly accurate or precise applications, one should perform the tests discussed below to determine the effect of the inhomogeneities in the wire.

No thermal voltage is developed when a loop of homogeneous wire is subjected to a temperature gradient. Similarly, no voltage is generated when two identical wires are joined and the pair of wires is placed in a temperature gradient. The problem in practical thermometry is, however, that the ideal characteristics, homogeneous and identical, are not sufficiently well approximated for real thermocouple materials. Actually, a loop of wire placed in a large temperature gradient will usually produce a thermoelectric voltage, sometimes as large as 10 μV for poor materials [27, 28, 29]. If wire from one spool is connected to wire from a different spool of the same nominal composition, their junction is placed in a thermometric fluid, and the free ends are held at room temperature, then a significant voltage may result: we have observed readings as large as hundreds of microvolts for poorly controlled alloys. These variable spurious voltages caused by inhomogeneities, physical imperfections, and chemical impurities are the main source of imprecision and inaccuracy in thermocouple systems.

Experimental methods described in this chapter allow an experimentalist to select materials that are most homogeneous and therefore have the smallest amount of spurious voltages. The tests also provide data necessary for making realistic error analyses.

For descriptive convenience we have divided inhomogeneities into four categories based on their distance of separation:

1. Short-range inhomogeneities occur in a single wire and are separated by less than 5 m, often being within a few cm of each other.
2. Medium-range inhomogeneities occur in wires that are from a single spool but are more than 5 m apart.
3. Long-range inhomogeneities are found in wires that are from the same general stock but are from different spools.
4. Inter-lot variations in chemical composition, thermal treatment, and handling occur in materials produced by different manufacturers, or even in wire produced by the same manufacturer at different times.

The latter categories of inhomogeneities lead to much larger spurious voltages in thermometer systems. Well-prepared thermocouple wire can have short-range inhomogeneity effects as low as 0.1 μV; poorly controlled alloys often have interlot variations as large as 100 μV.

Two types of probes, shown in Fig. 3.7, can be used to investigate the various effects of the four categories of inhomogeneities. The first probe configu-

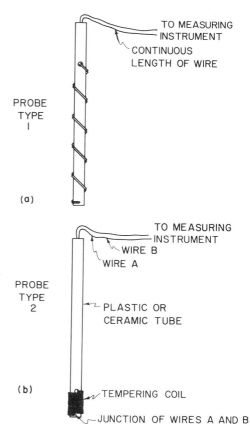

PROBE TYPE I

TO MEASURING INSTRUMENT

CONTINUOUS LENGTH OF WIRE

(a)

PROBE TYPE 2

TO MEASURING INSTRUMENT

WIRE B

WIRE A

PLASTIC OR CERAMIC TUBE

TEMPERING COIL

JUNCTION OF WIRES A AND B

(b)

FIG. 3.7 Probe configurations for inhomogeneity testing of thermocouple materials, (a) probe for short-range inhomogeneities; (b) probe for medium- and long-range inhomogeneities and interlot variations.

ration, shown in Fig. 3.7(a), consists of a single wire about 4 or 5 m long, part of which is attached to a plastic tube. It need not have a large number of coils, even straight lengths of wire are often satisfactory. Such a probe is used to test for short-range inhomogeneities. The second probe configuration, shown in Fig. 3.7(b), consists of two wires, each 2 or 3 m long, that are coiled on a plastic tube and joined at the bottom. This probe is used to test for the last three categories of inhomogeneities. The essential difference between the two types of probes (besides the junction) is in their manner of thermal tempering: the second type has tightly wound coils of wire near the junction in order to prevent a thermal gradient across the junction, which often contains dissimilar materials.

For both types of probes the ends of the wires are connected to a potentiometer or high-resistance voltmeter and the probes are then dipped into dewars containing cryogenic fluids, usually liquid helium or nitrogen, or into high-temperature furnaces. The first type of probe is dipped in two different manners, one way for static tests, another way for dynamic tests.

For static short-range inhomogeneity tests the probes are immersed to a given depth in the cryogenic fluid or furnace and the temperature gradient is allowed to come to equilibrium before readings are taken. In order to obtain more representative values, readings are usually taken at several different levels for each test.

For dynamic short-range inhomogeneity tests the probes are lowered into the fluid or furnace at a constant speed. Erratic output voltages are usually observed in these tests because large temperature gradients are developed over different, relatively short lengths of wire as the depth of immersion is changed. The magnitude of the output depends on the type of thermocouple wire, the specific specimen, and to some extent, the rate of immersion. Since comparable results are desired, a constant immersion rate should be used. We have arbitrarily selected 0.5 m/min for our tests.

The dynamic, short-range tests are sensitive because large temperature gradients are established over short sections of wire which may contain significant chemical and physical defects. However, the thermal gradients in the static short-range tests are smaller because of thermal diffusion. This results in spurious voltages which are correspondingly smaller. The static tests are therefore less sensitive indicators of inhomogeneities but more indicative of spurious voltages to be expected in practice. Short-range inhomogeneity tests give a good preliminary estimate of the imprecision that can be expected for temperature measurements in an actual system. Results from dynamic tests are most applicable to systems with rapidly fluctuating temperatures; results from static tests are most appropriate for stable systems like the typical laboratory cryostat or apparatus. Wire that exhibits unusually high spurious thermal voltages (typical values are given in the next section) can be detected and rejected before costly installation!

For tests on medium- or long-range inhomogeneities, or interlot variations, the probe configuration shown in Fig. 3.7(b) is used. The only differences are in the methods for selection of wires that will be assembled in the test rig. The selection criteria are simply those implied in the basic definitions of the three categories of inhomogeneities. The manner of joining the wires is not critical as long as good electrical contact is obtained and the materials are not mechanically or thermally strained more than a few centimeters away from the junction. The assembled probes are dipped into a cryogenic fluid or furnace in the same manner as in static short-range inhomogeneity tests.

Medium- and long-range inhomogeneity tests are useful for determining the variations that may occur in a selected lot of thermocouple wire. The medium-range inhomogeneity tests are especially useful for systems that include thermocouples made from consecutive lengths of wire. The deviations in voltage usually become progressively larger as the original positions of the wires become more widely separated. It is good practice to thermoelectrically compare the front and back ends of a spool, or lot of material, to determine if variations of the material are within acceptable limits. If the material is sufficiently homogeneous, then a single calibration will suffice for all thermocouples made from that spool. If the material varies beyond acceptable limits, then additional calibrations become necessary for each separate part of the spool or lot. In the limits of relatively inhomogeneous material or extremely accurate measurement systems, separate calibrations must be made for each length of wire.

Tests for interlot variations are absolutely necessary if wires from different melts or from different manufacturers are going to be used. Interlot variations can be very large: hundreds of microvolts have been observed for nominally identical material received from different manufacturers. Even material received from the same manufacturer at different times can have large variations, up to about 40 μV. Because of these variations, it is best to have all thermocouples in a system made from one lot, preferably one spool, of material. To insure this, enough wire should be ordered at the beginning of construction of a thermocouple system to allow repair or replacement of all initial thermocouples.

We have determined the spurious voltages caused by the first three categories of inhomogeneities for most types of thermocouple materials used at low temperatures. Results obtained with both types of probes on five different materials are summarized in Table 3.1. Short-range inhomogeneity results were obtained by using the Type 1, the other results by using the Type 2 probe with a junction in the wires. The numbers quoted in the columns are the maximum absolute values which were found for several spools from the same manufacturer. Similar results can be obtained on high temperature materials.

A close examination of Table 3.1 shows some interesting results. Material that is most homogeneous in short lengths is not always the most reproducible between spools (compare Type EP material from companies A and C). Material that is fairly reproducible at liquid nitrogen temperatures may be quite poor at liquid helium temperatures (for example, copper from company G). There is considerable variability in the homogeneity and interchangeability of material obtained from different manufacturers.

Temperature errors, as described earlier, can be caused by lack of calibration

TABLE 3.1 Thermocouple wire inhomogeneity voltages

Material	Company	Short-range Static Liq N$_2$	Liq He	Dynamic Liq N$_2$	Medium-range Liq N$_2$	Liq He	Different spools Liq N$_2$	Liq He
EP or KP	A	0.5	0.9	5.5	2.2	2.2	28.0	33.4
	B	1.0	1.2	6.9	1.0	1.0	38.0	39.1
	C	2.6	2.6	8.2	4.5	4.5	4.4	6.1
JP (iron)	D	7.5	8.1	10.1	20.0	20.0	55.0	57.3
	E	2.2	2.2	8.1	1.6	1.6	15.0	15.0
TP (copper)	F	0.4	0.4	1.0	0.4	0.7	0.5	0.5
	G	0.2	1.4	1.1	1.9	7.6	4.1	37.9
KN	H	0.7	0.7	13.7	3.8	4.9	42.0	45.4
	I	1.9	2.4	14.1	2.2	2.2	1.7	2.8
	J	2.0	2.0	21.4	2.6	2.6	4.6	4.6
EN or TN	K	0.9	1.2	1.3	1.6	1.7	6.6	6.6
(constantan)	L	3.0	3.0	5.1	2.4	2.5	26.0	27.7
	M	2.8	2.8	8.3	10.8	12.5	36.0	44.4
	N	1.6	2.1	6.8	5.8	5.8		

Note. All units μV.

and spurious voltages, but additional errors can be introduced by poor thermal contact and heat flow from the thermocouple wire to the object being measured. Poor thermal contact may result in a thermocouple which is not at the same temperature as the object. If there is significant heat flow down the thermocouple wires, the temperature of the object will be changed as well. The principal objective is to obtain a good thermal connection between the thermocouple and the object being measured and to achieve an isothermal region along the thermocouple leading up to the object being measured. The region of thermocouple which was strained in forming the junction should be isothermal. The problem of achieving adequate thermal anchors has been discussed recently by Hust [30], Kopp and Slack [31], and Allen et al. [32]. The first two papers are concerned with thermocouple thermal anchors to solids, while the latter discusses thermal anchoring to liquids. To compute the proper length of thermal anchoring, one needs data on the thermal conductivity of both the wire and the contact agent used. The thermal conductivity of solids has been compiled and reviewed by Childs et al. [33] and Powell [34].

The heat flow to the object being measured should be minimized. Heat flow can of course be reduced by using small diameter, low conductivity wires. In addition, one can also intercept the heat with an intermediate thermal shield whose temperature is controlled near that of the object being measured. This is generally accomplished by nulling the output of a differential thermocouple connected between the object being measured and the intercepting heat shield. This in essence creates an isothermal region along the thermocouple wires near the junction. This shield may also be used as a radiation shield surrounding the specimen. Thermal radiation can cause temperature errors by changing the temperature of the object being measured or by causing excessive heat flow to the thermocouple junction. To minimize the effects of this extraneous heat flow, one should provide adequate thermal radiation shielding in the system. This is often done with opaque baffles in pipes and good conducting, often reflective, shells surrounding the measurement region.

Emf measurements are accomplished with a deflection voltmeter for low accuracy measurements and a potentiometer (manual or self balancing) for accurate measurements. The digital voltmeter, which is now being used extensively, is often used as a substitute for a potentiometer. The deflection voltmeter measures the voltage at the terminals of the voltmeter which is dependent upon the IR drop of the external circuit. In general one would make R small compared to the input resistance of the voltmeter so as to minimize the effect of changes in R. The potentiometer is essentially a zero current device and therefore the measured voltage does not directly depend on R. However, the response time and sensitivity of potentiometers are dependent on the external circuit resistance. Thermocouple emfs are generally very small, often at the microvolt level. In the measurement of such voltages, one must be careful to avoid the effects of improper shielding or grounding. Probably the most common problem with circuit grounds is to have too many. If more than one ground is present, loop currents are set up introducing spurious voltages in the signal path. Most modern low level dc amplifiers have excellent 60 Hz ac rejection; however, because of the relatively large ac pickup possible, it is still necessary to design carefully. Excessive ac pickup may cause a low frequency beat with a chopper amplifier or may saturate the amplifier, resulting in a steady shift in its output. The following procedures for low level dc circuits are recommended:

1. Operate either at floating potential or with one ground preferably at the input of the amplifier.
2. Use extension cables that have a high rejection of electromagnetic, as well as static, noise [35]. The leads should be encased in a shield grounded at only one point. The normal copper shielding does not adequately shield low frequency noise, e.g., 60 c. However, there are cables that have ferromagnetic material for shielding; these are preferable. Cables and wire junctions can also be placed in iron conduits or boxes to provide additional electromagnetic shielding.

For further details about shielding and grounding the reader is referred to Morrison [36].

Care must be observed also in connecting either similar or dissimilar wires. Since the previously described homogeneity tests show that even nominally identical wires are thermoelectrically different, one should consider all junctions as if they were between dissimilar metals. To minimize the spurious thermo-electric voltages generated at these junctions, one should be careful to insure good electrical connections, as well as to minimize the temperature gradient across the junctions. One must be especially careful with electrical feed-throughs from one environment to another since it is likely that in such a situation one will encounter both vastly dissimilar materials and temperature gradients. When copper-to-copper junctions are to be soldered, special low thermal solder (70% cadmium, 30% tin) is recommended. One can also make thermocouple junctions by spot welding, either with an electrical discharge welder or with a small torch. In some apparatus, junctions are made by simply twisting the leads together. Each of these methods of forming junctions have unique applications and can be used successfully if done with care. The simple twist method is valuable when frequent disassembly is necessary. The welding method is valuable for wires which are not readily soldered. In any case, one must be careful to achieve an isothermal region over the part of the thermocouple near the junction which was strained in fabrication.

If any switches are used in thermocouple circuits, they need to be of high quality to prevent degradation of the signal-to-noise ratio. For manual operation, low-resistance rotary switches have proven completely satisfactory if enclosed in metal cabinets. For automatic switching operations, enclosed gold or silver plated switches have usually been satisfactory; they typically have about 1 μV transient noise and therefore introduce excessive noise if switched faster than about once per second. For either manual or automatic usage, switches should be thermally isolated since large thermal voltages may be generated at the junctions of the wires and the switch components.

Reviews of methods of testing thermocouples have been given by Roeser [37] and Hust et al. [38]. Kinzie [12], among others, has described the deviation of practical materials from the ideal. In separate brief sections he discusses the effects of inhomogeneities, uniformity, instability and drift, repeata-bility, accuracy, and response time.

3.3 STANDARDIZED MATERIALS

Most industrial and scientific thermocouple thermometry applications utilize standardized materials. Those standardized couples are now identified by a type letter rather than by proprietary names. Since many research workers are still unfamiliar with the type names, they are explained in Table 3.2. National Bureau of Standards Monograph 125 by Powell et al. [15] now serves as the standard reference for all typed materials.

Types S, R, and B (all platinum vs. platinum-rhodium combinations) are referred to as *noble metal types*, the others as *base metal types*. The noble metal types will be described first.

TABLE 3.2 Thermocouple characteristics

	Thermocouple combinations	
Type designation[a]	Temperature range, °C	Materials
B	0–1820	*Platinum*-30% rhodium vs. *platinum*-6% rhodium
E	− 270–1000	*Nickel*-chromium alloy vs. a *copper*-nickel alloy
J	− 210–1200	Iron vs. another slightly different *copper*-nickel alloy
K	− 270–1372	*Nickel*-chromium alloy vs. *nickel*-aluminum alloy
R	− 50–1767	*Platinum*-13% rhodium vs. platinum
S	− 50–1767	*Platinum*-10% rhodium vs. platinum
T	− 270– 400	Copper vs. a *copper*-nickel alloy

	Single-leg materials
. . . N	The negative wire in a combination
. . . P	The positive wire in a combination
BN	*Platinum*-nominal 6 wt% rhodium
BP	*Platinum*-nominal 30 wt% rhodium
EN or TN	A *copper*-nickel alloy, constantan: Cupron,[b] Advance,[d] ThermoKanthal JN[c]; nominally 55% Cu, 45% Ni; often referred to as Adams constantan
EP or KP	A *nickel*-chromium alloy: Chromel,[e] Tophel,[b] T-1,[d] ThermoKanthal KP[c]; nominally 90% Ni, 10% Cr
JN	A *copper*-nickel alloy similar to, but not always interchangeable with, EN and TN; SAMA specification
JP	Iron: ThermoKanthal JP[c]; nominally 99.5% Fe
KN	A *nickel*-aluminum alloy: Alumel,[e] Nial,[b] T-2,[d] ThermoKanthal KN[c]; nominally 95% Ni, 2% Al, 2% Mn, 1% Si
RN, SN	High-purity platinum
RP	*Platinum*-13 wt% rhodium
SP	*Platinum*-10 wt% rhodium
TP	Copper, usually Electrolytic Tough Pitch

[a]The letter designations used in this chapter follow the recommendations of Committee E-20 of the American Society for Testing and Materials. The letter type, e.g., Type T, designates the thermoelectric properties, not the precise chemical composition. Thermocouples of a given type may have variations in composition as long as the resultant thermoelectric properties remain within the established limits of error.

[b]Trademark—Wilbur B. Driver Co.

[c]Trademark—Kanthal Corp.

[d]Trademark—Driver-Harris Co.

[e]Trademark—Hoskins Manufacturing Co.

The use of trade names does not constitute an endorsement of any manufacturer's products. All materials manufactured in compliance with the established thermoelectric voltage standards are equally acceptable.

3.3.1   Type S—Platinum-10% Rhodium Alloy
vs. Platinum Thermocouples

This type is often referred to by the nominal chemical composition of its positive thermoelement: *platinum**-10% rhodium. Until 1973 the composition was somewhat different from the nominal 10% rhodium, usually about 9.91 wt%. However, as a result of the recent research by Bedford et al. [39], the composition of the positive thermoelement has been established to be *platinum*-10.00 ± 0.05 wt% rhodium. The negative thermoelement is commercially pure platinum. Differences between this commercial material and the platinum thermoelectric reference standard, Pt-67, are described later. The effect of differences in rhodium content of the positive thermoelement is described in this section.

The *platinum*-10% rhodium thermocouple was developed and tested by Le Chatelier [40] almost a century ago. Because of his pioneer work, this type of thermocouple was often referred to as the Le Chatelier couple. In this country the ASTM (E230-68) and ISA (C96.1-64) standards for this thermocouple type had been taken from NBS Circular 561 by Shenker et al. [41]. The reference tables given in NBS Circular 561 were based on functions given by Roeser and Wensel [42], as revised to conform with the IPTS-48. In Great Britain, the British Standards Institution (BSI) Table B.S. 1826:1952 for this thermocouple type was based on the research by Barber [43]. There were considerable differences between the American and British reference tables because of small differences in rhodium content and in different realizations of the temperature scales.

The early research by Le Chatelier [40] demonstrated the main advantages of the *platinum*-10% rhodium thermoelement: reproducibility, stability, and usefulness to moderately high temperature. An ASTM manual, STP-470 [8], and Bennett [44] have described many of the properties of noble metal thermocouples. For many industrial applications, the Type R or B noble metal thermocouples (described in the next two sections) have preferable characteristics and are often used instead of Type S systems. However, Type S thermocouples remain the standard for determining temperatures between 630.74°C and the freezing point of gold (1064.43°C). The CIPM [45] article details the temperature scale definition in this range and includes specifications on the limits of the thermocouple voltage. The other fixed point used for determining the constants in the specified quadratic interpolation formula of the thermoelectric voltage is the freezing point of silver, 961.93°C.

Because of the differences between the British and American standards for Type S thermocouples, an international program was begun several years ago to rectify the unsatisfactory disagreements and to establish a common set of standard reference tables. The program involved cooperation of three national laboratories: the National Bureau of Standards (USA), the National Physical

---

*Throughout this chapter, in the thermocouple designations an italicized word indicates the primary constituent of an alloy.

Laboratory (UK), and the National Research Council (Canada) and of seven manufacturers in Great Britain and the United States. The thoroughly documented results have been published by Bedford et al. [39].

Their research has confirmed that Type S thermocouples can be used from $-50°C$ up to the platinum melting point $1767.6°C$ (incidentally found to *not* be $1772°C$ as listed in the CIPM article). They may be used intermittently at temperatures up to the platinum melting point and continuously up to about $1400°C$. The ultimate useful life of the thermocouple is governed primarily by physical problems of impurity diffusion and grain growth which lead to mechanical failure. The thermocouple is most reliable when used in a clean oxidizing atmosphere (air) but may also be used in inert gaseous atmospheres or in a vacuum for short periods of time. However, as noted in Sec. 3.3.3, Type B thermocouples are generally more suitable for these applications. Continued use of Type S thermocouples at high temperatures, above about $1400°C$, leads to excessive grain growth in the platinum thermoelement. Stability of the thermocouple at high temperatures depends primarily upon the quality of the materials used for protecting and insulating the thermocouple. High purity alumina with low iron content appears to be the most suitable material for mechanically supporting and protecting the thermocouple wires.

The ASTM manual STP 470 [8] indicates the following restrictions on the use of Type S thermocouples at high temperatures: "They should not be used in reducing atmospheres, nor in those containing metallic vapor (such as lead or zinc), nonmetallic vapors (such as arsenic, phosphorus, or sulfur) or easily reduced oxides, unless suitably protected with nonmetallic protecting tubes. They should never be inserted directly into a metallic primary tube."

The positive thermoelement, *platinum*-10% rhodium, is unstable in a thermal neutron flux because the rhodium converts to palladium. The negative thermoelement, pure platinum, is relatively stable to neutron transmutation. However, fast neutron bombardment will cause physical damage, which will change the thermoelectric voltage unless it is annealed out.

McLaren and Murdock [46] have described several precautions that must be taken to insure accurate measurements in the intermediate temperature range, including the defined range from 630.74 to $1064.43°C$. They emphasized the importance of annealing techniques.

Both thermoelements of Type S thermocouples are sensitive to impurity contamination. In fact the Type R thermocouple described in the next section was developed essentially because of iron contamination effects on some British *platinum*-10% rhodium wires. Cochrane [47] and Aliotta [48] have recently given detailed descriptions of the effects of various impurities on the thermoelectric voltages of platinum based thermocouple materials. At present good alloy material typically contains less than 500 atomic ppm of impurities; good pure platinum, less than 100 atomic ppm of impurities. High temperature impurity contamination usually causes negative drifts in calibration, the extent of which will depend on the type and amount of chemical contaminant. Volatilization of rhodium from the positive thermoelement or diffusion of rhodium from the positive thermoelement into the pure platinum negative thermoelement will also cause negative drifts in the thermoelectric output.

At the gold point, 1064.43°C, the thermoelectric voltage of Type S thermocouples increases by about 340 μV ($\sim$3%) per weight percent increase in rhodium content; the Seebeck coefficient increases by about 4% per weight percent increase at the same temperature.

The thermoelectric voltages of platinum based thermocouples are sensitive to their heat treatments. In particular, quenching from high temperatures should be avoided. Before annealing, platinum and platinum-rhodium alloy wires should be degreased with solvents such as trichloroethane or trichloroethylene and then may be rinsed with distilled water. After cleaning, they should be annealed by the techniques recommended by Bedford et al.[39]: Anneal electrically in air at 1450°C for 30 min and then slowly cool them to room temperature. The air should be dust free, of course. After assembly in insulating and protecting tubes, the thermocouple should be reannealed for 1 hr at 1100°C or for 30 min at 1400°C and then slowly cooled (about 30 min) to room temperature.

The Office of Standard Reference Materials of the National Bureau of Standards has prepared a large lot of reference grade platinum wire approximately 0.5 mm in diameter. The material has been thoroughly characterized chemically and electrically and is available from the National Bureau of Standards as Standard Reference Material—SRM No. 680. A selected portion of that material has been set aside for use as a thermoelectric reference standard by the Temperature Section of the NBS. It is referred to as Pt-67.

ASTM Standard E230-72 in the Annual Book of ASTM Standards [49] specifies that the standard limits of error for Type S commercial thermocouples be ±1.4°C between 0 and 538°C and ±1/4% between 538 and 1482°C. Limits of error are not specified for Type S thermocouples below 0°C. The recommended upper temperature limit for continuous use of protected thermocouples, 1482°C, applies to AWG 24 (0.5 mm) wire.

The coefficients for the four sets of equations for the thermoelectric voltage of Type S thermocouples are given in Table 3.3. The reduced temperature expressions generated by Bedford et al. [39] are included. Values at selected fixed points are given in Table 3.4; values at 10°C intervals are given in the last section; a graph of the Seebeck coefficient is given in Fig. 3.8.

3.3.2 Type R–*Platinum*-13% Rhodium Alloy vs. Platinum Thermocouples

This type is also often referred to by the nominal chemical composition of its positive thermoelement: *platinum*-13% rhodium. Until this year the composition was somewhat different from the nominal 13% rhodium, usually about 12.85 wt%. However, as a result of the recent research by Bedford et al. [39], the composition of the positive thermoelement has been established to be platinum-

TABLE 3.3 Power series expansion for the thermoelectric voltage of Type S thermocouples

Temperature range	Degree	Coefficients	Term
−50–630.74°C	6	5.3995782346 ...	T
		1.2519770000 × 10^{-2}	T^2
		−2.2448217997 × 10^{-5}	T^3
		2.8452164949 × 10^{-8}	T^4
		−2.2440584544 × 10^{-11}	T^5
		8.5054166936 × 10^{-15}	T^6
630.74–1064.43°C	2	−2.9824481615 × 10^2	...
		8.2375528221 ...	T^1
		1.6453909942 × 10^{-3}	T^2
1064.43–1665°C	3	1.2766292175 × 10^3	...
		3.4970908041 ...	T^1
		6.3824648666 × 10^{-3}	T^2
		−1.5722424599 × 10^{-6}	T^3
1665–1767.6°C	3	9.7846655361 × 10^4	...
		−1.7050295632 × 10^2	T^1
		1.1088699768 × 10^{-1}	T^2
		−2.2494070849 × 10^{-5}	T^3
1064.43–1665°C $T^* = \dfrac{T-1365}{300}$	3	1.3943438677 × 10^4	...
		3.6398686553 × 10^3	T^{*1}
		5.0281206140 ...	T^{*2}
		−4.2450546418 × 10^1	T^{*3}
1665–1767.6°C $T^* = \dfrac{T-1715}{50}$	3	1.8113083153 × 10^4	...
		−5.6795375480 × 10^2	T^{*1}
		−1.2112492121 × 10^1	T^{*2}
		−2.8117588563 ...	T^{*3}

13.00 ± 0.05 wt% rhodium. The negative thermoelement is commercially pure platinum. Differences between this commercial material and the platinum thermoelectric reference standard, Pt-67, were described in the last section. The effect of differences in rhodium content on the positive thermoelement is described later in this section.

During the early years of this century the *platinum*-13% rhodium vs. platinum thermocouple was developed and tested in this country to give agreement with the British *platinum*-10% rhodium vs. platinum thermocouples which had been found to have significant iron contamination. Fairchild and Schmitt [50] discovered during prolonged high-temperature tests that American and British *platinum*-10% rhodium vs. platinum thermocouples differed significantly from each other in thermoelectric output and stability. The main chemical difference was traced to a 0.34% iron impurity in the British positive thermoelement which was presumably caused by the use of impure rhodium. Many instruments and systems had been calibrated on the basis of the thermoelectric voltages of the older, impure *platinum*-10% rhodium material. Therefore, when pure rhodium was used for alloying in order to improve the stability characteristics, the composition had to be changed to give thermoelectric values near the previous ones. When the more pure rhodium was used, it was found that about 13% rhodium had to be alloyed into the platinum to approximately match the previous *platinum*-10% rhodium British wire. That is the highly pragmatic reason for the development of the Type R thermocouple.

Type R thermocouples have a higher Seebeck coefficient than do Type S thermocouples, about 12% larger over much of the range. Type R thermocouples are not standard interpolating instruments on the IPTS–68 for the 630.74°C to gold freezing point range. Other than the above two items, and remarks on the history of development and the composition, all of the comments from the previous chapter on Type S also apply to Type R.

However, for emphasis the precautions and restrictions on usage are repeated: "They should not be used in reducing atmospheres, nor in those containing metallic vapor (such as lead or zinc), nonmetallic vapors (such as arsenic, phosphorus, or sulfur) or easily reduced oxides, unless suitably protected with nonmetallic protecting tubes. They should never be inserted directly into a metallic primary tube." Glawe [51] has described the effects on thermoelectric

TABLE 3.4 Thermoelectric values at the fixed points for Type S thermocouples

Fixed point	Temp., °C	$E, \mu V$	$S, \mu V/°C$
Mercury FP	− 38.862	− 189.54	4.318
Ice point	0.000	0.00	5.400
Ether TP	26.87	153.70	6.026
Water BP	100.000	645.34	7.333
Benzoic TP	122.37	812.88	7.640
Indium FP	156.634	1081.79	8.044
Tin FP	231.9681	1714.64	8.714
Bismuth FP	271.442	2063.97	8.977
Cadmium FP	321.108	2516.72	9.246
Lead FP	327.502	2575.94	9.276
Mercury BP	356.66	2848.34	9.406
Zinc FP	419.580	3447.87	9.643
Sulfur BP	444.674	3690.88	9.725
Cu-Al FP	548.23	4714.00	10.030
Antimony FP[a]	630.74	5552.10	10.295
Aluminum FP	660.37	5859.12	10.411
Silver FP	961.93	9148.20	11.403
Gold FP[a]	1064.43	10334.30	11.740
Copper FP	1084.5	10570.46	11.793
Nickel FP	1455	15033.80	12.085
Cobalt FP	1494	15504.28	12.040
Palladium FP	1554	16223.95	11.943
Platinum FP	1767.6	18693.89	10.663

[a]Junction point of different functions.

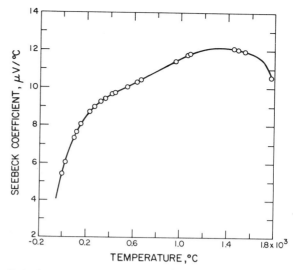

FIG. 3.8 Seebeck coefficient of Type S thermocouples.

voltages that occur from prolonged exposure at elevated temperatures in vacuum, air, and argon atmospheres.

ASTM Standard E230-72 in the Annual Book of ASTM Standards [49] specifies that the standard limits of error for Type R commercial thermocouples be ± 1.4°C between 0 and 538°C and ± ¼% between 538 and 1482°C. Limits of error are not specified for Type R thermocouples below 0°C. The recommended upper temperature limit for continuous service of protected Type R thermocouples is 1482°C and applies to AWG 24 (0.5 mm) wire.

The coefficients for the four sets of equations for the thermoelectric voltage of Type R thermocouples are given in Table 3.5. The reduced temperature expressions generated by Bedford et al. [39] are included. Values at selected fixed points are given in Table 3.6; values at 10°C intervals are given in the last section; a graph of the Seebeck coefficient is given in Fig. 3.9.

3.3.3 Type B—*Platinum*–30% Rhodium Alloy vs. *Platinum*–6% Rhodium Alloy Thermocouples

This type is often referred to by the nominal chemical composition of its thermoelements—*platinum*-30% rhodium vs. *platinum*-6% rhodium or "30-6." The actual composition is somewhat different however. The positive thermoelement, BP, is usually a platinum alloy with 29.0 ± 0.2 wt% rhodium; the negative thermoelement, BN, usually has 6.12 ± 0.02 wt% rhodium. The effect of differences in rhodium content are described later in this section. Because of its

TABLE 3.5 Power series expansion for the thermoelectric voltage of Type R thermocouples

Temperature range	Degree	Coefficients	Term
− 50–630.74°C	7	5.2891395059 ...	T
		1.3911109947 × 10^{-2}	T^2
		− 2.4005238430 × 10^{-5}	T^3
		3.6201410595 × 10^{-8}	T^4
		− 4.4645019036 × 10^{-11}	T^5
		3.8497691865 × 10^{-14}	T^6
		− 1.5372641559 × 10^{-17}	T^7
630.74–1064.43°C	3	− 2.6418007025 × 10^2	...
		8.0468680747 ...	T^1
		2.9892293723 × 10^{-3}	T^2
		− 2.6876058617 × 10^{-7}	T^3
1064.43–1665°C	3	1.4901702702 × 10^3	...
		2.8639867552 ...	T^1
		8.0823631189 × 10^{-3}	T^2
		− 1.9338477638 × 10^{-6}	T^3
1665–1767.6°C	3	9.5445559910 × 10^4	...
		− 1.6642500359 × 10^2	T^1
		1.0975743239 × 10^{-1}	T^2
		− 2.2289216980 × 10^{-5}	T^3
1064.43–1665°C $T^* = \dfrac{T - 1365}{300}$	3	1.5540414086 × 10^4	...
		4.2357772712 × 10^3	T^{*1}
		1.4693087343 × 10^1	T^{*2}
		− 5.2213889624 × 10^1	T^{*3}
1665–1767.6°C $T^* = \dfrac{T - 1715}{50}$	3	2.0416695016 × 10^4	...
		6.6850914082 × 10^2	T^{*1}
		− 1.2301472524 × 10^1	T^{*2}
		− 2.7861521235 ...	T^{*3}

favorable characteristics, the 30-6 thermocouple has rapidly gained acceptance and become widely used in this country. The Temperature Section of the National Bureau of Standards, as requested by the American Society for Testing and Materials, Committee E-20, prepared reference tables for the thermocouple to facilitate its use and calibration. Thermocouples were obtained from three manufacturers in the United States and from one European manufacturer and were calibrated by conventional methods. The results were published by Burns and Gallagher [52].

The 30-6 thermocouple was first introduced in Europe by Degussa of Hanau, Germany. Reference curves and tables were published for their thermocouple by Obrowski and Prinz [53]. The values of thermoelectric voltage were represented by a set of cubic equations developed from typical values at various thermometric fixed points.

Studies by Ehringer [54], by Walker, Ewing, and Miller [55, 56], and by

TABLE 3.6 Thermoelectric values at the fixed points for Type R thermocouples

Fixed point	Temp., °C	$E, \mu V$	$S, \mu V/°C$
Mercury FP	−38.862	183.05	4.090
Ice point	0.000	0.00	5.289
Ether TP	26.87	151.72	5.987
Water BP	100.000	647.23	7.476
Benzoic TP	122.37	818.57	7.837
Indium FP	156.634	1095.62	8.322
Tin FP	231.9681	1756.10	9.167
Bismuth FP	271.442	2125.04	9.517
Cadmium FP	321.108	2607.22	9.889
Lead FP	327.502	2670.59	9.933
Mercury BP	356.66	2962.99	10.121
Zinc FP	419.580	3611.34	10.479
Sulfur BP	444.674	3875.95	10.609
Cu-Al FP	548.23	5000.94	11.110
Antimony FP[a]	630.74	5933.08	11.477
Aluminum FP	660.37	6275.90	11.643
Silver FP	961.93	10003.09	13.052
Gold FP[a]	1064.43	11363.85	13.497
Copper FP	1084.5	11635.49	13.571
Nickel FP	1455	16811.06	14.102
Cobalt FP	1494	17360.36	14.065
Palladium FP	1554	18201.73	13.974
Platinum FP	1767.6	21103.11	12.668

[a]Junction point of different functions.

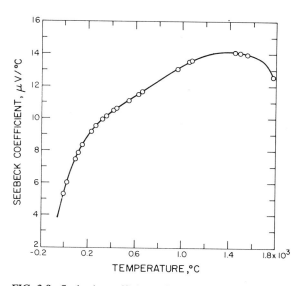

FIG. 3.9 Seebeck coefficient of Type R thermocouples.

Glawe [51] have demonstrated that thermocouples in which both legs are platinum-rhodium alloys are suitable for reliable temperature measurements at high temperatures. Such thermocouples have been shown to offer the following distinct advantages over the more familiar Type S (*platinum*-10% rhodium vs. platinum) and Type R (*platinum*-13% rhodium vs. platinum) thermocouples at high temperatures: (1) improved stability, (2) increased mechanical strength, and (3) higher possible operating temperatures. In addition to the *platinum*-rhodium thermocouple combinations which have standardized letter designations, Types S, R, and B, there are two other infrequently used combinations: *platinum*-20% rhodium vs. *platinum*-5% rhodium, and *platinum*-40% rhodium vs. *platinum*-20% rhodium, referred to respectively as 20-5 and 40-20 thermocouples. Reference tables for the 20-5 and 40-20 thermocouples (based on the IPTS–48) were published by Bedford [57, 58].

Of the three combinations where *platinum*-rhodium alloys are used in the negative leg, the 30-6 thermocouple offers the most favorable overall characteristics and is the only one with a standardized letter designation. Although the thermoelectric characteristics of the 30-6 thermocouple and the 20-5 thermocouple are similar, the 30-6 thermocouple offers a slightly greater Seebeck coefficient at the higher temperatures and it also has a somewhat higher tensile strength. Under similar conditions of temperature and environment, the 30-6 thermocouple shows less grain growth and less drift in calibration than either Type S or Type R thermocouples. Various physical properties of these types of thermocouples are described in the ASTM Manual STP 470 [8] on the use of thermocouples.

The research by Burns and Gallagher [52] indicated that the 30-6 thermocouple can be used intermittently (for several hours) up to 1800°C and continuously (for several hundred hours) at temperatures up to about 1750°C with only small changes in calibration. The maximum temperature limit for the thermocouple is governed, primarily, by the melting point of the Pt-6% Rh thermoelement which is estimated to be about 1820°C by Acken [59]. The thermocouple is most reliable when used in a clean oxidizing atmosphere (air) but has also been used successfully in neutral atmospheres or vacuum by Walker et al. [55], Hendricks and McElroy [60], Walker et al. [56], and Glawe [51]. The stability of the thermocouple at high temperatures has been shown by Walker et al. [55] to depend, primarily, upon the quality of the materials used for protecting and insulating the thermocouple. High purity alumina with low iron-content appears to be the most suitable material available today for the purpose.

The ASTM Manual STP 470 [8] indicates the following restrictions on the use of Type B thermocouples at high temperatures: "They should not be used in reducing atmospheres, nor those containing metallic or nonmetallic vapors, unless suitably protected with nonmetallic protecting tubes. They should never be inserted directly into a metallic primary protecting tube."

At temperatures below 450°C the Seebeck coefficient of Type B thermocouples becomes quite small and is almost negligible in the normal room temperature range. Consequently, in most applications the reference junction temperature of the thermocouple does not need to be controlled or even known, as long as it is between 0 and 50°C. For example, as shown by the reference tables, the voltage developed by the thermocouple with the reference junction at 0°C undergoes a reversal in sign at about 43°C, and between 0 and 50°C varies from a minimum of about -3 μV near 20°C to a maximum of about 2 μV at 50°C. Therefore, in use, if the reference junction of the thermocouple is within the range 0 to 50°C, then a 0°C reference junction temperature can be assumed and the error introduced will not exceed 3 μV. At high temperatures (above 1100°C) an additional error of 3 μV (about 0.3°C) in the measurements would be significant in most instances.

Burns and Gallagher [52] found significant differences in the purity of the platinum-rhodium thermoelements from the various manufacturers. Appreciable amounts of Fe, Ir, Au, Pd, Si, and Al were detected in some of the platinum-rhodium wires as well as small traces of Ca, Mg, B, and Cu. The amounts of impurities varied significantly from wire to wire.

In order to study the effect of varying the rhodium content of the alloys, Burns and Gallagher [52] measured the thermoelectric voltages of four wires near 29.60% rhodium and of five wires near 6.12% rhodium. They calculated that a 0.1% change in the rhodium content of the platinum-nominally 30% rhodium thermoelement produces a corresponding change in the thermocouple voltage of about 15 μV at 1500°C. In contrast a change of only 0.01% in the rhodium content of platinum-nominally 6% rhodium thermoelement also produces a voltage change of about 15 μV at this temperature. In both cases, a decrease in the rhodium content decreases the thermoelectric voltage of that thermoelement with respect to platinum.

The thermoelectric voltage of Type B thermocouples is sensitive to their history of annealing, heat treatment and quenching. Burns and Gallagher [52] recommend an electrical anneal in air for one hour at about 1450°C, followed by slow cooling. Calibration of Type B wires above 1600°C is undesirable in most circumstances.

ASTM Standard E230-72 in the Annual Book of ASTM Standards [49] specifies that the standard limits of error for Type B commercial thermocouples be ± 1/2% between 871 and 1705°C. Limits of error are not specified for Type B thermocouples below 871°C. The recommended upper temperature limit for protected thermocouples, 1705°C, applies to AWG 24 (0.5 mm) wire.

The information of Burns and Gallagher [52], after updating, can be summarized as follows:

1. Nearly all 30-6 thermocouples produced by manufacturers in this country will have thermoelectric voltages that agree with the values given in this chapter to within the equivalent of ± 0.5% of the temperature in the range 500 to 1800°C and to within ± 15 μV for temperatures below 500°C.

2. Calibration of a particular 30-6 thermocouple at four points (about 600, 1064, 1300, and 1554°C) will be sufficient to construct a deviation curve from the reference table such that the resulting calibration will be accurate to within ± 6 μV up to 1064°C, the equivalent of about ± 3°C up to 1554°C, and the equivalent of about ± 5°C above.

3. Actual calibration of the thermocouple above about 1600°C is not recommended, since some instability may result in the thermocouple. Values above 1600°C may be accurately determined by extrapolation.

4. High purity alumina is recommended for insulation and protection of the thermocouple but caution should be exercised at temperatures above about 1600°C for errors introduced by the electrical conductance of the insulators.

5. The use of large diameter wires (at least 0.5 mm or AWG 24) and larger size insulating tubes is recommended for operating temperatures above 1500 or 1600°C, so as to give the thermocouple added strength and to minimize errors due to electrical leakage.

6. In most instances the reference junction temperature need not be controlled since the emf and thermoelectric power of the thermocouple at normal room temperatures are very small.

The coefficients for the equation for the thermoelectric voltage of Type B thermocouple is given in Table 3.7. Values at selected fixed points are given in Table 3.8; values at 10°C intervals are given in the last section; a graph of the Seebeck coefficient is given in Fig. 3.10.

Base metal alloys are commonly used, especially at low temperatures. Some characteristic values for Types E, K, and T below room temperature are given in Fig. 3.11. The voltages of positive and negative legs for those materials are given in Fig. 3.12.

3.3.4 Type E–*Nickel*-Chromium Alloy vs. *Copper*-Nickel Alloy Thermocouples

This type, and the other base-metal types, do *not* have specific chemical compositions given in standards; rather, any material that fits the specified table within certain limits (see the end of this section) can be considered to be a Type E thermocouple. The positive thermoelement, EP, is the same material as KP. The negative thermoelement, EN, is the same material as TN.

The Type E thermocouple, as a combination of two specific thermoelements, does not have a well-documented history. The first officially recognized reference tables that we are aware of were those calculated by Shenker et al. [41] in NBS Circular 561. They based their tables upon a combination of data for KP vs. platinum and platinum vs. TN, where the primary data were taken from earlier NBS research papers.

Extensive research on the subzero properties of Type E thermocouples has been carried out by members of the Cryogenics Division in Boulder. That

TABLE 3.7 Power series expansion for the thermoelectric voltage of Type B thermocouples

Temperature range	Degree	Coefficients	Term
0–1820°C	8	$-2.4674601620 \times 10^{-1}$	T
		$5.9102111169 \times 10^{-3}$	T^2
		$-1.4307123430 \times 10^{-6}$	T^3
		$2.1509149750 \times 10^{-9}$	T^4
		$-3.1757800720 \times 10^{-12}$	T^5
		$2.4010367459 \times 10^{-15}$	T^6
		$-9.0928148159 \times 10^{-19}$	T^7
		$1.3299505137 \times 10^{-22}$	T^8

research has been summarized and tabulated by Sparks et al. [14] in NBS Monograph 124. They showed that Type E thermocouples are very useful down to about liquid hydrogen temperatures (n.b.p. 20.28K) where their Seebeck coefficient is about 8 $\mu V/°C$. They may even be used down to liquid helium temperatures (4.2K) though their Seebeck coefficient becomes quite low at 4K, only about 2 $\mu V/°C$. Both thermoelements of Type E thermocouples have a relatively low thermal conductivity, good resistance to corrosion in moist atmo-

TABLE 3.8 Thermoelectric values at the fixed points for Type B thermocouples

Fixed point	Temp., °C	$E, \mu V$	$S, \mu V/°C$
Ice point	0.000	0.00	−0.247
Ether TP	26.870	−2.39	0.068
Water BP	100.000	33.18	0.900
Benzoic TP	122.370	56.09	1.148
Indium FP	156.634	101.88	1.524
Tin FP	231.9681	247.37	2.334
Bismuth FP	271.442	347.74	2.750
Cadmium FP	321.108	497.15	3.265
Lead FP	327.502	518.24	3.331
Mercury BP	356.660	619.69	3.627
Zinc FP	419.580	867.78	4.256
Sulfur BP	444.674	977.66	4.501
Cu-Al FP	548.23	1495.10	5.483
Antimony FP	630.74	1978.43	6.227
Aluminum FP	660.37	2166.77	6.485
Silver FP	961.93	4490.76	8.848
Gold FP	1064.43	5433.59	9.539
Copper FP	1084.5	5626.33	9.667
Nickel FP	1455	9576.58	11.441
Cobalt FP	1494	10024.76	11.539
Palladium FP	1554	10720.56	11.646
Platinum FP	1772	13262.22	11.543

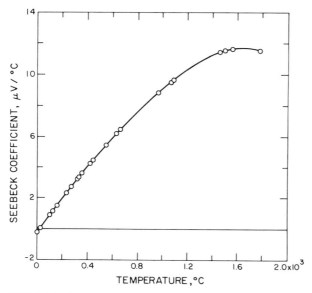

FIG. 3.10 Seebeck coefficient of Type B thermocouples.

spheres, and reasonably good homogeneity. Because of these three reasons, and their relatively high Seebeck coefficients, Type E thermocouples have been recommended by Sparks et al. [14] to be the most useful of the commercially standardized thermocouple combinations for subzero temperature measurements.

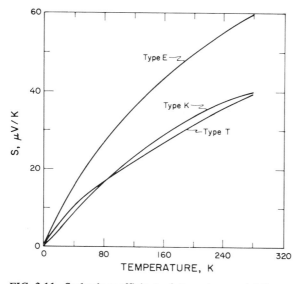

FIG. 3.11 Seebeck coefficient of three base metal thermo-couples.

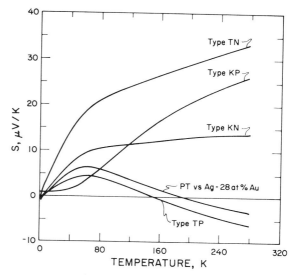

FIG. 3.12 Seebeck coefficients of some thermoelements.

For operation below 20K the nonstandardized combination KP versus *gold*-0.07 at % iron is recommended. The properties of this combination have been described by Sparks et al. [13].

Type E thermocouples also have the largest Seebeck coefficient above 0°C of any of the standardized thermocouples. For that reason they are being used more often whenever environmental conditions permit.

Type E thermocouples are recommended by the ASTM Manual [8] for use in the temperature range from −250 to 871°C in oxidizing or inert atmospheres. The negative thermoelement is subject to deterioration above about 871°C, but the thermocouple may be used up to 1000°C for short periods. The ASTM Manual [8] indicates the following restrictions on the use of Type E thermocouples at high temperatures: "They should not be used in sulfurous, reducing, or alternately reducing and oxidizing atmospheres unless suitably protected with protecting tubes. They should not be used in vacuum (at high temperatures) for extended times because the chromium in the positive thermoelement vaporizes out of solution and alters the calibration. They should also not be used in atmospheres that promote 'green-rot' corrosion (those with low, but not negligible, oxygen content)."

The negative thermoelement, a *copper*-nickel alloy, is subject to composition changes under thermal neutron irradiation since the copper is converted to nickel and zinc.

Neither thermoelement of Type E thermocouples is very sensitive to minor changes in composition or impurity level because both are already heavily alloyed. Similarly they are also not extremely sensitive to minor differences in heat treatment (provided that the treatment does not violate any of the restrictions mentioned above). For most general applications they may be used with the heat treatment given by the wire manufacturers. However, when the highest accuracy is sought, additional preparatory heat treatments may be desirable in order to enhance their performance. Details on this and other phases of the use and behavior of Type KP thermoelements (EP is the same as KP) are given in publications by Potts and McElroy [61], and N. A. Burley [62, 63].

ASTM Standard E230-72 in the Annual Book of ASTM Standards [49] specifies that the standard limits of error for Type E commercial thermocouples be ± 1.7°C between 0 and 316°C and ± 1/2% between 316 and 871°C. Limits of error are not specified for Type E thermocouples below 0°C. Type E thermocouples can also be supplied to meet special limits of error, which are less than the standard limits of error given above: ± 1 1/4°C between 0 and 316°C and ± 3/8% between 316 and 871°C. The recommended upper temperature limit for protected thermocouples, 871°C, applies to AWG 8 (3.3 mm) wire. For smaller wires the recommended upper temperature decreases to 649°C for AWG 14 (1.6 mm), 538°C for AWG 20 (0.8 mm), and 427°C for AWG 24 or 28 (0.5 or 0.3 mm).

The coefficients for the equations for the thermoelectric voltage of Type E thermocouples are given in Table 3.9. Values at selected fixed points are given in Table 3.10; values at 10°C are given in the last section; a graph of the Seebeck coefficient is given in Fig. 3.13.

TABLE 3.9 Power series expansion for the thermoelectric voltage of Type E thermocouples

Temperature range	Degree	Coefficients	Term
−270–0°C	13	$5.8695857799 \times 10^{1}$	T
		$5.1667517705 \times 10^{-2}$	T^2
		$-4.4652683347 \times 10^{-4}$	T^3
		$-1.7346270905 \times 10^{-5}$	T^4
		$-4.8719368427 \times 10^{-7}$	T^5
		$-8.8896550447 \times 10^{-9}$	T^6
		$-1.0930767375 \times 10^{-10}$	T^7
		$-9.1784535039 \times 10^{-13}$	T^8
		$-5.2575158521 \times 10^{-15}$	T^9
		$-2.0169601996 \times 10^{-17}$	T^{10}
		$-4.9502138782 \times 10^{-20}$	T^{11}
		$-7.0177980633 \times 10^{-23}$	T^{12}
		$-4.3671808488 \times 10^{-26}$	T^{13}
0–1000°C	9	$5.8695857799 \times 10^{1}$	T
		$4.3110945462 \times 10^{-2}$	T^2
		$5.7220358202 \times 10^{-5}$	T^3
		$-5.4020668085 \times 10^{-7}$	T^4
		$1.5425922111 \times 10^{-9}$	T^5
		$-2.4850089136 \times 10^{-12}$	T^6
		$2.3389721459 \times 10^{-15}$	T^7
		$-1.1946296815 \times 10^{-18}$	T^8
		$2.5561127497 \times 10^{-22}$	T^9

TABLE 3.10 Thermoelectric values at the fixed points for Type E thermocouples

Fixed point	Temp., °C	$E, \mu V$	$S, \mu V/°C$
Helium NBP	− 268.935	− 9833.09	2.084
Hydrogen TP	− 259.340	− 9792.66	6.180
Hydrogen NBP	− 252.870	− 9744.75	8.608
Neon TP	− 248.595	− 9704.62	10.181
Neon NBP	− 246.048	− 9677.58	11.073
Oxygen TP	− 218.789	− 9249.87	20.018
Nitrogen TP	− 210.002	− 9062.90	22.507
Nitrogen NBP	− 195.802	− 8716.78	26.181
Oxygen NBP	− 182.962	− 8360.81	29.230
Carbon dioxide SP	− 78.476	− 4227.53	48.541
Mercury FP	− 38.862	− 2193.03	54.039
Ice point[a]	0.000	0.00	58.606
Ether TP	26.870	1609.1	61.099
Water BP	100.000	6371.1	67.512
Benzoic TP	122.370	7846.8	89.229
Indium FP	156.634	10259.9	71.566
Tin FP	231.9681	15809.4	75.505
Bismuth FP	271.442	18820.7	77.012
Cadmium FP	321.108	22683.9	78.483
Lead FP	327.502	23186.2	78.643
Mercury BP	356.660	25489.1	79.293
Zinc FP	419.580	30512.6	80.302
Sulfur BP	444.674	32531.2	80.565
Cu-Al FP	548.23	40901.4	80.899
Antimony FP	630.74	47561.4	80.449
Aluminum FP	660.37	49941.2	80.177
Silver FP	961.93	73495.5	75.489

[a]Junction point of different functions.

It should be stressed that Types E, J, K and T thermocouple materials that conform closely to the high temperature tabular values may not necessarily conform closely at low temperatures (below 0°C) and vice versa. If thermocouples are to be used for accurate measurements both above and below 0°C, then the material must be calibrated in the full temperature range, both above and below 0°C. Special selection of material will usually be required.

3.3.5 Type J–Iron vs. *Copper*-Nickel Alloy (SAMA) Thermocouples

This is one of the most common types of industrial thermocouples because of its relatively high Seebeck coefficient and low cost. It has been reported that more than 200 tons of Type J materials are supplied annually to industry in this country. However, it is least suitable for accurate thermometry because there are

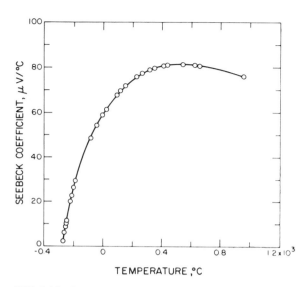

FIG. 3.13 Seebeck coefficient of Type E thermocouples.

significant nonlinear deviations in the thermoelectric output from different manufacturers. These irregular deviations lead to difficulties in obtaining accurate
calibrations based on a limited number of calibration points. The positive thermoelement is commercially pure (99.5% Fe) iron, usually containing significant
impurity levels of carbon, chromium, copper, manganese, nickel, phosphorus,
silicon, and sulfur. Thermocouple wire represents such a small fraction of the
total production of commercial iron wire that the producers do not control the
chemical composition to maintain constant thermoelectric properties. Instead,
instrument companies and thermocouple fabricators select material most suitable
for their thermocouple usage. The total and specific types of impurities that
occur in commercial iron change with time, location of primary ores, and
methods of smelting. Many unusual lots have been selected in the past, for
example spools of industrial iron wire and even scrapped rails from an elevated
train line. At present, iron wire that most closely fits these tables has about
0.25% Mn and 0.12% copper plus other minor impurities.

The negative thermoelement for Type J thermocouples is a *copper*-nickel
alloy known ambiguously as constantan. The word *constantan* has commonly
referred to copper-nickel alloys with anything from 45 to 60% of copper, often
with minor impurities of carbon, iron, or manganese. Constantan for Type J
thermocouples usually contains about 55% copper, 45% nickel, and a small but
thermoelectrically significant amount of iron and manganese, about 0.1% or
more. It should be emphasized that Type JN thermoelements are NOT generally
interchangeable with Type TN (or EN) thermoelements, although they are all
referred to as *constantan*. In order to provide some differentiation in nomenclature,
Type JN is often referred to as *SAMA constantan*.

Suppliers of Type J thermocouples usually select heats of iron and well-matched batches of constantan so that the total output of the combination closely fits the Type J table up to 760°C. In fact, with care in selection, Type J thermocouples can be produced that will fit calibration tables as accurately as the more expensive Type K thermocouples. While the overall thermocouple will conform to the limits of error published in ASTM or ISA standards (or even have closer limits of error), it should be emphasized that Type JP and JN thermoelements as supplied by different companies are *not* generally interchangeable.

The grandfather of the Type J thermocouple tables was the commercial table printed by Leeds and Northrup in 1913 and incorporated in a later NBS paper, Foote, Fairchild, and Harrison [64] and the NAS-NRC International Critical Tables [65]. They covered the range from 0 to 760°C. The usefulness of an iron vs. *copper*-nickel alloy thermocouple had been shown as early as 1892 by Lindeck [66]. By the 1930s a number of different tables had been published for iron vs. constantan thermocouples because of lack of standardization and differences in the irons used by various thermocouple manufacturers. Although other tables for iron vs. constantan thermocouples came into existence, the 1913 table was the most commonly used for instrument calibrations. Until the late 1930s however, thermocouples did not conform accurately to this curve. Therefore Roeser and Dahl [67] obtained a representative selection of both materials and carefully remeasured their thermoelectric properties between −200 and 1000°C. Their results were representative of material used by the military at that time, but deviated significantly from the 1913 table. Because the new tables differed by up to 2% from the previous tables, they were not generally accepted as a replacement for the earlier iron vs. constantan tables. To somewhat ease the confusion, the tables generated by Roeser and Dahl [67] wer referred to as Type Y or RP 1080 iron vs. constantan. Their tables were used by several military groups but were not generally used in civilian applications.

In an effort to promote uniformity, a section of the Scientific Apparatus Makers of America (SAMA) initiated a new program in 1948. A research project was established at the National Bureau of Standards and the results were published in 1953 by Corruccini and Shenker. The resultant tables were very close to the 1913 ones and have generally been accepted as the Type J, iron vs. constantan, thermocouple standards (NBS Circular 561, Shenker et al. [41]). After modifications for temperature scale changes, their research results were used for generating the functions given here.

Type J thermocouples are recommended by the ASTM [8] for use in the temperature range from 0 to 760°C in vacuum, oxidizing, reducing, or inert atmospheres. If used for extended times above 500°C, heavy gage wires are recommended because the oxidation rate is rapid at elevated temperatures. The ASTM Manual STP 470 [8] indicates the following restrictions on the use of Type J thermocouples: "They should not be used in sulfurous atmospheres above 500°C. Because of potential rusting and embrittlement, they are not recommended for sub-zero temperatures. They should not be cycled above 760°C even for a short time if accurate readings below 760°C are desired at a later time."

The positive thermoelement, iron, is relatively insensitive to composition changes under thermal neutron irradiation, but does exhibit a slight increase in manganese content. The negative thermoelement, a *copper*-nickel alloy, is subject to substantial composition changes under thermal neutron irradiation since copper is converted to nickel and zinc.

Both thermoelements of Type J thermocouples are variable in thermoelectric output because of compositional variations in the iron and in the *copper*-nickel alloy. Corruccini and Shenker [68] found an order of magnitude variation in both the manganese (0.03 to 0.38%) and copper (0.02 to 0.15%) impurities in the iron thermoelements, even though the materials were presumably specially selected lots of material. Not only were the thermoelectric voltages of different iron thermoelements different by as much as 2%, the output curves were sometimes different in shape. The negative thermoelements also differed by as much as 2%, but their deviations tended to be much more linear. At present the manufacturers are controlling the compositions and the matching of thermoelements more carefully and therefore deviations from the standards should be considerably less than those observed by Corruccini and Shenker [68].

Finch [69] has shown that the Seebeck coefficient of iron at $500°C$ is increased by additions of Cr, Mn, or S but decreased by additions of Ni, Si, Sn, P, and Cu (for impurity level above 0.1%). Manufacturers select the negative thermoelement to match a given lot of positive material. The composition of the *copper*-nickel alloy therefore also varies significantly. The average composition is around 55% copper and there is usually a significant impurity amount of iron.

Annealing below $760°C$ for short periods does not significantly alter the thermoelectric properties of Type J thermocouples.

Commercial iron undergoes a magnetic transformation near $769°C$ and an α-γ crystal transformation near $910°C$ (see Hansen and Anderko [70]). Both of these transformations, especially the latter, seriously affect the thermoelectric properties of iron, and therefore of Type J thermocouples. It is for that reason that iron vs. constantan thermocouples are not recommended as a standardized type above $760°C$. If Type J thermocouples are taken to high temperatures, especially above $900°C$, they will lose the accuracy of their calibration when they are recycled to lower temperatures. If Type J thermocouples are used at temperatures above $760°C$, only the largest wire, AWG 8 (3.3 mm) should be used and they should be held at the measured temperature for 10 to 20 min before readings are taken. The output of the Type J thermocouples may change by as much as 40 μV (or $1°C$ equivalent) per minute when first brought up to temperatures near $900°C$.

ASTM Standard E230-72 in the Annual Book of ASTM Standards [49] specifies that the standard limits of error for Type J commercial thermocouples be $\pm$ $2.2°C$ between 0 and $277°C$ and $\pm$ 3/4% between 277 and $760°C$. Limits of error are not specified for Type J thermocouples below $0°C$ or above $760°C$. Type J thermocouples can also be supplied to meet special limits of error, which are equal to one-half the limits given above. The recommended upper temperature limit for protected thermocouples, 760°C, applies to AWG 8 (3.3 mm) wire. For smaller wires the recommended upper temperature decreases to 593°C for AWG 14 (1.6 mm), 482°C for AWG 20 (0.8 mm), and 371°C for AWG 24 or 28 (0.5 or 0.3 mm).

The coefficients for the equations for the thermoelectric voltage of Type J thermocouples are given in Table 3.11. Values at selected fixed points are given in Table 3.12; values at 10°C intervals are given in the last section; a graph of the Seebeck coefficient is given in Fig. 3.14.

Values above 760°C are given as a guide only: thermoelectric properties of Type J thermocouples are not stable above 760°C and the thermocouple should NOT be considered to be a standardized type above this temperature.

3.3.6 Type K–*Nickel*-Chromium Alloy vs. *Nickel*-Aluminum Alloy Thermocouples

This type is more resistant to oxidation at elevated temperatures than the Types E, J or T thermocouples and consequently it finds wide application at temperatures above 500°C. The positive thermoelement, KP, which is the same as EP, is an alloy that typically contains about 89 to 90% nickel, 9 to nearly 9.5% chromium, both silicon and iron in amounts up to about 0.5%, plus smaller amounts of other constituents such as carbon, manganese, cobalt and niobium. The negative thermoelement, KN, is typically composed of about 95 to 96% nickel, 1 to 1.5% silicon, 1 to 2.3% aluminum, 1.6 to 3.2% manganese, up to about 0.5% cobalt and smaller amounts of other constituents such as iron, copper and lead. Type KN thermoelements with modified compositions are also available for use in special applications. These include alloys in which the

TABLE 3.11 Power series expansion for the thermoelectric voltage of Type J thermocouples

Temperature range	Degree	Coefficients	Term
− 210–760°C	7	5.0372753027×10^1	T
		$3.0425491284 \times 10^{-2}$	T^2
		$- 8.5669750464 \times 10^{-5}$	T^3
		$1.3348825735 \times 10^{-7}$	T^4
		$- 1.7022405966 \times 10^{-10}$	T^5
		$1.9416091001 \times 10^{-13}$	T^6
		$- 9.6391844859 \times 10^{-17}$	T^7
760–1200°C	5	2.9721751778×10^5	—
		$- 1.5059632873 \times 10^3$	T
		$3.2051064215 \quad \ldots$	T^2
		$- 3.2210174230 \times 10^{-3}$	T^3
		$1.5949968788 \times 10^{-6}$	T^4
		$- 3.1239801752 \times 10^{-10}$	T^5

TABLE 3.12 Thermoelectric values at the fixed points for Type J thermocouples

Fixed point	Temp., °C	$E, \mu V$	$S, \mu V/°C$
Nitrogen TP	− 210.002	− 8095.7	19.126
Nitrogen NBP	− 195.802	− 7796.3	22.972
Oxygen NBP	− 182.962	− 7480.7	26.148
Carbon dioxide SP	− 78.476	− 3718.7	43.721
Mercury FP	− 38.862	− 1906.3	47.586
Ice point	0.000	0.0	50.373
Ether TP	26.870	1373.9	51.832
Water BP	100.000	5267.7	54.348
Benzoic TP	122.370	6488.6	54.788
Indium FP	156.634	8374.3	55.238
Tin FP	231.9681	12551.7	55.537
Bismuth FP	271.442	14742.7	55.459
Cadmium FP	321.108	17492.8	55.280
Lead FP	327.502	17846.2	55.257
Mercury BP	356.660	19456.0	55.170
Zinc FP	419.580	22925.9	55.189
Sulfur BP	444.674	24312.3	55.321
Cu-Al FP	548.23	30109.5	56.985
Antimony FP	630.74	34910.8	59.594
Aluminum FP	660.37	36693.3	60.729
Silver FP	961.93	55669	60.246
Gold FP	1064.43	61716	58.100
Copper FP	1084.5	62880	57.926

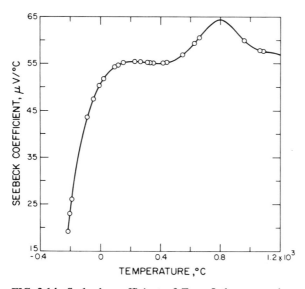

FIG. 3.14 Seebeck coefficient of Type J thermocouples.

manganese and aluminum contents are reduced or eliminated, while the silicon and cobalt contents are increased.

Type K thermocouples were developed in 1906 and were known originally as "Chromel"* vs. "Alumel"* thermocouples. The Hoskins Manufacturing Company was the sole manufacturer of these thermocouples in this country until the mid-1940s. After this time other Type K thermocouples such as Topel† vs. Nial†, T_1‡ vs. T_2‡, and Thermo Kanthal KP§ vs. ThermoKanthal KN§ were introduced commercially.

The first reference tables for Type K thermocouples to gain general industry-wide acceptance were the ones prepared by Roeser et al. in 1935 [71]. They based their tables upon the calibrations of 30 samples of No. 8 AWG Type KP and Type KN thermoelements, except that below 0°C they used only two samples each of the KP and KN thermoelements. The thermoelements were all furnished by the Hoskins Manufacturing Co. (the only manufacturer at the time), and they were selected by the manufacturer from 100 production tests. The tabular values of Roeser et al. [71] were corrected by Shenker et al. [41] to account for changes in the temperature scales and electrical units and presented in NBS Circular 561.

Extensive research on the subzero properties of Type K thermocouples was performed by members of the Cryogenics Division in Boulder. That research was summarized and tabulated by Sparks et al. [14] in NBS Monograph 124. While they found the Type E thermocouple to be the most satisfactory of the standardized letter-designated type thermocouples for measurements down to liquid hydrogen temperature (20.28K), Type K thermocouples may also be used at these temperatures. However, their Seebeck coefficient (about 4 μV/K at 20K) is only about one-half of that of Type E thermocouples. Furthermore, the thermoelectric homogeneity of KN thermoelements is generally not quite as good as that of EN thermoelements. Both the KP and the KN thermoelements do have a relatively low thermal conductivity and good resistance to corrosion in moist atmospheres at low temperatures.

Type K thermocouples are recommended by the ASTM [8] for continuous use at temperatures within the range -250 to 1260°C in oxidizing or inert atmospheres. Both the KP and the KN thermoelements are subject to oxidation when used in air above about 850°C, but even so, Type K thermocouples may be used at temperatures up to about 1350°C for short periods with only small changes in calibration. When oxidation occurs it normally leads to a gradual increase in the thermoelectric voltage with time. The magnitude of the change in the thermoelectric voltage and the physical life of the thermocouple will depend upon such factors as the temperature, time at temperature, diameter of the

---

*Trademark–Hoskins Manufacturing Company.
†Trademark–Wilbur B. Driver Company.
‡Trademark–Driver Harris Company.
§Trademark–Kanthal Corporation.

thermoelements and conditions of use. The thermoelectric instability of Type K thermocouples in air at elevated temperatures has been carefully studied by Dahl [72], Potts and McElroy [61], Burley and Ackland [73], and by Wang et al. [74] and their work should be consulted for details.

In addition, the ASTM Manual STP 470 [8] gives the following restrictions on the use of Type K thermocouples: "They should not be used in sulfurous, reducing, or alternately reducing and oxidizing atmospheres unless suitably protected with protecting tubes. They should not be used in vacuum (at high temperatures) for extended times because the chromium in the positive thermoelement vaporizes out of solution and alters the calibration. They should also not be used in atmospheres that promote 'green-rot' corrosion (those with low, but not negligible, oxygen content)."

Both thermoelements of Type K thermocouples are reasonably stable, thermoelectrically, under neutron irradiation since the resulting changes in their chemical compositions due to transmutation are small. The KN thermoelements are somewhat less stable than the KP thermoelements in that they experience a small increase in the iron content accompanied by a slight decrease in the manganese and cobalt contents.

Neither thermoelement of a Type K thermocouple is very sensitive to minor changes in composition or impurity level because both are already heavily alloyed. Similarly, they are also not extremely sensitive to minor differences in heat treatment (provided that the treatment does not violate any of the restrictions mentioned above). For most general applications they may be used with the heat treatment routinely given by the wire manufacturer. However, when extreme accuracy is sought, the thermoelements may require additional preparatory heat treatments in order to achieve the desired results. Details on this and other phases of the use and behavior of Type K thermocouples are given in the articles by Potts and McElroy [61], and Burley [62, 63].

ASTM Standard E230-72 in the Annual Book of ASTM Standards [49] specifies that the standard limits of error for Type K commercial thermocouples be $\pm 2.2°C$ between 0 and $277°C$ and $\pm 3/4\%$ between 277 and $1260°C$. Limits of error are not specified for Type K thermocouples below $0°C$. Type K thermocouples can also be supplied to meet special limits of error, which are equal to one half the standard limits of error given above. The recommended upper temperature limit for protected Type K thermocouples, $1260°C$, applies for AWG 8 (3.3 mm) wire. For smaller wires it decreases to $1093°C$ for AWG 14 (1.6 mm), $982°C$ for AWG 20 (0.8 mm), and $871°C$ for AWG 24 or 28 (0.5 or 0.3 mm).

While limits of error for single-leg thermoelements vs. platinum are not given in ASTM Standard E230-72, the Type KP and KN thermoelements are supplied, by common practice, to a voltage tolerance equivalent to one-half the tolerance specified for the Type K thermocouple.

The coefficients for the equations for the thermoelectric voltage of Type K thermocouples are given in Table 3.13. Values at selected fixed points are given

TABLE 3.13 Power series expansion for the thermoelectric voltage of Type K thermocouples

Temperature range	Degree	Coefficients	Term
-270–$0°C$	10	$3.9475433139 \times 10^{1}$	T
		$2.7465251138 \times 10^{-2}$	T^2
		$-1.6565406716 \times 10^{-4}$	T^3
		$-1.5190912392 \times 10^{-6}$	T^4
		$-2.4581670924 \times 10^{-8}$	T^5
		$-2.4757917816 \times 10^{-10}$	T^6
		$-1.5585276173 \times 10^{-12}$	T^7
		$-5.9729921255 \times 10^{-15}$	T^8
		$-1.2688801216 \times 10^{-17}$	T^9
		$-1.1382797374 \times 10^{-20}$	T^{10}
0–$1372°C$	$8+$ exp.	$-1.8533063273 \times 10^{1}$	$\cdots$
		$3.8918344612 \times 10^{1}$	T
		$1.6645154356 \times 10^{-2}$	T^2
		$-7.8702374448 \times 10^{-5}$	T^3
		$2.2835785557 \times 10^{-7}$	T^4
		$-3.5700231258 \times 10^{-10}$	T^5
		$2.9932909136 \times 10^{-13}$	T^6
		$-1.2849848798 \times 10^{-16}$	T^7
		$2.2239974336 \times 10^{-20}$	T^8

$$+ 125 \exp\left[-\frac{1}{2}\left(\frac{T-127}{65}\right)^2\right]$$

in Table 3.14; values at $10°C$ intervals are given in the last section; a graph of the Seebeck coefficient is given in Fig. 3.15.

3.3.7 Type T–Copper vs. *Copper*-Nickel Alloy Thermocouples

This type is one of the older and more popular thermocouples for determining temperatures within the range from about $300°C$ down to the hydrogen normal boiling point $(-252.87°C)$. It is at present the only one of the standardized letter-designated type thermocouples for which limits of error are specified below $0°C$. The positive thermoelement, TP, is typically electrolytic tough pitch copper that conforms to ASTM Specification B3 for soft or annealed bare copper wire. Such material is about 99.95% pure copper with an oxygen content varying from 0.02 to 0.07% (depending on sulfur content) and with other impurities totaling about 0.01%. Above about $-195°C$ the thermoelectric properties of Type TP thermoelements, which satisfy the above conditions, are exceptionally uniform and exhibit little variation between different lots. Below about 76K the thermoelectric properties are affected more strongly by the presence of dilute transition metal solutes, particularly iron.

The negative thermoelement, TN (or EN), is a *copper*-nickel alloy known ambiguously as constantan. As discussed in Sec. 3.3.5, the word *constantan* refers to a family of *copper*-nickel alloys containing anywhere from 45 to 60% copper. These alloys also typically contain small percentages of manganese and iron as well as other trace impurities such as carbon, magnesium, silicon, cobalt, etc. Constantan is also known by various trade names such as Advance* and Cupron†. The constantan for Type T thermocouples usually contains about 55% copper, 45% nickel, and small but thermoelectrically significant amounts of iron

*Trademark—Driver Harris Company.
†Trademark—Wilbur B. Driver Company.

TABLE 3.14   Thermoelectric values at the fixed points for Type K thermocouples

Fixed point	Temp., °C	E, μV	S, μV/°C
Helium NBP	− 268.935	− 6456.93	0.922
Hydrogen TP	− 259.340	− 6439.34	2.803
Hydrogen NBP	− 252.870	− 6416.69	4.212
Neon TP	− 248.595	− 6396.64	5.168
Neon NBP	− 246.048	− 6382.75	5.741
Oxygen TP	− 218.789	− 6144.56	11.630
Nitrogen TP	− 210.002	− 6034.65	13.373
Nitrogen NBP	− 195.802	− 5825.66	16.032
Oxygen NBP	− 182.962	− 5605.15	18.296
Carbon dioxide SP	− 78.476	− 2869.64	32.977
Mercury FP	− 38.862	− 1484.85	36.766
Ice point[a]	0.000	0.00	39.475
Ether TP	26.870	1076.0	40.564
Water BP	100.000	4095.3	41.371
Benzoic TP	122.370	5016.0	40.913
Indium FP	156.634	6403.9	40.142
Tin FP	231.9681	9420.1	40.399
Bismuth FP	271.442	11028.6	41.081
Cadmium FP	321.108	13084.9	41.670
Lead FP	327.502	13351.5	41.725
Mercury BP	356.660	14571.4	41.946
Zinc FP	419.580	17223.1	42.323
Sulfur BP	444.674	18286.7	42.441
Cu-Al FP	548.23	22696.5	42.638
Antimony FP	630.74	26207.0	42.393
Aluminum FP	660.37	27460.6	42.221
Silver FP	961.93	39779.4	39.317
Gold FP	1064.43	43755.0	38.245
Copper FP	1084.5	44520.4	38.020

[a]Junction point of different functions.

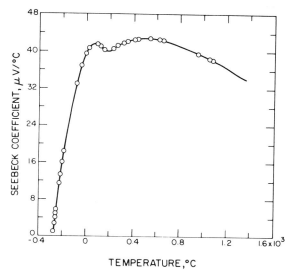

FIG. 3.15 Seebeck coefficient of Type K thermocouples.

and manganese, about 0.1% or larger. It should be emphasized that Type TN (or EN) thermoelements are NOT generally interchangeable with Type JN thermoelements, although they are all referred to as *constantan*. In order to provide some differentiation in nomenclature, Type TN (or EN) is often referred to as *Adams' constantan* and Type JN is usually referred to as *SAMA constantan*.

The thermoelectric relations for Type TN and EN thermoelements are the same, that is the voltage vs. temperature equations and tables for platinum vs. Type TN thermoelements is typical for both types of thermoelements over the temperature range recommended for each corresponding thermocouple type. However, it should *not* be assumed that Type TN and EN thermoelements are always interchangeable or that they have the same commercial limits of error.

The first reference tables for Type T thermocouples to be used on an industry-wide basis were those prepared by Roeser and Dahl [67]. They indicated that their tabular values were intended primarily for use above 0°C. They presented tabular values below 0°C as a matter of convenience only and the values were based upon measurements of only a relatively few samples. The values above 0°C were derived from measurements of the thermoelectric voltage on nine samples of copper vs. platinum, Pt-27, between 0 and 1000°C and on twenty-seven samples of constantan vs. platinum between 0 and 800°C. The samples were furnished by the various pyrometer instrument manufacturers at that time. Several years later, more representative tabular values were established by Scott [75] for the subzero temperature range. The tabular values of Roeser and Dahl [67] above 0°C were combined with those of Scott [75] and also corrected for changes in the temperature scales and electrical units by Shenker et al. [41]. The last revision of the tables in 1955 was presented in NBS Circular 561.

Extensive research on the subzero properties of Type T thermocouples was performed by members of the Cryogenic Division in Boulder. That research was summarized and tabulated by Sparks et al. [14] in NBS Monograph 124. They indicate that Type T thermocouples may be used down to liquid hydrogen temperatures (about 20K), but of the standardized letter-designated type thermocouples they emphasize that Type E thermocouples possess the most desirable characteristics in this range. The thermoelectric homogeneity of most Type TP and TN (or EN) thermoelements is reasonably good. However, the Seebeck coefficient of Type T thermocouples is moderately small at subzero temperatures (about 5.6 μV/K at 20K), being roughly two-thirds that of Type E thermocouples. This, together with the high thermal conductivity of Type TP thermoelements, is the major reason why Type T thermocouples are less suitable for use in the subzero range than Type E thermocouples. In addition, there is considerable variability in the thermoelectric properties of Type TP thermoelements at temperatures below about 76K caused by variations in the amounts and types of impurities present in these nearly pure materials. For these reasons, Type T thermocouples are generally unsuitable for use below about 20K.

Type T thermocouples are recommended by the ASTM [8] for use in the temperature range from -184 to $371°$C in vacuum or in oxidizing, reducing or inert atmospheres. The recommended upper temperature limit for continuous service of protected Type T thermocouples is set at $371°$C for AWG 14 (1.6 mm) thermoelements since Type TP thermoelements oxidize rapidly above this temperature. However, the thermoelectric properties of Type TP thermoelements are apparently not grossly affected by oxidation since Roeser and Dahl [67] observed negligible changes in the thermoelectric voltage of Nos. 12, 18 and 22 AWG Type TP thermoelements after heating for 30 hours in air at $500°$C. At this temperature the Type TN thermoelements have good resistance to oxidation and exhibit only small changes in thermal emf with long exposure in air, as shown by the studies of Dahl [72]. Higher operating temperatures up to at least $700°$C, are possible in vacuum or in inert atmospheres, where the deterioration of the Type TP thermoelement is no longer a problem. Operation of Type T thermocouples in hydrogen atmospheres at temperatures above about $370°$C is not recommended since severe embrittlement of the Type TP thermoelements may occur.

Type T thermocouples are not well suited for use in nuclear environments since both thermoelements are subject to significant changes in composition under thermal neutron irradiation. The copper in the thermoelements is converted to nickel and zinc.

Because of the high thermal conductivity of Type TP thermoelements, special care should be exercised in the use of the thermocouples to insure that both the measuring and reference junctions assume the desired temperatures. Caldwell [76] has determined the errors that arise from insufficient immersion of

Type TP thermoelements in ice baths, and his work should be consulted for details.

ASTM Standard E230-72 in the Annual Book of ASTM Standards [49] specifies that the standard limits of error for Type T commercial thermocouples be $\pm 2\%$ between -101 and $-59°C$, $\pm 0.8°C$ between -59 and $93°C$ and $\pm 3/4\%$ between 93 and $371°C$. Type T thermocouples can also be supplied to meet special limits of error, which are equal to one half the standard limits of error given above (plus a limit of error of $\pm 1\%$ is specified between -184 and $-59°C$). The recommended upper temperature limit for protected Type T thermocouples, $371°C$, applies to AWG 14 (1.6 mm) wire. For smaller wires it decreases to $260°C$ for AWG 20 (0.8 mm) and $240°C$ for AWG 24 or 28 (0.5 or 0.3 mm).

The coefficients for the equations for the thermoelectric voltage of Type T thermocouples are given in Table 3.15. Values at selected fixed points are given in Table 3.16; values at $10°C$ intervals are given in the last section; a graph of the Seebeck coefficient is given in Fig. 3.16.

3.4 NEWER THERMOCOUPLE MATERIALS

New materials have often been suggested, or even tested. See for example the Temperature Conference Proceedings of 1962 and 1972. Only three will be

TABLE 3.15 Power series expansion for the thermoelectric voltage of Type T thermocouples

Temperature range	Degree	Coefficients	Term
-270–$0°C$	14	$3.8740773840 \times 10^{1}$	T
		$4.4123932482 \times 10^{-2}$	T^2
		$1.1405238498 \times 10^{-4}$	T^3
		$1.9974406568 \times 10^{-5}$	T^4
		$9.0445401187 \times 10^{-7}$	T^5
		$2.2766018504 \times 10^{-8}$	T^6
		$3.6247409380 \times 10^{-10}$	T^7
		$3.8648924201 \times 10^{-12}$	T^8
		$2.8298678519 \times 10^{-14}$	T^9
		$1.4281383349 \times 10^{-16}$	T^{10}
		$4.8833254364 \times 10^{-19}$	T^{11}
		$1.0803474683 \times 10^{-21}$	T^{12}
		$1.3949291026 \times 10^{-24}$	T^{13}
		$7.9795893156 \times 10^{-28}$	T^{14}
0–$400°C$	8	$3.8740773840 \times 10^{1}$	T
		$3.3190198092 \times 10^{-2}$	T^2
		$2.0714183645 \times 10^{-4}$	T^3
		$-2.1945834823 \times 10^{-6}$	T^4
		$1.1031900550 \times 10^{-8}$	T^5
		$-3.0927581898 \times 10^{-11}$	T^6
		$4.5653337165 \times 10^{-14}$	T^7
		$-2.7616878040 \times 10^{-17}$	T^8

TABLE 3.16 Thermoelectric values at the fixed points for Type T thermocouples

Fixed point	Temp., °C	$E, \mu V$	$S, \mu V/°C$
Helium NBP	− 268.935	− 6256.29	1.408
Hydrogen TP	− 259.340	− 6229.19	4.07
Hydrogen NBP	− 252.870	− 6197.73	5.645
Neon TP	− 248.595	− 6171.38	6.682
Neon NBP	− 246.048	− 6153.58	7.298
Oxygen TP	− 218.789	− 5873.02	12.923
Nitrogen TP	− 210.002	− 5753.28	14.305
Nitrogen NBP	− 195.802	− 5535.59	16.328
Oxygen NBP	− 182.962	− 5314.72	18.067
Carbon dioxide SP	− 78.476	− 2740.70	30.828
Mercury FP	− 38.862	− 1434.94	35.022
Ice point[a]	0.000	0.0	38.741
Ether TP	26.870	1067.9	40.829
Water BP	100.000	4277.3	46.773
Benzoic TP	122.370	5341.4	48.343
Indium FP	156.634	7036.4	50.565
Tin FP	231.9681	11013.3	54.887
Bismuth FP	271.442	13218.8	56.821
Cadmium FP	321.108	16095.3	58.975
Lead FP	327.502	16473.3	59.240
Mercury BP	356.660	18217.9	60.415

[a]Junction point of different functions.

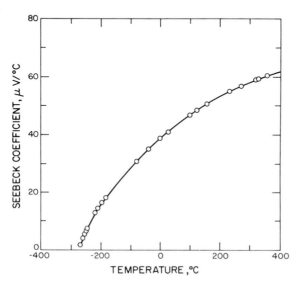

FIG. 3.16 Seebeck coefficient of Type T thermocouples.

briefly described here: (1) the gold-iron useful at low temperatures, (2) tungsten-rhenium useful at very high temperatures, and (3) Nicrosil/Nisil, a possible replacement for Type K.

Sparks and Powell [13] suggest that one of the best thermocouple combinations to use at low temperatures is Type KP for the positive leg vs. Au–0.07 at % Fe for the negative leg. The Seebeck coefficient of that combination is quite high, relative to other materials, at helium temperatures, approximately 16–17 μV/K. Although that combination is commonly used for cryogenic research, it has not become a standard.

Burns and Hurst [77] have carried out extensive research on various tungsten-rhenium combinations. Bare wire W–3% Re and W–25% Re thermoelements were exposed at 2400K in argon, hydrogen, or vacuum. They experienced a shift in their thermoelectric voltage upon initial exposure. But after the initial exposure, they experienced no significant further change for periods up to 1000 hr (except for the vacuum exposed ones, which had a continuous drift). More recent studies by them [78] have shown that carefully sheathed thermocouples of those compositions exhibit drifts less than 1°C/1000 hr exposure at 1800°C. No degradation was observed.

Burley et al. [79] are developing a new nickel alloy system that may replace Type K because of its superior stability and resistance to corrosion. Graphs of the Seebeck coefficient for the combined thermocouple and the two thermoelements are given in Figs. 3.17-3.19.

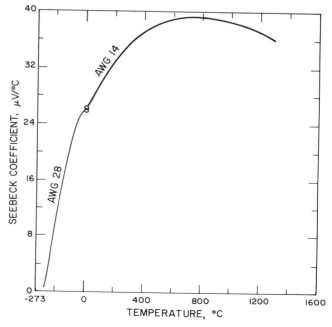

FIG. 3.17 Seebeck coefficient of Nicrosil vs. Nisil.

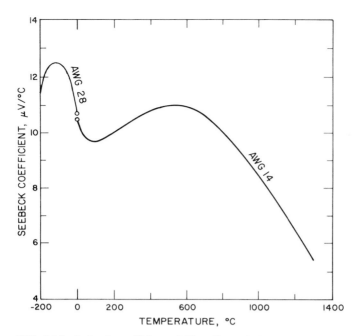

FIG. 3.18 Seebeck coefficient of Nicrosil vs. platinum.

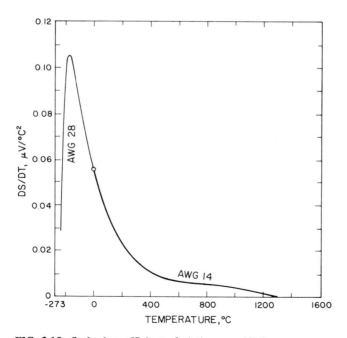

FIG. 3.19 Seebeck coefficient of platinum vs. Nisil.

REFERENCES

1. Seebeck, T. J., Magnetische Polarisation der Metalle und Erze durch Temperatur-Differenz, *Abh. Königl. Akad. Wiss. Berlin*, p. 265, 1822–1833.
2. Peltier, J. C. A., Nouvelles Expériences sur la Caloricité des Courants Électriques, *Ann. Chim. Phys.*, vol. 56, p. 371, 1834.
3. MacDonald, D. K. C. "Thermoelectricity: An Introduction to the Principles," Wiley, New York, 1962.
4. Pollack, D. D., "The Theory and Properties of Thermocouple Elements," ASTM, Philadelphia, 1971.
5. Barnard, R. D., "Thermoelectricity in Metals and Alloys," Taylor and Francis Ltd., London, 1972.
6. Bridgman, P. W., "The Thermodynamics of Electrical Phenomena in Metals," Dover, New York, 1961.
7. Harman, T. C. and Honig, J. M., "Thermoelectric and Thermomagnetic Effects and Applications," McGraw-Hill, New York, 1967.
8. American Society for Testing and Materials, "STP 470—Manual on the Use of Thermocouples in Temperature Measurement," ASTM, Philadelphia, 1970.
9. Benedict, R. P., "Fundamentals of Temperature, Pressure, and Flow Measurements," Wiley, New York, 1969.
10. Dike, P. H., "Thermoelectric Thermoelectricity," Leeds and Northrup, Philadelphia, 1954.
11. Heikes, R. R. and Vine, R. W., "Thermoelectricity: Science and Engineering," Interscience, New York, 1961.
12. Kinzie, P. A., "Thermocouple Temperature Measurement," Wiley, New York, 1973.
13. Sparks, L. L. and Powell, R. L., Low Temperature Thermocouples: KP, "Normal" silver, and Copper versus Au–0.02 at % Fe and Au–0.07 at % Fe, *J. Res. Nat. Bur. Stand.*, vol. 76A, p. 263, 1972.
14. Sparks, L. L., Powell, R. L. and Hall, W. J., Reference Tables for Low-temperature Thermocouples, *NBS Monograph 124*, National Bureau of Standards, Washington, D.C., 1972.
15. Powell, R. L., Hall, W. J., Hyink, C. H., Sparks, L. L., Burns, G. W., Scroger, M. C., and Plumb, H. H., Thermocouple Reference Tables Based on the IPTS–68, *NBS Monograph 125*, National Bureau of Standards, Washington, D.C., 1974.
16. Caldwell, F. R., in American Institute of Physics, "Temperature: Its Measurement and Control in Science and Industry," vol. 3, pt. 2, Reinhold, New York, 1962.
17. Caldwell, F. R., Thermocouple Materials, *NBS Monograph 40*, National Bureau of Standards, Washington, D.C., 1962.
18. Richards, D. B., Edwards, L. R. and Legvold, S., Thermocouples in Magnetic Fields, *J. Appl Phys.*, vol. 40, p. 3836, 1969.
19. Loscoe, C. and Mette, H., in American Institute of Physics, "Temperature: Its Measurement and Control in Science and Industry," vol. 3, pt. 2, p. 283, Reinhold, New York, 1962.
20. Kelly, M. J., Johnston, W. W., and Baumann, C. D., in American Institute of Physics, "Temperature: Its Measurement and Control in Science and Industry," vol. 3, pt. 2, p. 265, Reinhold, New York, 1962.
21. Browning, W. E. and Miller, C. E., in American Institute of Physics, "Temperature: Its Measurement and Control in Science and Industry," vol. 3, pt. 2, p. 271, Reinhold, New York, 1962.
22. Levy, G. F., Fouse, R. R. and Sherwin, R., in American Institute of Physics, "Temperature: Its Measurement and Control in Science and Industry," vol. 3, pt. 2, p. 277, Reinhold, New York, 1962.

23. Browning, W. E., "Progress in Nuclear Energy," Series IV, Pergamon, New York, 1963.
24. Heckelman, J. D., and Kozar, R. P., in American Institute of Physics, "Temperature: Its Measurement and Control in Science and Industry," vol. 4, pt. 3, p. 1935, Inst. Soc. America, Pittsburgh, 1972.
25. Carpenter, F. D., Sandsfur, N. L., Grenda, R. J., and Steibel, J. S., in American Institute of Physics, "Temperature: Its Measurement and Control in Science and Industry," vol. 4, pt. 3, p. 1927, Inst. Soc. America, Pittsburgh, 1972.
26. Martin, C. D., and Gabbard, C. H., in American Institute of Physics, "Temperature: Its Measurement and Control in Science and Industry," vol. 4, pt. 3, p. 1869, Inst. Soc. America, Pittsburgh, 1972.
27. White, W. P., The Constancy of "Thermoelements," *Phys. Rev.*, vol. 23, p. 449, 1906.
28. White, W. P., The Thermoelement as a Precision Thermometer, *Phys. Rev.*, vol. 31, p. 135, 1910.
29. Fuschillo, N., Inhomogeneity e.m.f.s in Thermoelectric Thermometers, *J. Sci. Instr.*, vol. 31, p. 133, 1954.
30. Hust, J. G., Thermal Anchoring of Wires in Cryogenic Apparatus, *Rev. Sci. Instr.*, vol. 41, p. 622, 1970.
31. Kopp, J., and Slack, G. A., Thermal Contact Problems in Low Temperature Thermocouple Thermometry, *Cryogenics*, vol. 11, p. 22, 1971.
32. Allen, L. D., Bedford, E. W., and Crabtee, R. D., Thermocouple Tempering in Cryogens, *Cry. Indus. Gases*, vol. 5, p. 19, April, 1970.
33. Childs, G. E., Ericks, L. J., and Powell, R. L., Thermal Conductivity of Solids at Room Temperature and Below, *NBS Monograph 131*, National Bureau of Standards, Washington, D.C., 1973.
34. Powell, R. L., "American Institute of Physics Handbook," 2nd Ed., sec. 4g, Am. Inst. of Physics, New York, 1963.
35. For example, the woven four conductor, magnetically shielded cable supplied by Magnetic Shield Div., Perfection Mica Co., Chicago, Ill.
36. Morrison, R., "Grounding and Shielding Techniques in Instrumentation," Wiley, New York, 1967.
37. Roeser, W. F., and Lonberger, S. T., Methods of Testing Thermocouples and Thermocouple Materials, *NBS Circular 590*, National Bureau of Standards, Washington, D.C., 1958.
38. Hust, J. G., Powell, R. L., and Sparks, L. L., in American Institute of Physics, "Temperature: Its Measurement and Control in Science and Industry," vol. 4, pt. 2, p. 1525, Inst. Soc. America, Pittsburgh, 1972.
39. Bedford, R. E., Ma, C. K., Barber, C. R., Chandler, T. R., Quinn, T. J., Burns, G. W., and Scroger, M., in American Institute of Physics, "Temperature: Its Measurement and Control in Science and Industry," vol. 4, pt. 3, p. 1585, Inst. Soc. America, Pittsburgh, 1972.
40. Le Chatelier, H., Sur la Variation Produite par une Élévation de la Température, dans la Force Electromotrice des Couples Thermo-électriques, *Compt. Rend.*, vol. 102, p. 819, 1886.
41. Shenker, H., Lauritzen, J. I., Corrucini, R. J., and Lonberger, S. T., Reference Tables for Thermocouples, *NBS Circular 561*, National Bureau of Standards, Washington, D.C., 1955.
42. Roeser, W. F., and Wensel, H. T., Reference Tables for Platinum to Platinum-Rhodium Thermocouples, *J. Res. Nat. Bur. Stand.*, vol. 10, p. 275, 1933.
43. Barber, C. R., Temperature calibration of platinum, 10% rhodium–platinum and platinum, 13% rhodium–platinum thermocouples over the temperature range 0°-1,760°C, *Proc. Phys. Soc.*, vol. 63B, p. 492, 1950.
44. Bennett, H. E., "Noble Metal Thermocouples," Johnson, Matthew, London, 1958.
45. CIPM–Comite Inst. des Poids et Mesures, International Practical Temperature Scale of 1968, *Metrologia*, vol. 5, p. 35, 1969.

46. McLaren, E. H., and Murdock, E. G., in American Institute of Physics, "Temperature: Its Measurement and Control in Science and Industry," vol. 4, pt. 3, p. 1543, Inst. Soc. America, Pittsburgh, 1972.

47. Cochrane, J., Relationship of Chemical Composition to the Electric Properties of Platinum, *Englhard Tech. Bull.*, vol. 11, p. 58, 1970.

48. Aliotta, J., Effects of Impurities on the Thermoelectric Properties of Platinum, *Instr. Control Sys.*, vol. 45, no. 3, p. 106, 1972.

49. Am. Soc. for Testing and Materials, "Annual Book of Standards," A.S.T.M., Philadelphia, 1974, 1975, etc.

50. Fairchild, C. O., and Schmitt, H. M., Life Tests of Platinum–Platinum-Rhodium Thermocouples, *Chem. Met. Eng.*, vol. 26, p. 158, 1922.

51. Glawe, G. E., in American Institute of Physics, "Temperature: Its Measurement and Control in Science and Industry," vol. 4, pt. 3, p. 1645, Inst. Soc. America, Pittsburgh, 1972.

52. Burns, G. W., and Gallagher, J. S., References Tables for the Pt-30 Percent Rh versus Pt-6 Percent Thermocouple, *J. Res. Nat. Bur. Stand.*, vol. 70C, p. 89, 1966.

53. Obrowski, W., and Prinz, W., Neu bestimmte Grundwerte für die Thermopaarkombination Pt 30% Rh-Pt 6% Rh, *Arch. Eisenhuttenwesen*, vol. 33, p. 1, 1962.

54. Ehringer, H., Über die Lebensdauer von Pt Rh-thermoelementen, *Mettall.*, vol. 8, p. 596, 1954.

55. Walker, B. E., Ewing, C. T., and Miller, R. R., Study of the Instability of Noble Metal Thermocouples in Vacuum, *Rev. Sci. Instr.*, vol. 33, p. 1029, 1962.

56. Walker, B. E., Ewing, C. T., and Miller, R. R., Study of the Instability of Noble Metal Thermocouples in Vacuum, *Rev. Sci. Instr.*, vol. 36, p. 601, 1965.

57. Bedford, R. E., Reference Tables for Platinum 20% Rhodium/Platinum 5% Rhodium Thermocouples, *Rev. Sci. Instr.*, vol. 35, p. 1177, 1964.

58. Bedford, R. E., Reference Tables for Platinum–40% Rhodium/Platinum–20% Rhodium Thermocouples, *Rev. Sci. Instr.*, vol. 36, p. 1571, 1965.

59. Acken, J. S., Some Properties of Platinum–Rhodium Alloys, *J. Res. Nat. Bur. Standards*, vol. 12, p. 249, 1934.

60. Hendricks, J. W., and McElroy, J. L., Report to *ASTM Comm. E-20*, Sub. IV, Sec. 2, June 25, 1964.

61. Potts, J. F., and McElroy, D. L., in American Institute of Physics, "Temperature: Its Measurement and Control in Science and Industry," vol. 3, pt. 2, p. 243, Reinhold, New York, 1962.

62. Burley, N. A., Solute Depletion and Thermo-emf Drift in Nickel-base Thermocouple Alloys, *J. Inst. Metals*, vol. 97, p. 252, 1969.

63. Burley, N. A., in American Institute of Physics, "Temperature: Its Measurement and Control in Science and Industry," vol. 4, p. 3, p. 1677, Inst. Soc. America, Pittsburgh, 1972.

64. Foote, P. D., Fairchild, C. O., and Harrison, T. R., Pyrometric Practice, *Tech. Paper NBS 14*, p. 306, 1920.

65. National Academy of Science/National Research Council, "International Critical Tables," McGraw-Hill, New York, 1926.

66. Lindeck, S., On Wire Standards of Electrical Resistance, *Report Brit. Assoc. Adv. Science*, p. 139, 1892.

67. Roeser, W. F., and Dahl, A. I., Reference Tables for Iron–Constantan and Copper–Constantan Thermocouples, *J. Res. Nat. Bur. Standards*, vol. 20, p. 337, 1938.

68. Corruccini, R. J., and Shenker, H., Modified 1913 Reference Tables for Iron-Constantan Thermocouples, *J. Res. Nat. Bur. Stand.*, vol. 50, p. 229, 1953.

69. Finch, D. I., U.S. Patent 2,325,759, Aug. 3, 1943.

70. Hansen, M., and Andreko, K., "Constitution of Binary Alloys," McGraw-Hill, New York, 1958.

71. Roeser, W. F., Dahl, A. I., and Gwens, G. J., Standard Tables for Chromel-Alumel Thermocouples, *J. Res. Nat. Bur. Standards*, vol. 14, p. 239, 1935.
72. Dahl, A. I., in American Institute of Physics, "Temperature: Its Measurement and Control in Science and Industry," vol. 2, p. 1238, Reinhold, New York, 1941.
73. Burley, N. A., and Acklund, R. G., Long Term Stability of Nickel-Base Thermocouple Alloys, *J. Austral. Inst. Metals*, vol. 12, p. 23, 1967.
74. Wang, T. P., Gottlieb, A. J., and Starr, C. D., The Emf Stability of Type K Thermocouple Alloys, *National Air Transport Meeting, paper 690426*, 1969.
75. Scott, R. B., Calibration of Thermocouples at Low Temperatures, *J. Res. Nat. Bur. Stand.*, vol. 25, p. 459, 1940.
76. Caldwell, F. R., Temperature of Thermocouple Reference Junctions in an Ice Bath, *J. Res. Nat. Bur. Stand.*, vol. 69C, p. 9S, 1965.
77. Burns, G. W., and Hurst, W. S., A Program in Refractory Metal Thermocouple Research, *J. Res. Nat. Bur. Stand.*, vol. 75C, p. 99, 1971.
78. Burns, G. W., and Hurst, W. S., Tungsten-Rhenium Alloy Thermocouple Assemblies, *International College on High-Temperature In-Pile Thermometry, Petten, Holland*, Dec. 13, 1974.
79. Burley, N. A., Burns, G. W., and Powell, R. L., Nicrosil and Nisil: Their Development and Standardization, *European Conference on Temperature Measurement, Teddington, England*, April 9, 1975.

APPENDIX: ABBREVIATED REFERENCE TABLES

See pp. 155–162.

TYPE S

PLATINUM-10 RHODIUM VS. PLATINUM THERMOCOUPLES

REFERENCE JUNCTION AT 0 DEGREES C.

TEMPERATURES IN DEGREES C (IPTS 1968).

THERMOELECTRIC VOLTAGE IN ABSOLUTE MILLIVOLTS

DEG C	0	10	20	30	40	50	60	70	80	90	100	DEG C
- 0	0.000	-0.053	-0.103	-0.150	-0.194	-0.236						- 0
+ 0	0.000	0.055	0.113	0.173	0.235	0.299	0.365	0.432	0.502	0.573	0.645	+ 0
100	0.645	0.719	0.795	0.872	0.950	1.029	1.109	1.190	1.273	1.356	1.440	100
200	1.440	1.525	1.611	1.698	1.785	1.873	1.962	2.051	2.141	2.232	2.323	200
300	2.323	2.414	2.506	2.599	2.692	2.786	2.880	2.974	3.069	3.164	3.260	300
400	3.260	3.356	3.452	3.549	3.645	3.743	3.840	3.938	4.036	4.135	4.234	400
500	4.234	4.333	4.432	4.532	4.632	4.732	4.832	4.933	5.034	5.136	5.237	500
600	5.237	5.339	5.442	5.544	5.648	5.751	5.855	5.960	6.064	6.169	6.274	600
700	6.274	6.380	6.486	6.592	6.699	6.805	6.913	7.020	7.128	7.236	7.345	700
800	7.345	7.454	7.563	7.672	7.782	7.892	8.003	8.114	8.225	8.336	8.448	800
900	8.448	8.560	8.673	8.786	8.899	9.012	9.126	9.240	9.355	9.470	9.585	900
1,000	9.585	9.700	9.816	9.932	10.048	10.165	10.282	10.400	10.517	10.635	10.754	1,000
1,100	10.754	10.872	10.991	11.110	11.229	11.348	11.467	11.587	11.707	11.827	11.947	1,100
1,200	11.947	12.067	12.188	12.308	12.429	12.550	12.671	12.792	12.913	13.034	13.155	1,200
1,300	13.155	13.276	13.397	13.519	13.640	13.761	13.883	14.004	14.125	14.247	14.368	1,300
1,400	14.368	14.489	14.610	14.731	14.852	14.973	15.094	15.215	15.336	15.456	15.576	1,400
1,500	15.576	15.697	15.817	15.937	16.057	16.176	16.296	16.415	16.534	16.653	16.771	1,500
1,600	16.771	16.890	17.008	17.125	17.243	17.360	17.477	17.594	17.711	17.826	17.942	1,600
1,700	17.942	18.056	18.170	18.282	18.394	18.504	18.612					1,700
DEG C	0	10	20	30	40	50	60	70	80	90	100	DEG C

PLATINUM-13 RHODIUM VS. PLATINUM THERMOCOUPLES

REFERENCE JUNCTION AT 0 DEGREES C.

TEMPERATURES IN DEGREES C (IPTS 1968).

THERMOELECTRIC VOLTAGE IN ABSOLUTE MILLIVOLTS

DEG C	0	10	20	30	40	50	60	70	80	90	100	DEG C
- 0	0.000	-0.051	-0.100	-0.145	-0.188	-0.226						- 0
+ 0	0.000	0.054	0.111	0.171	0.232	0.296	0.363	0.431	0.501	0.573	0.647	+ 0
100	0.647	0.723	0.800	0.879	0.959	1.041	1.124	1.208	1.294	1.380	1.468	100
200	1.468	1.557	1.647	1.738	1.830	1.923	2.017	2.111	2.207	2.303	2.400	200
300	2.400	2.498	2.596	2.695	2.795	2.896	2.997	3.099	3.201	3.304	3.407	300
400	3.407	3.511	3.616	3.721	3.826	3.933	4.039	4.146	4.254	4.362	4.471	400
500	4.471	4.580	4.689	4.799	4.910	5.021	5.132	5.244	5.356	5.469	5.582	500
600	5.582	5.696	5.810	5.925	6.040	6.155	6.272	6.388	6.505	6.623	6.741	600
700	6.741	6.860	6.979	7.098	7.218	7.339	7.460	7.582	7.703	7.826	7.949	700
800	7.949	8.072	8.196	8.320	8.445	8.570	8.696	8.822	8.949	9.076	9.203	800
900	9.203	9.331	9.460	9.589	9.718	9.848	9.978	10.109	10.240	10.371	10.503	900
1,000	10.503	10.636	10.768	10.902	11.035	11.170	11.304	11.439	11.574	11.710	11.846	1,000
1,100	11.846	11.983	12.119	12.257	12.394	12.532	12.669	12.808	12.946	13.085	13.224	1,100
1,200	13.224	13.363	13.502	13.642	13.782	13.922	14.062	14.202	14.343	14.483	14.624	1,200
1,300	14.624	14.765	14.906	15.047	15.188	15.329	15.470	15.611	15.752	15.893	16.035	1,300
1,400	16.035	16.176	16.317	16.458	16.599	16.741	16.882	17.022	17.163	17.304	17.445	1,400
1,500	17.445	17.585	17.726	17.866	18.006	18.146	18.286	18.425	18.564	18.703	18.842	1,500
1,600	18.842	18.981	19.119	19.257	19.395	19.533	19.670	19.807	19.944	20.080	20.215	1,600
1,700	20.215	20.350	20.483	20.616	20.748	20.878	21.006					1,700
DEG C	0	10	20	30	40	50	60	70	80	90	100	DEG C

TYPE B

PLATINUM 30 RHODIUM VS. PLATINUM 6 RHODIUM THERMOCOUPLES

TEMPERATURES IN DEGREES C (IPTS 1968).

REFERENCE JUNCTION AT 0 DEGREES C.

THERMOELECTRIC VOLTAGE IN ABSOLUTE MILLIVOLTS

DEG C	0	10	20	30	40	50	60	70	80	90	100	DEG C
0	0.000	-0.002	-0.003	-0.002	-0.000	0.002	0.006	0.011	0.017	0.025	0.033	0
100	0.033	0.043	0.053	0.065	0.078	0.092	0.107	0.123	0.140	0.159	0.178	100
200	0.178	0.199	0.220	0.243	0.266	0.291	0.317	0.344	0.372	0.401	0.431	200
300	0.431	0.462	0.494	0.527	0.561	0.596	0.632	0.669	0.707	0.746	0.786	300
400	0.786	0.827	0.870	0.913	0.957	1.002	1.048	1.095	1.143	1.192	1.241	400
500	1.241	1.292	1.344	1.397	1.450	1.505	1.560	1.617	1.674	1.732	1.791	500
600	1.791	1.851	1.912	1.974	2.036	2.100	2.164	2.230	2.296	2.363	2.430	600
700	2.430	2.499	2.569	2.639	2.710	2.782	2.855	2.928	3.003	3.078	3.154	700
800	3.154	3.231	3.308	3.387	3.466	3.546	3.626	3.708	3.790	3.873	3.957	800
900	3.957	4.041	4.126	4.212	4.298	4.386	4.474	4.562	4.652	4.742	4.833	900
1,000	4.833	4.924	5.016	5.109	5.202	5.297	5.391	5.487	5.583	5.680	5.777	1,000
1,100	5.777	5.875	5.973	6.073	6.172	6.273	6.374	6.475	6.577	6.680	6.783	1,100
1,200	6.783	6.887	6.991	7.096	7.202	7.308	7.414	7.521	7.628	7.736	7.845	1,200
1,300	7.845	7.953	8.063	8.172	8.283	8.393	8.504	8.616	8.727	8.839	8.952	1,300
1,400	8.952	9.065	9.178	9.291	9.405	9.519	9.634	9.748	9.863	9.979	10.094	1,400
1,500	10.094	10.210	10.325	10.441	10.558	10.674	10.790	10.907	11.024	11.141	11.257	1,500
1,600	11.257	11.374	11.491	11.608	11.725	11.842	11.959	12.076	12.193	12.310	12.426	1,600
1,700	12.426	12.543	12.659	12.776	12.892	13.008	13.124	13.239	13.354	13.470	13.585	1,700
1,800	13.585	13.699	13.814									1,800

DEG C	0	10	20	30	40	50	60	70	80	90	100	DEG C

TYPE E

NICKEL-CHROMIUM VS. COPPER-NICKEL THERMOCOUPLES

TEMPERATURES IN DEGREES C (IPTS 1968). REFERENCE JUNCTION AT 0 DEGREES C.

THERMOELECTRIC VOLTAGE IN ABSOLUTE MILLIVOLTS

DEG C	0	10	20	30	40	50	60	70	80	90	100	DEG C
-200	-8.824	-9.063	-9.274	-9.455	-9.604	-9.719	-9.797	-9.835				-200
-100	-5.237	-5.680	-6.107	-6.516	-6.907	-7.279	-7.631	-7.963	-8.273	-8.561	-8.824	-100
- 0	0.000	-0.581	-1.151	-1.709	-2.254	-2.787	-3.306	-3.811	-4.301	-4.777	-5.237	- 0
+ 0	0.000	0.591	1.192	1.801	2.419	3.047	3.683	4.329	4.983	5.646	6.317	+ 0
100	6.317	6.996	7.683	8.377	9.078	9.787	10.501	11.222	11.949	12.681	13.419	100
200	13.419	14.161	14.909	15.661	16.417	17.178	17.942	18.710	19.481	20.256	21.033	200
300	21.033	21.814	22.597	23.383	24.171	24.961	25.754	26.549	27.345	28.143	28.943	300
400	28.943	29.744	30.546	31.350	32.155	32.960	33.767	34.574	35.382	36.190	36.999	400
500	36.999	37.808	38.617	39.426	40.236	41.045	41.853	42.662	43.470	44.278	45.085	500
600	45.085	45.891	46.697	47.502	48.306	49.109	49.911	50.713	51.513	52.312	53.110	600
700	53.110	53.907	54.703	55.498	56.291	57.083	57.873	58.663	59.451	60.237	61.022	700
800	61.022	61.806	62.588	63.368	64.147	64.924	65.700	66.473	67.245	68.015	68.783	800
900	68.783	69.549	70.313	71.075	71.835	72.593	73.350	74.104	74.857	75.608	76.358	900
1,000	76.358											1,000

| DEG C | 0 | 10 | 20 | 30 | 40 | 50 | 60 | 70 | 80 | 90 | 100 | DEG C |

TYPE J

IRON VS. COPPER-NICKEL (SAMA) THERMOCOUPLES

TEMPERATURES IN DEGREES C (IPTS 1968). REFERENCE JUNCTION AT 0 DEGREES C.

THERMOELECTRIC VOLTAGE IN ABSOLUTE MILLIVOLTS

DEG C	0	10	20	30	40	50	60	70	80	90	100	DEG C
-200	-7.890	-8.096										-200
-100	-4.632	-5.036	-5.426	-5.801	-6.159	-6.499	-6.821	-7.122	-7.402	-7.659	-7.890	-100
-0	0.000	-0.501	-0.995	-1.481	-1.960	-2.431	-2.892	-3.344	-3.785	-4.215	-4.632	-0
+0	0.000	0.507	1.019	1.536	2.058	2.585	3.115	3.649	4.186	4.725	5.268	+0
100	5.268	5.812	6.359	6.907	7.457	8.008	8.560	9.113	9.667	10.222	10.777	100
200	10.777	11.332	11.887	12.442	12.998	13.553	14.108	14.663	15.217	15.771	16.325	200
300	16.325	16.879	17.432	17.984	18.537	19.089	19.640	20.192	20.743	21.295	21.846	300
400	21.846	22.397	22.949	23.501	24.054	24.607	25.161	25.716	26.272	26.829	27.388	400
500	27.388	27.949	28.511	29.075	29.642	30.210	30.782	31.356	31.933	32.513	33.096	500
600	33.096	33.683	34.273	34.867	35.464	36.066	36.671	37.280	37.893	38.510	39.130	600
700	39.130	39.754	40.382	41.013	41.647	42.283	42.922					700
DEG C	0	10	20	30	40	50	60	70	80	90	100	DEG C

TYPE J (EXTENDED RANGE)

IRON VS. COPPER-NICKEL (SAMA) THERMOCOUPLES

TEMPERATURES IN DEGREES C (IPTS 1968). REFERENCE JUNCTION AT 0 DEGREES C.

THERMOELECTRIC VOLTAGE IN ABSOLUTE MILLIVOLTS

DEG C	0	10	20	30	40	50	60	70	80	90	100	DEG C
700							42.922	43.563	44.207	44.852	45.498	700
800	45.498	46.144	46.790	47.434	48.076	48.716	49.354	49.989	50.621	51.249	51.875	800
900	51.875	52.496	53.115	53.729	54.341	54.948	55.553	56.155	56.753	57.349	57.942	900
1,000	57.942	58.533	59.121	59.708	60.293	60.876	61.459	62.039	62.619	63.199	63.777	1,000
1,100	63.777	64.355	64.933	65.510	66.087	66.664	67.240	67.815	68.390	68.964	69.536	1,100
1,200	69.536											1,200
DEG C	0	10	20	30	40	50	60	70	80	90	100	DEG C

TYPE K

NICKEL-CHROMIUM ALLOY VS NICKEL-ALUMINUM-SILICON ALLOY THERMOCOUPLES

REFERENCE JUNCTION AT 0 DEGREES C.

TEMPERATURES IN DEGREES C (IPTS 1968).

THERMOELECTRIC VOLTAGE IN ABSOLUTE MILLIVOLTS

DEG C	0	10	20	30	40	50	60	70	80	90	100	DEG C
-200	-5.891	-6.035	-6.158	-6.262	-6.344	-6.404	-6.441	-6.458				-200
-100	-3.553	-3.852	-4.138	-4.410	-4.669	-4.912	-5.141	-5.354	-5.550	-5.730	-5.891	-100
- 0	0.000	-0.392	-0.777	-1.156	-1.527	-1.889	-2.243	-2.586	-2.920	-3.242	-3.553	- 0
+ 0	0.000	0.397	0.798	1.203	1.611	2.022	2.436	2.850	3.266	3.681	4.095	+ 0
100	4.095	4.508	4.919	5.327	5.733	6.137	6.539	6.939	7.338	7.737	8.137	100
200	8.137	8.537	8.938	9.341	9.745	10.151	10.560	10.969	11.381	11.793	12.207	200
300	12.207	12.623	13.039	13.456	13.874	14.292	14.712	15.132	15.552	15.974	16.395	300
400	16.395	16.818	17.241	17.664	18.088	18.513	18.938	19.363	19.788	20.214	20.640	400
500	20.640	21.066	21.493	21.919	22.346	22.772	23.198	23.624	24.050	24.476	24.902	500
600	24.902	25.327	25.751	26.176	26.599	27.022	27.445	27.867	28.288	28.709	29.128	600
700	29.128	29.547	29.965	30.383	30.799	31.214	31.629	32.042	32.455	32.866	33.277	700
800	33.277	33.686	34.095	34.502	34.909	35.314	35.718	36.121	36.524	36.925	37.325	800
900	37.325	37.724	38.122	38.519	38.915	39.310	39.703	40.096	40.488	40.879	41.269	900
1,000	41.269	41.657	42.045	42.432	42.817	43.202	43.585	43.968	44.349	44.729	45.108	1,000
1,100	45.108	45.486	45.863	46.238	46.612	46.985	47.356	47.726	48.095	48.462	48.828	1,100
1,200	48.828	49.192	49.555	49.916	50.276	50.633	50.990	51.344	51.697	52.049	52.398	1,200
1,300	52.398	52.747	53.093	53.439	53.782	54.125	54.466	54.807				1,300
DEG C	0	10	20	30	40	50	60	70	80	90	100	DEG C

TYPE T

COPPER VS. COPPER-NICKEL THERMOCOUPLES

TEMPERATURES IN DEGREES C (IPTS 1968).　　　　REFERENCE JUNCTION AT 0 DEGREES C.

THERMOELECTRIC VOLTAGE IN ABSOLUTE MILLIVOLTS

DEG C	0	10	20	30	40	50	60	70	80	90	100	DEG C
-200	-5.603	-5.753	-5.889	-6.007	-6.105	-6.181	-6.232	-6.258				-200
-100	-3.378	-3.656	-3.923	-4.177	-4.419	-4.648	-4.865	-5.069	-5.261	-5.439	-5.603	-100
- 0	0.000	-0.383	-0.757	-1.121	-1.475	-1.819	-2.152	-2.475	-2.788	-3.089	-3.378	- 0
+ 0	0.000	0.391	0.789	1.196	1.611	2.035	2.467	2.908	3.357	3.813	4.277	+ 0
100	4.277	4.749	5.227	5.712	6.204	6.702	7.207	7.718	8.235	8.757	9.286	100
200	9.286	9.820	10.360	10.905	11.456	12.011	12.572	13.137	13.707	14.281	14.860	200
300	14.860	15.443	16.030	16.621	17.217	17.816	18.420	19.027	19.638	20.252	20.869	300
400	20.869											400
DEG C	0	10	20	30	40	50	60	70	80	90	100	DEG C

4 Temperature measurement in cryogenics

JOHN A. CLARK
University of Michigan, Ann Arbor

The word *cryogen* is derived from the two Greek words *kyros-* and *-gen,* meaning literally "the production of icy-cold." More simply, a cryogen is a refrigerant. Cryogenic is the adjective form of the noun, and signifies physical phenomena below −150°C (123K). This is the approximate temperature at which physical properties of many substances begin to show significant variation with temperature [1]. Of natural importance to research, engineering design, and operation at the low temperatures is the measurement of the temperature itself. The purpose of this chapter is to discuss this question in a reasonably broad context within the framework of the 1968 state of the art.

The selection of a suitable temperature sensing element depends on a number of important considerations. Perhaps most fundamental of all is the accuracy required in the measurement. Entirely different techniques will be employed, for example, if an accuracy of 0.001K is necessary or if 1.0K is sufficient. Other significant factors include the influence of transient effects, sensitivity, type of readout, nature of signal, availability or desirability of recording and control, durability, stability, and ruggedness of the sensing element, and, of course, replacement, interchangeability and cost.

Until 1968 no internationally accepted standard for temperature measurement existed below the defining fixed point temperature for oxygen (90.18K, −182.97°C). In 1960, the 11th General Conference on Weights and Measures dropped the notation "fundamental" and "primary" fixed points and adopted instead the terms defining fixed points and secondary reference points for the various two-phase reference states for temperature calibration [2, 3]. The adoption of a uniform scale extended the range of the present international scale to 13.8K [4]. Below this temperature no international scale will exist for some time although convenient and practical methods for measurement of temperatures to 0.2K and lower have been developed. These methods will be included in this discussion. The 1968 International Temperature Scale is included as an Appendix to this book.

This chapter will cover the following topics: The concept of temperature, the absolute (thermodynamic) temperature scale, the gas thermometer, the international temperature scales, the international practical temperature scales, temperature scales below 90K, thermocouples, resistance thermometry, and magnetic thermometry or adiabatic demagnetization. The basic principles as well as practical considerations of measurement will be presented.

The principal sources of reference for cryogenic temperature measurement are listed. Most of these will be cited from time to time in the body of this chapter and hence are also included among the references. However, in view of their value it is important to have them conveniently listed.

American Institute of Physics, Temperature, Its Measurement and Control in Science and Industry, Reinhold.

> Vol. I, 1941, 1343 pages.
> Vol. II, Hugh C. Wolfe (ed.), 1955, 451 pages.
> Vol. III, Charles M. Hertzfeld, (editor-in-chief), 1962.
> Part I, "Basic Concepts, Standards and Methods," F. G. Brickwedde (ed.), 1962, 838 pages.
> Part II, "Applied Methods and Instruments," A. I. Dahl (ed.), 1962, 1087 pages.
> Part III, "Biology and Medicine," J. D. Hardy (ed.).

R. B. Scott, Cryogenic Engineering, Van Nostrand, 1959.

R. W. Vance and W. M. Duke (eds.), Applied Cryogenic Engineering, John Wiley, 1962.

R. W. Vance (ed.), Cryogenic Technology, John Wiley, 1963.

Dirk De Klerk, "Adiabatic Demagnetization," S. Flügge (ed.), Encyclopedia of Physics, Vol. XV, Low Temperature Physics 2.

Journal of Research NBS, Section A (Physics and Chemistry); Section C, (Engineering and Instrumentation).

Advances in Cryogenic Engineering, K. D. Timmerhaus (ed.), Vol. 1-20, 1955-75 (to date).

Cryogenics, Vol. 1-15, 1960-75 (to date).

Metrologia, Vol. 1-11, 1964-75 (to date). Published under the auspices of the International Committee of Weights and Measures.

The International Temperature Scale (ITS)

> 1927 ITS
> G. F. Burgess, "International Temperature Scale," *J. Research NBS,* Vol. 1, 1925, pp. 635-637.
> See also, Temperature, Vol. 1, 1941, pp. 21-23.
> 1948 ITS
> J. A. Hall, Temperature, Vol. 1, 1955, pp. 115-141.

H. F. Stimson, "The International Temperature Scale of 1948," *J. Research NBS*, Vol. 42, 1949, p. 209.
1960 text revisions of the 1948 ITS
H. F. Stimson, "International Practical Temperature Scale of 1948. Text Revision of 1960," *J. Research NBS*, Vol. 65A, No. 3, 1961. See also, H. F. Stimson, "The Text Revision of the International Temperature Scale of 1948, Temperature, Vol. 3, Part I, 1962, pp. 59–67.
1968 IPTS (see Appendix to this book).

4.1 THE CONCEPT OF TEMPERATURE

The concept of temperature is old and doubtless stems from the desire to attach numerical quantities to a feeling of hotness or coldness. Galileo (1600) was one of the earliest to experiment with the design of an instrument to which a scale was attached for the purpose of indicating a numerical temperature. These early instruments were called "thermoscopes" and were said to have measured "degrees of heat." Today, 375 years later, we find the art of temperature measurement highly developed but under continued study. The basic standard instrument presently employed is the gas thermometer, in some respects similar to the first of Galileo, but registering in degrees of absolute temperature, not in degrees of heat. This distinction was made only after the discovery of latent heats by Joseph Black and James Watt (1764) and the enunciation of the Second Law of Thermodynamics (about 1850).

A sensation of warmth or cold is of little value to the physical world when measurement and reproducibility of temperature are required. While it is of small value in measurement, the subjective sense might be used to indicate an equivalent "hotness" or "coldness" of two separate bodies. It is common experience, for example, that two blocks of iron, one taken from an ice bath and the other from a furnace, will approach the same feeling of warmth if they are brought into thermal contact with one another. Of course, great doubt might reasonably be raised concerning the validity of the conclusion of equal hotness if the indications were taken by touching the blocks with the skin of the hands. However, other schemes could be used which are less subjective and would produce the same result. One might not use the hands but use, for instance, a small rod of silver and place it in intimate thermal contact with each iron block. The increase or decrease in the length of the silver rod could then serve as an indicating device since by experience it is known that this dimension will change as the rod is heated or cooled; also, when the heating or cooling ceases, the changes in length also cease. Hence, after some period of time following the bringing together of the two blocks very careful observations of the length (say with a powerful microscope) would show no subsequent change in length of the rod when it was placed successively in contact with each block. Furthermore, if the block and the silver rod were mutually in thermal contact with each other and with nothing else, the length of the rod would be the same when it was attached to each block.

This would define the measurable state known as "the equality of temperature." The silver rod might also be called a temperature meter or, more simply, a thermometer. The process just described has led to a generalization, or law called the Zeroth Law of Thermodynamics. This law, which is the logical basis of all temperature measurement, may be stated as follows: Two bodies (the iron blocks) at the state of equality of temperature with a third body (the silver rod) are in a state of equality of temperature with each other.

Returning to the silver rod that has been called a thermometer, one might be led to attempt to assign a sort of numerical scale to its length so as to convert its elongation or contraction into some definite, reproducible scale of temperature. It is apparent that this could be done with no particular mechanical difficulty, although some amplification of the changes in the length might be necessary for convenience in use. The selection of the type and magnitude of the units on this scale is wholly arbitrary. It is known that when the rod is placed into a bath of ice and water, or in a bath of saturated steam at constant pressure and allowed to reach a state of equality of temperature with the bath in each instance, it does not change in length, but has a greater length in the steam than in the ice. Because the length is greatest in the steam bath, it would seem reasonable to assign it the greatest level of temperature, although this is arbitrary and the reverse has been done, as in the Celsius scale (1740). The temperature of the ice bath, for convenience, then could be established as zero while that of the steam bath could be taken as 100. And this would define a difference of 100 degrees of temperature between these two fixed points.

The final decision to be made in the construction of a temperature scale is the selection of an interpolation device to be employed to obtain the level of temperature and the magnitude of the unit degree of temperature in this interval from observations of the length of the rod. Freedom of choice is to be had in this selection also, that is, any curve connecting the 0 and 100-degree points in Fig. 4.1 may be chosen as the interpolation device. A scale of temperature is established both by the curves and the scale of the ordinate, as is evident, but freedom of choice is preserved if one fixes the scale of the ordinate and allows the curves to have an arbitrary shape. Some curves obviously are more convenient than others, for example, one would find a multivalued curve quite inconvenient to use.

Three curves or interpolating devices are shown in Fig. 4.1. The ordinate is divided arbitrarily into 100 equal divisions between 0 and 100. The abscissa is the ratio Δl, the difference between the length of the rod at intermediate levels of temperature and at the ice bath temperature to Δl_{100}, the difference between its length in the steam bath and the ice bath. Clearly, there are an infinite number of different scales one could select. One could take curve B which is a straight line joining the fixed points of temperature and have a linear scale; if the silver rod were used as the thermometer, this could be called the "linear silver

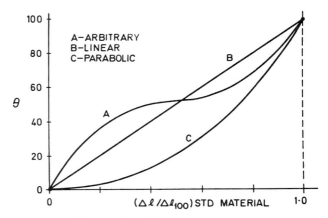

FIG. 4.1 Arbitrary scales of temperature.

scale of temperature." By denoting this as the θ scale the relationship between length changes and θ is the simple one

$$\frac{\Delta l}{\Delta l_{100}} = \frac{\theta}{100} \tag{4.1}$$

A question might be raised now: Is this θ scale fundamental, or if only the material in the rod is changed, will a different scale result? Indeed, it turns out that this scale is not fundamental, and different temperatures would be obtained with, say, a linear copper scale. This important consequence of physics may be demonstrated as follows.

Consider that a linear silver scale has been adopted for reference, with the silver rod being selected as the primary standard thermometer. All temperatures will be referred to this scale which we have called θ. We may take any property of matter which is measurable and which changes with heating and cooling as the indicating quantity. Changes in such properties may be called $\Delta p = p - p_0$ where p_0 is the magnitude of the property p at the level of temperature of the ice bath. In a rod of any other material the temperature coefficient of p based on the linear silver scale of temperature is defined as

$$\alpha = \frac{\partial p}{\partial \theta} \tag{4.2}$$

Hence

$$\Delta p = \int_0^\theta \alpha \, d\theta \tag{4.3}$$

also

$$\Delta p_{100} = \int_0^{100} \alpha \, d\theta \tag{4.4}$$

so

$$\frac{\Delta p}{\Delta p_{100}} = \frac{\displaystyle\int_0^{\theta} \alpha \, d\theta}{\displaystyle\int_0^{100} \alpha \, d\theta} \tag{4.5}$$

Each of the integrals in Eqs. (4.3) and (4.4) may be expressed as follows:

$$\int_0^{\theta} \alpha \, d\theta = \bar{\alpha}_\theta \cdot \theta \quad \text{and} \quad \int_0^{100} \alpha \, d\theta = \bar{\alpha}_{100} \cdot 100$$

where $\bar{\alpha}$ is the average value of α in a range of temperature.

Equation (4.5) is then written

$$\frac{\Delta p}{\Delta p_{100}} = \frac{\bar{\alpha}_\theta}{\alpha_{100}} \frac{\theta}{100} \tag{4.6}$$

It is a matter of experience that any property which changes with heating or cooling (such as length) does not do so at a constant rate of change in terms of the temperature of the scale of temperature employed to measure it. The exception, of course, is the property of the material used to define the linear scale of a selected standard temperature scale. But this is trivial.

A linear scale based on $\Delta p/\Delta p_{100}$ of any other material would be written

$$\frac{\Delta p}{\Delta p_{100}} = \frac{t}{100} \tag{4.7}$$

where t is the temperature on the linear scale pertaining to this other material.

Combining Eqs. (4.6) and (4.7) we have

$$\frac{\bar{\alpha}_\theta}{\bar{\alpha}_{100}} \frac{\theta}{100} = \frac{t}{100} \quad \text{or} \quad \frac{\theta}{t} = \frac{\bar{\alpha}_{100}}{\bar{\alpha}_\theta} \tag{4.8}$$

Equation (4.8) is the expression relating the linear scales of temperature (θ and t) of two different materials.

We find, therefore, that except in the rather improbable instance of $\bar{\alpha}_{100}/\bar{\alpha}_\theta$ being universally unity for all materials, a fundamental difference must be expected between scales of temperature defined in the manner outlined here. It is only for the case of a constant value of the temperature coefficient of change of a temperature dependent property that would produce a value of $\bar{\alpha}_{100}/\bar{\alpha}_\theta$ of unity and exact agreement between all linear scales of temperature. This circumstance cannot reasonably be expected in nature.

We conclude from these arguments that thermometers constructed after the fashion described, while useful as operational tools to measure and reproduce levels of temperature, would each produce different values of temperature when used to measure the state of a given system. This was roughly the state of thermometry for temperatures below 90K until the adoption of the IPTS of 1968. In this range there existed no accepted standard thermometer nor scale although many working groups had defined their own wire scales [5]. However, these scales all differed from each other. This is demonstrated by Hust [6] in Fig. 4.2, where temperature scales from several different laboratories in the United States, Canada,

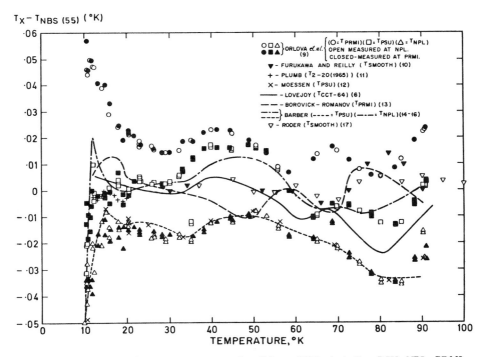

FIG. 4.2 Comparison of low-temperature scales (10 to 90K), including PSU, NPL, PRMI, CCT-64, and NBS (scales 2–20) 1965, with respect to NBS-55 scale. PSU = Pennsylvania State University; NPL = National Physical Laboratory (England); PRMI = USSR; CCT = Advisory Committee on Thermometry, International Committee on Weights and Measures; NBS = National Bureau of Standards (USA).

and Europe are compared with a scale created by the National Bureau of Standards (NBS-55) in the range 10K to 90K. This scale is formed by lowering all temperatures on a previous scale (NBS-39) [7, 8] by 0.01K.

4.2 THE ABSOLUTE (THERMODYNAMIC) TEMPERATURE SCALE

In 1848 Kelvin extended the reasoning of Carnot and demonstrated the existence of a scale of temperature which, unlike those shown in Fig. 4.1, would be completely independent of a thermometric substance. Such a scale is called an "absolute" scale and being deduced from the laws of thermodynamiccs only, it is known as the absolute thermodynamic temperature scale. Actually, there are an infinite number of such scales of temperature possible, the final one selected being a matter of convenience. The scale we employ today is not the first scale Kelvin proposed, but rather his second. His first scale had $+\infty$ and $-\infty$ as the upper and lower bounds of temperature, which is inconvenient, but his second scale remedied this by having bounds at 0 and $+\infty$.

From the Second Law of Thermodynamics, Kelvin was able to show that the rate of the heat quantities from a cyclically operating reversible heat engine could be written

$$\frac{Q_1}{Q_0} = \frac{f(t_1)}{f(t_0)} \tag{4.9}$$

where $f(t)$ denotes an unknown but arbitrary function of temperature alone. The form of this function is specified by the scale of temperature chosen and can, therefore, have an infinite number of possible forms. When a specific form is chosen, a scale of temperature is then defined that is independent in its definition of any thermometric substance. The reversible engine becomes the thermometer, but the nature of such a device, as well as the ratio of its heat quantities, is quite independent of the fluid—the thermometric substance—employed to operate the engine. Kelvin selected as his second scale the simple function $f(t) = T$ where T is called the absolute thermodynamic temperature and is given in degrees Kelvin. Hence Eq. (4.9) becomes

$$\frac{Q_1}{Q_0} = \frac{T_1}{T_0} \tag{4.10}$$

where T_1 and T_0 are the Kelvin temperatures of the heat source and the heat sink, respectively.

The complete definition of the Kelvin scale including the establishment of the size of the degree is dependent on a single arbitrary constant. Originally this was accomplished by defining the difference between the steam and ice points as 100K, exactly. The absolute temperature of the ice point was then determined experimentally using a gas thermometer. Since 1954, however, the arbitrary constant selected for defining the scale has been the temperature of the triple point of water, taken to be 273.16K exactly [2, 3]. In 1854, 100 years earlier, Kelvin had stated that the triple point of water must be adopted ultimately as a defining fixed point [3]. The triple point has been adopted to replace the ice point owing to a greater reliability in establishing its temperature experimentally and the fact that the triple-point temperature is not pressure sensitive as are two-phase states. The size of the degree in the interval between the ice and steam points now must be determined experimentally. This interval will certainly not be exactly 100K but will probably be within 0.001K or less of this value.

The dependence of the Kelvin scale on a single arbitrary constant may be shown by a simple argument. For fixed thermal states at say, T_1 (steam point) and T_0 (ice point) the ratio of the heat quantities in Eq. (4.10) is established by the constraints of nature. Thus

$$\frac{Q_1}{Q_0} = \Gamma = \frac{T_1}{T_0} = \text{constant} \tag{4.11}$$

If we define ΔT_{01} as $T_1 - T_0$, then the temperature of the ice point T_0 may be expressed in terms of ΔT_{01} as

$$T_0 = \frac{\Delta T_{01}}{\Gamma - 1} \tag{4.12}$$

Hence, since Γ is fixed by the thermal states (T_0, T_1, whose numerical temperatures may be yet unknown), T_0 is determined by the value of ΔT_{01}, or ΔT_{01} is determined by the value of T_0. One constant only, either T_0 or ΔT_{01}, is sufficient to define the scale. Prior to 1954, ΔT_{01} was chosen as 100K and T_0 determined by the gas thermometer, but since 1954 the reverse procedure has been adopted and T_0 is defined as the triple point of water at 273.16K. (It is defined by the international practical temperature scale as $+ 0.01°C$ which gives the ice point as absolute temperature of 273.15K [3].) Now, the thermodynamic temperatures of all defining fixed points and secondary reference points must be determined by the gas thermometer.

4.3 THE ABSOLUTE (GAS) TEMPERATURE SCALE

The practical use of the Kelvin scale would require the operation of reversible engines. This, of course, is quite impossible and it is necessary to find an approximation to the absolute thermodynamic scale. This is found from the properties of an ideal gas for if such a fluid is employed in a reversible engine it may be shown that

$$\frac{Q_1}{Q_0} = \frac{\theta_1}{\theta_0} \tag{4.13}$$

where an ideal gas is defined as a substance having an internal energy U, a function only of temperature and an equation of state written as $pv = R\theta$, and θ is a temperature on a new scale called the absolute (gas) temperature scale which would be determined from pressure and volume measurements on the gas, as

$$\theta = \frac{pv}{R} \tag{4.14}$$

Further, if the difference $\theta_1 - \theta_0$ or θ_0 is defined as identically the same as $T_1 - T_0$ or T_0, respectively, then it follows from Eqs. (4.10) and (4.11) that

$$\theta_1 = T_1$$
$$\theta_0 = T_0 \tag{4.15}$$
or $$\theta = T$$

The absolute (gas) temperature scale is identical with the absolute (thermodynamic) temperature scale if an ideal gas is employed in a reversible engine or any other device permitting the measurement of p and v. Such a device (which now replaces the reversible engine) is known as a gas thermometer and is the primary standard thermometer used to determine and establish the absolute thermodynamic temperature scale. Beattie [18] and Barber [19] describe the gas thermometer and the techniques of its use. Gas thermometers are either of constant pressure or constant volume and although both types give essentially identical results independent of the type of gas used, the constant volume type appears to be the more widely used. The MIT thermometer used by Beattie [18, 20] and by Beattie, Benedict, and Kaye [21] is of the constant volume type and uses nitrogen gas in a very pure state as the thermometric substance. Gas thermometry also has used H_2, He, Ne, A and air as thermometric substances.

It was found by Eq. (4.15) that the requirement of a reversible engine as a thermometer was unnecessary to realize the Kelvin scale, if one could substitute a gas thermometer using an ideal gas. However, owing to the departure of the properties of a real gas from ideal conditions, it becomes necessary to empirically correct the readings taken from a gas thermometer in order to determine the absolute temperature. Such temperatures are not acutally Kelvin temperatures, but because of the necessary corrections must be looked upon as approximations. They are called absolute (gas) temperatures and represent the closest approximations to the absolute scale of Kelvin. The nature of these corrections is well established and is outlined by Keyes [22]. However, it will be appropriate to indicate the nature of the result for a constant-volume gas thermometer. From thermodynamic theory and the properties of real gases it may be shown that for $V^* = V$

$$\frac{p - p^*}{(T - T^*)p^*} = \frac{1}{T^*} + \frac{T}{(T - T^*)p^*} \int_{T^*}^{T} C_v \left(\frac{\partial T}{\partial v}\right)_u d\, \frac{1}{T} \qquad (4.16)$$

The integral is the gas scale correction and is evaluated from measurements of the properties of real gases. For example, having once obtained the absolute temperature T^* of the triple point of water by definition at 273.16K, then Eq. (4.16) can be used to obtain the absolute temperature T on the gas scale of any other thermodynamic state from gas thermometer measurements of p and p^* and the real properties of the gas. A simple constant-volume gas thermometer for cryogenic application is illustrated by Barber [19] and shown in Fig. 4.3. This gas thermometer is used for the calibration of platinum resistance thermometers in the liquid hydrogen temperature range. A typical gas bulb with platinum thermometer receptacles and auxiliary apparatus is shown in detail in Fig. 4.4 taken from Moessen, Astonand, and Ascah [12]. Figure 4.5 shows temperature scale corrections for the nonideality of the helium gas scale and the absolute temperature scale. This indicates that gas thermometer temperatures from various laboratories may differ from each other in the cryogenic temperature range by as much as 0.03K. These differences are attributed to the various methods used to account for gas imperfectibility and do not include the influence of dead space volume, gravity, and other deviations from ideal measuring conditions.

A negative absolute temperature may seem to be a suitable subject for a discussion of cryogenic temperatures. The concept of a negative absolute temperature is presented by Ramsey [24] and Hertzfeld [25]. Such a condition may be described whenever the population of an energy state of higher energy is greater than the population of a lower energy state [25] under conditions which permit a statistical mechanical interpretation of thermodynamics. Hertzfeld [25] lists these conditions as: (1) The system must be fairly well isolated from its surroundings, but must come to internal equilibrium rapidly; (2) the states of the system must be quantized; and (3) the system must have a highest state in the same sense that a system has a lowest state. Examples of such systems are the nuclear magnetic moments of the constituents of certain crystals. No violation of the principles of thermodynamics is envisioned in this phenomenon. Interestingly, states at negative absolute temperatures are "hotter" than those at infinite temperature. This results from the fact that at T equal to ∞ all states are equally populated. However, if higher energy states have larger populations than lower energy states the system must be "hotter" than those states having a temperature of infinity. This naturally removes the discussion from the cryogenic range!

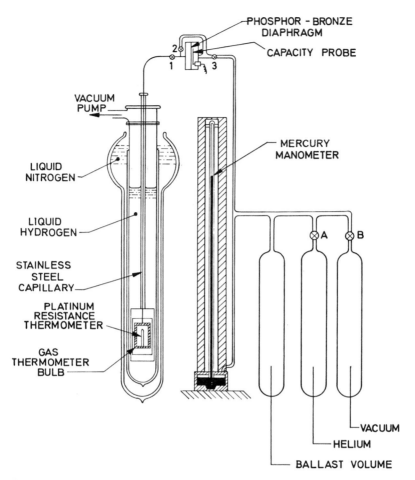

FIG. 4.3 Constant volume thermometer used for calibration of a platinum resistance thermometer. (From Barber, 1962, Ref. [19].)

4.4 THE INTERNATIONAL TEMPERATURE
SCALE (ITS)

Owing to its size and complexity, the gas thermometer is impractical to use for laboratory or industrial measurement of temperature, as most gas thermometers occupy a room-sized space and require elaborate and time-consuming preparations for their use. For these reasons a simple, reproducible, and convenient secondary temperature standard is required even for precise laboratory measurements. The gas thermometer remains, of course, the primary standard and is employed to determine the gas scale absolute temperature of the various defining

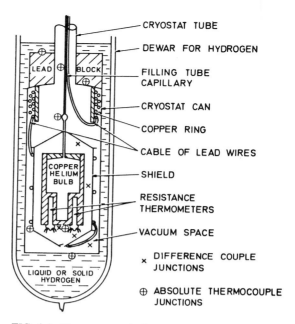

FIG. 4.4 Thermometer bulb, shields, and cryostat.

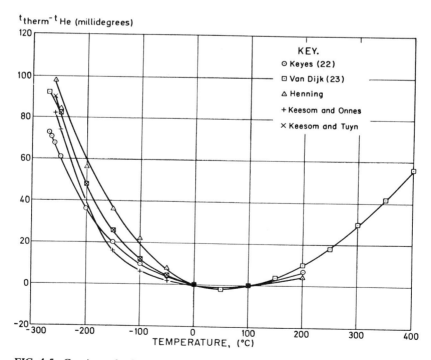

FIG. 4.5 Gas imperfection corrections for constant volume helium gas thermometer.

fixed points and to calibrate precision laboratory secondary standard thermometers.

The definition of the secondary standard thermometers is established by international agreement by the General Conference on Weights and Measures, a body consisting of scientific representatives from 36 nations which meets every 6 years. This parent body is assisted by a smaller executive group known as the International Committee on Weights and Measures which consists of 18 elected members from the various nations and normally meets every 2 years. It is this group that supervises the International Bureau of Weights and Measures at Sévres, near Paris. This committee also oversees the publication of *Metrologia*, a journal devoted to original papers on research "directed towards the significant improvement of fundamental measurements in any field of physics." An Advisory Committee on Thermometry (Comité Consultatif de Thermométrie), the CCT, actively assists the International Committee on Weights and Measures in all matters relating to temperature measurement.

International temperature scales (ITS) have been adopted in 1927 [26] and in 1948 [27,28]. In 1960 a textual revision of the 1948 scale was made [2,3] but this revision did not significantly affect the numerical values on the scale. The principal changes were to:

Replace the ice point with the triple point of water (0.01°C, exactly),

Use the zinc point (pressure insensitive) instead of the sulfur point (pressure sensitive),

Change the name to the international practical temperature scale (IPTS).

The ITS (1927 and 1948) is a Celsius (°C, 1948) scale and no mention was made in the text of an international practical Kelvin temperature scale (IPKTS). However, in practice this conversion was made on the basis of the accepted value of T_0, the Kelvin scale temperature of the ice point. The 1960 revision of the ITS (1948) specifically mentions the IPKTS and relates it to the IPTS by

$$T\text{K(IPKTS)} - t°\text{C(IPTS)} + T_0 \qquad (4.17)$$

where T_0 is now defined to be 273.15K. This was done so that the IPKTS will have the same value (273.16K) for the triple point of water as the Kelvin absolute temperature scale. The IPTS and the Celsius scale both are defined to have a value of 0.01°C as the triple point of water. Stimson [2] observes that for precision no greater than 0.001°C, the zero on the 1948 Celsius scale (ITS 1948) may be realized with an ice bath as described in the 1948 ITS.

The international practical temperature scale (IPTS) specified four things: (a) the gas scale temperatures of reproducible defining fixed points and the secondary reference points at which instruments are calibrated, (b) the types of instruments to be used in realizing the scale, (c) the equations to be used for interpolating or extrapolating from the fixed points, and (d) the experimental procedures recommended for both measurement and calibration. A summary of the IPTS and a comparison of the 1927, 1948, and 1960 revision of the 1948 scale are given in Tables 4.1, 4.2, and 4.3, taken from Hust [6]. The complete range of temperatures from −182.97°C to 1063°C is included for completeness.

It is difficult to determine exact differences of the ITS and IPTS in the radiation law range because of the variability of λ; the wavelength of the radiation on the 1927 scale is restricted only to the visible spectrum and is not restricted at all on the 1948 scale. Table 4.4 contains the differences calculated at $\lambda_1 = 0.4738 \times 10^{-4}$ cm and $\lambda_2 = 0.65 \times 10^{-4}$ cm according to Corruccini [29].

Twenty-two secondary reference points from −78.5°C to 3380°C including their vapor temperature-pressure relations are given by Stimson [3]. The relationship between the international practical temperature scale (1948) and the thermodynamic scale is shown in Table 4.5 [2].

Because of the polynomial form of the interpolating equations used to describe the international practical temperature scale (IPTS) between the defining fixed points, an inherent difference exists between the IPTS and the thermodynamic temperature scale (TTS). This difference is determined by comparing gas thermometer readings with those from standard thermometers as prescribed by

TABLE 4.1   Defining fixed points for the international temperature scales

Description	ITS Int 1927 (°C)	ITS Int 1948 (°C)	IPTS Int 1948 1960 text rev. (°C)
Oxygen point:[a] equilibrium between liquid and gaseous oxygen	− 182.97	− 182.97	− 182.97
Ice point:[a] equilibrium between ice- and air-saturated liquid water	0.000	0.000	
Triple point of water: equilibrium between ice liquid water and gaseous water			0.010
Steam point:[a] equilibrium between liquid and gaseous water	100.00	100.000	100.000
Sulfur point:[a] equilibrium between liquid and gaseous sulfur	444.60	444.60	(444.60)
Zinc point:[a] equilibrium between solid and liquid zinc (recommended to replace the sulfur point)			419.505
Silver point:[a] equilibrium between solid and liquid silver	960.5	960.8	960.8
Gold point:[a] equilibrium between solid and liquid gold	1063.0	1063.0	1063.0

Source.   Reference [6].
[a]At 1 atm.

TABLE 4.2 Official interpolation procedures for the 1927 and 1948 international temperature scales

Item	1927 scale	1948 scale
Temperature range	−190 to 0°C	−182.97 to 0°C
Defined by	Platinum resistance thermometer	Platinum resistance thermometer
Quartic interpolation formula	$R_t = R_0 [1 + At + Bt^2 + C(t - 100)t^3]$	$R_t = R_0 [1 + At + Bt^2 + C(t - 100)t^3]$
Calibrate at	O_2, ice, steam, S points	O_2, ice, steam, S points
Temperature range	0 to 660°C	0 to 630.5°C
Quartic interpolation formula	$R_t R_0 (1 + At + Bt^2)$ $R_{100}/R_0 \geq 1.390;\ R_{444.6}/R_0 \geq 2.645$	$R_t = R_0 (1 + At + Bt^2)$ $R_{100}/R_0 > 1.3920$
Calibrate at	ice, steam, S points	ice, steam, S points
Temperature range	660 to 1063°C	630.5 to 1063°C
Defined by	Pt-Pt 10% Rh thermocouple	Pt-Pt 10% Rh thermocouple
Quartic interpolation formula	$e = a + bt + ct^2$	$e = a + bt + ct^2$
Calibrate at	Sb, Ag, Au points	Sb, Ag, Au points
Temperature range	Above 1063°C	Above 1063°C
Defined by	Monochromatic optical pyrometer	Monochromatic optical pyrometer
Quartic interpolation formula	$\log \dfrac{J_t}{J_{Au}} = \dfrac{c_2}{\lambda}\left(\dfrac{1}{1336} - \dfrac{1}{t+T_0}\right)$	$\log \dfrac{J_t}{J_{Au}} = \dfrac{\exp\left[\dfrac{c_2}{tAu + T_0}\right] - 1}{\dfrac{c_2}{t + T_0} - 1}$
	$c_2 = 1.432$ cm deg	$c_2 = 1.438$ cm deg

Source. Reference [6].

Note. t, in °C; T, in °K; T_0 is temperature of ice point, $T_0 = 273.15$°K.

TABLE 4.3 Differences between ITS (1927) and
IPTS (1948) in the platinum thermometer range

$t°C$ (Int) 1948	$\Delta t°C$ (Int) 1948 − °C (Int) 1927
630.5	0.00[a]
650	0.08[a]
700	0.24
750	0.35
800	0.42
850	0.43
900	0.40
950	0.32
1000	0.20
1050	0.05
1063	0.00

Source. Reference [6].

[a]These values are uncertain since platinum thermometers are defined only up to 630.5°C on 1948
scale (see Corruccini [29]).

the IPTS. Hust [6] reports the data of several investigators who examined the
differences between the TTS and the IPTS in the range −190°C to 0°C. These
results are shown in Fig. 4.6. As may be noted, fairly large discrepancies exist
between these various results probably because of differences in the gas thermometer measurements as suggested by the data in Fig. 4.5. In order to examine
and define these differences systematically, Preston-Thomas and Kirby [35]
redetermined part of the TTS, in terms of platinum resistance thermometer
readings, by means of a constant volume helium gas thermometer of reasonably
high accuracy. These authors expect to extend similar measurement to −219°C,
the triple point of oxygen. Their measurements in the range −183°C to 100°C
are given in Fig. 4.7. As may be seen by the data in Figs. 4.6 and 4.7, the IPTS
and the TTS differ by a maximum of about 0.04°C in the cryogenic range.

TABLE 4.4 Differences between the ITS and IPTS in
radiation law range

$t°C$ (Int) 1948	$\Delta t°C$ (Int) 1948 λ_1	°C (Int) 1927 λ_2
1063	0	0
1500	− 2	− 2
2000	− 6	− 6
2500	− 12	− 12
3000	− 19	− 20
3500	− 28	− 30
4000	− 38	− 43

TABLE 4.5 Relation between International Practical Temperature Scale and Thermodynamic Scale (1960)

	International Practical Temperature Scale		Thermodynamic Scale	
Item	Celsius	Absolute	Celsius	Absolute
Name	International practical temperature	International practical Kelvin temperature	Thermodynamic celsius temperature	Thermodynamic Kelvin temperature
Symbol	t_{int}	$T_{int} = t_{int} + T_0$ [a]	$t = T - T_0$	T
Designation	°C (Int 1948) Degrees Celsius International Practical 1948	°K (Int 1948) Degrees Kelvin International Practical 1948	°C (therm.) Degrees Celsius Thermodynamic	°K Degrees Kelvin

Note. For the international practical temperature the subscript "int" after t and T may be omitted if there is no possibility of confusion.

[a] $T_0 = 172.15°$.

4.5 THE INTERNATIONAL PRACTICAL TEMPERATURE SCALE OF 1968

For detailed information concerning the International Practical Temperature Scale of 1968, it will be necessary to refer to the Appendix appearing at the end of this book. However, it might be of value here to indicate the nature of that scale in the range of temperatures considered in this chapter, that is, the cryogenic range.

The standard instrument used from 13.81K to 630.74°C is the strain-free

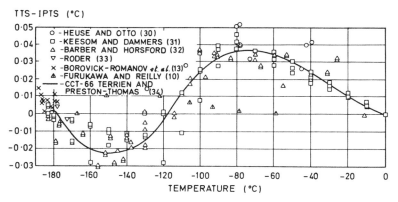

FIG. 4.6 Temperature differences between thermodynamic and international temperature scales.

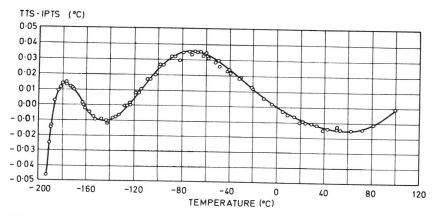

FIG. 4.7 Comparison of measurements of TTS-IPTS. (From Preston-Thomas and Kirby, 1968, Ref. [35].)

platinum resistance thermometer made from annealed pure platinum. Interpolation between the fixed point temperatures is provided by formula used to describe the relation between indications of the standard instruments and values of the International Practical Temperature. From 13.81K to 273.15K the temperature T_{68} (International Practical Temperature of 1968) is defined by the relation

$$W(T_{68}) = W_{CCT-68}(T_{68}) + \Delta W(T_{68})$$

where $W(T_{68})$ is the resistance ratio of the platinum resistance thermometer and $W_{CCT-68}(T_{68})$ is a resistance ratio as given by a set of tabular values provided in the above reference. The last term in the above relationship is a deviation in the resistance ratio and is determined at the temperatures of the defining fixed points from measured values of $W(T_{68})$ and from tabular values which are provided. At intermediate temperatures, interpolation formula are provided to determine values of the deviations. These deviations are determined by separate polynomial functions in the following ranges of temperature, 13.81K to 20.28K, 20.28K to 54.361K, 54.361K to 90.188K and 90.188K to 273.15K.

A detailed study of the relationships between the IPTS-68 and the NBS-55, NPL-61, PRMI-54 and PSU-54 temperatures scales in the range 13.81K to 90.188K is provided in the Appendix at the end of the book. All scale differences are less than 40×10^{-3}K in this range of temperature.

The important range below 13.8K has not been defined by an international standard and since this includes the entire region of the He^4 and He^3, it is hoped that similar efforts by the CCT will bring about a standard scale at these very low temperatures.

4.6 EARLY TEMPERATURE SCALES
BELOW 90K

When there was no international standard below 90K several "national" or "laboratory" scales were developed, each of which differed from the others and from the thermodynamic scale. They were based on the resistance characteristics of platinum calibrated against a gas thermometer and several of these scales were compared in Fig. 4.2. Scott [36] compared several other scales as indicated in Fig. 4.8. The NBS (1939) scale was superseded by the NBS (1955) scale formed by lowering all temperatures on the NBS (1939) scale by 0.01°C and these scales have been the basis for all NBS calibrations in the interval 12K to 90K from 1939 until 1968. The agreement of the NBS (1939) scale with the thermodynamic scale was ± 0.02°C in the range 12K to 90K. It is interesting to note that the scale identified as "Calif (1927)" in Fig. 4.8 is formulated on the basis of a copper-Constantan thermocouple which was stable over a period of 3 years having an estimated accuracy of 0.05K in the range 12K to 90K [37].

In 1964 the CCT established a provisional temperature scale to be considered as a replacement for the IPTS below 273.15K. This scale is referred to as "CCT-64" and is in the form of a resistance-temperature table for platinum thermometers, extending from 10K to 273.15K. The derivation for the range 90K to 273.15K is given by Barber and Hayes [38]. Certain modifications were considered in it prior to its recommendation as an international scale [5] but, modification in CCT-64 was small, and the low temperature part is assumed [6] to be the best approximation to the thermodynamic scale in this region. Calibration of thermometers below 90K may be accomplished using a number of multiphase equilibrium states, called "fixed points" and Timmerhaus [39] lists several of these states which are given in Table 4.6.

These are not presently "secondary reference points" as prescribed by an international temperature scale, but represent the best literature values. An

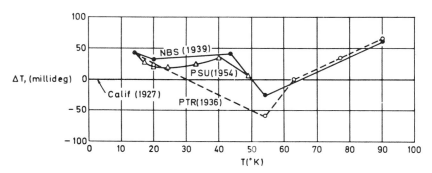

FIG. 4.8 Comparison of temperature scales [36]: Calif = University of California [37]; NBS = National Bureau of Standards (F18); PSU = Pennsylvania State University [12]; PTR = Physikalishe Technistlae Reichsanstalt [36].

TABLE 4.6 Fixed points below 90K

Point	Temperature ($^\circ$K)
Lambda point of helium	2.173
Boiling point of helium (1 atm)	4.215
Triple point of equilibrium hydrogen	13.81
Triple point of normal hydrogen	13.95
Boiling point of equilibrium hydrogen (1 atm)	20.27
Boiling point of normal hydrogen (1 atm)	20.39
Triple point of neon	24.57
Boiling point of neon (1 atm)	27.17
Triple point of oxygen	54.36
Triple point of nitrogen	63.14
Boiling point of nitrogen (1 atm)	77.35

Source. Reference [39].

equilibrium cell or vapor pressure thermometer is employed to determine these states and would be similar to the oxygen vapor-pressure thermometer described by Timmerhaus [40] and shown in Fig. 4.9.

For precise calibration of thermometers in the range 0.20K to 5.2K the vapor pressure scales of He^4 and its light isotope He^3 are available. For temperatures between 1K and 5.2K the International Committee on Weights and Measures in October 1958 recommended for international use a scale based on equilibrium between He^4 liquid and its vapor now known as the "1958 He^4 scale of temperatures." This scale is described by Brickwedde et al. [41] where the vapor pressure of He^4 is tabulated for intervals of 0.001K from 0.50K to 5.22K. Clement [42] concludes that the He^4 1958 scale is accurate within 0.001K to 0.002K with a roughness less than 0.0001K. Values of the vapor pressure of He^4 in microns (10^{-3} mm Hg) for intervals of 0.01K are given in the Appendix at the end of this chapter, taken from [41].

A comparison of the 1958 He^4 scale with previous scales is shown in Fig. 4.10, taken from Hust [6]. The identification of the various scales is given by Brickwedde et al. [41]. The acoustical thermometer of Plumb and Cataland [11] provides an interesting comparison with the results of the vapor pressure scale.

The phase diagram for He^4 is shown in Fig. 4.11. Below the λ point at 2.173K liquid helium experiences a transition to its superfluid state. The tendency of superfluid helium to flow makes vapor pressure measurements difficult below the λ point and furthermore, below 1K the vapor pressure of He^4 is less than 120 μm adding further problems in measurement. To overcome both of these drawbacks the vapor pressure of the light isotope He^3 was determined and developed into the He^3 scale. A comparison of the vapor pressures of He^3 and He^4 is given in Fig. 4.12, taken from Arp and Kropschot [43]. For comparison it will be noted that the vapor pressure of He^3 at 1K is 8,842 μm and that of He^4 at the same temperature is only 120 μm. The International Committee on

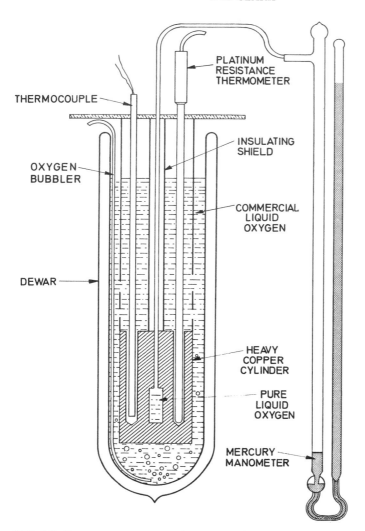

FIG. 4.9 Oxygen vapor-pressure thermometer for calibrating working thermometers. (From Timmerhaus, 1963, Ref. [40].)

Weights and Measures in 1962 recommended the use of the He[3] vapor pressure temperature data for international use. This scale, which is known as the "1962 He[3] scale of temperatures," is tabulated in intervals of 0.001K from 0.20K to 3.324K by Sherman, Sydoriak, and Roberts [44]. Vapor pressure data in intervals of 0.01K for He[3] are given in Table 4.7, taken from the summary table of Sydoriak, Sherman, and Roberts [45].

Preston-Thomas and Bedford [5] have examined the reproducibilities of various actual and postulated temperature scales in the range 1K to 1063K. Their

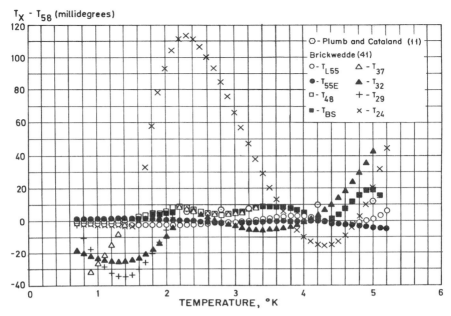

FIG. 4.10 Deviations of earlier helium vapor-pressure scales from 1958 He⁴ scale. (From Hust, 1968, Ref. [6].)

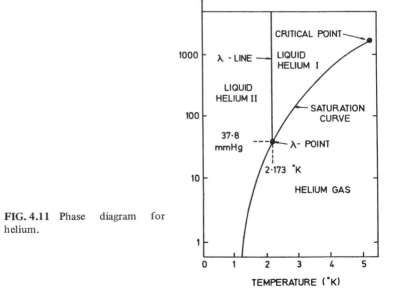

FIG. 4.11 Phase diagram for helium.

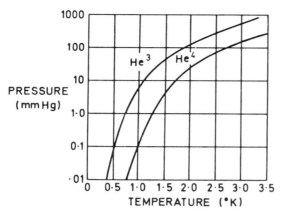

FIG. 4.12 Vapor pressures of He³ and He⁴. (From
Arp and Kropschot, Ref. [43].)

results are given in Fig. 4.13 indicating a general reproducibility of 10^{-2} to 10^{-3}
over the full range.

4.7 THERMOMETERS FOR CRYOGENIC TEMPERATURES

The remaining sections of this chapter will consider the principal devices which
are employed to measure cryogenic temperatures, including thermoelectric, electrical resistance, and magnetic thermometers. A survey of available low temperature thermometers is illustrated in Fig. 4.14 for the temperature range 0.05K to
300K. This is an extension of a similar chart presented by Timmerhaus [40]. A
comparison of the performance of several of these devices has been prepared by
Corruccini [63] in a survey of temperature measurements at cryogenic temperatures. His results are given in Table 4.8.

4.7.1 Thermocouples

The familiar thermoelectric circuit—the thermocouple—in which an emf is produced by subjecting the junctions of dissimilar metallic combinations to different
temperatures is commonly used in the cryogenic temperature range. In circumstances where measurement accuracy is from 0.25K to 0.50K, a thermocouple
may even be the preferred temperature-sensing element. There are several reasons
for this. A thermocouple is easily made, is small, and can be mounted relatively
simply in remote and fairly inaccessible locations; requires only standard laboratory or industrial measuring instruments; can be made rugged and relatively
insensitive to environmental disturbances; and is inexpensive. Other desirable
characteristics that can be obtained using thermocouples are: A large net thermal

TABLE 4.7 He^3 vapor pressure on the 1962 He^3 scale at 0°C and standard gravity, 980.665 cm/sec²

T	0.00	0.01	0.02	0.03	0.04	0.05	0.06	0.07	0.08	0.09
0.20	0.012	0.024	0.046	0.084	0.144	0.239	0.382	0.592	0.891	1.308
0.30	1.877	2.636	3.633	4.921	6.561	8.619	11.173	14.304	18.105	22.673
0.40	28.11	34.54	42.08	50.86	61.01	72.68	86.02	101.17	118.31	137.61
0.50	159.2	183.3	210.1	239.8	272.5	308.5	347.9	391.1	438.0	489.1
0.60	544.4	604.3	668.9	738.4	813.0	893.0	978.7	1070.1	1167.6	1271.4
0.70	1381	1498	1622	1753	1892	2038	2192	2355	2525	2704
0.80	2892	3089	3295	3511	3736	3971	4216	4472	4739	5016
0.90	5304	5603	5914	6237	6572	6918	7277	7649	8034	8431
1.00	8.842	9.267	9.704	10.156	10.622	11.102	11.597	12.106	12.631	13.170
1.10	13.725	14.295	14.881	15.484	16.102	16.737	17.388	18.056	18.741	19.443
1.20	20.163	20.900	21.655	22.428	23.220	24.029	24.857	25.704	26.571	27.456
1.30	28.360	29.285	30.229	31.193	32.177	33.181	34.206	35.252	36.319	37.407
1.40	38.516	39.646	40.799	41.973	43.169	44.388	45.629	46.893	48.179	49.489
1.50	50.822	52.178	53.558	54.961	56.389	57.840	59.316	60.817	62.342	63.892
1.60	65.467	67.068	68.694	70.345	72.022	73.726	75.455	77.211	78.993	80.802
1.70	82.638	84.501	86.391	88.309	90.254	92.228	94.229	96.258	98.315	100.402
1.80	102.516	104.660	106.833	109.035	111.266	113.527	115.818	118.138	120.489	122.870
1.90	125.282	127.724	130.197	132.701	135.236	137.803	140.401	143.031	145.692	148.386
2.00	151.112	153.870	156.661	159.485	162.342	165.232	168.155	171.112	174.102	177.126
2.10	180.184	183.276	186.403	189.564	192.760	195.990	199.256	202.557	205.894	209.266
2.20	212.673	216.117	219.597	223.113	226.665	230.255	233.881	237.544	241.244	244.982
2.30	248.757	252.570	256.420	260.309	264.236	268.202	272.206	276.249	280.331	284.452
2.40	288.613	292.813	297.053	301.333	305.653	310.013	314.414	318.855	323.337	327.861
2.50	332.425	337.031	341.679	346.368	351.100	355.874	360.690	365.549	370.450	375.395
2.60	380.383	385.414	390.489	395.608	400.771	405.978	411.230	416.526	421.868	427.254
2.70	432.686	438.164	443.687	449.256	454.872	460.534	466.242	471.998	477.801	483.651
2.80	489.549	495.495	501.488	507.531	513.622	519.762	525.951	532.189	538.477	544.815
2.90	551.203	557.642	564.131	570.672	577.264	583.907	590.602	597.349	604.149	611.002
3.00	617.907	624.866	631.879	638.945	646.066	653.241	660.472	667.757	675.098	682.496
3.10	689.949	697.459	705.026	712.650	720.332	728.072	735.871	743.728	751.644	759.620
3.20	767.656	775.753	783.910	792.128	800.408	808.750	817.155	825.622	834.153	842.747
3.30	851.406	860.130	868.918	877.773						

Source. Reference [45].

Note. The units of pressure are microns (10^{-3} mm) of mercury below 1K and millimeters of mercury at higher temperatures.

emf, a monotonic or linear emf-temperature characteristic, a stable emf-temperature characteristic, resistance to chemical corrosion, including the effects of both oxidizing and reducing atmospheres, uniformity of the wire material in large batches, and high thermal response. The influence of the environment must often be reduced by the use of protection tubes. The common thermoelectric elements in use today at temperatures from 4K to 300K are the gold-cobalt vs. copper and the Constantan vs. copper junctions, described below. Measurements made at the liquid nitrogen, liquid oxygen, and liquid hydrogen temperature probably have used the Constantan vs. copper thermocouple with greater frequency than any other single combination. As was mentioned earlier, the Calif (1927) scale in Fig. 4.8 was formulated on the basis of a copper-Constantan thermocouple [37] and this particular thermocouple was found to be stable for a period of 3 years in the temperature range 12K to 90K with an accuracy of ± 0.05K.

Although thermocouples are commonly used at temperatures below 300K the International Practical Temperature Scale is not specified in terms of thermo-

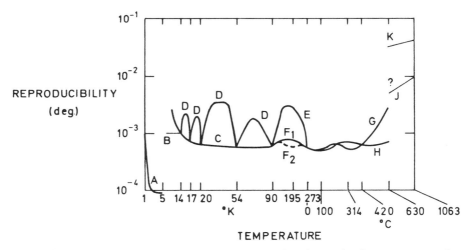

FIG. 4.13 Estimated reproducibilities of various actual and postulated temperature scales. (A) Vapor pressure measurement, 1958 He^4 scale; (B) Estimated reproducibility of a Ge resistance thermometer "wire scale"; (C) Estimated reproducibility of any of the Pt resistance thermometer "national" scales; (D) Reproducibility of the Barber and Van Dijk scale [46] on the (optimistic) assumption of the fixed points being realizable to 0.2 millikelvin; (E) IPTS in the 90K to 273K range assuming the indeterminacy shown by Barber and Horsford [32]; (F) IPTS in the 90K to 273K range assuming the indeterminacy shown by Lovejoy [47] (F_1) or alternatively the use of an additional fixed point (CO_2 point) in a modification of the IPTS (F_2); (G) IPTS in the range 0°C to 630°C; (H) Reproducibility of a modification of the IPTS in the manner suggested by McLaren and Murdock [48]; (J) Estimated reproducibility of a Pt resistance thermometer scale in the 630°C to 1063°C range; (K) Reproducibility of the IPTS (using Pt 10 Rh/Pt thermocouples) in the 630°C to 1063°C range. (From Preston-Thomas and Bedford, 1968, Ref. [5].)

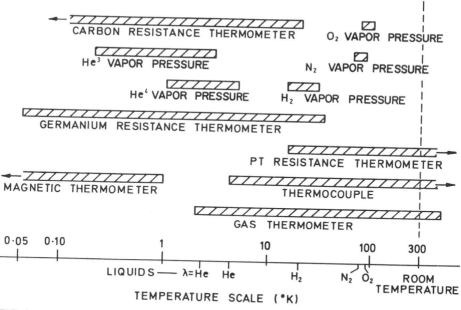

FIG. 4.14 Temperature ranges normally associated with various low-temperature thermometers.

electric systems in the cryogenic range of temperatures. The principal reasons for this are accuracy and reproducibility as compared with the platinum resistance thermometers.

The general principles of thermoelectric thermometry and the various thermoelectric circuits and instrumentation are, of course, important to design for the installation of thermocouples. In view of space limitations here and the generally wide availability of this kind of information, it will not be included in this discussion. Finch [49] has given a thorough presentation of the principles of thermoelectricity. Some improved reference tables for iron-Constantan, Chromel-Alumel, copper-Constantan and Chromel-Constantan thermocouples are presented by Benedict and Ashby [50]. Caldwell [51] discusses the properties of various

TABLE 4.8 Comparison of cryogenic temperature-measuring devices

Type	Range ($^\circ$K)	Best reproducibility ($^\circ$K)	Best accuracy ($^\circ$K)
Platinum resistance	10 to 900	10^{-3} to 10^{-4}	10^{-2} to 10^{-4}
Carbon	1 to 30	10^{-2} to 10^{-3}	10^{-2} to 10^{-3}
Germanium	1 to 100	10^{-3} to 10^{-4}	10^{-2} to 10^{-3}
Gold-cobalt vs. copper thermocouple	4 to 300	10^{-1} to 10^{-2}	0.10

materials that could be used as thermocouple elements at temperatures above 0°C and the use of thermocouples in engineering measurements and their circuits are given by Weber [52], Baker, Ryder, and Baker [53], and Dike [54], among others.

The behavior of a thermocouple element is usually characterized by its thermoelectric potential of emf, "E," and its thermoelectric power dE/dT. Its emf, "E," is always related to an arbitrarily selected reference temperature. At cryogenic temperatures, the common thermoelectric combinations are gold-cobalt (Au + 2.11 at. % Co) vs. copper, copper vs. Constantan (60% Cu and 40% Ni), gold-cobalt vs. normal silver (Ag + 0.37 at. % Au), Iron vs. Constantan and Chromel-P (90% Ni and 10% Cr) vs. Alumel (95% Ni and 5% (Al, Si, Mn). The more frequently used thermocouples, however, are the gold-cobalt vs. copper and the copper vs. Constantan combinations. These thermocouples have been used to temperatures as low as 0.2K. Their best accuracy in the temperature range 4K to 300K is 0.10K for gold-cobalt vs. copper and 0.50K for copper vs. Constantan.

The thermoelectric potential differences for these five thermocouples is given in Table 4.9, taken from Powell, Bunch, and Caywood [55].

The principal advantage of the use of the gold-cobalt vs. copper combination is evident from these data as it has a significantly higher thermal emf. However, owing to inhomogeneities in its chemical constituency this combination produces irregular emf's that are uncompensated for in its calibration and, therefore, give rise to measurement errors. In fact, this lack of homogeneity in composition is the greatest single defect in the Au-Co vs. Cu thermocouple. When the chemical metallurgy of the Au-Co wires can produce a product having a constant, controllable and stable composition this thermocouple will come into much wider use at very low temperatures. The effects of inhomogeneity are usually greatest when the measuring and reference junctions are at widely different temperatures. In such cases the thermocouple lead wires are subjected to steep temperature gradients and at the points of greatest temperature change in the wire chemical inhomogeneity will produce an emf. Thus, the Au-Co vs. Cu combination is best used where temperature differences to be measured are small. A convenient and practical application is the use of the Au-Co vs. Cu combination as a differential

TABLE 4.9 Thermoelectric potential differences in microvolts for several thermocouple combinations

Temperature (°K)	Constantan vs. copper	Gold-cobalt vs. copper	Normal silver vs. copper	Iron vs. constantan	Chromel P vs. alumel
4–20	57.8	171.4	0.2	59	41
20–76	646.9	1562.5	37.9	805	616
76–273	5545.6	8123.2	133.7	8252	6182

Source. Reference [55].

TABLE 4.10 Inhomogeneity of thermoelectric voltages obtained from dip tests

	Bath Temperatures			
	4K–300K		76K–300K	
	Voltage (μV)		Voltage (μV)	
Samples	Maximum	Average	Maximum	Average
Cu[a]	4.5	2.5	2.0	0.8
Cu[b]	1.8	0.7	1.0	0.3
Constantan[c]	0.5	0.2	0.5	0.2
Au-Co[d]	5.0	3.0	4.0	2.5
Au-Co[e]	5.5	3.5	4.0	2.5
Ag-Au[f]	2.2	1.2	1.2	0.8

Source. Reference [50].
[a] Instrument grade copper, 32 AWG.
[b] Thermocouple grade copper, 36 AWG.
[c] Thermocouple grade Constantan, 36 AWG.
[d] Gold-cobalt, Bar 9, 36 AWG (1960).
[e] Gold-cobalt, Bar 5, 36 AWG (1958).
[f] "Normal" silver, 36 AWG.

thermocouple in an installation where the temperature differences are small or zero, as in constant temperature baths, cryostats, and equilibrium cells for temperature calibration.

The Fe vs. Con and Ch vs. Al thermocouples are infrequently used at low temperatures principally because of voltage uncertainties resulting from inhomogeneities in the wires.

An estimate of the inhomogeneity of thermoelectrical voltages obtained by placing one section of a wire sample in a cryogen while the two ends of the wire are attached to a potentiometer is reported by Powell et al. [56] and shown in Table 4.10. As is evident from this sampling, the gold-cobalt wire is subject to the greatest voltage uncertainty. Thermocouple grade copper and Constantan exhibit the least voltage uncertainty owing to inhomogeneity.

Powell et al. [55, 56] have studied the thermoelectric characteristics of several thermocouple combinations in the temperature range 4K to 300K. A summary of their results for the Au-Co vs. Cu and Cu vs. Con thermocouples is given in Table 4.11 for a 0($^\circ$K) reference temperature. An extensive tabulation of the thermoelectric potentials and thermoelectric power in 1K intervals for these combinations from 0($^\circ$K) to 300K is found in [56]. These results represent the best average or smoothed data for a family of thermocouple combinations and thus, they may be used as the standard reference data for each thermocouple. Such data are of great value in thermocouple calibration, as is discussed later.

The thermoelectric power, dE/dT, in microvolts (μV) per degree Kelvin ($^\circ$K) for the five thermocouple combinations in Table 4.12 is given in Fig. 4.15 as a

TABLE 4.11 Thermoelectric potential differences in microvolts for gold-cobalt and constantan vs. copper thermocouples

Temperature (°K)	Au-Co	Constantan	Temperature (°K)	Au-Co	Constantan
0	0.00	0.00	90	2246.8	946.7
2	2.09	0.66	95	2433.3	1038.5
4	8.22	2.62	100	2622.6	1133.7
6	18.20	5.83	110	3008.5	1333.7
8	31.83	10.26	120	3402.8	1546.4
10	48.93	15.88	130	3804.1	1771.7
12	69.30	22.64	140	4211.2	2009.5
14	92.75	30.50	150	4623.2	2260.0
16	119.1	39.43	160	5039.1	2522.7
18	148.1	49.40	170	5458.4	2797.1
20	179.6	60.40	180	5880.4	3083.1
25	269.1	92.31	190	6304.6	3380.3
30	372.5	130.3	200	6730.6	3688.6
35	483.0	173.9	210	7158.0	4007.7
40	614.2	222.9	220	7586.4	4337.4
45	749.9	276.8	230	8015.7	4677.5
50	893.9	335.6	240	8445.5	5027.8
55	1045.2	398.8	250	8875.6	5388.0
60	1202.9	466.2	260	9305.9	5757.9
65	1366.2	537.5	270	9736.2	6137.3
70	1534.5	612.7	280	10166.3	6526.0
75	1707.1	691.2	290	10596.1	6923.7
80	1883.5	773.0	300	11025.5	7330.2
85	2063.4	858.1			

Source. Reference [56].

function of temperature. At temperatures from 4K to about 200K, the Au-Co vs. Cu thermocouple is clearly the superior combination from the standpoint of thermoelectric power. Below about 40K the Fe vs. Con, Ch vs. Al and Cu vs. Con thermocouples have about the same thermoelectric power. Normal silver vs. copper produces an almost insignificant thermoelectric power at low temperatures. The thermoelectric power, dE/dT, in Fig. 4.15 may be used to give an indication of the sensitivity of temperature measurement when it is related to the measurement sensitivity of the potentiometer ΔE^* in microvolts, used to measure the thermal emf of the thermocouple. That is, the uncertainty in temperature indication, ΔT^*, may be written

$$\Delta T^* = \frac{\Delta E^*}{dE/dT} \qquad (4.18)$$

Thus, thermocouples having large thermoelectric power will enable a smaller measurement uncertainty for a given measuring instrument.

TABLE 4.12 August 1965 table for platinum resistance thermometer 1653433

Temperature (°K)	Resistance (abs Ω)	Inverse difference	Temperature (°K)	Resistance (abs Ω)	Inverse difference
90	6.20974		140	11.66916	9.305
91	6.32084	9.001	141	11.77656	9.311
92	6.43186	9.008	142	11.88390	9.316
93	6.54279	9.015	143	11.99118	9.322
94	6.65364	9.022	144	12.09840	9.327
95	6.76440	9.029	145	12.20556	9.332
96	6.87507	9.035	146	12.31265	9.337
97	6.98567	9.042	147	12.41969	9.343
98	7.09617	9.049	148	12.52667	9.348
99	7.20660	9.056	149	12.63358	9.353
100	7.31695	9.063	150	12.74044	9.358
101	7.42721	9.069	151	12.84724	9.363
102	7.53739	9.076	152	12.95399	9.368
103	7.64749	9.083	153	13.06067	9.373
104	7.75751	9.089	154	13.16730	9.379
105	7.86745	9.096	155	13.27387	9.384
106	7.97732	9.102	156	13.38038	9.389
107	8.08710	9.109	157	13.48684	9.393
108	8.19681	9.115	158	13.59324	9.398
109	8.30644	9.122	159	13.69959	9.403
110	8.41599	9.128	160	13.80588	9.408
111	8.52546	9.135	161	13.91212	9.413
112	8.63486	9.141	162	14.01830	9.418
113	8.74419	9.147	163	14.12443	9.423
114	8.85343	9.153	164	14.23050	9.427
115	8.96261	9.160	165	14.33653	9.432
116	9.07171	9.166	166	14.44249	9.437
117	9.18074	9.172	167	14.54841	9.441
118	9.28969	9.178	168	14.65428	9.446
119	9.39857	9.184	169	14.76009	9.451
120	9.50738	9.190	170	14.86585	9.455
121	9.61612	9.196	171	14.97156	9.460
122	9.72479	9.202	172	15.07722	9.464
123	9.83338	9.208	173	15.18283	9.469
124	9.94191	9.214	174	15.28839	9.473
125	10.05036	9.220	175	15.39390	9.478
126	10.15875	9.226	176	15.49936	9.482
127	10.26707	9.232	177	15.60477	9.484
128	10.37532	9.238	178	15.71013	9.491
129	10.48350	9.244	179	15.81545	9.495
130	10.59162	9.249	180	15.92071	9.500
131	10.69967	9.255	181	16.02593	9.504
132	10.80765	9.261	182	16.13110	9.508
133	10.91556	9.267	183	16.23623	9.513
134	11.02341	9.272	184	16.34130	9.517
135	11.13120	9.278	185	16.44633	9.521
136	11.23892	9.283	186	16.55132	9.525
137	11.34657	9.289	187	16.65625	9.529
138	11.45417	9.294			
139	11.56169	9.300			

THERMOELECTRIC
POWER (µv/°K)

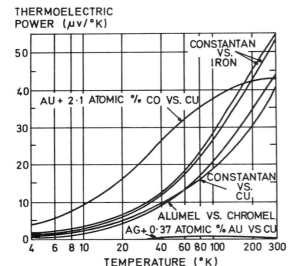

FIG. 4.15 Thermoelectric power as a function of temperature for various thermocouple combinations. (From Corruccini, 1963, Ref. [63].)

The calibration of a thermocouple is conveniently done by establishing its deviation characteristic as compared with standard thermocouple emf. Standard thermocouple potentials at cryogenic temperatures are given in Table 4.11 in summary form for Au-Co vs. Cu and Cu vs. Con and may be found in much greater detail in [56]. The thermocouple "deviation" is defined

$$\Delta E_{\text{dev}} = E_{\text{std}} - E_{\text{obs}} \tag{4.19}$$

where E_{std} is the standard emf corresponding to the temperature of the thermocouple for which E_{obs} is its observed emf. A deviation plot is usually constructed by obtaining several corresponding values of ΔE_{dev} and E_{obs} over a range of temperature and generally, when these data are plotted as ΔE_{dev} vs. E_{obs} a smooth, frequently almost linear, curve may be used to join the data points. This is a result of the fact that while each individual thermocouple wire combination will differ slightly from others of similar composition, the emf-temperature characteristics of a "family" of similar wires are essentially parallel. Thus, their deviations will be almost linear and exactly zero at a common reference temperature.

The use of a deviation plot provides a very convenient method for making accurate temperature measurements with thermocouples. Owing to the essential linearity of a deviation curve, interpolation between a minimum number of calibration points may be done with confidence. The determination of an unknown temperature in measurement is made by computing E_{std} from Eq. (4.19) using ΔE_{dev} taken from the deviation plot corresponding to the E_{obs} for the thermocouple. The unknown temperature is found from the "standard" table of E_{std} vs. temperature. The "standard" table is usually formulated in great detail and represents the best, smoothed data for a family of thermocouple combinations.

A potentiometer is probably the most satisfactory instrument for precision temperature measurement using thermocouples. These instruments are described in detail elsewhere [52, 53, 57]. Instruments presently available by the Leeds and Northrup Co., the K-5 (facility) and K-3, have sensitivities 0.02-0.1 μV and 0.5 μV, respectively. The Wenner potentiometer has a sensitivity of 0.1 μV and a potentiometer manufactured by the Minneapolis Honeywell Co. also has a sensitivity of 0.1 μV. Similar potentiometers are produced by other companies. Comparing these sensitivities with the thermoelectric power of the Au-Co vs. Cu thermocouple, temperature sensitivities will range from 0.002K to 0.01K at a level of temperature of 10K.

4.7.2 Resistance Thermometry

The variation of electrical resistance with temperature provides a very convenient, accurate, and practical method for temperature measurement. This method is enhanced when the material from which the thermometer is made has a stable and easily reproducible composition. Otherwise, the method becomes impractical owing to inherent instabilities in the resistance-temperature characteristic and consequent uncertainties in the temperature. The basic measurement required is that of electrical resistance and this can be done with great precision using available resistance bridges or potentiometers. Hence, with a stable material and present instrumentation a resistance thermometer can be used to measure temperature to a high degree of accuracy [52, 53, 58-60]. For precision measurements the platinum resistance thermometer is the most widely used temperature measuring device in the range 1K to 300K. As mentioned earlier, the international practical temperature scale (1948) is defined in terms of the resistance characteristics of platinum from $-183°C$ to $630.5°C$. The international temperature scale, introduced [4-6] in 1968, also employs the resistance characteristic of platinum as the standard below 300K, as well as above that temperature. The reason for this of course, is the unusually high degree of purity that can be achieved in the production of platinum, the reproducibility of the purity from batch to batch, its monotonic resistance-temperature curve in the strain-free, annealed state, and its inertness to chemical contamination. Its cost is high which is a factor in its use. Other materials which are also used include copper, nickel, carbon, germanium and certain semiconductors known as *thermistors*. These will be discussed later.

The resistance-temperature characteristic of platinum is shown in Fig. 4.16. Above 50K this relationship is essentially linear. The 1948 IPTS requires that the resistance ratio in Fig. 4.16 be equal to or greater than 1.3920 at 373.15K (100°C) to ensure purity in the platinum wire. The resistance-temperature characteristics of platinum film on a nonconducting substrate, nickel, tungsten, and copper are shown in Fig. 4.17.

Precision platinum resistance thermometers are made of a fine coil of highly purified, strain-free platinum wire wound around a nonconducting frame. A typical method of construction is shown in Fig. 4.18. The ice-point resistance of these thermometers is commonly set at approximately 25.5 absolute ohms. The platinum thermometer is usually manufactured as a capsule (Fig. 4.18) or as a cane. In each case four lead wires are provided for resistance measurement. Precision resistance is best measured using a Mueller bridge with a four-lead-wire thermometer as shown in Fig. 4.19. The accuracy of this bridge circuit is 10^{-5} ohms [39] which would correspond to approximately 0.003K at 12K and 0.00009K at 100K. Except at very low temperatures, accuracies from 0.001K to 0.0001K can be obtained using a platinum thermometer and Mueller bridge. As indicated in Fig. 4.19, the four-lead-wire circuit provides a means for reversing lead wire connections during a measurement, this technique permitting the complete cancellation of lead wire resistance so that the net measured resistance is that of the platinum resistance thermometer wire itself. Potentiometric methods for resistance measurement are summarized by Dauphinee [61].

Calibration of a platinum thermometer can be made using a gas thermometer, another standard thermometer or using the defining fixed points and a polynomial equation between resistance and temperature, such as the Callendar or Callendar-VanDusen equations [28] which are equivalent to those given in Table 4.2. In the United States, calibration is frequently done by the National Bureau of Standards, Institute for Basic Standards. Typical calibration data for a platinum thermometer is given in Table 4.13. The constants α, δ, and β were found from the Callendar-VanDusen formula

R/R (273·15 °K)

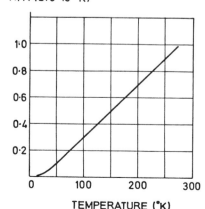

FIG. 4.16 Resistance ratio of platinum as a function of temperature.

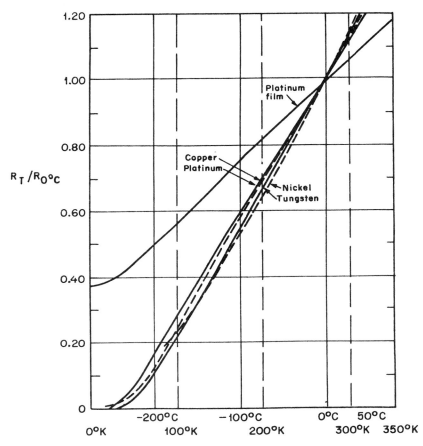

FIG. 4.17 Resistance-temperature relationships for various resistance-type temperature sensors—medium range. (From Rosemount Engineering Company, Platinum Resistance Temperature Sensors, *Bulletin 9612*, Rev. B, p. 4, Rosemount Engineering Company, Minneapolis, Minn., 1962. Reprinted courtesy of Rosemount, Inc.)

$$t = \frac{R_t - R_0}{\alpha R_0} + \delta\left(\frac{t}{100} - 1\right)\left(\frac{t}{100}\right) + \beta\left(\frac{t}{100} - 1\right)\left(\frac{t}{100}\right)^3 \qquad (4.20)$$

Additional qualifying information was provided by the NBS for this calibration as follows [62]:

The value of δ was estimated using the assumption, based on experience with similar thermometers, that the product $\alpha \cdot \delta$ is a constant. The uncertainty in the estimated value of δ is equivalent to an uncertainty at the sulphur point of less than ± 0.01 deg C. The other values given are determined from measurements at the triple point of water, the steam

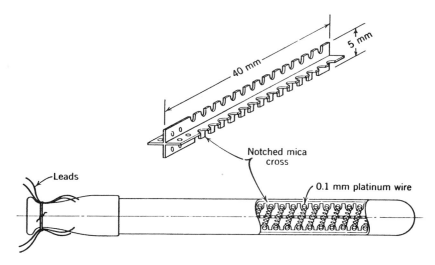

FIG. 4.18 Capsule-type, strain-free resistance thermometer. (From Timmerhaus, 1963, Ref. [40].)

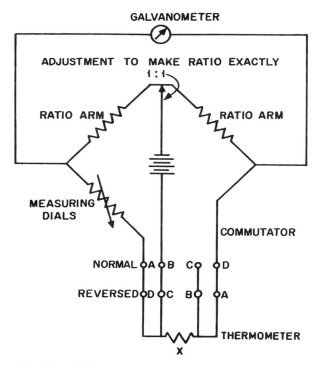

FIG. 4.19 Mueller bridge with a four-lead platinum resistance thermometer.

TABLE 4.13 Report of calibration of platinum resistance thermometer, L & N No. 1653433

Constant	Value
α	0.003925780
δ	1.49168
β	0.11116 (t below 0°C)
β	0 (t above 0°C)
R_0	25.5510 abs Ω

Source. Submitted by the University of Michigan. Reference [62].

point, and the oxygen point. The uncertainty of the measurements at these points, expressed in temperature, is less than ± 0.0003, ± 0.0015 and ± 0.005 deg C respectively. About one-half of each of these uncertainties is an allowance for systematic errors, including the differences among national laboratories, the remaining part representing the effect of random errors in the measurement process. The effect of these uncertainties on other measured temperatures are discussed in "Intercomparison of Platinum Resistance Thermometers between −190 and 445°C," *J. Research NBS* 28, 217 (1942). During calibration the value of R_0 changed by the equivalent of 5×10^{-4} deg C. These results indicate that this thermometer is satisfactory for use as a defining standard in accordance with the text of the International Practical Temperature Scale.

Resistance-temperature data on the thermometer described in Table 4.14 are listed in Table 4.12 for a small range of temperatures above 90K. These data are computed from the following equation and represent a few of the numerical results abstracted from the original calibration

TABLE 4.14 Short-range stability of a Western Electric 14A Thermistor

Time (min)	Thermistor A	Thermistor B
0	96,234.0	96,234.7
5	96,234.6	96,234.6
10	96,234.2	96,234.9
15	96,235.0	96,234.4
20	96,234.8	96,234.4
25	96,234.8	96,234.8
30	96,234.7	96,234.2
35	96,234.8	96,234.7
40	96,234.6	96,234.0
45	96,235.0	96,234.6

$$\frac{R_t}{R_0} = 1 + \alpha t \left[1 + \delta \left(1 - \frac{t}{100} \right)^{10^{-6}} + \beta t^2 \left(1 - \frac{t}{100} \right)^{10^{-6}} \right] \qquad (4.21)$$

The first column is the temperature in K (IPTS 1948), the second column is the thermometer resistance in absolute ohms (abs Ω) and the third column gives the inverse (reciprocal) of the difference between each two successive values in the second column. These reciprocal first differences are included to facilitate interpolation. The error introduced by using linear interpolation will be less than 0.0001°C. The third column may also be expressed as dT/dR, K/Ω, as the tabular difference in the first column is 1.0K. The thermometer described in Table 4.14 was also calibrated and the results tabulated in 0.1K intervals from 11K to 92K by the NBS using the NBS-1955 temperature scale. This temperature scale was referred to earlier in Fig. 4.2 and defines the temperature in terms of the electrical resistance of platinum in the range 10K to 90K.

An important class of low temperature thermometers are those whose electrical resistance increases with decrease in temperature, rather than the opposite, as is the case with platinum. Below 20K these thermometers become most practical. This class of thermometers included carbon, germanium, and the semiconductors (thermistors) and are the most sensitive resistance elements to temperature changes at low temperatures available. The electrical resistance characteristics of these materials is shown in Fig. 4.20 in comparison with platinum, tungsten and indium.

The most common resistance element, which also is readily available and inexpensive, is the conventional carbon radio resistor. In addition to its high thermal sensitivity at low temperatures, the carbon resistor can be made small, is rather insensitive to magnetic fields and has a small heat capacity for rapid thermal response. It is slightly pressure sensitive, having temperature changes of 0.31K at 20K and 0.02K at 4K for an increase in pressure of 1000 psi [64], and is subject to thermal instabilities or aging. This lack of reproducibility is particularly significant after the resistor has been exposed to thermal cycling. Carbon in the form of thin graphite coatings has been used as a thermometer [65], this type of thermometer being especially useful where high response is required, as in low-temperature (0.1K) adiabatic demagnetization experiments.

Lindenfeld [66] reports on the use of carbon and germanium thermometers between 0.30K and 20K. One problem in the use of carbon radio resistors below 1K is the difficulty in measuring their high resistance. Maximum power dissipated in these resistors is about 10^{-8} W for temperatures 1K and higher and using a Wheatstone bridge temperature changes of 10^{-5}K to 10^{-6}K can be detected. The use of the carbon resistor in measurement is greatly aided if a reasonably simple and accurate formula can be written relating resistance to temperature. Clement and Quinnell [67] found that Allen-Bradley Company cylindrical carbon radio resistors has a resistance temperature relationship below

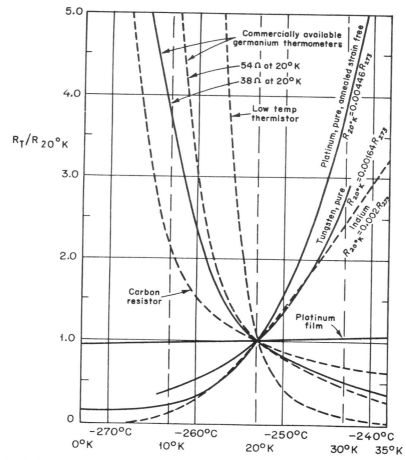

FIG. 4.20 Resistance-temperature relationship for various resistance-type temperature sensors—low range. (From Rosemount Engineering Company, Platinum Resistance Temperature Sensors, *Bulletin 9612*, Rev. B, p. 4, Rosemount Engineering Company, Minneapolis, Minn., 1962. Reprinted courtesy of Rosemount, Inc.)

20K which could be expressed to within ± 1/2% by a semiempirical expression of the form

$$\log_{10} R + \frac{K}{\log_{10} R} = A + \frac{B}{T} \tag{4.22}$$

The constants K, A, and B are determined by a calibration of the resistor at a minimum of three known temperatures. Typical resistance-temperature curves for two Allen-Bradley carbon resistors are shown in Fig. 4.21. Schulte [68] calibrated an Allen-Bradley 0.1 W, 270 Ω carbon resistor between 4K and 296K

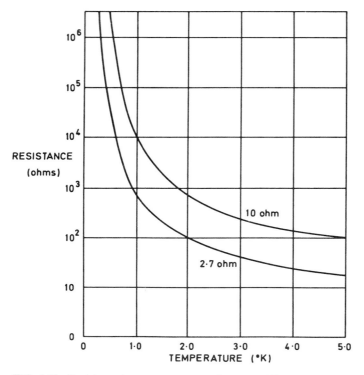

FIG. 4.21 Resistance-temperature curve for two Allen-Bradley carbon resistors. (From Timmerhaus, 1962, Ref. [39].)

and found his results to correlate within 7% of Eq. (4.22). For a range of temperatures from 2K to 20K, Mikhailov and Kaganouskii [69] also found that Eq. (4.22) gave satisfactory results for carbon thermometers. In this case the constants in the equation were determined from calibrations at 2K, 4.2K, and 20.4K. This permitted temperatures to be calculated with an accuracy of a few hundredths of a degree in the range 2K to 4.2K. After 100 heating and cooling cycles between 300K and 77K, uncertainty in the temperature measurements in the same 2.2K interval did not exceed 0.01K.

Measurement of the resistance of a carbon thermometer may be made with a resistance bridge, as in Fig. 4.19, or with a potentiometer using an accurately calibrated monitoring resistor of known resistance, a schematic diagram of which is shown in Fig. 3.22 as used by Greene [70]. He calibrated a carbon resistor having a nominal resistance of 82Ω with a measuring current of 10 μA. The results of this calibration are given in Fig. 4.23 which illustrates the influence of thermal cycling, the reproducibility of the calibration before and after a calibration run, and heat conduction along the thermometer lead wires. The ordinate in Fig. 4.23 is the voltage drop across the resistor for a 10 μA current. During any

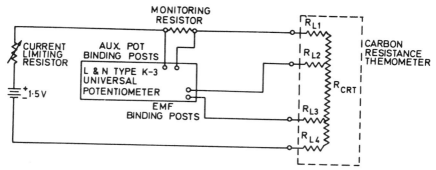

FIG. 4.22 Schematic diagram of the L & N-type K-3 universal potentiometer circuit (From Greene, 1966, Ref. [70].)

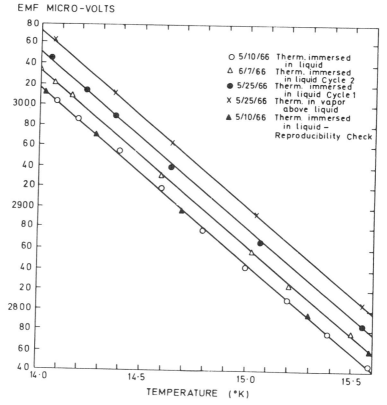

FIG. 4.23 Typical calibration curves for carbon resistance thermometer (nominal resistance 82 Ω). (From Greene, 1966, Ref. [70].)

one calibration the accuracy amounts to ± 0.22K and is within the precision of the measurements. Although thermal cycling did produce a shift in the calibration curve, its slope remains constant and thermal conduction along the lead wires raised the calibration curve by approximately 0.10K in this instance.

The use of carbon resistors for field measurement where laboratory precision is not demanded has been studied by Herr, Terbeek, and Tieferman [71]. Allen-Bradley Co. 0.1-W, 100-Ω (± 5% at 300K) resistors were found to be reproducible within ± 1% of the absolute temperature in the range 19.5K to 55.5K (35°R to 100°R). The measurement of resistance ratio rather than absolute resistance was found to be a more satisfactory method owing to drift in the resistance values of the carbon resistor.

The word *thermistor* is a trade name for a class of semiconducting solids having a large negative temperature coefficient of electrical resistance. It is a name derived from the word combination thermal-sensitive-resistor. In a physical description these substances are classed as electronic semiconductors whose characteristics have been given much theoretical and experimental examination since World War II.

Semiconductors may be classed with those substances having electronic conductivities in the range 10^{-5} to 10^3 $(\Omega\ cm)^{-1}$, or resistivities falling between 10^{-3} and 10^5 Ω cm [72]. This can be compared with the pure metals and metallic alloys whose resistivities [73] are generally less than 10^{-4} Ω cm or with the electrical insulators, as mica and quartz, having resistivities above 10^6 Ω cm at ordinary temperatures. Figure 4.24 shows these relationships. The important difference between semiconductors and metals for thermal-sensitive uses is not,

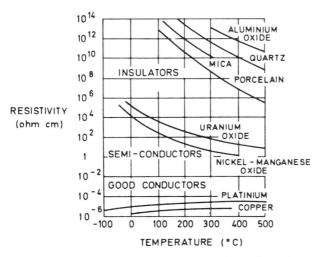

FIG. 4.24 Temperature-resistivity relationship of insulators, semiconductors, and good conductors. (From Clark and Kobayashi, 1967, Ref. [84].)

however, their orders of magnitude of resistivity but the great differences in the change of resistivity with temperature as compared with the metals. This may be illustrated by a typical thermistor which will increase in resistance from 780 to 17,800 Ω for a temperature change from $+ 30$ to $- 30°$C. This is a total change of approximately 17,000 Ω or a percentage change of about 2,000%. Compared with standard platinum and copper resistance thermometers the corresponding change is about 6 and 100 Ω, respectively, over the same range of temperature, both changing about 20%. The thermistor, then, undergoes a percentage change in resistivity of about 100 times that of the metals in this range of temperature. Should a greater interval of temperature be examined, as in Fig. 4.24, the percentage change for the thermistor might be as large as 2×10^6.

Possibly of greater significance in the field of thermal measurements is dR/dT, the rate of change of resistance with temperature, of this thermistor as a function of temperature. At $25°$C, for example, dR/dT is 44 $\Omega/°$C and at $-30°$C it is 1120 $\Omega/°$C, while for a standard 25-Ω platinum resistance thermometer, dR/dT is about 0.10 $\Omega/°$C in this same range of temperature. This means that if one is able to measure changes in resistance, say, to 0.01 Ω, the temperature change capable of detection with this thermistor is $0.0002°$C at $25°$C and $0.000009°$C at $-30°$C but some commercially available thermistors have sensitivities 100 to 1,000 times greater than this. The ordinary resistance thermometer would detect a temperature change of $0.1°$C under these same circumstances. It is quite generally true that thermistors have greatest sensitivity at lower temperatures. For absolute temperature measurement other considerations, naturally, are necessary, not least among which is the thermal stability of the thermistor element, a property possessed in the highest degree by an annealed, strain-free platinum resistance thermometer.

Thermistors are available from the manufacturers in a variety of shapes and sizes: Discs, beads, rods, washers, and wafers. The shape selected depends on the use to be made of the element and sizes range from 0.0152 mm to 2.54 mm diameter for beads, 5.08 mm to 19.05 mm diameter and 1.02 mm to 12.7 mm thick for discs, wafers, and washers, and from 0.0254 mm to 12.7 mm diameter, 6.35 mm to 50.8 mm long for rods. Lead wires of various lengths and diameters consist of platinum, platinum-iridium alloys, or copper which can be butt-soldered, wrapped and soldered or fired in place on the thermistor element. Silver paste contacts are available to which the user can soft-solder lead wires, if desired. Washer-type elements have terminals that may be mechanically clamped into place against the faces of the element. Protective coatings are frequently placed over the thermistor to prevent or retard atmospheric attach. These consist of a thin or thick layer of glass or enamel coating. For certain applications the element can be placed in an evacuated or gas-filled bulb.

The recommended maximum temperature for continuous service varies but it can be as high as $300°$C; however, some manufacturers recommend a temperature no greater than $150°$C. To a large extent this will depend upon such things as the accuracy required, the atmosphere surrounding the element and the melting point of the solder, if any, used to fasten the lead wires to the element. In any event, the thermistor is used to its greatest advantage, from a thermal-sensitive consideration, at lower temperatures.

Most thermal-sensitive semiconductors (thermistors) are manufactured by sintering various mixtures and combinations of metallic oxides; the common materials are the oxides of manganese, nickel, cobalt, copper, uranium, iron, zinc, titanium, and magnesium. For the commercial thermistors, the oxides of manganese, nickel, and cobalt, however, are the most commonly used substances for the mixtures. The result of this type of manufacturing process is a hard, dense ceramic type of material. Other materials [72, 74] which may be classed with the semiconductors and which possess a large negative temperature coefficient of electrical resistance include chlorides such as NaCl, some sulfides like Ag_2S, CuS, PbS, CaS, and some iodides, bromides, and nitrides. Lead sulfide has been used as a detector of infrared radiation in a radiation pyrometer and is marketed commercially. Its response is high (10,000 cps) and it can detect temperatures as low as 89K. The uses of thermistors in a radiation-type pickup is reported [77] for measurement of subzero temperatures.

Some pure materials such as silicon, tellurium, germanium, and selenium [74] which are monatomic become semiconductors in the presence of certain impurities. This effect is shown qualitatively in Figs. 4.25 and 4.26 for silicon containing an unknown impurity and for cuprous oxide with varying amounts of oxygen in excess of the stoichiometric. Figure 4.25 taken from Becker, Green,

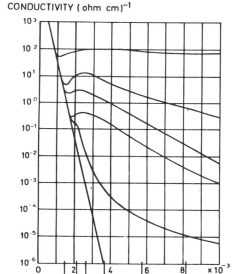

CONDUCTIVITY (ohm cm)$^{-1}$

INVERSE ABSOLUTE TEMPERATURE$\left(\frac{1}{°K}\right)$

FIG. 4.25 Logarithm of the conductivity of various specimens of silicon as a function of inverse absolute temperature. (From Clark and Kobayashi, 1967, Ref. [84].)

CONDUCTIVITY (ohm cm)$^{-1}$

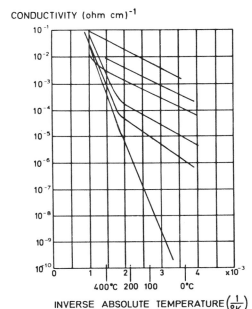

FIG. 4.26 Logarithm of the conductivity of various specimens of cuprous oxide as a function of inverse absolute temperature.

INVERSE ABSOLUTE TEMPERATURE $\left(\frac{1}{°K}\right)$

and Pearson [75], shows a 10^7 increase in the conductivity of pure silicon by the addition of a foreign impurity. A similar large increase in conductivity is seen in the case of cuprous oxide, Fig. 4.26, also taken from [75] where the increase is due to an excess of oxygen up to 1%. These effects vary greatly with the type of impurity, its amount, its dispersion within the solid, and the heat treatment of the solid.

Generally speaking, a thermistor can be considered for use in any application requiring a thermal-sensitive electrical resistance element; the obvious and perhaps most widely employed application is temperature measurement. As was pointed out earlier, it is possible to detect very minute changes in temperature with a thermistor owing to the large change in its electrical resistance with temperature. Brown [76] employed a Western Electric 17A thermistor to measure small changes in air temperature. The device was used in a bridge circuit; the output was amplified and fed into a recording oscillograph and during the initial measurement it was found that the thermistor was so sensitive that it recorded with fidelity the fluctuation in air temperature resulting from atmospheric turbulence. A typical oscillograph is shown in Fig. 4.27. Changes in temperature could be measured to an estimated $0.0007°C$.

Theoretical work of Wilson [78, 80] and others has led to the following expression for the electronic conductivity of a semiconductor

$$\rho = Ae^{-B/T} \tag{4.23}$$

Since the conductivity δ is the reciprocal of the resistivity ζ we may write

$$\rho = Ce^{B/T}$$

or

$$\rho = p_0 e^{B[(1/T)-(1/T)]} \tag{4.24}$$

Also, since the electrical resistance is a geometric extension of the resistivity, Eq. (4.24) may be written

$$R = R_0 e^{B[(1/T)-(1/T_0)]} \tag{4.25}$$

Because of the form of Eqs. (4.24) and (4.25), the logarithm of the resistivity or resistance is frequently plotted against the reciprocal of the absolute temperature, as shown in Fig. 4.28, in order to demonstrate the electrical characteristics of a thermistor and to compare it with others. These data are experimental and are taken from Becker, Green, and Pearson [75].

The experimental curves in Fig. 4.28 are almost straight, as required by Eq. (4.24). However, close inspection will disclose a slight curvature which may be shown to increase linearly with increase in level of temperature [75]. Hence, the equation is sometimes modified as

$$\rho = ET^{-c}e^{D/T} \tag{4.26}$$

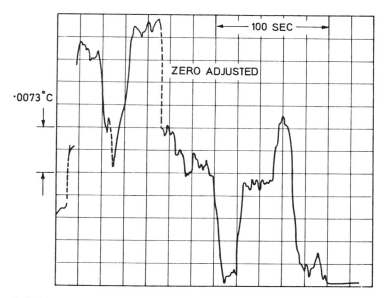

FIG. 4.27 Thermistor response to room temperature variations. (From Clark and Kobayashi, 1967, Ref. [84].)

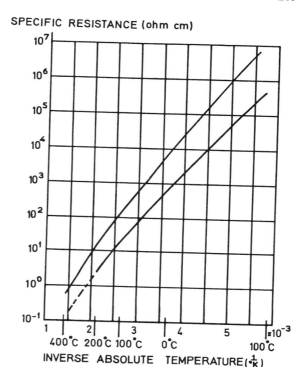

SPECIFIC RESISTANCE (ohm cm)

INVERSE ABSOLUTE TEMPERATURE$(\frac{1}{\cdot K})$

FIG. 4.28 Logarithm of the specific resistance of two thermistor materials as a function of inverse absolute temperature. (From Clark and Kobayashi, 1967, Ref. [84].)

where c is a small number compared with D or B and may be positive, negative, or zero depending on the material [75]. For our present purpose we shall employ Eq. (4.25), since if the interval $T - T_0$ is not too great, this equation will adequately represent the data and it is somewhat easier to handle mathematically.

As was mentioned above, the relationship of resistance to absolute temperature, given by Eq. (4.25), has the same shape as the curve shown in Fig. 4.28, from which several important characteristics may be obtained relative to the suitability of a thermistor as a temperature-sensing element. A curve of R vs. $1/T$ is also a convenient chart for comparing several different thermistors for use in temperature measurement. By taking logarithms and differentiating Eq. (4.25) the following equations are obtained

$$\frac{dR}{R} = -\frac{B}{T^2}\,dT \tag{4.27}$$

or

$$\frac{1}{R}\frac{dR}{dT} = -\frac{B}{T^2} \tag{4.28}$$

and
$$\frac{dR}{dT} = - B \frac{R}{T^2} \tag{4.29}$$

Equation (4.29) may be interpreted in relation to a curve similar to Fig. 4.28 or log R vs. $1/T$. It will be noted that the slope of a curve on such a chart is written

$$\frac{d(\log R)}{d(1/T)} = \frac{dR/R}{-dR/T^2} = \left(\text{slope of log } R - \frac{1}{T} \text{ curve}\right) \tag{4.30}$$

Comparison of Eqs. (4.29) and (4.30) disclosed that the right-hand side of Eq. (4.30), the slope of a curve plotted as log R vs. $1/T$, is equal to the parameter B in Eq. (4.25). Hence

$$B = \left(\text{slope of log } R \text{ vs. } \frac{1}{T} \text{ curve}\right) \tag{4.31}$$

Equation (4.29) is then rearranged to

$$\frac{dR}{dT} = - (\text{slope}) \frac{R}{T^2} \tag{4.32}$$

Interpretation of Eq. (4.32) is as follows. For use as a temperature-sensing element it is desirable that a thermistor have as large a value of dR/dT as possible in order that it be sensitive and capable of detecting small changes in temperature for any given resistance measuring system. From Eq. (4.32) it follows that at any given temperature, the thermistor which has the greatest slope on a log R vs. $1/T$ plot and the greatest resistance will also be the most sensitive as a temperature-sensing element. In this way, therefore, a series of thermistors can be very rapidly evaluated as to their thermal sensitivity.

Another method for evaluation of thermistors consists of plotting log R_0 vs. B, where B is determined from experimental thermistor data in the region of T_0, which may be taken to be $0°C$; R_0 is then the resistance of the thermistor at $0°C$. Because most thermistors have similar characteristics it will be generally true that a thermistor with superior thermal sensitivity at $0°C$ will also have superior sensitivity at other temperatures. In any event the resistance-temperature characteristics of the thermistor can be obtained approximately from Eq. (4.25) or from the manufacturer's published data. Equation (4.25) is approximate owing to the nonlinear nature of log R vs. $1/T$, as mentioned above in connection with Eq. (4.26) and Fig. 4.28.

The technical literature does not contain a large body of data on the stability or aging effects of thermistors so what is reported here are heterogeneous results of a number of observers on a few isolated tests. It may be generally concluded, however, that an aging effect may be expected which usually is of the nature of an increase with time of the electrical resistance which is not linear but logarithmic, resulting in smaller percentage changes in resistance with increased time. Preaging may be accomplished by heating or by the passage of higher than service current through the thermistor [79]. These have the effect also of accelerating the aging if the temperature is high enough.

The change of electrical resistance is sometimes attributed to a rearrangement in the distribution of the components of the mixture of oxides making up a semiconductor. Heat treatment is believed to play a major role in the dispersion of the components so that aging and preaging usually involve some kind of heat treatment. Muller and Stolen [81] tested two Western Electric 14A thermistors at 25°C over a period of 6 months and they report a decrease in resistance of about 50 Ω out of a total of approximately 100,000 Ω. This corresponds to an aging effect of about 0.012°C.

Figure 4.29 shows aging data [75] taken on 3/4-in. diameter discs of materials No. 1 and No. 2 (No. 1 is composed of manganese, nickel oxides; No. 2 is composed of oxides of manganese, nickel, and cobalt) with silver contacts and soldered leads. These discs were measured soon after production, were aged in an oven at 105°C, and were periodically tested at 24°C. The percentage change in resistance over its initial value is plotted versus the logarithm of the time in the aging oven. It is to be noted that most of the aging takes place in the first day or week so that if these discs were preaged for a week or a month and the subsequent change in resistance referred to the resistance after preaging, they would age only about 0.2% in one year. In a thermistor thermometer, this change in resistance would correspond to a temperature change of 0.05°C while thermistors mounted in an evacuated tube, or coated with a thin layer of glass age even less than those shown in the figure. For some applications such high

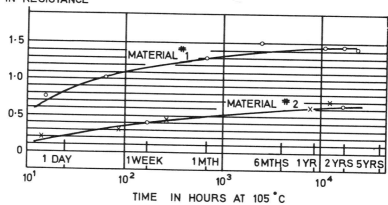

FIG. 4.29 Effect of aging in 105°C oven on thermistor characteristics; materials 1 and 2. (From Clark and Kobayashi, 1967, Ref. [84].)

stability is not essential and it is not necessary to give the thermometers special treatment. Thermistors have been used at high temperatures with satisfactory aging characteristics. Extruded rods of material No. 1 have been tested for stability by treating them for 2 months at a temperature of 300°C and − 75°C for a total of 700 temperature cycles, each lasting 1/2 hr. The resistance of typical units changed by less than 1%.

In order to determine the life of a 1A thermistor, Pearson [82] placed it in a circuit where an off-and-on current of 10 mA ac was repeated 30 sec over an extended period of time. Resistance measurements were made on the units periodically in order to determine their stability with time and the general trend was a rise in resistance during the first part of its life, after which the resistance became quite constant. Over a period of 15 months, during which time the thermometer was put through 650,000 heating cycles, the cold resistance did not increase by more than 7%. The resistance of the thermistor when hot was found to be equally stable.

The characteristics of both thermistors and thermocouples shift when exposed to high temperatures for lengthy periods of time [83]. For thermistors the resistance change varies logarithmically with time with higher temperatures accelerating the change. This suggests that if thermistors are subjected for several days or weeks to temperatures somewhat higher than those encountered in actual use, the major portion of the change would have occurred. For thermocouples, the change in voltage output becomes greater as the exposure time to high temperature is increased. Over a 3-month period in which thermistors and thermocouples were exposed to 365K for about 15 hr, the thermistor shifted a maximum of 0.11K, while the thermocouple shifted 0.17K. However, when new elements were tested and aged for 100 hr at 530K, the thermistors still shifted only 0.11K while the thermocouples shifted twice as much or 0.33K.

It was found by Muller and Stolen [81] that if the exciting potential is left impressed across the thermistor, a steady state is reached. This implied a resistance change of less than 1 Ω on daily measurement, the cold resistance of this thermistor at 0°C being 350,000 Ω. Short-range stability of a Western Electric 14A thermistor measured at 5-min intervals at 25°C (in Ω) is shown in Table 4.14. The authors [81] used the thermistor to measure small temperature difference in a laboratory experiment. The conclusion is that no significant change in resistance was detected which could not be attributed to measurement uncertainty.

To obtain a stable thermistor the following steps are generally thought to be necessary [75]. By these precautions remarkably good stabilities can be attained.

1. Select only semiconductors which are pure electronic conductors.
2. Select those which do not change chemically when exposed to the atmosphere at elevated temperatures.
3. Select one which is not sensitive to impurities likely to be encountered in manufacture or in use.
4. Treat it so that the degree of dispersion of the critical impurities is in

equilibrium or else that the approach to equilibrium is very slow at operating temperatures.

5. Make a contact which is intimate, sticks tenaciously, has an expansion coefficient compatible with the semiconductor, and is durable in the atmosphere to which it will be exposed.

6. In some cases, enclose the thermistor in a thin coating of glass or a material impervious to gases and liquids; the coating should have a suitable expansion coefficient.

7. Preage the unit for several days or weeks at a temperature somewhat higher than that to which it will be subjected.

Clark and Kobayashi [84, 85] have studied the general characteristics of thermistors to be used for temperature measurement. This includes the theory of their conductance properties, the dynamic response and steady state error of the thermistor temperature-sensing element, their stability and the resistance-temperature characteristics of approximately 300 commercially available thermistors from eight different manufacturers. Friedberg [79] describes a semiconducting film of ZnO used as a thermometer at 2K which had an electrical resistance of $5(10^5)$ Ω at liquid helium temperatures, and a sensitivity of approximately $5(10^4)$ Ω/K at 2K.

Germanium, with impurities consisting variously of arsenic, gallium, or indium, has become one of the most satisfactory materials for thermal resistance elements in the range 0.2K to 20K. This material possesses a negative temperature coefficient of resistance, a moderate level of resistance, high sensitivity of resistance change to temperature change, high reproducibility and stability to thermal cycling, and is readily manufactured and fabricated. The impurities are included in the germanium in controlled quantities to influence both the resistance-temperature characteristics and the sensitivity; a typical resistance-temperature curve for germanium "doped" with 0.001 at.% indium is shown in Fig. 4.30 for the temperature range 1K to 5K. This particular element was

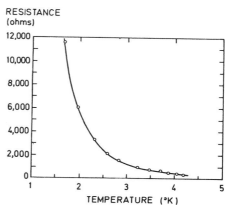

FIG. 4.30 Calibration curve for a germanium thermometer. (From Friedberg, 1955, Ref. [79].)

found to be highly reproducible over a period of several months and the thermometer was subjected to a number of warming and cooling cycles following which its resistance-temperature (R-T) characteristic could be reproduced to within ± 0.001K. The measuring current used was 0.01 ma although the author reports an increase of current to 0.1 ma did not appreciably influence the R-T characteristic [79].

Edlow and Plumb [86, 87] studied the reproducibility and temperature-resistance characteristics of a number of commercially available germanium thermometers; the germanium had either arsenic or gallium as the impurity. Their purpose was to find out if a germanium thermometer was sufficiently stable to be used as a basic secondary standard thermometer. As a consequence of their study the NBS adopted the germanium resistance thermometer as the basis for the NBS scale from 2K to 20K and used it for basic temperature calibration in this range. The determination of reproducibility was made by cycling the resistance element from 4.2K to 300K and measuring the resistance change at 4.2K which was then related to the corresponding temperature change. Two typical heating-cooling cycle tests are shown in Figs. 4.31 and 4.32. In each case the reproducibility is within ± 0.001K. In the case of resistor D (Fig. 4.32), the reproducibility is within 0.0005K after 86 cycles. Because of this high degree of stability the resistor of Fig. 4.32 became one of the NBS standard thermometers. This result is quite typical of that found by others. Kunzler, Geballe, and Hall [88], for example, cycled arsenic-"doped" germanium encapsulated in helium-filled thermometers as many as 50 times and found no evidence of calibration

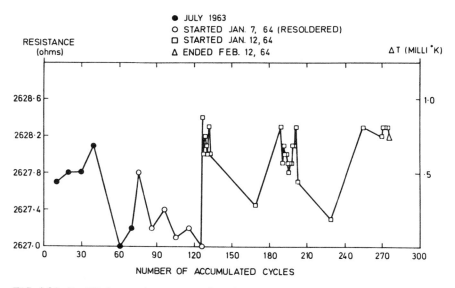

FIG. 4.31 Equilibrium resistance as a function of the number of accumulated cycles. $T = 4.2K$. (From Edlow and Plumb, 1966, Ref. [87].)

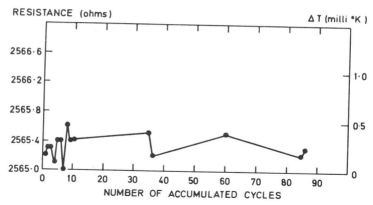

FIG. 4.32 Equilibrium resistance as a function of the number of accumulated cycles for a resistor D. $T = 4.2$K. (From Edlow and Plumb, 1966, Ref. [87].)

change of as much as 0.0001K. Furthermore, they report two such thermometers in use for 3 years on low-temperature experimental apparatus with no observable change in calibration. From results such as these it seems safe to conclude that germanium "doped" with a selected impurity is a suitable material for low-temperature thermometers below 20K.

The resistance-temperature calibration data for a number of encapsulated, hermetically sealed, arsenic-"doped" germanium thermometers was determined by Edlow and Plumb [86] in the range 2.1K to 5.0K. The resistance was measured at temperature intervals of 0.1K in a pressure-controlled helium liquid-vapor equilibrium cell and other measurements were made in a calibration comparator apparatus. The results agreed to within 0.001K; the basic standard temperature reference was the NBS 1958 He[4] scale (Appendix to this chapter), and some typical data are given in Fig. 4.33. A polynomial function was derived for each thermometer to represent its resistance-temperature calibration in the range 2.1K to 5.0K. The sensitivity of a germanium thermometer, dR/dT, manufactured by Cryo Cal, Inc. [89], is shown in Fig. 4.34 for the temperature range 2K to 28K. At 20K the sensitivity of this thermometer is 3 Ω/K which can be compared with a sensitivity of 0.0185 Ω/K for a platinum thermometer at the same temperature. The very large increase in sensitivity for germanium at temperatures below 20K is characteristic of this type of resistance thermometer.

The use of arsenic-doped germanium prepared from a single germanium crystal is reported by Kunzler et al. [88]; the germanium element is cut into the form of a bridge of dimensions 0.06 X 0.05 X 0.52 cm with side arms near each end for electrical connections. An encapsulated thermometer design is illustrated in Fig. 4.35. When covered with a platinum case it is filled with helium gas which limits its lowest useful temperature to about 0.25K. Bare bridges have also been used in applications such as adiabatic demagnetization experiments

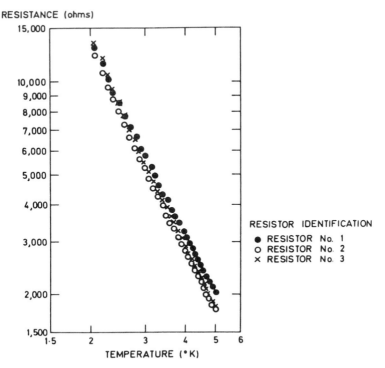

RESISTANCE (ohms)

RESISTOR IDENTIFICATION
● RESISTOR No. 1
○ RESISTOR No. 2
× RESISTOR No. 3

TEMPERATURE (°K)

FIG. 4.33 A plot of the resistance-temperature calibration data for resistors 1, 2, and 3. Temperatures were derived from liquid He⁴ vapor pressures. (From Edlow and Plumb, 1967, Ref. [86].)

where a thermometer with minimum heat capacity is required to give high response and in this case a lag time between the thermometer and sample was 0.1 sec. Higher currents are permitted with the bare bridge than with encapsulated models owing to the improved cooling permitted by the exposed germanium element. The resistance-temperature characteristics of the encapsulated model were found to be unaffected (within ± 0.0001K) by thermal cycling or aging over a period of several years. Cycling of the bare bridge between 4.2K and 293K several times produced only a few thousandths of a degree change in its calibration. The resistance-temperature characteristics of four typical encapsulated thermometers are shown in Table 4.15 and Fig. 4.36.

The R-T characteristics of a carbon thermometer is shown for comparison. At low temperatures the germanium thermometers have widely different electrical properties because the arsenic impurity concentration is not the same in each sample even though they were cut from the same germanium crystal. This is a good example of the extreme sensitivity of the electrical properties of these resistors to impurity concentration. However, the resistivity-temperature characteristics of an element are defined approximately by its resistivity at 4.2K [88]

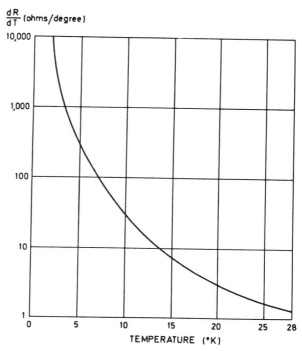

FIG. 4.34 Sensitivity dR/dT typical standard cryoresistor. (Courtesy of Cryo Cal, Inc.) (From Herder, 1968, Ref. [89].)

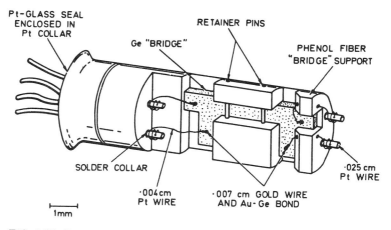

FIG. 4.35 Encapsulated germanium thermometer—model II, with cover removed. (From Kunzler, Geballe, and Hall, 1962, Ref. [88].)

TABLE 4.15 Resistance-temperature Characteristics of Germanium Thermometers Fabricated from Arsenic-doped Crystal 8-899-N

T (°K)	Sample no.							
	9-2R		15-2R		21-2R		27-2R	
	$R(\Omega)$	dR/dT	$R(\Omega)$	dR/dT	$R(\Omega)$	dR/dT	$R(\Omega)$	dR/dT
273	~ 1		~ 1		~ 1		~ 1	
77	2.0		1.9		1.8		1.7	
35	4.6	0.16	4.3	0.13	3.9	0.11	3.4	0.07
20	8.0	0.5	7.0	0.3	5.9	0.22	4.9	0.15
15	10.5	0.8	9.0	0.6	7.3	0.4	5.9	0.25
10	18.3	2.8	14.1	1.8	10.7	1.1	8.0	0.67
4.2	101	50	53	15.5	29	7.7	16.7	2.6
2	789	1000	216	200	77	46	29.5	11.2
1.5	2300		450		120		36.5	

Source. Reference [88].

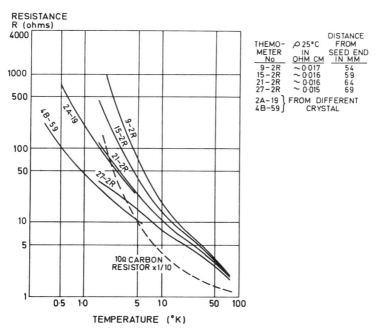

FIG. 4.36 Resistance-temperature characteristics of germanium thermal-sensing elements.

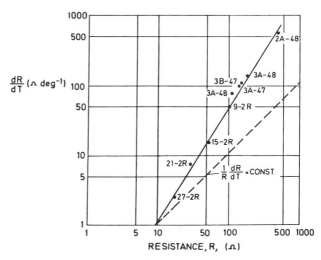

FIG. 4.37 Variation of dR/dT with resistance at 4.2K. (From Kunzler et al., 1962, Ref. [88].)

and in the range 2K to 35K, the ratio $(dR/R)(dT/T)$ is of the order of unity for all samples, a result found with germanium bridges cut from other crystals as well. This characteristic is responsible for the tremendous temperature sensitivity of the resistance thermometers. The sensitivity dR/dT at 4.2K of several thermometers, including those in Table 4.15, is shown in Fig. 4.37. An increased sensitivity may be achieved roughly according to $R^{3/2}$, by selecting a thermometer of higher resistance providing the instrumentation is compatible with the selected resistor.

The resistance of germanium is influenced by a magnetic field. The variation of the magnetoresistance with temperature was studied by Kunzler et al. [88] using a germanium bridge have a zero field resistance $(R_{H=0})$ of 200 Ω at 4.2K. The magnetoresistance $(\Delta R/R_{H=0})$ at 18 kG was found to be 0.16 at 4.2K and gradually increased with decreasing temperature, reaching a maximum value of 0.28 at 1.9K. Below 1.9K the magnetoresistance decreased to a value of $\Delta R/R_{H=0}$ equals 0.21 at 1.2K. At 4.2K the corresponding change in temperature calibration as a result of the magnetic field would be approximately 0.20K. The magnetoresistance of germanium is also slightly anisotropic, being somewhat less than 10% of the total magnetoresistance over 180 angular degrees at 4.2K and in an 18-kG field.

Other papers treating the germanium resistance thermometers have been published by Low [90] and Orlova, Astrov, and Medvedeva [91]. Antcliffe et al. [92] report the use of germanium thermometers below 1K. Their lowest temperature was 0.40K obtained by a He³ bath and the range of temperatures investigated was 4.20K to 0.40K. Between 4.20K and 1.20K the data were fitted to

$$R = CT^{-A} e^{-B/T} \tag{4.33}$$

with representative values of the constants as $A = 0.507\text{-}1.004$, $-B = 1.245\text{-}1.399$ and $C = 71.89\text{-}112.8$ for three resistors calibrated.

4.7.3 Magnetic Thermometry

The practical minimum temperature which may be produced by pumping helium is about 1K for He^4 and 0.5K for He^3 as below these temperatures the vapor pressure is too low to be maintained for most useful experimental purposes. To produce as well as measure temperatures below 0.5K the properties of paramagnetic substances, usually paramagnetic alums, are used. The low temperatures are achieved by demagnetizing these salts adiabatically from an initial state of a high magnetic field and a temperature of approximately 1K as the process of adiabatic demagnetization produces a rapid drop in the temperature of the salt owing to a decrease in its energy by the work of demagnetization. This is analagous to the drop in temperature of a compressed gas as it expands isentropically or adiabatically while doing work on its environment. Garrett [93] reports the limit of cooling by this process (electronic) is about 10^{-4} K but if the energy of nuclear spin is involved a temperature of 10^{-6} K is thought possible. A two-state demagnetization of a diluted chromium alum has produced a final temperature of 10^{-3} K with a field of 9,000 gauss (G) but to reach this temperature in a single demagnetization from 1K a field of 25,000 G would be required. Temperatures of a few hundredths of a degree absolute can be achieved without exceptional difficulties and those of the order of .001 degree can be obtained with somewhat greater effort [94, 95]. Table 4.16, prepared by Zemansky [96], summarizes some results of adiabatic demagnetization experiments and identified the paramagnetic salts used. Certain properties of commonly used paramagnetic salts are given in Table 4.17 [96].

At liquid helium temperatures, the orientation of the magnetic ions in the paramagnetic salts are influenced by a magnetic field in a significant way and contribute to both the energy and the entropy of the salt. Lattice vibrations also have energy and entropy contributions but at the low temperatures (< 1K) associated with adiabatic demagnetization experiments these effects are small. The partial spatial ordering of the paramagnetic ions in the presence of a magnetic field at constant temperature results in a decrease of the system entropy, as would be expected in an isothermal transition from a less ordered to a greater ordered state. Thus, the effect of an increase in the magnetic field on a paramagnetic salt is exactly analogous to the isothermal compression of a fluid or the isothermal extension of an elastic substance. This is illustrated in Fig. 4.38 which shows the temperature-entropy diagram of a paramagnetic salt for magnetic fields of strength H. States of lower entropy at a given temperature correspond to the magnetic fields of greater strength, i.e., $H_4 > H_3$, etc.

The process of magnetic cooling consists first of cooling a sample of

TABLE 4.16 Temperatures attained by adiabatic demagnetization of various paramagnetic salts

Experimenter	Date	Paramagnetic salt	Initial field (Oe)	Initial temp. ($^\circ$K)	Final magnetic temp., T^* ($^\circ$K)
Giauque and MacDougall	1933	Gadolinium sulfate	8,000	1.5	0.25
De Haas, Wiersma, and Kramers	1933	Cerium fluoride	27,600	1.35	0.13
		Dysprosium ethyl sulfate	19,500	1.35	0.12
		Cerium ethyl sulfate	27,600	1.35	0.085
De Haas and Wiersma	1934	Chromium potassium alum	24,600	1.16	0.031
	1935	Iron ammonium alum	24,075	1.20	0.018
		Alum mixture	24,075	1.29	0.0044
		Cesium titanium alum	24,075	1.31	0.0055
Kurti and Simon	1935	Gadolinium sulfate	5,400	1.15	0.35
		Manganese ammonium sulfate	8,000	1.23	0.09
		Iron ammonium alum	14,100	1.23	0.038
		Iron ammonium alum	8,300	1.23	0.072
		Iron ammonium alum	4,950	1.23	0.114
MacDougall and Giauque	1936	Gadolinium nitrobenzene sulfonate	8,090	0.94	0.098
Kurti, Laine, Rollin, and Simon	1936	Iron ammonium alum	32,000	1.08	0.010
Kurti, Laine, and Simon	1939	Iron ammonium alum	28,800	9.5	0.36
Ashmead	1939	Copper potassium sulfate	35,900	1.17	0.005
De Klerk[a]	1956	Chromium potassium alum	–	–	0.0029

Source. Reference [96].
[a]Reference [95].

paramagnetic salt to as low a temperature as possible in the absence of any significant magnetic field ($H \cong 0$) shown as state A, Fig. 4.38. This is usually accomplished in a helium cryostat pumped to a temperature T_1 of approximately 1K. While maintained at this temperature by the helium bath a magnetic field is introduced into the system which causes the entropy to decrease to state B, Fig. 4.38. In state B the paramagnetic salt is removed from the immediate influence of its cooling bath, usually accomplished by pumping away the helium surrounding the salt, and the magnetic field switched off. With the removal of the field the paramagnetic ions are reoriented to a state of greater disorder in a reversible-

TABLE 4.17 Properties of paramagnetic salts

Paramagnetic salt	Gram-ionic weight, M (g)	Density $\left(\dfrac{g}{cm^3}\right)$	Curie constant $C\left(\dfrac{cm^3\ deg}{g\ ion}\right)$
Cerium magnesium nitrate $2Ce(NO_3)_3\,3Mg(NO_3)_2\,24H_2O$	765	–	0.318
Chromium potassium alum $Cr_2(SO_4)_3\,K_2SO_4\,24H_2O$	499	1.83	1.86
Chromium methylammonium alum $Cr_2(SO_4)_3\,CH_3NH_3SO_4\,24H_2O$	492	1.645	1.87
Copper potassium sulfate $CuSO_4\,K_2SO_4\,6H_2O$	442	2.22	0.445
Iron ammonium alum $Fe_2(SO_4)_3(NH_4)_2SO_4\,24H_2O$	482	1.71	4.35
Gadolinium sulfate $Gd_2(SO_4)_3\,8H_2O$	373	3.010	7.85
Manganese ammonium sulfate $MnSO_4(NH_4)_2SO_4\,6H_2O$	391	1.83	4.36
Titanium cesium alum $Ti_2(SO_4)_3\,Cs_2SO_4\,24H_2O$	589	~ 2	0.118

Source. Reference [96].

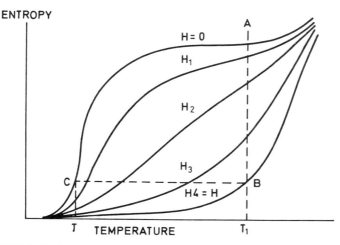

FIG. 4.38 Temperature-entropy diagram for a paramagnetic salt under the influence of different magnetic fields. (From Timmerhaus, 1963, Ref. [40].)

adiabatic process with a corresponding flow of work to the environment by virtue of the magnetic rearrangement. The consequence of this is an isentropic drop in energy of the salt to a state of zero magnetic field and lower temperature T, shown as state C, Fig. 4.38. A cryostat for doing this is described by De Klerk and Steenland [97] and shown in Fig. 4.39. The time required to reduce the magnetic field is about 1 sec which may be compared with the spin-spin relaxation time of 10^{-9} sec and the spin-lattice relaxation time of 10^{-3} sec [93, 98].

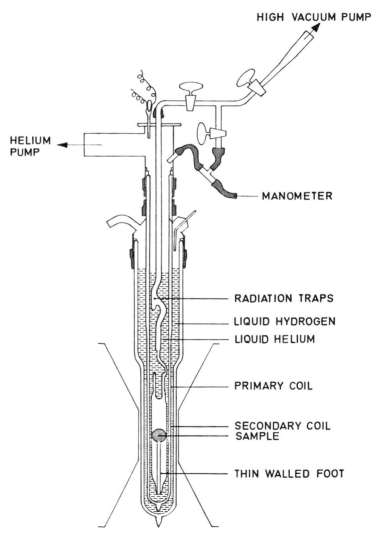

FIG. 4.39 Typical Leiden demagnetization cryostat, one-fifth of real size. (From De Klerk and Steenland, 1955, Ref. [97].)

The temperature T to which the paramagnetic salt was cooled is computed from Curie's law:

$$M = \frac{C}{T} H \tag{4.34}$$

or

$$M = \chi H \tag{4.35}$$

with

$$\chi = \frac{C}{T} \tag{4.36}$$

where M is the magnetization or magnetic moment, C is Curie's constant, H is the magnetic field strength and χ is the magnetic susceptibility. The magnetic susceptibility is related to the permeability μ and the magnetic flux intensity B by

$$B = \mu H \tag{4.37}$$

and

$$4\pi\chi = \mu - 1 \tag{4.38}$$

Substances are classified according to the value of χ: *diamagnetic* for $\chi < 0$, *paramagnetic* for $\chi > 0$, and *ferromagnetic* for $\chi \gg 0$.

Departures from Curie's law result from the effects of the shape of the paramagnetic salt and are expressed by the Curie-Weiss law

$$\chi = \frac{C}{T - \Delta} \tag{4.39}$$

where Δ is the Curie-Weiss constant and is equal to zero for a spherical sample. For this reason spherical samples are used, as in Fig. 4.39, if possible or, if not, the results are corrected to that of a spherical sample. Typical spherical and spheroidal sample tubes are illustrated by De Klerk [94] in Fig. 4.40. The magnetic temperature computed from Eq. (4.36) or (4.39) is not a true thermodynamic temperature owing to the empirical constants C and Δ which do not follow from considerations of the second law of thermodynamics. These temperatures will be denoted as T^* and will be related to the thermodynamic temperature T later.

For either spherical or spheroidal samples the defined magnetic temperature is written [94] as

$$T^*_{\text{sphere}} = \frac{C}{\chi_{\text{sphere}}} \tag{4.40}$$

where

$$\chi_{\text{sphere}} = \frac{\chi_{\text{spheroid}}}{1 + (4\pi/3 - \alpha)\chi_{\text{spheroid}}} \tag{4.41}$$

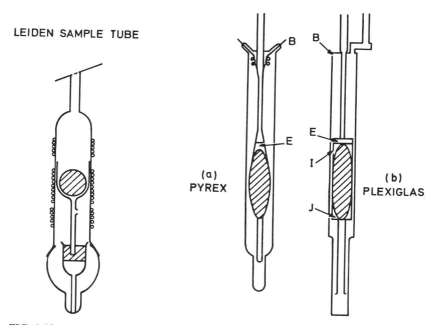

FIG. 4.40 Typical paramagnetic salt samples. (From De Klerk, 1956, Ref. [94].)

$$\alpha = 4\pi\left(\frac{1 - e^2}{e^2}\right)\left[\frac{1}{2e}\ \log_e\left(\frac{1 + e}{1 - e}\right) - 1\right] \tag{4.42}$$

and
$$e = (1 - \epsilon^2)^{1/2} \tag{4.43}$$

ϵ is defined as the eccentricity of the spheroid as outlined by Maxwell [99].

The magnetic behavior of four paramagnetic salts showing their conformance with Curie's law at liquid helium temperatures and higher is given in Figs. 4.41 and 4.42, taken from Kittel [100]. At low temperatures saturation effects cause departure from Curie's law at a certain magnetic field strength, as illustrated in Fig. 4.42.

The determination of T^*_{sphere} or $T - \Delta$ is made by measuring the magnetic inductance produced by the paramagnetic salt in an electrical measuring circuit, as shown in Fig. 4.43 and the magnetic inductance is proportional to the magnetic susceptibility χ. The magnetic susceptibility of the salt is determined at 1K and 4K using the 1958 He^4 scale and then extrapolated to lower temperatures for use during demagnetization experiments. T^*_{sphere} is a good approximation to T for a few tenths of a degree below 1K but at lowest temperatures it

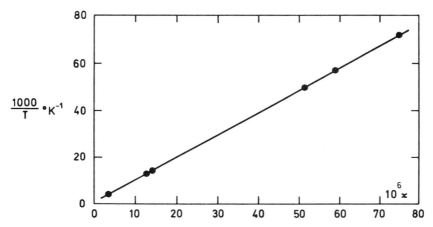

FIG. 4.41 Plot of susceptibility per gram vs. reciprocal temperature for powdered $CuSO_4 \cdot K_2SO_4 \cdot 6H_2O$, showing the Curie law temperature dependence. (From Kittel, 1956, Ref. [100].)

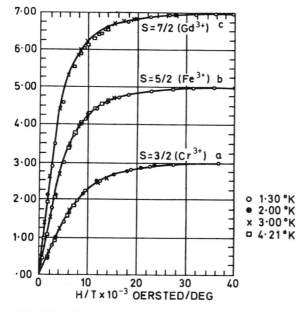

FIG. 4.42 Plot of magnetic moment H/T for spherical samples of (a) potassium chromium alum, (b) ferric ammonium alum, and (c) gadolinium sulfate octalhydrate. (From Kittel, 1956, Ref. [100].)

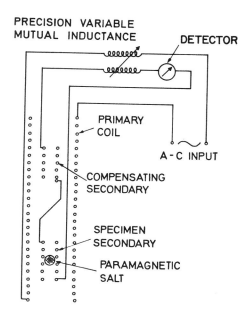

PRECISION VARIABLE
MUTUAL INDUCTANCE

DETECTOR

PRIMARY
COIL

A - C INPUT

COMPENSATING
SECONDARY

SPECIMEN
SECONDARY

PARAMAGNETIC
SALT

FIG. 4.43 Electrical circuit for magnetic thermometer. (From Timmerhaus, 1963, Ref. [40].)

fails to represent T since for all known salts a maximum in χ has been found which produces a minimum in T^*_{sphere}. Demagnetization from still higher fields results in progressively lower values of χ and higher T^*_{sphere} and T may differ by an order of magnitude. The final temperature T^*_{sphere} for a chromium potassium alum as a function of magnetic field strength is shown in Fig. 4.44. In this case a minimum T^*_{sphere} is approached asymptotically for progressively higher fields.

The thermodynamics of a paramagnetic salt indicate [93, 94] that the entropy S is a function of T and H. Thus

$$S = f(T, H) \tag{4.44}$$

This may also be written [93, 94] as

$$\text{T}ds = C_H dT + T\left(\frac{\partial M}{\partial T}\right)_H dH \tag{4.45}$$

where

$$C_H = \left(\frac{\partial E}{\partial T}\right)_H = T\left(\frac{\partial S}{\partial T}\right)_H \tag{4.46}$$

For the isentropic demagnetization, state B to C, Fig. 4.38, then Eq. (4.45) indicates that

$$T - T_1 = \int_0^{H_4} \frac{T}{C_H}\left(\frac{\partial M}{\partial T_H}\right) dH \tag{4.47}$$

FINAL MAGNETIC
TEMPERATURE (T_f^*)

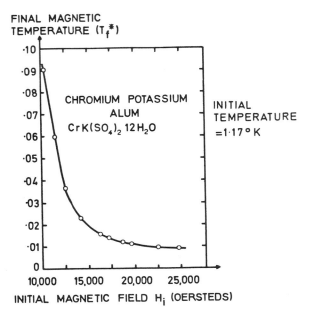

INITIAL MAGNETIC FIELD H_i (OERSTEDS)

FIG. 4.44 De Klerk's results in the adiabatic demagnetization of chromium potassium alum. (From Zemansky, 1957, Ref. [96].)

Since $(\partial M/\partial T)_H < 0$ by Eq. (4.34), the temperature T will always be less than T_1 for adiabatic demagnetization processes.

The relation between T_{sphere}^* (to be identified as T^* hereafter) and the thermodynamic temperature T is determined from the thermodynamic definition

$$T = \left(\frac{dQ}{dS}\right)_{rev} \tag{4.48}$$

This may also be written

$$T = \frac{(dQ/dT^*)_{H=0}}{(dS/dT^*)_{H=0}} \tag{4.49}$$

Now (dQ/dT^*) may be found from heating experiments in which

$$\left(\frac{dQ}{dT^*}\right)_{H=0} = m\left(\frac{dh}{dT^*}\right)_{H=0} \tag{4.50}$$

in which m is the mass of a sample of paramagnetic salt and h is its enthalpy per unit mass.

The quantity $(dS/dT^*)_{H=0}$ is determined from a series of demagnetization experiments from T_1 (Fig. 4.38) and a number of different magnetic fields such as H_1, H_2, H_3, and H_4. The entropies corresponding to the isotherm T_1 and the various fields is determined from Eq. (4.45) as

$$S(H, T_1) - S(O, T_1) = \int_0^H \left(\frac{\partial M}{\partial T}\right)_H dH \qquad (4.51)$$

The corresponding temperatures are then computed from Eq. (4.47), giving a curve of S vs. T^* for a zero field ($H = 0$). From this the slope (dS/dT^*) may be derived and T computed from Eq. (4.49).

The relationship between T and T^* for several paramagnetic salts is shown in Figs. 4.45 and 4.46.

NOMENCLATURE

A constant
B constant
C constant, Curie constant, Eq. (4.33)

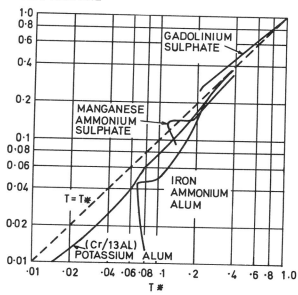

FIG. 4.45 Deviation of Curie temperature T^* from thermodynamic temperature for several paramagnetic sales. (From Timmerhaus, 1963, Ref. [40].)

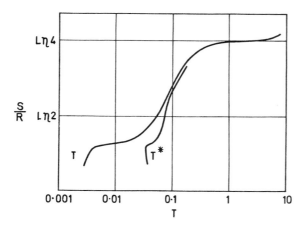

FIG. 4.46 Entropy of potassium chrome alum as a function of absolute temperature and of T^*. (From Kittel, 1956, Ref. [100].)

D	constant
dE/dF	thermoelectric power
E	thermocouple emf; constant, Eq. (4.26); internal energy
$F(t)$	arbitrary function of temperature
H	magnetic field strength
h	enthalpy
K	constant
l	length
M	magnetic moment
m	mass
P	arbitrary temperature dependent property, pressure
Q_0	quantity of heat rejected by a Carnot engine to a heat reservoir at temperature t_0
Q_1	quantity of heat absorbed by a Carnot engine from a heat reservoir at temperature t_1
R	resistance
R_0	resistance at the ice point
R_t	resistance at temperature t
r	ratio of two thermodynamic temperatures
T	absolute thermodynamic temperature (Kelvin scale)
T^*	thermodynamic temperature of the triple point of water, magnetic temperature
t	temperature according to an arbitrary temperature scale, platinum resistance temperature
V	volume

Greek Symbols

α	temperature coefficient of any arbitrary property P, constant
β	constant
Δ	Curie-Weiss constant
ΔE^*	potentiometer sensitivity
ΔT^*	uncertainty in temperature indication
δ	constant
θ	temperature according to an arbitrary temperature scale
ρ	electrical resistivity
σ	electrical conductivity
χ	magnetic susceptibility

REFERENCES

1. Clark, J. A., Advances in Heat Transfer, in T. F. Irvine, Jr., and J. P. Hartnett (eds), "Cryogenic Heat Transfer," vol. 5, Academic, New York, 1968.
2. Stimson, H. F., The Text Revision of the International Temperature Scale of 1948, in American Institute of Physics, "Temperature, Its Measurement and Control in Science and Industry," vol. 3, pt. 1, pp. 59–66, Reinhold, New York, 1962.
3. Stimson, H. F., International Practical Temperature Scale of 1948: Text Revision 1960, Nat. Bur. Stand. (U.S.) Monogr. 37, September 8, 1961. (See also J. Res. Nat. Bur. Stand. (U.S.) vol. 65A, no. 3, 1961.
4. Terrien, J., News from the International Bureau of Weights and Measures, Metrologia, vol. 4, no. 1, p. 41, 1968.
5. Preston-Thomas, H., and Bedford, R. E., Practical Temperature Scales Between 11°K and 273°K, Metrologia, vol. 4, no. 1, pp. 14–30, 1968.
6. Hust, J. G., Personal communication, Cryogenics Division, National Bureau of Standards, Boulder, Colo., April 4, 1968.
7. Hoge, H. J., and Brickwedde, F. G., Establishment of a Temperature Scale for the Calibration of Thermometers between 14 and 83°K, J. Research NBS, vol. 22, pp. 351–373, 1939.
8. Hoge, H. J., A Practical Temperature Scale Below the Oxygen Point and a Survey of Fixed Points in this Range, in American Institute of Physics, "Temperature, Its Measurement and Control in Science and Industry," vol 1, pp. 141–156, Reinhold, New York, 1941.
9. Orlova, M. P., Sharevskaya, D. I., Astrov, D. N., and Krutikova, S., The Derivation of the Provisional Reference Table CCT-64, $T = f(W)$ for Platinum Resistance Thermometers for the Range from 12 to 273.15°K, Pt. I. The Derivation of the Table for the Range from 12 to 95°K, Metrologia, vol. 2, no. 1, pp. 6–10, 1966.
10. Furukawa, G. T., and Reilly, M. L., Application of Precise Heat Capacity Data to the Analysis of the Temperature Intervals of the NBS (1955) and the International Practical Temperature Scales in the Region of 90°K, J. Res. Nat. Bur. Stand. (U.S.), vol. 69A, no. 1, pp. 5–12, 1965.
11. Plumb, H. H., and Cataland, G., Acoustical Thermometer and the National Bureau of Standards Provisional Temperature Scale 2–20 (1965), Metrologia, vol. 2, no. 4, pp. 127–139, 1966.
12. Moessen, G. W., Astonand, J. G., and Ascah, R. G., The Pennsylvania State University Thermodynamic Temperature Scale Below 90°K and the Normal Boiling Points of Oxygen and Normal Hydrogen on the Thermodynamic Scale, in American Institute of

Physics, "Temperature, Its Measurement and Control in Science and Industry," vol. 3, pp. 90–102, Reinhold, New York, 1962.

13. Borovick-Romanov, A. S., Strediclov, P. R., Orlova, M. P., and Astrov, D. N., The I.M.P.R. Temperature scale for 10 to 90°K Region, in American Institute of Physics, "Temperature, Its Measurement and Control in Science and Industry," vol. 3, pp. 113–128, Reinhold, New York, 1962.

14. Barber, C. R., New Gas-Thermometer Measurements over the Range from 10° to 90°K and the Extension of the International Temperature Scale below 90°K, in "Progress in Refrigeration Science and Technology," vol. 1, pp. 174–178, Pergamon, New York, 1960.

15. Barber, C. R., The Construction of a Practical Scale of Temperature at Sub-Zero Temperatures, *Proc. Inst. Refrig.*, vol. 58, pp. 153–168, 1961.

16. Barber, C. R., The Establishment of a Practical Scale of Temperature for the Range 10–90°K, *J. Appl. Phys. (Brit.)*, vol. 13, pp. 235–241, 1962.

17. Roder, H. M. Irregularities in the NBS (1955) Provisional Temperature Scale. *J. Nat. Bur. Stand. (U.S.)*, vol. 69A, pp. 527–530, 1965.

18. Beattie, J. A., Gas Thermometry, in American Institute of Physics, "Temperature, Its Measurement and Control in Science and Industry," vol. 2, pp. 63–97, Reinhold, New York, 1955.

19. Barber, C. R., Helium Gas Thermometry at Low Temperatures, in American Institute of Physics, "Temperature, Its Measurement and Control in Science and Industry," vol. 3, pp. 103–112, Reinhold, New York, 1962.

20. Beattie, J. A., The Thermodynamic Temperature of the Ice-Point, in American Institute of Physics, "Temperature, Its Measurement and Control in Science and Industry," vol. 1, pp. 74–88, Reinhold, New York, 1941.

21. Beattie, J. A., Benedict, M., and Kaye, J., An Experimental Study of the Absolute Temperature Scale, *Proc. Amer. Acad. Arts Sci.*, vol. 74, no. 11, pp. 327–398, December 1941.

22. Keyes, F. G., Gas Thermometer Scale Corrections Based on an Objective Correlation of Available Data for Hydrogen, Helium and Nitrogen, in American Institute of Physics, "Temperature, Its Measurement and Control in Science and Industry," vol. 1, pp. 45–59, Reinhold, New York, 1941.

23. Van Dijk, H., Concerning Temperature Units and Temperature Scales, *Z. Angew. Phys.*, vol. 15, pp. 561–566, 1963.

24. Ramsey, N. F., Thermodynamics and Statistical Mechanics of Negative Absolute Temperatures, in American Institute of Physics, "Temperature, Its Measurement and Control in Science and Industry," vol. 3, pt. 1, Reinhold, New York, 1962.

25. Hertzfeld, C. M., The Thermodynamic Temperature Scale, Its Definition and Realization, in American Institute of Physics, "Temperature, Its Measurement and Control in Science and Industry," vol. 3, pt. 1, p. 49, Reinhold, New York, 1962.

26. Burgess, G. F., International Temperature Scale, *J. Res. Nat. Bur. Stand. (U.S.)*, vol. 1, pp. 635–637, 1928.

27. Stimson, H. F., The International Temperature Scale of 1948, *J. Res. Nat. Bur. Stand. (U.S.)*, vol. 42, p. 209, 1949.

28. Hall, J. A., The International Temperature Scale, in American Institute of Physics, "Temperature, Its Measurement and Control in Science and Industry," vol.2, pp. 115–139, Reinhold, New York, 1955.

29. Corruccini, R. J., Differences between the International Temperature Scales of 1948 and 1927, *J. Res. Nat. Bur. Stand. (U.S.)*, vol. 43, pp. 133–136, 1949.

30. Heuse, W., and Otto, J., A New Gas Thermometer Determination of Some Fixed Points below 0° in Connection with Pressure and Resistance Thermometers, *Ann. Phys.*, vol. 9, no. 5, pp. 486–504, 1931.

31. Keesom, W. H., and Dammers, B. G., Comparison of Some Platinum Thermometers with the Helium Thermometer between 0 and − 183°C, *Physica*, vol. 2, pp. 1080–1090, 1935.

32. Barber, C. R., and Horsford, A., Differences between the Thermodynamic Scale and the International Practical Temperature Scale from 0°C to − 183°C, *Metrologia*, vol. 1, no. 3, pp. 75–80, 1956.

33. Roder, H. M., Irregularities in the NBS (1955) Provisional Temperature Scale, *J. Res. Nat. Bur. Stand. (U.S.)*, vol. 69A, pp. 527–530, 1965.

34. Terrien, J., and Preston-Thomas, H., Progress in the Definition and in the Measurement of Temperature, *Metrologia*, vol. 3, no. 1, pp. 29–31, 1967.

35. Preston-Thomas, H., and Kirby, C. G. M., Gas Thermometer Determinations of the Thermodynamic Temperature Scale in the Range − 183°C to 100°C., *Metrologia*, vol. 4, no. 1, pp. 30–40, 1968.

36. Scott, R. B., Low Temperature Scales from 90° to 5°K, in American Institute of Physics, "Temperautre, Its Measurement and Control in Science and Industry," vol. 2, pp. 179–184, Reinhold, New York, 1955.

37. Giauque, W. F., Buffington, R. M., and Schulze, W. A., Copper-Constantan Thermocouples and the Hydrogen Thermometer Compared from 15° to 283° Absolute, *J. Amer. Chem. Soc.*, vol. 49, p. 2343, 1927.

38. Barber, C. R., and Hayes, J. G., The Derivation of the Original Reference Table CCT-64, $T = f(W)$ for Platinum Resistance Thermometers for the Range from 12 to 273.15°K. Pt. II. The Derivation of the Table for the Range from 90 to 273.15°K, *Metrologia*, vol. 2, no. 1, pp. 11–13, 1966.

39. Timmerhaus, K. D., Low Temperature Thermometry, in R. W. Vance and W. M. Duke (eds.), "Applied Cryogenic Engineering," Wiley, New York, 1962.

40. Timmerhaus, K. D., Measurement of Low Temperatures, in R. W. Vance (ed.) "Cryogenic Technology," Wiley, New York, 1963.

41. Brickwedde, F. G., The 1958 He⁴ Scale of Temperatures. Pt. I. Introduction; Van Dijk, H., Durieux, M., Clement, J. R., Logan, J. K. Pt. 2. Table for the 1958 Temperature Scale, *Nat. Bur. Stand. (U.S.) Monogr.* 10, June 17, 1960. (See also *J. Res. Nat. Bur. Stand. (U.S.)*, vol. 64A, no. 1, 1960.

42. Clement, J. R. The 1958 He⁴ Temperature Scale, in American Institute of Physics, "Temperature, Its Measurement and Control in Science and Industry," vol. 3, pt. 1, pp. 67–74, Reinhold, New York, 1962.

43. Arp, V., and Kropschot, R. H., Helium in R. W. Vance and W. M. Duke (eds.) "Applied Cryogenic Engineering," Wiley, New York, 1962.

44. Sherman, R. H., Sydoriak, S. G., and Roberts, T. R., The 1962 He³ Scale of Temperatures. 4. Tables, *J. Res. Nat. Bur. Stand. (U.S.)* vol. 68A, no. 6, pp. 579–588, 1964.

45. Sydoriak, S. G., Sherman, R. H., and Roberts, T. R., The 1962 He³ Scale of Temperatures, pts. 1–3, *J. Res. Nat. Bur. Stand. (U.S.)* vol. 68A, no. 6, pp. 547–578, 1964.

46. Barber, C. R., and Van Dijk, H., The Provisional International Practical Temperature Scale of Temperatures of 1967, Document no. 3, presented to the CCT, 1967.

47. Lovejoy, D. R., Quelque points d'ebullition et point triples au-dessons de 0°C, CCT 6th Session, 23, 1962.

48. McLaren, E. H., and Murdock, E. G., The Freezing Points of High Purity Metals as Precision Temperature Standards VIII b. Sb: Liquidus Points and Alloy Melting Ranges of Seven Samples of High Purity Antimony; Temperature Scale Realization and Reliability in the Range 0–631°C, *Can. J. Phys.* (in press).

49. Finch, D. I., General Principles of Thermoelectric Thermometry, in American Institute of Physics, "Temperature, Its Measurement and Control in Science and Industry," vol. 3, pp. 3–32, Reinhold, New York, 1962.

50. Benedict, R. P., Ashby, H. F., Improved Reference Tables for Thermocouples, in American Institute of Physics, "Temperature, Its Measurement and Control in Science and Industry," vol. 3, pt. 2, pp. 51–64, Reinhold, New York, 1962.

51. Caldwell, F. R., Thermocouple Materials, in American Institute of Physics, "Temperature, Its Measurement and Control in Science and Industry," vol. 3, pp. 81–134, Reinhold, New York, 1962.

52. Weber, R. L., "Heat and Temperature Measurement," Prentice-Hall, Englewood, New Jersey, 1950.

53. Baker, H. D., Ryder, E. A., and Baker, N. H., "Temperature Measurement in Engineering," vol. 1, Wiley, New York, 1955.

54. Dike, P. H., "Thermoelectric Thermometry," Leeds and Northrup, Philadelphia, 1954.

55. Powell, R. L., Bunch, M. D., and Caywood, L. P., Low Temperature Thermocouple Thermometry, in K. D. Timmerhaus (ed.), "Advances in Cryogenic Engineering," vol. 6, pp. 537–542, Plenum, New York, 1961. (See also, "Temperature, Its Measurement and Control in Science and Industry," vol. 3, pp. 65–67, Reinhold, New York, 1962.)

56. Powell, R. L., Bunch, M. D., and Corruccini, R. J., Low Temperature Thermocouples – 1. Gold Cobalt or Constantan vs. Copper or "Normal Silver," *Cryogenics*, pp. 139–150, March 1961.

57. White, W. P., Potentiometers for Thermoelectric Measurements, in American Institute of Physics, "Temperature, Its Measurement and Control in Science and Industry," vol. 1, pp. 265–279, Reinhold, New York, 1941.

58. Mueller, E. F., Precision Resistance Thermometry, in American Institute of Physics, "Temperature, Its Measurement and Control in Science and Industry," vol. 1, pp. 162–179, Reinhold, New York, 1941.

59. Stimson, H. F., Precision Resistance Thermometry and Fixed Points, in American Institute of Physics, "Temperature, Its Measurement and Control in Science and Industry," vol. 2, pp. 141–168, Reinhold, New York, 1955.

60. Werner, F. D., Some Recent Developments in Applied Platinum Resistance Thermometry, in American Institute of Physics, "Temperature, Its Measurement and Control in Science and Industry," vol. 3, pt. 2, pp. 299–304, Reinhold, New York, 1962.

61. Dauphinee, T. M., Potentiometric Methods of Resistance Measurement, in American Institute of Physics, "Temperature, Its Measurement and Control in Science and Industry," vol. 3, pt. 1, pp. 269–283, Reinhold, New York, 1962.

62. Sonntag, R. E., Report of Calibration, National Bureau of Standards, Text no. G35936A, completed Sept. 24, 1965. Communication to J. A. Clark, University of Michigan, Ann Arbor.

63. Corruccini, R. J., Temperature Measurements in Cryogenic Engineering, in K. D. Timmerhaus (ed.), "Advances in Cryogenic Engineering," vol. 8, pp. 315–334, Plenum, New York, 1963.

64. Dean, J. W., and Richards, R. J., Hydrostatic Pressure Effects in Carbon and Germanium Thermometers, in K. D. Timmerhaus (ed.), "Advances in Cryogenic Engineering," vol. 14, Plenum, New York, 1969.

65. Dupre, A., Itterbeek, A., Michiels, L., and Van Neste, L., The Use of Graphite Thermometers in Heat Conductivity Experiments below 1°K, *Cryogenics*, vol. 4, no. 6, pp. 354–357, December 1964.

66. Lindenfeld, P., Carbon and Semiconductor Thermometers for Low Temperatures, in American Institute of Physics, "Temperature, Its Measurement and Control in Science and Industry," vol. 3, pt. 1, pp. 339–405, Reinhold, New York, 1962.

67. Clement, J. R., and Quinell, E. H., Low Temperature Characteristics of Carbon-Composition Thermometers, *Rev. Sci. Instrum.*, vol. 23, no. 5, pp. 213–217, May 1952.

68. Schulte, E. H., Carbon Resistors for Cryogenic Temperature Measurement, *Cryogenics*, vol. 6., no. 6, pp. 321–323, December 1966.

69. Mikhailov, M. N., and Kaganovskii, A. Ya., Carbon Resistance Thermometers for Low Temperatures, *Cryogenics*, vol. 2, no. 6, pp. 98–100, December 1961.

70. Greene, N. E., "Three-Phase, Solid-Liquid-Vapor, Equilibrium in the System Hydrogen-Helium," Ph.D. thesis, Department of Mechanical Engineering, University of Michigan, Ann Arbor, December 1966.

71. Herr, A. C., Terbeek, H. G., and Tieferman, M. W., Temperature, Suitability of Carbon Resistors for Field Measurement of Temperatures in the Range 30 to 100°R, in American Institute of Physics, "Temperature, Its Measurement and Control in Science and Industry," vol. 3, pt. 2, pp. 355–359, Reinhold, New York, 1962.

72. Maurer, R. J., The Electrical Properties of Semi-Conductors, *J. Appl. Phys.*, vol. 16, p. 563, October 1945.

73. "Handbook of Physics and Chemistry," 34th ed., pp. 2185–2193, Chemical Rubber Publishing, New York, 1952–1953.

74. Seitz, F., The Basic Principles of Semi-Conductors, *J. Appl. Phys.*, vol. 16, p. 553, October 1945.

75. Becker, J. A., Green, C. B., and Pearson, G. L., Properties and Uses of Thermistors—Thermally Sensitive Resistors, *Elec. Eng.* vol. 65, p. 711, 1946.

76. Brown, L. G., Air Rate Measurement in Vertical Down Flow of Fluidized Solids, S. B. thesis, Department of Mechanical Engineering, Massachusetts Institute of Technology, Cambridge, June 1955.

77. Stoll, A. M., Wide Range Thermistor Radiometer for Measurement of Skin Temperatures and Environmental Radiant Temperature, *Rev. Sci. Instrum.*, vol. 25, no. 2, p. 184, 1954.

78. Wilson, A. H., Theory of Electronic Semi-Conductors, *Proc. Roy. Soc. (London)*, vol. 133, p. 458, 1931.

79. Friedberg, S. A., Semi-Conductors as Thermometers, in American Institute of Physics, "Temperature, Its Measurement and Control in Science and Industry," vol. 2, pp. 359–382, Reinhold, New York, 1955.

80. Wilson, A. H., "Semi-Conductors and Metals," Cambridge, New York, 1939.

81. Muller, R. H., and Stolen, H. J., Use of Thermistors in Precise Measurement of Small Temperature Differences, *Anal. Chem.*, vol. 25, pp. 1103–1106, July 1953.

82. Pearson, G. L., Life Characteristics of a 1A Thermistor, *Bell Tel. Lab. Rec.*, vol. 19, p. 107, December 1940.

83. Benedict, R. P., Thermistors vs. Thermocouples, *Elec. Manuf.*, August 1954, p. 120.

84. Clark, J. A., and Kobayashi, Y., Property of Thermistors, *Tech. Rep.* 188, Contract DA-19-016-ENG-3204, U.S. Army Materiel Command, New Hampshire, May 1967.

85. Clark, J. A., and Kobayashi, Y., Resistance-Temperature Characteristics of Some Commercially Available Thermistors, *Suppl. Tech. Rep.* 188 (Ref. [84]), May 1967.

86. Edlow, M. H., and Plumb, H. H., Germanium Resistance Thermometry in the Range 2.1 to 5.0°K, *J. Res. Nat. Bur. Stand. (U.S.)*, vol. 71C, no. 1, pp. 29–41, January–March 1967.

87. Edlow, M. H., and Plumb, H. H., Reproducibility of Germanium Resistance Thermometers at 4.2°K, *J. Res. Nat. Bur. Stand. (U.S.)*, vol. 70C, no. 4, pp. 245–255, October–November 1966.

88. Kunzler, J. E., Geballe, T. H., and Hall, G. W., Jr., Temperature, Germanium Resistance Thermometers, in American Institute of Physics, "Temperature, Its Measurement and Control in Science and Industry," vol. 3, pt. 1, pp. 391–397, Reinhold, New York, 1962.

89. Herder, T., Properties of Germanium Thermometers, private communication to J. A. Clark, Cryo Cal, Inc. Riviera Beach, Fla., October 1968.

90. Low, F. J., Gallium-Doped Germanium Resistance Thermometers, in K. D. Timmerhaus (ed.), "Advances in Cryogenic Engineering," vol. 7, pp. 514–516, Plenum, New York, 1962.

91. Orlova, M. P., Astrov, D. N., and Medvedeva, L. A., A Germanium Thermometer for Low Temperatures, *Cryogenics*, pp. 165–167, June 1965.
92. Antcliffe, G. A., Einspruch, N. G., Pinatti, D. G., and Rorschauch, H. E., Jr., Germanium Resistance Thermometry at Temperatures below 1°K, *Rev. Sci. Instrum.*, vol. 39, pp. 254–255, February 1968.
93. Garrett, C. G. B., "Magnetic Cooling," Harvard, Cambridge, Massachusetts, 1954.
94. De Klerk, D., Adiabatic Demagnetization, Low Temperature Physics II, in S. Flugge (ed.), "Encyclopedia of Physics," vol. 15, pp. 38–209, Springer-Verlag, Berlin, 1956.
95. De Klerk, D., Temperature, Thermometry below 1°K, in American Institute of Physics, "Temperature, Its Measurement and Control in Science and Industry," vol. 2, pp. 251–264, Reinhold, New York, 1955.
96. Zemansky, M. W., "Heat and Thermodynamics," 4th ed., McGraw-Hill, New York, 1957.
97. De Klerk, D., and Steenland, M. J., Adiabatic Demagnetization, in C. J. Gorter (ed.) "Progress in Low Temperature Physics," vol. 1, chap. 14, Interscience, New York, 1955.
98. Jackson, L. C., "Low Temperature Physics," 4th ed., Wiley, New York, 1955.
99. Maxwell, J. C., "A Treatise on Electricity and Magnetism," 3rd ed., vol. 2, p. 69, Oxford, London, 1904.
100. Kittel, C., "Introduction to Solid State Physics," 2nd ed., Wiley, New York, 1956.

APPENDIX: VAPOR PRESSURE OF ^{4}He (1958 SCALE) IN MICRONS (10^{-3} mm Hg) AT 0°C AND STANDARD GRAVITY (980.665 cm/sec^2)

T (K)	μ	T (K)	μ	T (K)	μ
0.50	0.016342	.66	1.0574	.81	13.187
		.67	1.2911	.82	15.147
.51	.022745	.68	1.5682	.83	17.348
.52	.031287	.69	1.8949	.84	19.811
.53	.042561			.85	22.561
.54	.057292	0.70	2.2787		
.55	.076356			.86	25.624
				.87	29.027
.56	.10081	.71	2.7272	.88	32.800
.57	.13190	.72	3.2494	.89	36.974
.58	.17112	.73	3.8549		
.59	.22021	.74	4.5543		
		.75	5.3591	0.90	41.581
0.60	0.28121				
		.76	6.2820	.91	46.656
.61	.35649	.77	7.3365	.92	52.234
.62	.44877	.78	8.5376	.93	58.355
.63	.56118	.79	9.9013	.94	65.059
.64	.69729			.95	72.386
.65	.86116	0.80	11.445	.96	80.382

T (K)	μ	T (K)	μ	T (K)	μ
.97	89.093	1.36	1724.91	1.75	10395.9
.98	98.567	1.37	1825.58		
.99	108.853	1.38	1930.79	1.76	10788.2
		1.39	2040.67	1.77	11191.2
1.00	120.000			1.78	11605.1
		1.40	2155.35	1.79	12030.1
1.01	132.070				
1.02	145.116	1.41	2274.99	1.80	12466.1
1.03	159.198	1.42	2399.73		
1.04	174.375	1.43	2529.72	1.81	12913.7
1.05	190.711	1.44	2665.09	1.82	13372.8
		1.45	2805.99	1.83	13843.6
1.06	208.274			1.84	14326.1
1.07	227.132	1.46	2952.60	1.85	14820.7
1.08	247.350	1.47	3105.04		
1.09	269.006	1.48	3263.48	1.86	15327.3
		1.49	3428.07	1.87	15846.3
1.10	292.169			1.88	16377.7
		1.50	3598.97	1.89	16921.7
1.11	316.923				
1.12	343.341	1.51	3776.32	1.90	17478.2
1.13	371.512	1.52	3960.32		
1.14	401.514	1.53	4151.07	1.91	18047.7
1.15	433.437	1.54	4348.79	1.92	18630.1
		1.55	4553.58	1.93	19225.5
1.16	467.365			1.94	19834.1
1.17	503.396	1.56	4765.68	1.95	20455.9
1.18	541.617	1.57	4958.18		
1.19	582.129	1.58	5212.26	1.96	21091.1
		1.59	5447.11	1.97	21739.7
1.20	625.025			1.98	22402.0
		1.60	5689.88	1.99	23077.9
1.21	670.411				
1.22	718.386	1.61	5940.76	2.00	23767.4
1.23	769.057	1.62	6199.90		
1.24	822.527	1.63	6467.42	2.01	24470.9
1.25	878.916	1.64	6743.57	2.02	25188.1
		1.65	7028.47	2.03	25919.2
1.26	938.330			2.04	26664.2
1.27	1000.87	1.66	7322.31	2.05	27423.3
1.28	1066.67	1.67	7625.21		
1.29	1135.85	1.68	7937.40	2.06	28196.3
1.30	1208.51	1.69	8259.02	2.07	28983.2
		1.70	8590.22	2.08	29784.2
1.31	1284.81			2.09	30599.1
1.32	1364.83	1.71	8931.18		
1.33	1448.73	1.72	9282.06	2.10	31428.1
1.34	1536.61	1.73	9643.02	2.11	32271.1
1.35	1628.62	1.74	10014.3	2.12	33128.0

T (K)	μ	T (K)	μ	T (K)	μ
2.13	33998.6	2.51	79022.2	2.90	156204
2.14	34882.8	2.52	80572.2		
2.15	35780.3	2.53	82142.9	2.91	158671
		2.54	83734.6	2.92	161164
2.16	36690.9	2.55	85347.2	2.93	163684
2.17	37614.3			2.94	166230
2.18	38550.2	2.56	86981.2	2.95	168802
2.19	39500.3	2.57	88636.7		
		2.58	90313.8	2.96	171402
2.20	40465.6	2.59	92012.6	2.97	174028
				2.98	176682
2.21	41446.6	2.60	93733.4	2.99	179364
2.22	42443.5				
2.23	43456.5	2.61	95476.0	3.00	182073
2.24	44485.7	2.62	97240.8		
2.25	45531.3	2.63	99028.2	3.01	184810
		2.64	100838	3.02	187574
2.26	46593.5	2.65	102669	3.03	190366
2.27	47672.5			3.04	193187
2.28	48768.6	2.66	104525	3.05	196037
2.29	49881.8	2.67	106403		
		2.68	108304	3.06	198914
2.30	51012.3	2.69	110228	3.07	201820
				3.08	204755
2.31	52160.2	2.70	112175	3.09	297719
2.32	53325.8				
2.33	54509.2	2.71	114145	3.10	210711
2.34	55710.5	2.72	116139		
2.35	56930.0	2.73	118156	3.11	213732
		2.74	120198	3.12	216783
2.36	58167.8	2.75	122263	3.13	219864
2.37	59423.8			3.14	222975
2.38	60698.8	2.76	124353	3.15	226115
2.39	61992.0	2.77	126465		
		2.78	128603	3.16	229285
2.40	63304.3	2.79	130765	3.17	232484
				3.18	235714
2.41	64635.2	2.80	132952	3.19	238974
2.42	65985.4				
2.43	67354.8	2.81	135164	3.20	242266
2.44	68743.5	2.82	137401		
2.45	70152.0	2.83	139663	3.21	245587
		2.84	141949	3.22	248939
2.46	71580.2	2.85	144260	3.23	252322
2.47	73028.1			3.24	255736
2.48	74496.0	2.86	146597	3.25	259182
2.49	75984.2	2.87	148961		
		2.88	151349	3.26	262658
2.50	77493.1	2.89	153763	3.27	266166

T (K)	μ	T (K)	μ	T (K)	μ
3.28	269706	3.66	428968	4.04	641700
3.29	273278	3.67	438846	4.05	648099
		3.68	438760		
3.30	276880	3.69	443713	4.06	654541
				4.07	661026
3.31	280516	3.70	448702	4.08	667554
3.32	284183			4.09	674125
3.33	287883	3.71	453729		
3.34	291615	3.72	458794	4.10	680740
3.35	295380	3.73	463897		
		3.74	469038	4.11	687399
3.36	299178	3.75	474218	4.12	694103
3.37	303008			4.13	700851
3.38	306871	3.76	479435	4.14	707643
3.39	310768	3.77	484691	4.15	714479
		3.78	489985		
3.40	314697	3.79	495317	4.16	721360
				4.17	728185
3.41	318659	3.80	500688	4.18	735255
3.42	322654			4.19	742269
3.43	326684	3.81	506098		
3.44	330747	3.82	511547	4.20	749328
3.45	334845	3.83	517036		
		3.84	522564	4.21	756431
3.46	338976	3.85	528132	4.22	763579
3.47	343141			4.23	770772
3.48	347341	3.86	533739	4.24	778010
3.49	351575	3.87	539387	4.25	785294
		3.88	545075		
3.50	355844	3.89	550805	4.26	792623
				4.27	799999
3.51	360147	3.90	556574	4.28	807422
3.52	364485			4.29	814893
3.53	368860	3.91	562383		
3.54	373269	3.92	568234	4.30	822411
3.55	377714	3.93	574126		
		3.94	580059	4.31	829978
3.56	382194	3.95	586034	4.32	837592
3.57	386710			4.33	845255
3.58	391262	3.96	592051	4.34	852966
3.59	395849	3.97	598110	4.35	860725
		3.98	604210		
3.60	400471	3.99	610352	4.36	868533
				4.37	876390
3.61	405130	4.00	616537	4.38	884296
3.62	409825			4.39	892252
3.63	414556	4.01	622764		
3.64	419324	4.02	629033	4.40	900258
3.65	424128	4.03	635345		

T (K)	μ	T (K)	μ	T (K)	μ
4.41	908313	4.69	1154761	4.96	1433533
4.42	916418			4.97	1444690
4.43	924573	4.70	1164339	4.98	1455911
4.44	932778			4.99	1467191
4.45	941033	4.71	1173972		
		4.72	1183662	5.00	1478535
4.46	949338	4.73	1193407		
4.47	957693	4.74	1203209		
4.48	966099	4.75	1213066	5.01	1489940
4.49	974556			5.02	1501409
		4.76	1222981	5.03	1512940
4.50	983066	4.77	1232955	5.04	1524535
		4.78	1242983	5.05	1536192
4.51	991628	4.79	1253069		
4.52	1000239			5.06	1547912
4.53	1008905	4.80	1263212	5.07	1559698
4.54	1017621			5.08	1571546
4.55	1026390	4.81	1273414	5.09	1583458
		4.82	1283673		
4.56	1035213	4.83	1293991	5.10	1595437
4.57	1044087	4.84	1304367	5.11	1607481
4.58	1053014	4.85	1314802	5.12	1619589
4.59	1061995			5.13	1631761
		4.86	1325297	5.14	1644000
4.60	1071029	4.87	1335850	5.15	1656305
		4.88	1346462		
4.61	1080114	4.89	1357136		
4.62	1089254			5.16	1668673
4.63	1098449	4.90	1367870	5.17	1681108
4.64	1107699			5.18	1693612
4.65	1117002	4.91	1378662	5.19	1706180
		4.92	1389516		
4.66	1126359	4.93	1400429	5.20	1718817
4.67	1135772	4.94	1411404	5.21	1731521
4.68	1145239	4.95	1422438	5.22	1744290

Source. Brickwedde, F. G., "The 1958 He[4] Scale of Temperatures," Part 1, Introduction; Van Dijk, H., Durieux, M., Clement, J. R., Logan, J. K. Part 2, "Table for the 1958 Temperature Scale," NBS Monograph 10, June 17, 1960. (See also *J. Research NBS, A Physics and Chemistry*, vol. 67A, No. 1, 1960.)

5 Optical techniques for temperature measurement

R. J. GOLDSTEIN
University of Minnesota, Minneapolis

Many optical techniques have been used in the measurement of temperature. Among these are (a) spectroscopic methods in which the emitted or absorbed electromagnetic radiation from a gas (possibly with a tracer added) is measured, (b) total or spectral radiation methods in which the temperature of an opaque surface is measured by comparison with the Stefan-Boltzmann or Planck radiation laws, (c) a scattered radiation technique in which the Doppler broadening of a light beam is measured to determine the temperature level (usually of scattering electrons), and (d) what can be called index of refraction methods in which the index of refraction or spatial derivatives of the index of refraction of a medium are measured and from this the temperature field is inferred.

Only the methods falling within the last category are examined in this chapter. These include schlieren, shadowgraph, and interferometer techniques. References [1–7] have considerable information and extensive bibliographies on these methods which are used to study the temperature fields in transparent media (usually gases or liquids). Although all three methods depend on variation of the index of refraction in a transparent medium and the resulting effects on a light beam passing through the test region, quite different quantities are measured in each one. Shadowgraph systems are used to indicate the variation of the second derivative (normal to the light beam) of the index of refraction. With a schlieren system the first derivative of the index of refraction (in a direction normal to the light beam) is determined. Interferometers permit direct measurement of differences in optical path length essentially giving the index of refraction field directly.

Optical measurement of a temperature field has many advantages over other techniques but perhaps the major one is the absence of an instrument probe which could influence the temperature field. The light beam can also be considered as essentially inertialess so that very rapid transients can be studied. The sensitivities of the three optical methods are quite different so that they can be used to study a variety of systems. Thus interferometers are often used to study free convection boundary layers where temperature gradients are very small while schlieren and shadowgraph systems are often employed in studying shock and flame phenomena where very large temperature and density gradients are present.

Shadowgraph, schlieren, and interferometric measurements are essentially integral ones in that they integrate the quantity measured over the length of the light beam. For this reason they are best suited to measurements in two- (or one-) dimensional fields where there is no index of refraction or density variation in the field along the light beam, except at the beam's entrance to and exit from the test (disturbed) region. These latter variations can be considered as sharp discontinuities or appropriate end corrections can be made. Axisymmetric fields can also be studied, as is demonstrated in the Appendix to this chapter specifically for interferometric measurements. If the field is three dimensional, an average of the measured quantity (along the light beam) can still be determined. Since both schlieren and shadowgraph systems are primarily used for qualitative studies this is often acceptable and even in interferometric studies the averaging done by the light beam can sometimes be quite valuable [8].

Since the three methods to be studied really measure the index of refraction (or one of its spatial derivatives) the relationship between this property and the temperature must be known. Actually, the index of refraction of a homogeneous medium is a function of the thermodynamic state and not necessarily of the temperature alone. According to the Lorenz-Lorentz relation the index of refraction of a homogeneous transparent medium is primarily a function of density

$$\frac{1}{\rho} \frac{n^2 - 1}{n^2 + 2} = \text{a constant} \tag{5.1}$$

In particular, when $n \sim 1$, this reduces to the Gladstone-Dale equation

$$\frac{n - 1}{\rho} = C$$

or
$$\rho = \frac{n - 1}{C} \tag{5.2a}$$

which holds quite well for gases. The constant C, called the Gladstone-Dale constant, is a function of the particular gas and varies slightly with wavelength. Usually instead of using C directly, the index of refraction at standard temperature and pressure n_0 is given

$$n - 1 = \frac{\rho}{\rho_0} (n_0 - 1) \tag{5.2b}$$

When the first or second derivative (say, with respect to y) is determined as in a schlieren or shadowgraph apparatus then, for gases, from Eq. (5.2a)

$$\frac{\partial \rho}{\partial y} = \frac{1}{C} \frac{\partial n}{\partial y} \tag{5.3}$$

$$\frac{\partial^2 \rho}{\partial y^2} = \frac{1}{C} \frac{\partial^2 n}{\partial y^2} \tag{5.4}$$

If the pressure can be assumed constant and the ideal gas equation of state ($\rho = P/RT$) holds then

$$\frac{\partial n}{\partial y} = -\frac{CP}{RT^2} \frac{\partial T}{\partial y} = -\frac{n_0 - 1}{T} \frac{\rho}{\rho_0} \frac{\partial T}{\partial y} \tag{5.5a}$$

or

$$\frac{\partial T}{\partial y} = -\frac{T}{n_0 - 1} \frac{\rho_0}{\rho} \frac{\partial n}{\partial y} \tag{5.5b}$$

and

$$\frac{\partial^2 n}{\partial y^2} = C \left[-\frac{\rho}{T} \frac{\partial^2 T}{\partial y^2} + \frac{2\rho}{T^2} \left(\frac{\partial T}{\partial y} \right)^2 \right] \tag{5.6}$$

Note that Eq. (5.5b) which would apply in a schlieren study shows a relatively simple relationship between the gradient of the temperature and the gradient of the index of refraction which is measured. For a shadowgraph the equivalent relation, Eq. (5.6), is more complicated, although under many conditions the second term may be small.

The index of refraction of a gas as measured in an interferometer can indicate the temperature directly. From Eqs. (5.2a) and (5.2b) assuming constant pressure and the perfect gas equation of state

$$T = \frac{C}{n-1} \frac{P}{R} = \left(\frac{n_0 - 1}{n - 1} \right) \frac{P}{P_0} \times T_0 \tag{5.7}$$

The index of refraction of a liquid is primarily a function of temperature and for accurate results should be obtained from direct measurement. For comparison, Table 5.1, derived from References [9] and [10], cites values at 20°C and 1 atm for air and water. The two wavelengths chosen are a commonly used mercury line (546.1 nm) and the visible line from a CW He-Ne laser (632.8 nm).

Of the three systems to be discussed, two of them—schlieren and shadow-

TABLE 5.1 Index of refraction for air and water at 20°C and 1 atmosphere

λ	$n_{air} - 1$	n_{H_2O}	dn/dT_{air} (°C)$^{-1}$	dn/dT_{H_2O} (°C)$^{-1}$
546.1 nm	2.733×10^{-4}	1.3345	-0.932×10^{-6}	-0.895×10^{-4}
632.8 nm	2.719×10^{-4}	1.3317	$-0.927/\times 10^{-6}$	-0.880×10^{-4}

graph—can be described by geometrical or ray optics although under certain conditions diffraction effects can be significant. Interferometers, as the name implies, depend on the interference of coherent light beams and some discussion of physical (wave) optics will be required.

5.1 SCHLIEREN SYSTEM

To study both schlieren and shadowgraph systems, the path of a light beam in a medium whose index of refraction is a function of position must be analyzed. Consider Fig. 5.1 where a light beam, traveling initially in the z direction, passes through a medium whose index of refraction varies (for simplicity) only in the y direction. At time τ the beam is at position z and the wavefront (surface normal to the path of the light) is as shown. After a time interval $\Delta\tau$ the light has moved a distance of $\Delta\tau$ times the velocity of light which, in general, is a function of y so the wavefront or light beam may have turned an angle $d\alpha'$. The local value of the speed of light is c_0/n. With reference to Fig. 5.1, and assuming that only small deviations occur, the distance that the light beam (Δz) travels during time interval $\Delta\tau$ is

$$\Delta z = \left(\frac{c_0}{n}\right)\Delta\tau$$

Now

$$\Delta^2 z = \Delta z_y - \Delta z_{y+\Delta y}$$

or

$$\Delta^2 z = -c_0\,\frac{\Delta(1/n)}{\Delta y}\,\Delta\tau\Delta y$$

and the angular deflection of the ray is

$$\Delta\alpha' \approx \frac{\Delta^2 z}{\Delta y} = -n\,\frac{\Delta(1/n)}{\Delta y}\,\Delta z$$

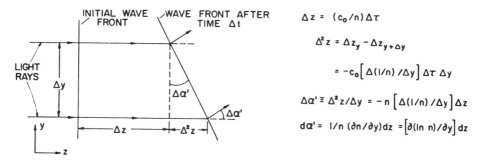

FIG. 5.1 Bending of a light ray in an inhomogeneous medium.

In the limit if Δy and Δz are considered to be very small

$$d\alpha' = \frac{1}{n}\frac{\partial n}{\partial y}\,dz = \frac{\partial(\ln n)}{\partial y}\,dz \tag{5.8}$$

Since α' is also the slope dy/dz of the light beam for small deviations

$$\frac{\partial^2 y}{\partial z^2} = \frac{1}{n}\frac{\partial n}{\partial y} \tag{5.9}$$

If the angle α' remains small this expression will hold over the light path through the disturbed region and the angle at the exit of the test region is then

$$\alpha' = \int\frac{1}{n}\frac{\partial n}{\partial y}\,dz = \int\frac{\partial(\ln n)}{\partial y}\,dz \tag{5.10}$$

where the integration is performed over the entire length of the light beam in the test region.

If the test region is enclosed by glass walls and the index of refraction within the test section is considerably different from the ambient air, then, from Snell's law, an additional angular deflection is present. If α is the angle of the light beam after it has passed through the test section and emerged into the surrounding air $n_a \sin \alpha = n \sin \alpha'$. Assuming the test section windows are plane and of uniform thickness, for small values of α and α', $\alpha = (n/n_a)\alpha'$. Using Eq. (5.10)

$$\alpha = \frac{n}{n_a}\int\frac{1}{n}\frac{\partial n}{\partial y}\,dz$$

Assuming $1/n$ within the integrand does not change greatly through the test section

$$\alpha = \frac{1}{n_a}\int\frac{\partial n}{\partial y}\,dz$$

$$\tag{5.11}$$

$$\alpha \cong \int\frac{\partial n}{\partial y}\,dz$$

since $n_a \approx 1$. Note that if a gas (not at extremely high density) is the test fluid, $\alpha' \cong \alpha$.

The angle α in Eq. (5.11) is in the $y - z$ plane. If there is also a variation of index of refraction in the x direction then, again assuming small angular devia-tion, a similar expression would give the angle in the $x - z$ plane proportional to $\partial n / \partial x$.

If variations in the x and z direction, as well as the effect of significant angular deflection are included, the resulting equation for the path of the light beam, equivalent to Eq. (5.9), is [2]

$$y'' = \frac{1}{n} [1 + x'^2 + y'^2] \left[\frac{\partial n}{\partial y} - y' \frac{\partial n}{\partial z} \right] \qquad (5.12)$$

where primes refer to differentiation with respect to z. A similar equation could be written for x''.

Note that the light beam is turned in the direction of increasing index of refraction. In most media this means that the light is bent toward the region of higher density or lower temperature. Basically a schlieren system is a device to measure or indicate this small angle (typically of the order of 10^{-6} to 10^{-3} rad) as a function of position in the $x - y$ plane normal to the light beam.

Consider the simple system shown in Fig. 5.2. A light source which we shall assume to be rectangular (dimensions a_s by b_s) is at the focus of lens L_1. The resulting parallel light beam enters the field of disturbance in the test section. The deflected rays, when the disturbance is present, are indicated by cross-hatched lines. The light is collected by a second lens L_2 at whose focus a knife edge is placed and passes onto a screen located at the conjugate focus of the test section. As will be shown below, if the screen is not at the focus of the disturbance, shadowgraph effects will be superimposed on the schlieren pattern.

If no disturbance is present then ideally the light beam at the focus of L_2 would be as shown in Fig. 5.3, with dimensions $a_0 - b_0$ which are related to the initial dimensions by

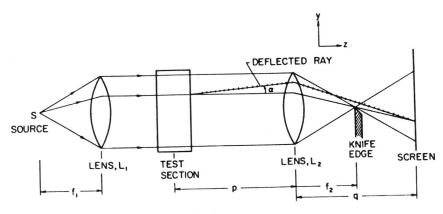

FIG. 5.2 Typical schlieren system using lenses.

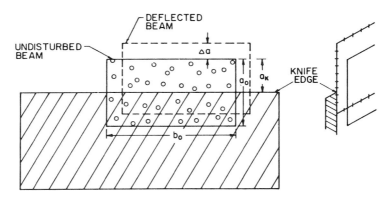

FIG. 5.3   View of deflected and undisturbed beam at knife edge of schlieren system.

$$\frac{a_0}{a_s} = \frac{b_0}{b_s} = \frac{f_2}{f_1} \tag{5.13}$$

where f_1 and f_2 are the focal lengths of L_1 and L_2 respectively. As shown in the figure, the source is usually adjusted so that the shorter dimension a_0 is at right angles to the knife edge to maximize sensitivity.* The knife edge (typically a razor blade) is adjusted, when no disturbance is present, to cut off all but an amount a_K (typically $a_K = a_0/2$) of the height a_0. When the knife edge is moved across the beam exactly at the focus, the illumination at the screen will decrease uniformly but if the knife edge is not right in the focal plane the image at the screen will not darken uniformly. The illumination at the screen when no knife edge is present is I_0, and with the knife edge inserted in the focal plane the illumination is

$$I_K = \frac{a_K}{a_0} I_0 \tag{5.14}$$

The light passing through any part of the test region comes from all parts of the source. Thus, at the focus not only is the image of the source composed of light coming from the whole field of view, but light passing through every point in the field of view gives an image of the source at the knife edge. If the light beam at a position x, y in the test region is deflected by an angle α, then, from

---

*If a light source with sharply defined edges as in Fig. 5.3 is not available, a condensing lens is sometimes used with the source. At the conjugate focus of this lens, an auxiliary knife edge is aligned so that the image of the source adjacent to the main knife edge (in the beam following the test section) is as straight and sharply defined as possible.

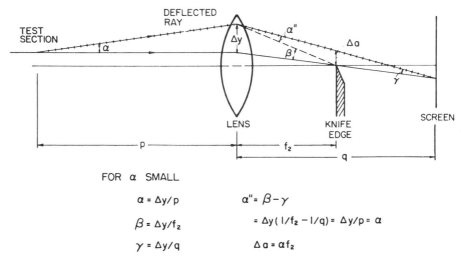

FOR α SMALL

$$\alpha = \Delta y / p \qquad\qquad \alpha'' = \beta - \gamma$$

$$\beta = \Delta y / f_2 \qquad\qquad = \Delta y (1/f_2 - 1/q) = \Delta y/p = \alpha$$

$$\gamma = \Delta y / q \qquad\qquad \Delta a = \alpha f_2$$

FIG. 5.4 Ray displacement at knife edge for a given angular deflection.

Fig. 5.4 the image of the source coming from that position will be shifted at the knife edge by an amount

$$\Delta a = \pm f_2 \cdot \alpha \tag{5.15}$$

where the sign is determined by the position of the knife edge; it is positive when (as in Fig. 5.4) $\alpha > 0$ gives $\Delta a > 0$ and negative if the knife edge were reversed so that $\alpha > 0$ leads to $\Delta a < 0$. The illumination at the image of position x, y on the screen will be (see Fig. 5.3)

$$I_d = I_K \left(\frac{a_K + \Delta a}{a_K} \right) = I_K \left(1 + \frac{\Delta a}{a_K} \right) \tag{5.16}$$

where Δa is positive if the light is deflected away from the knife edge and negative if the light is deflected toward the knife edge. The relative intensity or contrast is

$$\text{Contrast} = \frac{\Delta I}{I_K} = \frac{I_d - I_K}{I_K} = \frac{\Delta a}{a_K} = \pm \frac{\alpha f_2}{a_K} \tag{5.17}$$

using Eq. (5.15). Note that the sensitivity of the schlieren system for measuring the deflection is

$$\frac{d \, \text{Contrast}}{d\alpha} = \frac{f_2}{a_K} \tag{5.18}$$

or proportional to f_2 and inversely proportional to a_K. For a given optical system minimizing a_K by movement of the razor blade would maximize the contrast. However, this would limit the range for deflection of the beam towards the knife edge to

$$\alpha_{max} = \frac{a_K}{f_2} \tag{5.19}$$

as all deflections this large or larger would give (neglecting diffraction) no illumination. The maximum angle of deflection away from the knife edge that could be measured is

$$\alpha_{max'} = \frac{a_0 - a_K}{f_2} \tag{5.20}$$

as a deflection of this magnitude would permit all the source illumination to pass to the screen. For equal range in both directions $a_K = a_0/2$ and

$$\alpha_{max} = \alpha_{max'} = \frac{a_0}{2f_2} = \frac{a_s}{2f_1} \tag{5.21}$$

Note from Fig. 5.3 that deflections in the x direction are parallel to the knife edge and will not affect the illumination at the screen so if density gradients in the x direction within the test region are to be studied, the knife edge must be turned at right angles. For maximum sensitivity (since $a_s < b_s$) the source should also be rotated 90 deg.

Combining Eqs. (5.11) and (5.17)

$$\text{Contrast} = \frac{\Delta I}{I_K} = \pm \frac{f_2}{a_K n_a} \int \frac{\partial n}{\partial y} \, dz \tag{5.22}$$

Assuming a two-dimensional field with $\partial n/\partial y$ constant at a given x, y position over the length L in the z direction,

$$\text{Contrast} = \pm \frac{f_2}{a_K} \frac{1}{n_a} \frac{\partial n}{\partial y} L \tag{5.23}$$

This equation holds for every x, y position in the test section and gives the contrast at the equivalent position in the image on the screen.

If the deflection is toward the knife edge the field will darken and the contrast will be negative. Using the coordinate system of Fig. 5.3, if the knife edge covers up the region $y < 0$ (i.e., knife edge pointing upwards) at the focus

$$\frac{\Delta I}{I_K} = + \frac{f_2}{a_K} \frac{1}{n_a} \frac{\partial n}{\partial y} L \tag{5.24a}$$

While if the knife edge is reversed and covers the region $y > 0$

$$\frac{\Delta I}{I_K} = - \frac{f_2}{a_K} \frac{1}{n_a} \frac{\partial n}{\partial y} L \tag{5.24b}$$

Changing the knife edge reverses the dark and light images on the screen. The brighter areas of the image represent regions in the test section where the index of refraction (and thus usually density) increases in the direction away from the knife edge (Fig. 5.5). Dark areas represent regions where the index of refraction increases in the direction opposite to that in which the knife edge points (Fig. 5.5).

Equation (5.22) can be rewritten in the case of a gas at constant pressure, using Eq. (5.5a)

$$\frac{\Delta I}{I_K} = \pm \frac{f_2}{a_K n_a} \int \frac{n_0 - 1}{T} \frac{\rho}{\rho_0} \frac{\partial T}{\partial y} dz \tag{5.25}$$

and equivalent to Eq. (5.23)

$$\frac{\Delta I}{I_K} \approx \pm \frac{f_2}{a_K} \left(\frac{n_0 - 1}{\rho_0} \right) \frac{P}{RT^2} \frac{\partial T}{\partial y} L \tag{5.26}$$

since $n_a \approx 1$.

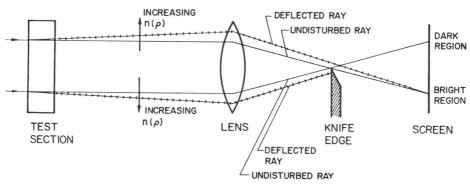

FIG. 5.5 Effect of index of refraction gradient on illumination at screen.

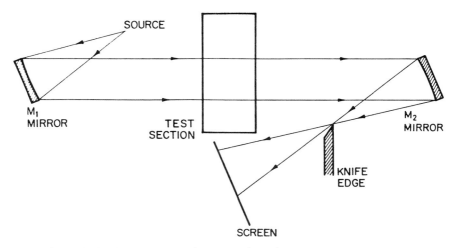

FIG. 5.6 Typical schlieren system using converging mirrors.

For a liquid

$$\frac{\Delta I}{I_K} = \pm \frac{f_2}{a_K n_o} \int \frac{\partial T}{\partial y} \frac{dn}{dT} dz \qquad (5.27)$$

If the field is two dimensional and n does not change greatly

$$\frac{\Delta I}{I_K} = \pm \frac{f_2}{a_K n_a} \frac{\partial T}{\partial y} \frac{dn}{dT} L \qquad (5.28a)$$

$$\approx \pm \frac{f_2}{a_K} \frac{\partial T}{\partial y} \frac{dn}{dT} L \qquad (5.28b)$$

In a quantitative study, measurements of the illumination or contrast, usually of the image on a photographic negative, must be made. These are quite time consuming and the resulting accuracy has not usually warranted the effort. Thus standard schlieren systems are usually employed for qualitative studies of a temperature or density field although quantitative measurements of shock wave angles or positions can be made. The minimum value of the contrast that can easily be observed is of the order of 0.05 which can be used in determining the overall sensitivity of the system. Since dn/dT varies slightly with wavelength it is preferable that the light source be relatively monochromatic, although for both schlieren and shadowgraph systems this is usually not a major criterion.

The high cost of large aberration-free lenses usually precludes construction of the system shown in Fig. 5.2, but the optically similar system using concave mirrors shown in Fig. 5.6 is widely used. The source and knife edge should be in the same plane and on opposite sides of the axes of the two mirrors in the "Z" arrangement shown. This eliminates the aberration coma although in the off-axis system astigmatism is still present. The legs of the "Z" should each be at the same angle to the line between the two mirrors; this angle should be as small as possible to reduce astigmatism [11].

Schlieren photographs taken with a system similar to that of Fig. 5.6 are shown in Figs. 5.7 and 5.8. The light source is a zirconium arc lamp with a source size of about 0.76 mm. Each mirror has an aperture of about 20 cm and a focal length of about 107 cm. The disturbed region is 3.5 m from the second mirror. In place of a screen a 35-mm camera with a 135-mm focal length lens is used to record the image. To focus on the test section a −1 auxiliary lens is also used.

Figure 5.7 is a schlieren image of a burning gas flame and jet leaving a standard propane torch. Figure 5.8 shows different images of the free convection field around a heated cylinder (3.2-cm diameter, 20 cm long) in air. Although

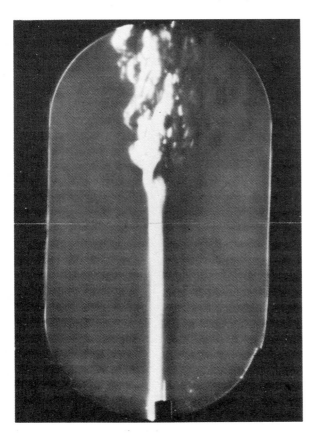

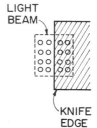

LIGHT
BEAM

KNIFE
EDGE

FIG. 5.7 Schlieren image of propane gas flame.

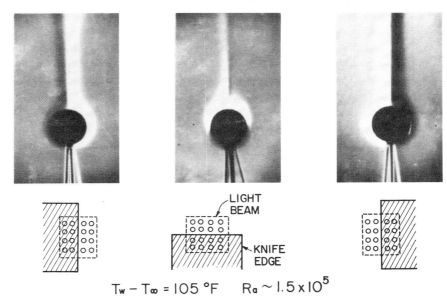

$$T_w - T_\infty = 105\ °F \qquad R_a \sim 1.5 \times 10^5$$

FIG. 5.8 Schlieren image of the natural convection field around a heated cylinder.

the temperature of the cylinder and thus the Rayleigh number varies only slightly for the different photographs, the difference in knife-edge orientation, shown for each view, dramatically changes the image. As mentioned above the schlieren image lightens when the index of refraction increases in the direction away from the knife edge. Thus the lighter regions on the photo show that the local temperature gradient is positive in the direction towards the knife edge while the reverse is true for the darker regions.

A number of variations on the schlieren systems shown in Figs. 5.2 and 5.6 have been used, and two of these are shown in Fig. 5.9. In Fig. 5.9a one plane and one converging mirror are used. Since plane mirrors are easier to make than are converging ones this apparatus would be somewhat less expensive than the one in Fig. 5.6. The main reason for its use, however, is the amplification of the angle representative of the disturbance. As the beam passes through the test section twice, the deflection angle will be doubled and, all other parameters being the same, so will the sensitivity. However, this double passage will in general cause a slight blurring of the image as the beam does not go through the exact same part of the test section on each passage. In addition, since the screen cannot be at the focus of both views of the test section, some shadowgraph effects will be present. A single-mirror system, shown in Fig. 5.9b, is still simpler although the blurring of the image would be still more serious than in Fig. 5.9a because the light is not parallel. The source and knife edge are at conjugate foci and for convenience they are often kept close together and thus at a distance

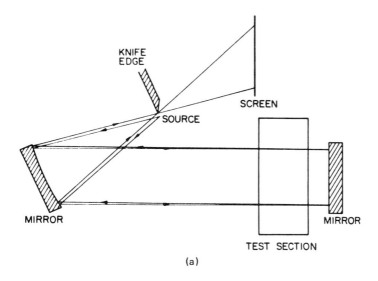

(a)

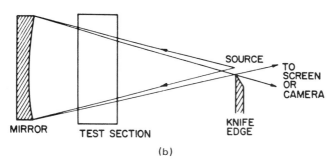

(b)

FIG. 5.9 Alternative schlieren systems. (*a*) One converging and one plane mirror. (*b*) One converging mirror.

twice the focal length f from the mirror. If α is the deflection of the beam after a single transit of the test section the sensitivity, equivalent to Eq. (5.18), would be $4f/a_K$. Since the source and knife edge in the two systems shown in Fig. 5.9 are in such close proximity, a splitter plate is sometimes used to give them a larger physical separation. It should be apparent that in all of these systems a camera placed in the beam after the knife edge and focused on the test section can be used in place of the screen.

Other schlieren systems have used one of the optical arrangements shown above, usually that of Fig. 5.6, but without the knife edge. In a color schlieren the knife edge is replaced by colored filters held at the focus and the deflection of the light beam (necessarily nonmonochromatic) then gives rise to different colors. In another system an aperture is placed at the focus giving a darker image for any deflection or temperature gradient irrespective of the direction of the gradient. An opaque disc at the focus gives a brightening of the image for a light deflection in any direction.

Of particular interest for quantitative studies is the Ronchi or grid-schlieren system [12–14]. A grid is shown as part of a focusing lens schlieren system (Fig. 5.10) which could be used in any of the optical arrangements previously described, although quantitative studies are best performed using parallel light which passes through the test section only once. The grid has equally spaced opaque lines (whose widths are usually set equal to the spacing) on a transparent sheet. It is placed before the focus as shown in Fig. 5.10. The resulting schlieren image is a series of lines, parallel if no disturbance is present. When a disturbance is present the displacement of these fringes is directly proportional to the local angular deflection α. The grid could also be placed right at the focus where the knife edge is located in a standard schlieren system; in this system the beam should just pass through one of the gaps between two opaque lines when no disturbance is present. Thus, the spacing between the lines would equal a_0 (Fig. 5.3) and the screen would be uniformly illuminated. In the presence of a disturbance the deflection of the light beam at the focus Δa will cause the beam to traverse the grid producing a light or dark image depending on the magnitude of Δa. The resulting image is a series of fringes, called "isophotes," representing regions of constant angular deflection (often constant density or temperature gradient). Placement of the grid at the focus of the second lens, however, is often impractical. The required line spacing on the grid may be so small as to cause significant diffraction effects.

Other systems which may be of interest in heat transfer studies include: a self-illuminated schlieren system for study of plasma system jets [15], a sharp-focusing schlieren system [16], and stereoscopic schlieren [17]. The latter two, though cumbersome to use, have application to the study of three-dimensional fields. Schlieren systems using lasers as light sources have also been studied [18]. The Schmidt-schlieren system, because of its similarity to a shadowgraph, is outlined in Sec. 5.2 while schlieren interferometers are described in the section on interferometry.

5.2 SHADOWGRAPH SYSTEM

In a shadowgraph system the linear displacement of the perturbed light beam is measured, rather than the angular deflection as in a schlieren system. The

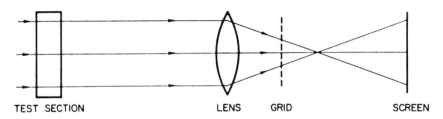

TEST SECTION   LENS GRID   SCREEN

FIG. 5.10 Grid-schlieren system.

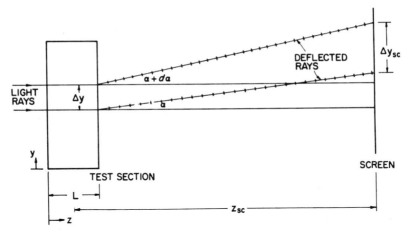

FIG. 5.11 Displacement of light beam for shadowgraph evaluation.

shadowgraph image can be understood by referring to Fig. 5.11 in which a parallel light beam enters a nonuniform test section. To simplify the derivation, variations are assumed to exist only in the y direction. At the exit of the test section the beam is not usually parallel having been deflected by an angle α which is a function of y. Consider the illumination at the exit of the test section. The linear displacement of the light beam is probably not large there because of the relatively short distance the light has traveled; if the illumination is uniform entering the test section it should still be closely uniform there. The light, however, is bent through an angle α. The illumination within the region defined by Δy at this position will be within the region defined by Δy_{sc} at the screen. If the initial intensity is I_T then at the screen

$$I_0 = \frac{\Delta y}{\Delta y_{sc}} I_T \tag{5.29}$$

If z_{sc} is the distance to the screen, then from Fig. 5.11,

$$\Delta y_{sc} = \Delta y + z_{sc} d\alpha \tag{5.30}$$

The contrast is

$$\frac{\Delta I}{I_T} = \frac{I_0 - I_T}{I_T} = \left(\frac{\Delta y}{\Delta y_{sc}} - 1 \right) \approx -z_{sc} \frac{\partial \alpha}{\partial y} \tag{5.31}$$

Combining this with Eq. (5.11)

$$\frac{\Delta I}{I_T} = -\frac{z_{sc}}{n_a}\int \frac{\partial^2 n}{\partial y^2}\, dz \tag{5.32}$$

If the index of refraction is only a function of temperature

$$\frac{\Delta I}{I_T} = -\frac{z_{sc}}{n_a}\int \frac{\partial^2 T}{\partial y^2}\frac{dn}{dT}\, dz \approx -z_{sc}\int \frac{\partial^2 T}{\partial y^2}\frac{dn}{dT}\, dz \tag{5.33}$$

assuming dn/dT is constant. For a gas, Eq. (5.6) could be substituted into Eq. (5.32).

If there are additional variations of the index of refraction in the x direction, then equivalent to Eq. (5.32),

$$\frac{\Delta I}{I_T} = -\frac{z_{sc}}{n_a}\int \left(\frac{\partial^2 n}{\partial x^2} + \frac{\partial^2 n}{\partial y^2}\right) dz \tag{5.34}$$

The shadowgraph, like the schlieren and interferometer methods, is best utilized in a two-dimensional system where there is no variation in the z direction aside from the sharp change at entrance and exit of the test region. Note that variations of the index of refraction in both x and y directions are obtained from a single image, while schlieren systems usually only indicate variations normal to the knife edge.

Different optical geometries are possible for shadowgraph systems as shown in Fig. 5.12. Parallel light systems, as in Fig. 5.12a, are easiest to understand although the lensless and mirrorless system of Fig. 5.12b is also usable if the distance from the (small) source to the test region is large. Other combinations of mirrors and lenses analogous to the schlieren systems of Figs. 5.2, 5.6, and 5.9 have been used. It should be noted that if a mirror or lens is used after the test region in a shadowgraph system it should not be placed so that the conjugate focus of the test section is in the plane of the screen. At the conjugate focus the parts of the beam deflected at different angles in the test region are all brought back together so there is no linear displacement there and thus no shadowgraph effect.

The standard shadowgraph is rarely used for quantitative density measurements. The contrast would have to be measured accurately and Eq. (5.33), for example, integrated twice to get the density or temperature distribution. Even the temperature gradient which is of interest in heat transfer studies would require one integration. If, however, very large gradients of density or temperature are present as in a shock wave or a flame, shadowgraph pictures can be very useful. As with a schlieren system quantitative measurements such as shock

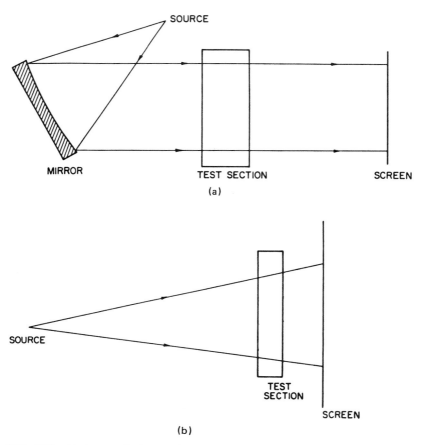

FIG. 5.12 Alternative shadowgraph systems. (*a*) One converging mirror. (*b*) No lens or mirror.

angles can be made. In addition, a shadowgraph is quite useful in indicating a boundary layer transition [3] (p. 28).

Figure 5.13 shows a shadowgraph of the flame whose schlieren photo is shown in Fig. 5.7. The same optical system was used as described in reference to Fig. 5.7 and Fig. 5.8 (i.e., a system similar to the one used in Fig. 5.6) with the knife edge removed and the camera set out of focus to get the shadowgraph image.

One particular modification of the shadowgraph has been used in heat transfer studies to obtain surface heat fluxes. It is sometimes called the Schmidt-schlieren system and is relatively simple to use and evaluate [19, 20]. Although a shadow picture is taken, a deflection that is proportional to the index of refraction gradient at a solid-fluid interface is measured. From this deflection the wall temperature gradient, and thus the surface heat flux, can be obtained.

Consider the light path in Fig. 5.14a, where the light enters the disturbed or test region parallel to the test surface. There is no variation of properties in the z direction within or outside the test section and at first only deflections in the $y - z$ plane are considered. From Eq. (5.9) the path of the light beam for small angular deflections is described by

$$\frac{\partial^2 y}{\partial z^2} = \frac{1}{n} \frac{\partial n}{\partial y}$$

Integrating once gives the value of the slope of a light ray at the exit of the test section

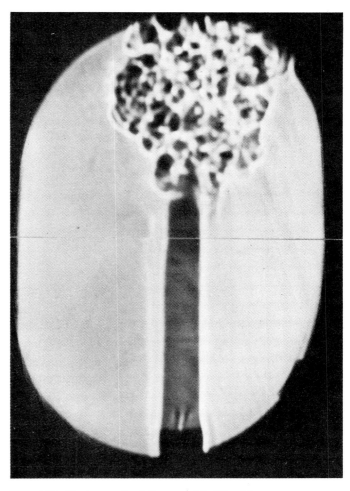

FIG. 5.13 Shadowgraph of flame used in Fig. 5.7.

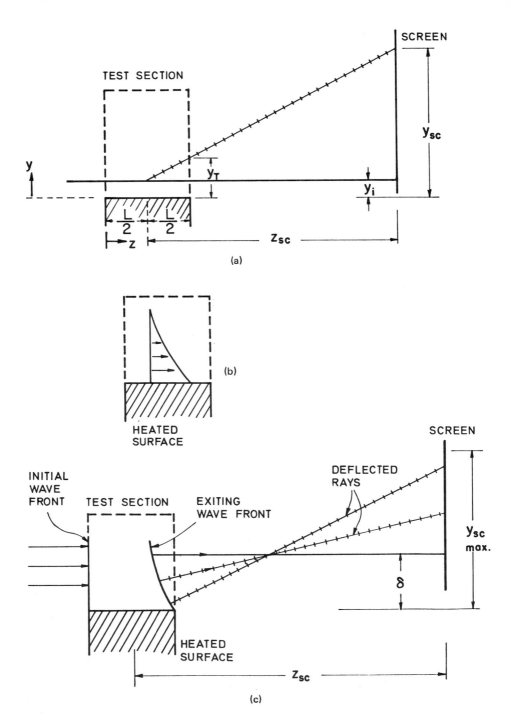

FIG. 5.14 Schmidt-schlieren system. (*a*) Path of light ray passing heated surface. (*b*) Temperature profile in boundary layer. (*c*) Light beam passing heated surface.

$$\left(\frac{\partial y}{\partial z}\right)_T = \frac{1}{n}\frac{\partial n}{\partial y}L \tag{5.35}$$

Integrating again to get the value of y at this position for a ray that entered at y_i

$$y_T - y_i = \frac{1}{n}\frac{\partial n}{\partial y}\frac{L^2}{2} \tag{5.36}$$

In these integrations we have assumed that $(1/n)(\partial n/\partial y)$ is constant along the beam even though the y position changes slightly. This is valid if $\partial^2 n/\partial y^2$ is small [20]. If no other variation in index of refraction is present, after leaving the test section the ray travels in a straight line to the screen. The ray height at the screen is given by

$$y_{sc} = y_T + \left(\frac{\partial y}{\partial z}\right)_T \left(z_{sc} - \frac{L}{2}\right)$$

or

$$y_{sc} - y_i = \frac{1}{n}\frac{\partial n}{\partial y}z_{sc}L \tag{5.37a}$$

$$\approx \frac{\partial n}{\partial y}z_{sc}L \tag{5.37b}$$

when $n \approx 1.0$.

If the test fluid has an index of refraction differing greatly from the surrounding air then upon emerging from the test section window the slope of the light beam would be

$$\frac{1}{n_a}\frac{\partial n}{\partial y}L$$

assuming near-normal incidence.

Neglecting displacement of the beam during passage through the window

$$y_{sc} - y_i = \frac{1}{n}\frac{\partial n}{\partial y}\frac{L^2}{2} + \frac{1}{n_a}\frac{\partial n}{\partial y}L\left(z_{sc} - \frac{L}{2}\right) \tag{5.38}$$

and if $L \ll z_{sc}$

$$y_{sc} - y_i = \frac{1}{n_a}\frac{\partial n}{\partial y}Lz_{sc} \tag{5.39a}$$

$$\approx \frac{\partial n}{\partial y}Lz_{sc} \tag{5.39b}$$

In most thermal boundary layers the maximum value of the temperature gradient, and thus the maximum of the index of refraction gradient, is at the fluid-solid interface. With heat flow from the solid to the fluid the light will normally be bent away from the surface, and the ray that passes just adjacent to the surface will undergo maximum bending. If the screen is placed far enough away from the test section this ray will also have the maximum deflection $y_{sc, max}$. Since the value of $(\partial^2 T)/(\partial y^2)$ and thus $\partial^2 n/\partial y^2$ is usually small or even zero near the surface, a significant light bundle adjacent to the surface will be deflected this maximum amount, producing a bright contour on the screen whose position can be measured fairly accurately. Then from Eq. (5.37b) or (5.39b)

$$\left(\frac{\partial n}{\partial y}\right)_w = \frac{y_{sc, max}}{z_{sc} L}$$

If the fluid is a gas this can be combined with Eq. (5.5b), and assuming that T does not vary greatly along the light ray

$$\left(\frac{\partial T}{\partial y}\right)_w = -\frac{T_w}{(n_0 - 1)} \frac{\rho_0}{\rho} \frac{y_{sc, max}}{z_{sc} L} \tag{5.40}$$

The heat flux at the wall in the y direction is

$$q_w = -k\left(\frac{\partial T}{\partial y}\right)_w$$

and the Nusselt number can be calculated directly,

$$\text{Nu} = \frac{q_w D}{k(T_w - T_\infty)} = \frac{T_w}{n_0 - 1} \frac{\rho_0}{\rho} \frac{D}{z_{sc} L} \frac{y_{sc, max}}{T_w - T_\infty} \tag{5.41}$$

With a liquid

$$\left(\frac{\partial T}{\partial y}\right)_w = \frac{y_{sc, max}}{(dn/dT)_w z_{sc} L} \tag{5.42}$$

and

$$\text{Nu} = -\frac{y_{sc, max}}{T_w - T_\infty} \frac{D}{z_{sc} L (dn/dT)_w} \tag{5.43}$$

It should be noted that to measure the displacement on the screen the zero position for y must be known. This must usually be determined when no disturbance is in the test region so the rays passing the wall are not deflected. If a slit is placed in the light beam before the test section, so that primarily the solid-fluid interface is illuminated, a somewhat sharper image of the maximum deflection can be obtained, although care must be taken to avoid diffraction effects.

If a thermal boundary layer (as in Figs. 5.14b and 5.14c) is present, all light within the boundary layer will be deflected at least to some extent. If the distance to the screen is large a shadow will appear which is representative of the thickness of the boundary layer. Since the boundary layer does not really have a finite thickness, however, this shadow height at the screen is somewhat a function of the screen position even for large values of z_{sc}.

Shadowgraphs of a heated horizontal circular cylinder (the same as in Fig. 5.8) in air are shown in Fig. 5.15. The same optical system as described in relation to Figs. 5.7 and 5.8, and modified for Fig. 5.13, was used. The first photo shows the cylinder and plume; the second photo is an enlargement of the region near the cylinder. The heart-shaped halo around the cylinder is indicative of the local heat transfer coefficient variation around the periphery. For quantitative measurements, the position of the solid surface itself could be put on the figure and measurements of $y_{sc,\,max}$ (in this case in the radial direction) could be taken using that surface as the datum. From the figure one can observe the relatively large heat flux and thus large heat transfer coefficient (since the cylinder wall temperature was uniform) at the bottom of the cylinder. The heat transfer coefficient gradually decreases as one goes up around the cylinder reaching a minimum at the top.

A number of other modifications of the basic shadow system can be used for quantitative studies. Several designs use a narrow inclined slit in the light beam before the test section [1 (Chap. 6)]. The distortion of the slit image on the screen is particularly useful in studying one-dimensional temperature or density fields.

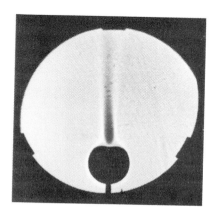

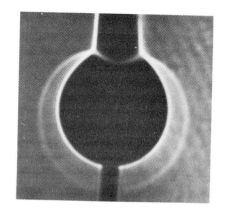

$T_w - T_\infty = 80\ °F$ $Ra \sim 1.1 \times 10^5$

FIG. 5.15 Shadowgraphs of cylinder used in Fig. 5.8.

5.3 INTERFEROMETER

5.3.1 Basic Principles

The third optical device for measurement of temperature (or density) is the interferometer which is often used in quantitative studies. Interferometry, unlike the schlieren and shadowgraph systems, does not depend upon the deflection of a light beam to determine density. In fact refraction effects are usually of second order and undesirable in interferometry as they introduce deviations or errors in the evaluating equations. To understand interferometric measurements one must consider the wave nature of light, and this is perhaps best done by first examining a particular system that is widely used.

The Mach-Zehnder interferometer is often employed in heat transfer and aerodynamic studies. One of the main advantages of the Mach-Zehnder system over other interferometers is the large displacement of the reference beam from the test beam. In this way the reference beam can pass through a uniform field. In addition, since the test beam passes through the disturbed region only once the image is sharp and optical paths can be clearly defined. References [7, 21-27] discuss some of the details of the optics of Mach-Zehnder interferometers.

Figure 5.16 is a schematic diagram of a Mach-Zehnder interferometer. A monochromatic light source is used in conjunction with a lens to obtain a parallel beam of light. The requirement of a very narrow spectral width for the light source is more critical in an interferometer than with schlieren or shadowgraph systems. The parallel light beam strikes the first splitter plate Sp_1 which is a partially silvered mirror permitting approximately half the impinging light to pass directly through it. This transmitted light follows path 1 to mirror M_1 where it is reflected toward the second splitter plate Sp_2. The light reflected by Sp_1 follows path 2 to mirror M_2 where it is also reflected towards Sp_2. The second splitter plate also transmits about half the impinging light and reflects most of the rest, and the recombined beams pass on to the screen. Note that there would also be another recombined beam leaving Sp_2, but in general only one beam from the final splitter is used. The mirrors and splitter plates are usually set at corners of a rectangle [25]. They should then be all closely parallel and at an angle of $\pi/4$ to the initial parallel beam.

Let us assume that in Fig. 5.16 the mirrors are perfectly parallel and that in both paths 1 and 2 there are no variations in optical properties normal to either beam. This would require uniform properties not only between the mirrors but also for the splitter plates. (Note that the effect of a variation in the thickness of a splitter plate can be corrected by a slight rotation of one of the mirrors if both surfaces of the plate are flat.) The two beams (from paths 1 and 2) emerging from Sp_2 would then be parallel.

In general the amplitude of a plane light wave in a homogeneous medium can be represented by

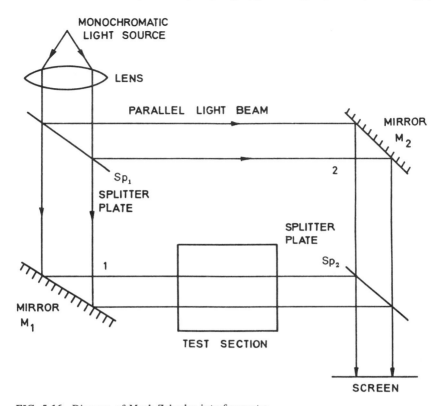

FIG. 5.16   Diagram of Mach-Zehnder interferometer.

$$A = A_0 \sin \frac{2\pi}{\lambda} (c\tau - z) \qquad (5.44)$$

where A_0 is the peak amplitude, c the speed of light, τ the time, z the distance, and λ the wavelength. Consider the amplitude of beam 1 at a fixed position past Sp_2 to be

$$A_1 = A_{01} \sin \frac{2\pi c\tau}{\lambda} \qquad (5.45)$$

The other beam could be represented at the same position by

$$A_2 = A_{02} \sin \left(\frac{2\pi c\tau}{\lambda} - \Delta \right) \qquad (5.46)$$

where the phase difference Δ appears since the two beams will probably not be exactly in phase because of a difference in their optical path length. Before the first splitter plate the two beams (1 and 2) were, of course, one beam and in phase.

Since beams 1 and 2 come from the same source they are coherent and can interfere with each other. This is really implicit in Eqs. (5.45) and (5.46) if Δ is not a function of time but only of the path difference. Summing Eqs. (5.45) and (5.46) and assuming that $A_{01} = A_{02} = A_0$,

$$A_T = A_1 + A_2 = A_0\left[\sin\left(\frac{2\pi c\tau}{\lambda} - \Delta\right) + \sin\left(\frac{2\pi c\tau}{\lambda}\right)\right] \tag{5.47}$$

which can be rewritten

$$A_T = 2A_0 \cos\left(\frac{\Delta}{2}\right) \sin\left(\frac{2\pi c\tau}{\lambda} - \theta\right) \tag{5.48}$$

where θ is a new phase difference. Thus the sum of the two waves is a new wave of the same frequency and wavelength.

Of particular importance is the intensity of the combined beam, which is the quantity observed visually or measured by a photographic plate. The intensity I is proportional to the square of the peak amplitude or

$$I \sim 4A_0^2 \cos^2\left(\frac{\Delta}{2}\right) \tag{5.49}$$

Note that when $\Delta/2\pi$ is an integer (say j), the peak intensity is four times that of either of the two beams, but when $\Delta/2\pi$ is a half integer $(j + 1/2)$ the intensity is zero. The interesting yet not too difficult paradox in this latter case is where did the energy go?

The optical path along a light beam is defined by

$$PL = \int n \, dz \tag{5.50}$$

$$PL = \int \frac{c_0}{c} \, dz = \lambda_0 \int \frac{dz}{\lambda} \tag{5.51}$$

Thus the optical path length is the vacuum wavelength times the real light path in wavelengths (which can vary along the path). In Fig. 5.16 the difference between paths 1 and 2 would be

$$\overline{\Delta PL} = PL_1 - PL_2 = \lambda_0 \left(\int_1 \frac{dz}{\lambda} - \int_2 \frac{dz}{\lambda}\right) \tag{5.52}$$

or

$$\overline{\Delta PL} = \int_1 n\, dz - \int_2 n\, dz \tag{5.53}$$

The phase difference between two points a distance dz apart—see Eq. (5.44)—is

$$\frac{2\pi\, dz}{\lambda}$$

Thus the difference in phase of the two beams upon recombination is, from Eqs. (5.45), (5.46), and (5.52),

$$\Delta = 2\pi \left(\int_1 \frac{dz}{\lambda} - \int_2 \frac{dz}{\lambda} \right)$$

or

$$\frac{\Delta}{2\pi} = \frac{\overline{\Delta PL}}{\lambda_0} \tag{5.54}$$

If $\overline{\Delta PL}/\lambda_0$ is zero or an integer then from Eqs. (5.49) and (5.54) there will be constructive interference and the field on the screen in Fig. 5.16 would be uniformly bright.

5.3.2 Fringe Pattern with Mach–Zehnder Interferometer

Consider a Mach-Zehnder interferometer with beams 1 and 2 passing through homogeneous media so that initially the recombined beam is uniformly bright ($\overline{\Delta PL}$ assumed to be zero). If a disturbance (inhomogeneity) were put in part of the field of light beam 1, then the path difference $\overline{\Delta PL}$ would no longer be zero nor would the field be uniform. At any position on the cross section of the beam (neglecting refraction) Eq. (5.53) can be written to give ϵ, the path length difference in terms of vacuum wavelengths

$$\epsilon = \frac{\overline{\Delta PL}}{\lambda_0} = \frac{1}{\lambda_0} \int (n - n_{ref})\, dz \tag{5.55}$$

where n_{ref} is the reference value of the index of refraction that was initially present throughout beam 1 and is in the reference beam 2. If $\overline{\Delta PL}/\lambda_0$ is an integer the field will be bright while if $\overline{\Delta PL}/\lambda_0$ is a half integer the field will be dark. Thus the initially uniformly bright field will have a series of bright and dark regions (fringes) each one representative of a specific value of $\overline{\Delta PL}/\lambda_0$ and differing in magnitude from the adjacent fringe of the same intensity by a value $\epsilon = \overline{\Delta PL}/\lambda_0 = 1$. If a gas in light beam 1 is causing the variation in optical path the Gladstone-Dale relation, Eq. (5.2), can be used in Eq. (5.55):

$$\epsilon = \frac{C}{\lambda_0} \int (\rho - \rho_{\text{ref}}) \, dz \qquad (5.56)$$

If the field is two dimensional in that the only variations in the index of refraction along the light beam (i.e., in the z direction), are the sharp discontinuities at the entrance and exit of the test section (see Appendix at the end of this chapter for a treatment of axisymmetric density fields) and ρ only varies over a length L, the fringe shift ϵ is given by

$$\epsilon = \frac{n - n_{\text{ref}}}{\lambda_0} L \qquad (5.57)$$

For a gas

$$\epsilon = \frac{C}{\lambda_0} (\rho - \rho_{\text{ref}}) L \qquad (5.58)$$

or

$$\rho = \frac{\lambda_0 \epsilon}{CL} + \rho_{\text{ref}} \qquad (5.59)$$

If the pressure is constant and the ideal gas law is used

$$\frac{1}{T} = \frac{\lambda_0 R}{PCL} \epsilon + \frac{1}{T_{\text{ref}}} \qquad (5.60)$$

$$T = \frac{PCL \cdot T_{\text{ref}}}{PCL + \lambda_0 R \epsilon T_{\text{ref}}} \qquad (5.61)$$

or

$$T - T_{\text{ref}} = \left[\frac{-\epsilon}{(PCL/\lambda_0 R T_{\text{ref}}) + \epsilon} \right] T_{\text{ref}} \qquad (5.62)$$

For a two-dimensional field in a liquid Eq. (5.57) would be written

$$n = \frac{\lambda_0 \epsilon}{L} + n_{\text{ref}} \qquad (5.63)$$

where n and n_{ref} would have to be known as functions of temperature. For small temperature differences

$$\epsilon = \frac{L}{\lambda_0} \frac{dn}{dT} (T - T_{\text{ref}}) \qquad (5.64)$$

and
$$T - T_{\text{ref}} = \frac{\epsilon \lambda_0}{L} \frac{1}{dn/dT} \qquad (5.65)$$

If $\lambda_0 = 546.1$ nm and L is 30 cm, then each fringe (i.e., $\epsilon = 1$) represents a temperature difference of about $2°C$ in air at $20°C$ and 1 atm. In water under the same conditions each fringe represents a temperature difference of about $0.02°C$.

If the initial optical path lengths of 1 and 2 are not exactly equal (but their difference is less than the coherence length of the light source), the above equations are still valid in determining temperature and density differences between different parts of the cross section in the disturbed beam 1. This is the usual manner in which interferometers are employed requiring a reference point in the cross section of the test beam 1 where the properties are known and, of course, uniform optical path length across the cross section of the reference beam. Then n_{ref}, ρ_{ref}, and T_{ref} refer to this specific, hopefully uniform, portion of the test region, and all properties at other locations in the test section are measured relative to the properties there.

When the beams are recombined parallel to each other (which is often called "infinite fringe setting," as discussed below), each fringe is the locus of points in a two-dimensional field where the density or temperature is constant. When the fringes represent isotherms in heat transfer studies they delineate thermal boundary layers and are very useful for qualitative temperature-field visualization as well as for quantitative studies.

In practice a Mach-Zehnder interferometer is not always used with the beams perfectly parallel upon recombination as in the discussion above. Consider two beams each of which is uniform (in phase) normal to the direction of its propagation, although diverging slightly, at a small angle θ, from each other as represented by the two wave trains shown in Fig. 5.17. Lines are shown drawn through the crests (maxima of amplitude) for each wave train to represent the planes (wave fronts) normal to the direction of propagation. Constructive interference occurs where the maxima of the two beams coincide; dashed lines representing the locus of these positions are also shown. If a screen is placed approximately normal to the two beams, the intensity distribution on the screen would follow a cosine-squared law Eq. (5.49) as shown on the figure. Thus parallel, equally spaced, alternately dark and bright fringes (called "wedge fringes") appear on the screen when there are no disturbances in either field. The difference in optical path length between the two beams varies linearly across the field of view with wedge fringes so that only one fringe in the field (the "zero-order" fringe) represents exactly equal path lengths. From Fig. 5.17 the spacing between the fringes is

$$d = \frac{\lambda/2}{\sin \theta/2} \qquad (5.66)$$

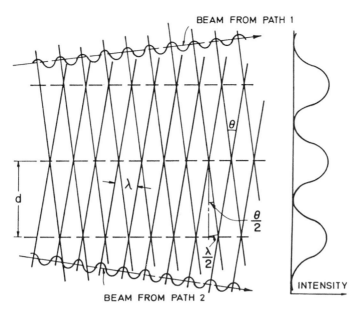

FIG. 5.17 Intensity pattern from two intersecting plane light beams.

or
$$d \sim \frac{\lambda}{\theta}$$

for small θ. To observe fringes θ must be very small. For example, if d is about 5 mm then θ (using the green mercury line) is about 10^{-4} rad. As θ is decreased to zero the fringes get further and further apart approaching the "infinite fringe" pattern found when the two beams are parallel.

When a Mach-Zehnder interferometer is adjusted to give wedge fringes, the fringes are localized as shown in Fig. 5.18. Only a pair of rays is shown, which after diverging are brought to a focus on the screen (or film in a camera) by a focusing lens. The angular separation of the beams is greatly exaggerated on the figure. The dashed lines represent the paths from the virtual object of the beams in 1 and 2 as they would appear along the other path. The fringes are localized where, tracing backwards along the real and image paths, the rays intersect. The plane of localization can be adjusted, once the beams are close enough to parallel to see fringes (i.e., d not too small) by rotation primarily of Sp_2 (and also of M_2 to keep the fringes in view) about two orthogonal axes in the plane of Sp_2. If the fringes are localized at M_2 then the rotation of M_2 will not affect the plane of localization and will only change the fringe spacing and orientation. To have both the fringes (localized at M_2) and the test section (actually the center of the test section—as discussed below) in focus on the screen or in the camera, the interferometer mirrors can be placed on the corners of a 2-1 rectangle, the distance from Sp_2 to M_2 then being the same as the distance from Sp_2 to the middle of the test section.

When a disturbance is present within the test section (which is in beam 1 in Fig. 5.18) the optical path is no longer uniform in this beam 1. The fringes then are no longer straight, but curved as in Fig. 5.19. In this figure the original (undisturbed) positions of the fringes are shown by dashed lines. In general the undisturbed fringes should be aligned in a direction in which the expected temperature (index of refraction) gradient will be large. The difference in optical path length from the original value or from the reference position in the field of view where the fringes have not changed is shown on the figure in terms of the fringe shift ϵ. If the total fringe shift is large, only integral values of ϵ are usually measured; for small differences in optical path length, fractional values of ϵ can be measured as shown. One of the major advantages of wedge fringes is the possibility of measuring fractional values of ϵ. In addition it is difficult to be certain with infinite fringe spacing that the undisturbed (or reference) field is at the maximum brightness so that there is an uncertainty in the reference position. This problem does not occur with wedge fringes as long as there is a region of known uniform properties in the field of view. Contours of constant optical path length can also be obtained with wedge fringes by superposing the disturbed interference pattern over the undisturbed pattern. The resulting moiré fringe pattern gives grey lines which can represent isotherms in a two-dimensional test region. If there are irregularities in the undisturbed image due to faulty optical parts this superposition method can still be used to obtain quantitative results.

The versatility of interferometry in heat transfer studies can be observed in

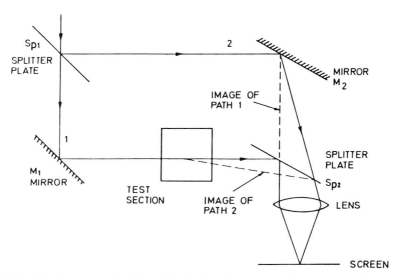

FIG. 5.18 Light rays for Mach-Zehnder interferometer indicating the preferred position of focus.

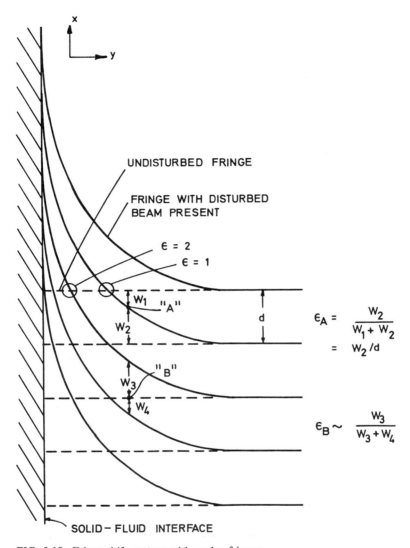

FIG. 5.19 Fringe shift pattern with wedge fringes.

Figs. 5.20 to 5.24. Figures 5.20 to 5.22 are obtained from an initially infinite fringe spacing. Figures 5.20 and 5.21 indicate the free convection isotherms in air between two horizontal cylinders maintained at different temperatures and around a single heated horizontal cylinder, respectively. The single cylinder is the one for which schlieren and shadowgraph (Schmidt-schlieren) patterns are presented in Figs. 5.8 and 5.13. Figure 5.22 shows how transient boundary layer development, in this case with free convection on a vertical foil in air, can be followed with an interferometer. Figure 5.23 shows the moiré fringe pattern obtained when a double exposure is made, indicating the free convection boundary layers on both sides of a thin foil in water. Figure 5.24 shows the temperature field in a forced flow using wedge fringes. The flow is of air over a rearward facing step with an approximately uniform heat flux surface.

5.3.3 Design and Adjustment

A diagram of an operating Mach-Zehnder interferometer is shown in Fig. 5.25. Instead of lenses, mirrors are used to obtain the initial parallel light beam and to focus the final combined beam. Light from a low pressure mercury vapor lamp is focused, through a filter to have a more monochromatic beam, on a small (0.25-mm diameter) illuminating mirror D which acts as a near point light source for paraboloidal mirror F. The parallel light from this mirror goes to the first splitter plate Sp_1. The light that passes directly through the splitter plate goes to mirror M_1 where it is turned at right angles and directed through the test

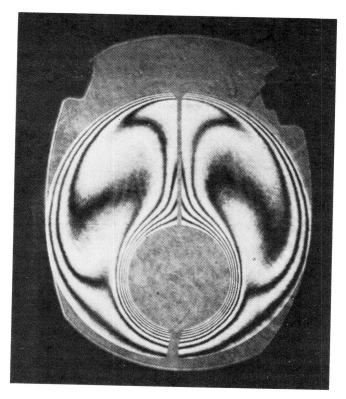

FIG. 5.20 Natural convection isotherms in the annulus between a heated inner cyclinder and a cooled outer cylinder.

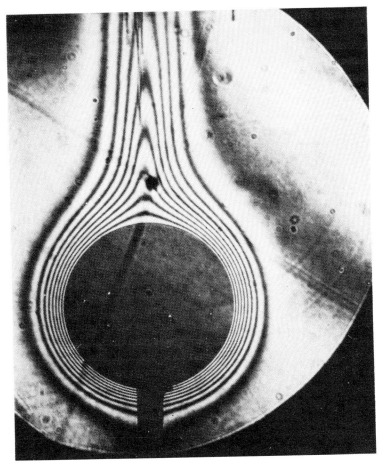

FIG. 5.21 Isotherms about a heated horizontal cylinder.

section and then to the second splitter plate Sp_2. The reference beam, after being reflected at Sp_1 goes to mirror M_2 and is there reflected to Sp_2. The two beams, after recombining at Sp_2, are reflected by plane mirror M_3 to the spherical mirror M_4, then onto a small plane mirror H and finally into a camera or onto a viewing plate. If thick windows must be used on the test section or if the fluid in the test section has a refractive index very different from that of air, it may be necessary to have a compensating tank Q to ensure that the path lengths of the beams are not too different. Using a laser light source with its large coherence length could eliminate the need for such a compensating tank.

Note that the test section should be placed as shown rather than in the other light beam. In this way the beam representing the shadow image of the test section does not pass through the last splitter plate which could cause considerable astigmatism [28].

The alignment of an interferometer such as shown in Figs. 5.16 and 5.25 though somewhat complicated and time consuming is not so horrendous a task as is often stated. A number of methods have been described [21, 25, 29, 30] which greatly simplify the task. The chief concern is to align the reference and test beams so that they are closely parallel. If they are not, the fringe spacing is so small that the fringes cannot be detected.

When first aligning the interferometer the two path lengths are set approximately equal. The difference in length must be less than the coherence length of

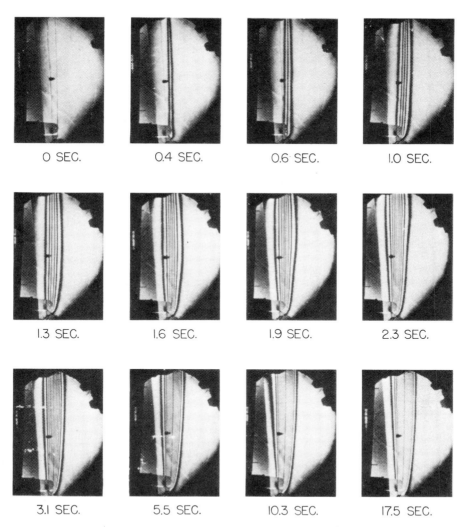

O SEC.   0.4 SEC.   0.6 SEC.   1.0 SEC.

1.3 SEC.   1.6 SEC.   1.9 SEC.   2.3 SEC.

3.1 SEC.   5.5 SEC.   10.3 SEC.   17.5 SEC.

FIG. 5.22 Transient natural convection boundary layer development on both sides of a heated vertical sheet.

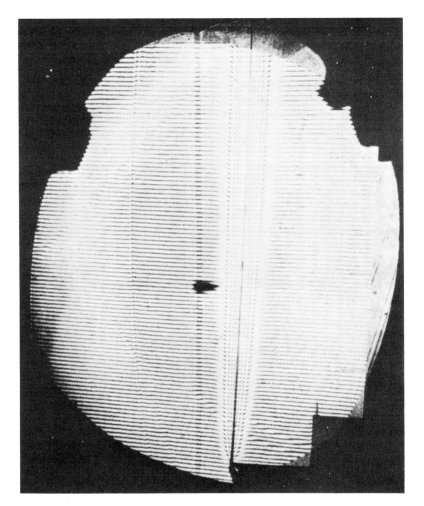

FIG. 5.23 Moiré fringe pattern of interferogram showing temperature fields on both sides of a heated vertical sheet.

the source-filter combination which increases as the light becomes more mono-chromatic. If the light intensity variation with wavelength is Gaussian with a bandwidth of $\Delta\lambda$, the coherence length or optical path difference over which fringes can still be observed is approximately $\lambda^2/\Delta\lambda$ [31]. The parallelism of the beam leaving the paraboloidal mirror F can be determined by measuring its dimensions at various positions along its path or, more accurately, by having it reflected back to F by a plane mirror and observing the proximity of the focus of this returned beam to the small illuminating mirror D.

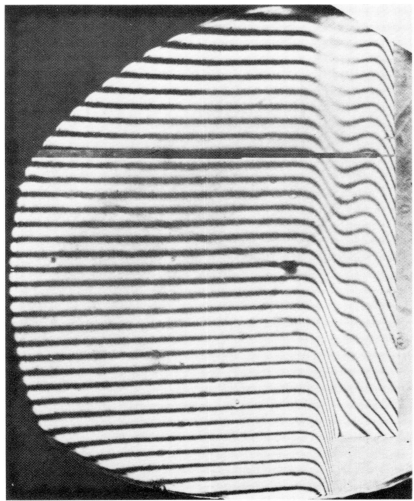

AIR FLOW →

FIG. 5.24 Interferogram of temperature field in a forced flow of air using wedge fringes.

A - LIGHT SOURCE
B - CONDENSER
C - COLOR FILTER
D - ILLUMINATING MIRROR
F - PARABOLOIDAL MIRROR
Sp_1, Sp_2 - SPLITTER PLATES 15cm dia
M_1, M_2, M_3 - PLANE MIRRORS 15cm dia
G - MIRROR TRANSLATION MOUNT
AND MOTOR
M_4 - SPHERICAL MIRROR

H - SMALL MIRROR
I - CAMERA
L - ROTATION CONTROLS FOR
M_2 AND Sp_2
N - TRANSLATION MOTOR CONTROLS
P_1, P_2 - LIGHT PATHS
Q - COMPENSATING CHAMBER
WHEN NECESSARY
T - TEST SECTION
U - WINDOW

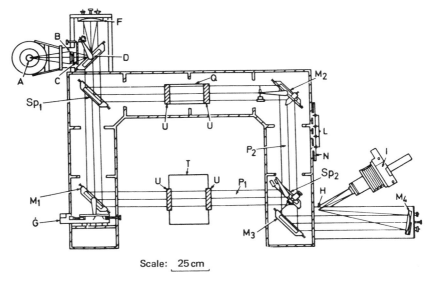

Scale: 25 cm

FIG. 5.25 University of Minnesota Heat Transfer Laboratory Mach-Zehnder interferometer.

To obtain fringes beams 1 and 2 (Fig. 5.18) must be nearly parallel to each other following Sp_2. This can be accomplished by aligning the images of two objects observed by looking back through Sp_2 preferably with a small telescope. The two objects examined should be far apart and must both be located before the first splitter plate Sp_1 so that images are obtained for paths 1 and 2. In practice the illuminating mirror (D in Fig. 5.25) or point source (Fig. 5.16), which can be considered at infinity, is used as well as some observable object (perhaps the illuminated cross hairs holding the mirror) placed in the field of view before Sp_1. Rotation of M_2 and Sp_2 about two orthogonal axes moves each pair of virtual images of the two objects. When both images of each object are superimposed the two beams leaving Sp_2 are closely parallel and fringes should appear in the field of view. Focusing on the center of the test section and M_2 with the telescope, camera, or screen, fringes are made as sharp as possible by further rotation of Sp_2 meanwhile rotating M_2 to keep the fringe spacing from getting too small.

When the plane of focus of the fringes is in M_2 (and the center of the test section), the final adjustment of optical path length can be made. It is advantageous to keep the interferometer set close to the zero-order fringe (zero optical path length between the two light beams) since that is the position for sharpest fringes even for the filtered light. In addition, white light and the zero-order fringe are useful for measuring the index of refraction for fluids whose optical properties are unknown or in tracing fringes through regions of large density gradients. In the apparatus shown in Fig. 5.25 the path length of the test beam can be altered by translation of the mirror (M_1). If white light (an incandescent bulb suffices) replaces the filtered mercury light on half the field of view the mirror can be translated until the zero-order white light fringe is observed. In practice it is often helpful, in particular if the initial setting is far from the zero-order fringe, to use light of varying coherence lengths from the most monochromatic to white to make each set of fringes as sharp as possible while translating the mirror.

Once adjusted the interferometer, if properly constructed, usually needs only minor readjustment. Placing the unit in a vibration-free constant-temperature area helps to maintain alignment.

5.3.4 Interferometer Error Analysis in a Two-Dimensional Field

Since the Mach-Zehnder interferometer is of great value for quantitative studies in two-dimensional fields, considerable attention [32–40] has been directed toward the corrections that must be applied when the idealizations assumed in the derivation of Eqs. (5.62) and (5.65) are not strictly met. The two most significant errors usually encountered are due to refraction and end effects. Refraction occurs when there is a density (really index of refraction) gradient normal to the light beam causing the beam to "bend." The resulting error increases with increasing density gradient and with increasing path length in the disturbed region L. It is refraction that usually prevents accurate interferometric measurements in thin forced convection boundary layers. End effects are caused by deviation from two dimensionality of the actual density field, in particular where the light beam enters and leaves the disturbed region. The end effects are usually large when the disturbed region is large normal to the light beam direction and the test section's length along the light beam is relatively short. End effects are often significant with thick thermal boundary layers. Thus if an experimental apparatus is designed to minimize refraction error, the end effect error may be large and vice versa.

The end effects will be considered first. At the two ends of the test section, where the light enters and leaves, the density field may no longer be truly two dimensional, and some correction to the calculation of the two-dimensional field

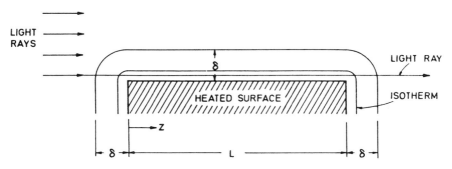

FIG. 5.26 Assumed temperature distribution to determine end effect correction.

in the center of the test piece must be made. The exact correction depends on the design of the test apparatus including the presence or absence of windows at the edges of the test section. For the purposes of demonstration consider Fig. 5.26 where a horizontal cross section of a constant temperature heated vertical plate losing heat by natural convection is shown. In the region near the center of the plate the temperature field is two dimensional (one dimensional in the section shown). As an approximation of the effect of the ends assume that the isotherms or constant density lines are arcs of circles at the corners in the figure. The fringe shift of the ray at the wall is, from Eq. (5.55),

$$\epsilon_w = \frac{1}{\lambda_0} \int (n - n_{\text{ref}}) \, dz$$

$$= \frac{2}{\lambda_0} \left[\int_{L/2}^{L} (n_w - n_{\text{ref}}) \, dz + \int_{L}^{L+\delta} (n - n_{\text{ref}}) \, dz \right]$$

integrated along $y = 0$. To a first approximation

$$\epsilon_w = \frac{2}{\lambda_0} \frac{dn}{dT} \left[\int_{L/2}^{L} (T_w - T_{\text{ref}}) \, dz + \int_{L}^{L+\delta} (T - T_{\text{ref}}) \, dz \right] \qquad (5.67)$$

where n_{ref} and T_{ref} are respectively the index of refraction and temperature outside the boundary layer. A reasonable polynomial approximation to the temperature distribution in the boundary layer (assumed laminar) is

$$T - T_{\text{ref}} = (T_w - T_{\text{ref}}) \left(1 - \frac{y}{\delta} \right)^2 \qquad (5.68)$$

Then at $y = 0$, for $L < z < L + \delta$

$$T - T_{ref} = (T_w - T_{ref})\left(1 - \frac{z - L}{\delta}\right)^2 \tag{5.69}$$

and from Eq. (5.67)

$$\epsilon_w = 2\frac{dn}{dT}\frac{1}{\lambda_0}(T_w - T_{ref})\left(\frac{L}{2} + \frac{\delta}{3}\right) \tag{5.70}$$

The error in the fringe shift, $\Delta\epsilon$, due to the end effect is the difference between the values calculated using Eqs. (5.70) and (5.64) (i.e., the effect of finite δ)

$$\frac{\Delta\epsilon}{\epsilon} \approx +\frac{2}{3}\frac{\delta}{L} \tag{5.71}$$

The error as calculated above is positive in that a temperature larger than the true wall temperature would be indicated if no correction were used. More complicated temperature distributions near the edge of a test section have also been studied [32, 38, 41]. In some cases the end-effect error in the outer region of a boundary layer may be negative.

Refraction of the light beam leads to two different deviations from the evaluating equations. When the light ray is bent in the assumed two-dimensional field [say $n = n(x, y)$] no longer does it go through a fixed value of $x - y$ along its path. Equation (5.55) should be used along the true optical path rather than assuming that the index of refraction is constant along the path and can be taken outside the integral sign. Another refraction effect is the apparent displacement of the beam which leads to an error in the $x - y$ position assigned to a particular fringe displacement. If the interferometer is focused at the center of the test section the displacement error is essentially zero whether the index of refraction is close to unity [37] or not [42], and this is done in most precision measurements.

To determine the remaining (i.e., first) part of the refraction error consider, for simplicity, a one-dimensional problem where the index of refraction varies only in the y direction as in Fig. 5.14. This is often a reasonable assumption. Additionally, let us assume that $\partial n/\partial y$ is constant over the path of a given light ray. Although other variations can be chosen this is a good approximation in a thermal boundary layer close to the solid-fluid interface. Then Eq. (5.9), which applied to Fig. 5.14 as long as the angles were small, can be integrated twice—see Eq. (5.36)—yielding

$$y - y_i = \frac{1}{2}\frac{1}{n}\frac{\partial n}{\partial y}z^2 \tag{5.72}$$

Since $\partial n/\partial y$ is constant

$$n = n_i + \frac{\partial n}{\partial y}(y - y_i) \tag{5.73}$$

where n_i is the value of the index of refraction at y_i. Then from Eq. (5.55)

$$\epsilon = \frac{1}{\lambda_0} \int_0^L \left[(n_i - n_{\text{ref}}) + \frac{\partial n}{\partial y} \frac{1}{2n} \frac{\partial n}{\partial y} z^2 \right] dz \tag{5.74}$$

or

$$\epsilon = \frac{(n_i - n_{\text{ref}})L}{\lambda_0} + \frac{1}{\lambda_0} \left(\frac{\partial n}{\partial y} \right)^2 \frac{1}{6n} L^3 \tag{5.75}$$

for a two-dimensional field with only small changes in n. Subtracting the value of ϵ obtained by neglecting refraction, i.e., $(n_i - n_{\text{ref}})L/\lambda_0$, the error in fringe shift is

$$\Delta\epsilon = \frac{1}{\lambda_0} \left(\frac{\partial n}{\partial y} \right)^2 \frac{L^3}{6n} \tag{5.76}$$

From Eq. (5.65) the error in temperature is

$$\Delta T = \frac{\lambda_0 \, \Delta\epsilon}{L \, dn/dT} \tag{5.77}$$

Taking

$$\frac{\partial n}{\partial y} = \frac{dn}{dT} \frac{\partial T}{\partial y}$$

$$\Delta T = \left(\frac{dn}{dT} \right) \left(\frac{\partial T}{\partial y} \right)^2 \frac{L^2}{6n} \tag{5.78}$$

This is the normal refraction correction that should be considered in an experiment. According to more exact analyses the refraction error is a function of the location of the focal plane. When the interferometer is focused at the center of the test section the refraction error is half of the value given by Eq. (5.78) [33, 40, 42]. Additional refraction effects appear when the test beam passes through a test section window, but this correction is usually quite small [38].

An additional error may appear when measuring the fringe positions. Fringe shifts are usually determined visually with a traveling microscope which permits quite accurate measurements [43]. Other refinements [44, 45] have led to more exotic measuring systems.

5.3.5 Other Interferometers

A number of interferometric systems other than the Mach-Zehnder have been used in heat transfer and aerodynamic studies. These in general are used to produce interferograms which can be evaluated in a similar manner to the Mach-Zehnder patterns. In some systems gratings have been used to divide the initial beam into two coherent beams, one of which traverses the test region. The beams are recombined on another grating yielding an interference pattern. One system [46, 47] uses two gratings but the reference and test beams then pass very close to each other. In a four-grating apparatus [48] the beams are further apart but residual fringes are often superimposed on the pattern. A laser light source has also been used with a grating interferometer [49].

Several laser light source interferometers have been used or suggested for temperature measurement [50–52]. Direct use of a laser as the light source in a Mach-Zehnder interferometer increases the coherence length to the point where a compensating tank may not be required even for a relatively large difference in path length of the two beams in the interferometer.

Other systems that are used with laser sources include schlieren or shearing interferometers. In one such system [53–55] a very small wire or stop is placed at the focus of a standard schlieren system to block the central maximum of the Fraunhofer diffraction pattern. Then the phase distribution produced when the beam passes through a disturbance in the test region can be observed in reference to an undisturbed part of the test beam. The result is a fringe pattern similar to that observed with a Mach-Zehnder interferometer set for infinite fringe spacing. Another system [56] uses a small glass shearing plate in place of the knife edge. When the angle between the incoming light beam and the normal to the plate is approximately 50 degrees, the first two reflected beams will interfere, resulting in fairly straight parallel fringes. A Wollaston prism [57–59] and a grating [60] have also been used to produce finite fringe interference patterns. Shearing interferometers in which the reference beams are sheared slightly [52, 61] in a lateral direction have been used with conventional and laser light sources. A polarization interferometer [7, 62] is a wave-shearing interferometer in which two coherent beams polarized at right angles to one another are produced from a single incoming beam. The two beams often diverge at a finite angle, and when recombined after passage through the test section, a fringe pattern results. If the shear is very small the fringe pattern represents the first derivative of the index of refraction field rather than the index of refraction directly. Either a finite fringe or infinite fringe field can be obtained using three Wollaston prisms [63, 64]. An advantage of a shearing interferometer over a Mach-Zehnder interferometer is that the temperature gradient near the surface of an object can be measured directly, giving the heat flux. The sensitivity is proportional to the shear spacing as well as to the length of the test object in the light beam direction. With a small shear spacing the errors due to refraction and thick window effects tend to vanish. Among the disadvantages is the need to integrate the fringe displacement to obtain density distributions. Only the density gradient in the direction of the beam shear can be measured at any one time. The closest to a surface that an accurate measurement of the density gradient can be made is on the order of one half the shear displacement.

A gas laser has been used in a Twyman-Green interferometer for heat transfer studies [65] as well as in a tilted plate interferometer producing Fizeau fringes [66]. In addition, lasers in holographic systems can be specifically applied to heat transfer studies [67–71].

5.4 CONCLUSION

The properties of shadowgraph, schlieren, and interferometric systems have been described. The basic operating equations for all three systems have been derived with emphasis on their use in determining density and temperature distributions in two-dimensional fields.

Shadowgraph and schlieren systems are principally used for qualitative descriptions of a density or temperature field. However, they can be used for quantitative studies in particular if a grid-schlieren or Schmidt-schlieren apparatus is used. Because schlieren and shadowgraph photographs yield information on the first and second derivatives of density (or temperature), their widest application can be found in systems where there are steep gradients of density and temperature as with flame fronts and shock waves.

Interferometers, usually in the form of a Mach-Zehnder system, are often used in quantitative studies of two-dimensional (including axisymmetric) density and temperature fields. For this reason details of the evaluating equations and the possible error corrections have been presented. Interferometers are often employed with systems in which density or temperature gradients are relatively small, such as free convection boundary layers.

The advent of lasers permits novel and useful interferometric designs. Lasers can be used in schlieren interferometers and in holography for temperature and density measurements.

ACKNOWLEDGMENT

I am grateful to W. Aung and T. Y. Chu for their aid during the preparation of the photographic plates and the review of the original manuscript. T. Kuehn was instrumental in checking the revised manuscript including new developments and references.

NOMENCLATURE

A amplitude of light beam
A_1 amplitude of light beam from path 1
A_2 amplitude of light beam from path 2

A_0	maximum of amplitude
A_{01}	maximum of amplitude of beam from path 1
A_{02}	maximum of amplitude of beam from path 2
A_T	amplitude of recombined light beams from paths 1 and 2
Δa	deflection of light beam away from schlieren knife edge
a_s, a_0, a_K	dimension of schlieren beam normal to knife edge at source, at knife edge, and above knife edge when no disturbance is present respectively
b_0, b_s	dimension of schlieren beam parallel to knife edge at source and knife edge respectively
C	Gladstone-Dale constant
c	speed of light
c_0	speed of light in vacuum
D	diameter of cylinder and dimension in Nusselt number
d	fringe spacing
f	focal length of lens or mirror
h	heat transfer coefficient
I	light intensity
I_d	illumination at screen of disturbed field when knife edge is present
I_K	illumination at screen of undisturbed field when knife edge is present
I_0	illumination at screen with no knife edge
I_T	illumination at exit of test section (and at screen if no deflection) in shadowgraph system
K	knife edge
k	thermal conductivity
L	length of test section in light beam direction z
n	index of refraction
n_a	index of refraction of air outside test section
n_0	index of refraction at standard conditions
n_{ref}	index of refraction in reference region
Δn	$(n - n_{\text{ref}})$
Nu	Nusselt number
P	pressure
p	object distance from lens or mirror
$\overline{\text{PL}}$	optical path length
$\overline{\Delta\text{PL}}$	difference in optical path length
q	image distance from lens or mirror
q_w	wall heat flux
R	gas constant in terms of mass
r	radial position in axisymmetric field
r_0	value of r such that for $r > r_0$ the index of refraction is that of the reference region
Ra	Rayleigh number based on diameter for circular cylinder
S	source
Sp	splitter plate

T	temperature
T_{ref}	temperature in reference region
T_∞	free stream temperature
x	direction normal to y and z
y	direction perpendicular to z, is usually the direction in which the gradient of density and temperature lies
y_i	height of light ray at entrance to test section
y_{sc}	height of light ray at screen
$y_{sc, max}$	maximum value of y_{sc}
y_T	height of light ray at exit from test section
$(\partial y/\partial z)_T$	slope of light ray at exit of test section
z	direction along light beam
z_{sc}	distance from test section to screen

Greek Symbols

α	angular deflection of light ray as measured in air outside test section; is same as α' if $n \approx n_a$
α_{max}	maximum deflection angles that can be measured with schlieren system
α_{max}'	maximum deflection angles that can be measured with schlieren system
α'	angular deflections of light ray within test fluid; $\alpha' \cong \alpha$ if $n \approx n_a$ (i.e., if test fluid is gas)
α''	angle defined and used in Fig. 5.4
β	angle defined and used in Fig. 5.4
γ	angle defined and used in Fig. 5.4
δ	boundary layer thickness
Δ	phase difference; also "change of"
ϵ	interferometer fringe shift, optical path length difference in vacuum wavelengths
θ	angle between interferometer beams when recombined
λ	wavelength of light
λ_0	vacuum wavelength
λ_{sc}	wavelength at screen or when beams are recombined
$\Delta\lambda$	spectral width of light source
ρ	density
ρ_0	density at standard condition
ρ_{ref}	density in reference region
σ	dummy variable introduced into Eq. (5A.5), Appendix to this chapter
τ	time; dummy variable used in Eq. (5A.7), Appendix to this chapter

Subscripts

w	refers to condition at the wall
sc	screen

REFERENCES

1. Weinberg, F. J., "Optics of Flames," Butterworth, London, 1963.
2. Ladenburg, R. W., Lewis, B., Pease, R. N., and Taylor, H. S., Physical Measurements in Gas Dynamics and Combustion, in "High-Speed Aerodynamics and Jet Propulsion," vol. 9, chaps. A1, A2, A3, Princeton University Press, Princeton, N.J., 1954.
3. Holder, D. W., North, R. J., and Wood, G. P., Optical Methods for Examining the Flow in High-Speed Wind Tunnels, pts. I and II, *AGARD*, 1965.
   Holder, D. W., and North, R. J., Schlieren Methods, *Notes Appl. Sci. No.* 31, National Physical Laboratory, London, 1963.
4. Barnes, N. F., Optical Techniques for Fluid Flows, *J. Soc. Mot. Pict. Tel. Eng.*, vol. 61, p. 487, 1953.
5. Dean, R. C., Jr., Aerodynamic Measurements, Gas Turbine Lab., Massachusetts Institute of Technology, Cambridge, 1953.
6. Shardin, H., Toepeler's Schlieren Method; Basic Principles for Its Use and Quantitative Evaluation, *Navy Translation* 156, 1947.
7. Hauf, W. and Grigull, U., Optical Methods in Heat Transfer, in J. P. Hartnett and T. F. Irvine, Jr. (eds.), "Advances in Heat Transfer," vol. 6, p. 133, Academic Press, New York, 1970.
8. Chu, T. Y., and Goldstein, R. J., Turbulent Convection in a Horizontal Layer of Water, *J. Fluid Mech.*, vol. 60, part 1, p. 141, 1973.
9. Landolt-Bornstein, Physikalisch-Chemische Tabellen, Suppl. #3, p. 1677, 1935.
10. Tilton, L., and Taylor, J., Refractive Index and Dispersion of Distilled Water for Visible Radiation at Temperatures 0 to 60°C, *J. Res. Nat. Bur. Stand. (U.S.)*, vol. 20, p. 419, 1938.
11. Speak, G. S., and Walters, D. J., Optical Considerations and Limitations of the Schlieren Method, *ARC Tech. Rep.*, R and M No. 2859, London, 1954.
12. Ronchi, V., Due Nuovi Metodi per lo Studio delle Superficie e dei Sistemi Ottici, *Ann. R. Scuola Normale Sup. Pisa*, vol. 15, 1923 (bound 1927).
13. Darby, P. F., The Ronchi Method of Evaluating Schlieren Photograph, *Tech. Conf. Opt. Pheno. Supersonic Flow, NAVORD Rep.* 74–46, 1946.
14. Didion, D. A., and Oh, Y. H., A Quantitative Schlieren-Grid Method for Temperature Measurement in a Free Convection Field, *Tech. Rep.* 1, Mechanical Engineering Department, Catholic University of America, Washington, D.C., 1966.
15. Watermeier, L. A., Self-Illuminated Schlieren System, *Rev. Sci. Instrum.*, vol. 37, p. 1139, 1966.
16. Kantrowitz, A., and Trimpi, R. L., A Sharp Focusing Schlieren System, *J. Aero. Sci.*, vol. 17, p. 311, 1950.
17. Hett, J. H., A High Speed Sterioscopic Schlieren System, *J. Soc. Mot. Pict. Tel. Eng.*, vol. 56, p. 214, 1951.
18. Ackerman, J. A., and Brill, G. A., Jr., Final Report on Research on the Adaptability of Lasers to Schlieren Systems, *ARL* 65-139, 1965.
19. Schmidt, E., Schlierenaufnahmen des Temperaturfelds in der Nähe Wärmeabgebender Körper, *Forschung. Ing. Wes.*, vol. 3, p. 181, 1932.
20. Boelter, L. M. K., and Cherry, V. H., Measurement of Heat Transfer by Free Convection from Cylindrical Bodies by the Schlieren Methods, *ASHVE Trans.*, vol. 44, p. 499, 1938.
21. Eckert, E. R. G., Drake, R. M., Jr., and Soehngen, E., Manufacture of a Zehnder-Mach Interferometer, *Wright-Patterson AFB, Tech. Rep.* 5721, ATI-34235, 1948.
22. Bennett, F. D., and Kahl, G. D., A Generalized Vector Theory of the Mach-Zehnder Interferometer, *J. Opt. Soc. Amer.*, vol. 43, p. 71, 1953.
23. Zobel, T., The Development and Construction of an Interferometer for Optical Measurement of Density Fields, *NACA Tech. Note* 1184, 1947.

24. Shardin, H., Theorie und Anwendung des Mach-Zehnderschen Interferenz-Refrakto-meters, *Z. Instrumentenk.*, vol. 53, pp. 396 and 424, 1933; *DRL Trans. No. 3*, University of Texas.

25. Tanner, L. H., The Optics of the Mach-Zehnder Interferometer, *ARC Tech. Rep.*, R and M No. 3069, London, 1959.

26. Tanner, L. H., The Design and Use of Interferometers in Aerodynamics, ARC Tech. Rep. R and M, No. 3131, London, 1957.

27. Wilkie, D., and Fisher, S. A., Measurement of Temperature by Mach-Zehnder Interfero-metry, *Proc. Inst. Mech. Eng.*, vol. 178, p. 461, 1963–1964.

28. Prowse, D. B., Astigmatism in the Mach-Zehnder Interferometer, *Appl. Opt.*, vol. 6, p. 773, 1967.

29. Price, E. W., Initial Adjustment of the Mach-Zehnder Interferometer, *Rev. Sci. Instrum.*, vol. 23, p. 162, 1952.

30. Prowse, D. B., A Rapid Method of Aligning the Mach-Zehnder Interferometer, *Aust. Def. Sci. Serv., Tech. Note* 100, 1967.

31. Born, M. and Wolf, E., "Principles of Optics," 3rd ed., p. 322, Pergamon, New York, 1965.

32. Eckert, E. R. G., and Soehngen, E. E., Studies on Heat Transfer in Laminar Free Convection with the Zehnder-Mach Interferometer, *Air Force Tech. Rep.* 5747, ATI-44580, 1948.

33. Wachtell, G. P., Refraction Effect in Interferometry of Boundary Layer of Supersonic Flow along Flat Plate, *Phys. Rev.*, vol. 78, p. 333, 1950.

34. Blue, R. E., Interferometer Corrections and Measurements of Laminar Boundary Layers in Supersonic Stream, *NACA Tech. Note* 2110, 1950.

35. Eckert, E. R. G., and Soehngen, E. E., Distribution of Heat-Transfer Coefficients around Circular Cylinders in Crossflow at Reynolds Numbers from 20 to 500, *Trans. ASME*, vol. 74, p. 343, 1952.

36. Kinsler, M. R., Influence of Refraction on the Applicability of the Zehnder-Mach Interferometer to Studies of Cooled Boundary Layers, *NACA Tech. Note* 2462, 1951.

37. Howes, W. L., and Buchele, D. R., A Theory and Method for Applying Interferometry to the Measurement of Certain Two-Dimensional Gaseous Density Fields, *NACA Tech. Note* 2693, 1952.

38. Howes, W. L., and Buchele, D. R., Generalization of Gas-Flow Interferometry Theory and Interferogram Evaluation Equations for One-Dimensional Density Fields, *NACA Tech. Note* 3340, 1955.

39. Howes, W. L., and Buchele, D. R., Practical Considerations in Specific Applications of Gas-Flow Interferometry, *NACA Tech. Note* 3507, 1955.

40. Howes, W. L., and Buchele, D. R., Optical Interferometry of Inhomogeneous Gases, *J. Opt. Soc. Amer.*, vol. 56, p. 1517, 1966.

41. Goldstein, R. J., Interferometric Study of the Steady State and Transient Free Convec-tion Thermal Boundary Layers in Air and in Water about a Uniformly Heated Vertical Flat Plate, Ph.D. thesis, University of Minnesota, Minneapolis, 1959.

42. Chu, T. Y., Personal communication, 1968.

43. Howes, W. L., and Buchele, D. R., Random Error of Interference Fringe Measurements Using a Mach-Zehnder Interferometer, *Appl. Opt.*, vol. 5, p. 870, 1966.

44. Werner, F. D., and Leadon, B. M., Very Accurate Measurement of Fringe Shifts in an Optical Interferometer Study of Gas Flow, *Rev. Sci. Instrum.*, vol. 24, p. 121, 1953.

45. Dew, G. D., A Method for the Precise Evaluation of Interferograms, *J. Sci. Instrum.*, vol. 41, p. 160, 1964.

46. Kraushaar, R., A Diffraction Grating Interferometer, *J. Opt. Soc. Amer.*, vol. 40, p. 480, 1950.

47. Sterrett, J. R., and Erwin, J. R., Investigation of a Diffraction Grating Interferometer for use in Aerodynamic Research, *NACA Tech. Note* 2827, 1952.

48. Weinberg, F. J., and Wood, N. B., Interferometer Based on Four Diffraction Gratings, *J. Sci. Instrum.*, vol. 36, p. 227, 1959.

49. Sterrett, J. R., Emery, J. C., and Barber, J. B., A Laser Grating Interferometer, *AIAA J.*, vol. 3, p. 963, 1965.

50. Goldstein, R. J., Interferometer for Aerodynamic and Heat Transfer Measurements, *Rev. Sci. Instrum.*, vol. 36, p. 1408, 1965.

51. Oppenheim, A. K., Urtiew, P. A., and Weinberg, F. J., On the Use of Laser Light Sources in Schlieren-Interferometer Systems, *Proc. Roy. Soc. A.*, vol. 291, p. 279, 1966.

52. Tanner, L. H., The Design of Laser Interferometers for Use in Fluid Mechanics, *J. Sci. Instrum.*, vol. 43, p. 878, 1966.

53. Gayhart, E. L., and Prescott, R., Interference Phenomenon in the Schlieren System, *J. Opt. Soc. Amer.*, vol. 39, p. 546, 1949.

54. Temple, E. B., Quantitative Measurement of Gas Density by Means of Light Interference in a Schlieren System, *J. Opt. Soc. Amer.*, vol. 47, p. 91, 1957.

55. Brackenridge, J. B., and Gilbert, W. P., Schlieren Interferometry. An Optical Method for Determining Temperature and Velocity Distributions in Liquids, *Appl. Opt.*, vol. 4, p. 819, 1965.

56. Wick, C. J., and Winnikow, S., Reflection Plate Interferometer, *Appl. Opt.*, vol. 12, p. 841, 1973.

57. Small, R. D., Sernas, V. A., and Page, R. H., Single Beam Schlieren Interferometer Using a Wollaston Prism, *Appl. Opt.*, vol. 11, p. 858, 1972.

58. Merzkirch, W., and Erdmann, W., Evaluation of Axisymmetric Flow Patterns with a Shearing Interferometer, *Appl. Opt.*, vol. 12, p. 119, 1973.

59. Merzkirch, W., Generalized Analysis of Shearing Interferometers as Applied for Gas Dynamic Studies, *Appl. Opt.*, vol. 13, p. 409, 1974.

60. Yokozeki, S., and Suzuki, T., Shearing Interferometer Using the Grating as the Beam Splitter, *Appl. Opt.*, vol. 10, p. 1575, 1971.

61. Bryngdahl, O., Applications of Shearing Interferometry, in E. Wolf (ed.), "Progress in Optics," vol. 4, Wiley, New York, 1965.

62. Chevalerias, R., Latron, Y., and Veret, C., Methods of Interferometry Applied to the Visualization of Flows in Wind Tunnels, *J. Opt. Soc. Amer.*, vol. 47, p. 703, 1957.

63. Black, W. Z., and Carr, W. W., Application of a Differential Interferometer to the Measurement of Heat Transfer Coefficients, *Rev. Sci. Instrum.*, vol. 42, p. 337, 1971.

64. Black, W. Z., and Norris, J. K., Interferometric Measurement of Fully Turbulent Free Convective Heat Transfer Coefficients, *Rev. Sci. Instrum.*, vol. 45, p. 216, 1974.

65. Grigull, U., and Rottenkolber, H., Two-Beam Interferometer Using a Laser, *J. Opt. Soc. Amer.*, vol. 57, p. 149, 1967.

66. Marsters, G. F., and Advani, A., A Tilted Plate Interferometer for Heat Transfer Studies, *Rev. Sci. Instrum.*, vol. 44, p. 1015, 1973.

67. Horman, M. H., An Application of Wavefront Reconstruction to Interferometry, *Appl. Opt.*, vol. 4, p. 333, 1965.

68. Heflinger, L. O., Wuerker, R. F., and Brooks, R. E., Holographic Interferometry, *J. Appl. Phys.*, vol. 37, p. 642, 1966.

69. Tanner, L. H., Some Applications of Holography in Fluid Mechanics, *J. Sci. Instrum.*, vol. 43, p. 81, 1966.

70. Bryngdahl, O., Shearing Interferometry by Wavefront Reconstruction, *J. Opt. Soc. Amer.*, vol. 58, p. 865, 1968.

71. Kupper, F. P., and Dijk, C. A., A Method for Measuring the Spatial Dependence of the Index of Refraction with Double Exposure Holograms, *Rev. Sci. Instrum.*, vol. 43, p. 1492, 1972.

APPENDIX: INTERFEROGRAM ANALYSIS
OF AXISYMMETRIC FIELDS

Although the optical techniques described in this paper have found their widest use in two-dimensional rectangular coordinate systems, they are also applicable to other geometries. In particular, interferograms of axisymmetric density or temperature distributions can be quantitatively evaluated using the Abel inversion [A1].

Consider Fig. 5A.1(a) which is a cross section of a field in which the index of refraction is a function only of r, and possibly position x normal to the section, but not of the angular position. We shall only consider the field in this particular section (i.e., constant x). A light beam from an interferometer passes through the test region in the direction z. At radial positions greater than r_0 the field is assumed to be uniform with an index of refraction of n_{ref}. This is no real limitation as we can make r_0 as large as we want. Neglecting refraction [A2] the fringe shift from the light ray at a particular position y is, from Eq. (5.55),

$$\epsilon(y) = \frac{1}{\lambda_0} \int_{-z_0}^{z_0} [n(r) - n_{ref}] \, dz \qquad (5A.1)$$

where the integration is carried out at constant y. The integration limits are functions of y but could with full generality be extended to $\pm \infty$. Since

$$z = \sqrt{r^2 - y^2} \qquad (5A.2)$$

and at constant y

$$dz = \frac{r \, dr}{\sqrt{r^2 - y^2}} \qquad (5A.3)$$

Eq. (5A.1) can be written as

$$\epsilon(y) = \frac{2}{\lambda_0} \int_{y}^{r_0} \frac{\Delta n(r) r \, dr}{\sqrt{r^2 - y^2}} \qquad (5A.4)$$

Multiplying both sides of Eq. (5A.4) by

$$\frac{y \, dy}{\sqrt{y^2 - \sigma^2}}$$

and integrating between σ and r_0

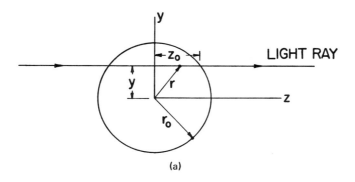

(a)

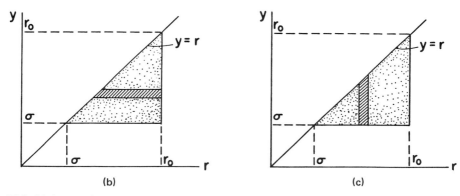

(b)   (c)

FIG. 5A.1 Interferogram analysis. (*a*) Light beam passing through axisymmetric field. (*b*) Region of integration for Eq. (5A.5). (*c*) Region of integration for Eq. (5A.6).

$$\int_\sigma^{r_0} \frac{\epsilon(y)y \, dy}{\sqrt{y^2 - \sigma^2}} = \frac{2}{\lambda_0} \int_\sigma^{r_0} \left[\int_y^{r_0} \frac{\Delta n(r)ry \, dr}{\sqrt{y^2 - \sigma^2} \sqrt{r^2 - y^2}} \right] dy \qquad (5A.5)$$

Note that the integration on the right-hand side is first over r between y and r_0 and then over y between σ and r_0. This is shown in Fig. 5A.1(b) where the integration is carried out over the whole shaded area by first integrating to obtain the horizontal element shown and then in the second integration (over y) to sweep out the finite area. Integration of the integrand over this same region can be performed as shown in Fig. 5A.1(c) by first getting the vertical element (i.e., integrating over y between σ and r) and then the whole area (a second integration over r between σ and r_0). Thus

$$\int_\sigma^{r_0} \frac{\epsilon(y)y \, dy}{\sqrt{y^2 - \sigma^2}} = \frac{2}{\lambda_0} \int_\sigma^{r_0} \Delta n(r) \cdot r \left(\int_\sigma^r \frac{y \, dy}{\sqrt{y^2 - \sigma^2} \sqrt{r^2 - y^2}} \right) dr \quad (5A.6)$$

The significance of this change in the order of integration can be observed by considering the inner integral on the right-hand side of Eq. (5A.6). Let

$$\tau = \frac{y^2 - \sigma^2}{r^2 - \sigma^2}$$

and since r and σ are constant over the inner integration

$$d\tau = \frac{2y \, dy}{r^2 - \sigma^2}$$

Thus

$$\int_\sigma^r \frac{y \, dy}{\sqrt{y^2 - \sigma^2} \sqrt{r^2 - y^2}} = \frac{1}{2} \int_0^1 \frac{d\tau (r^2 - \sigma^2)}{\sqrt{y^2 - \sigma^2} \sqrt{r^2 - y^2}}$$

$$= \frac{1}{2} \int_0^1 \frac{d\tau}{\sqrt{\tau(1 - \tau)}} \tag{5A.7}$$

The right-hand side of Eq. (5A.7) is in the form of a beta function and is equal to $\pi/2$.

Equation (5A.6) thus becomes

$$\int_\sigma^{r_0} \frac{\epsilon(y) y \, dy}{\sqrt{y^2 - \sigma^2}} = \frac{\pi}{\lambda_0} \int_\sigma^{r_0} \Delta n(r) r \, dr \tag{5A.8}$$

If the left-hand side is integrated by parts

$$\int_\sigma^{r_0} \frac{\epsilon(y) y \, dy}{\sqrt{y^2 - \sigma^2}} = \epsilon(y) \sqrt{y^2 - \sigma^2} \Bigg]_\sigma^{r_0} - \int_\sigma^{r_0} \sqrt{y^2 - \sigma^2} \frac{d\epsilon(y)}{dy} \, dy \tag{5A.9}$$

$$= - \int_\sigma^{r_0} \sqrt{y^2 - \sigma^2} \frac{d\epsilon(y)}{dy} \, dy \tag{5A.10}$$

since the first term on the right-hand side of Eq. (5A.9) is zero at both limits. Combining Eqs. (5A.8) and (5A.10)

$$\int_{\sigma}^{r_0} \Delta n(r) r \, dr = -\frac{\lambda_0}{\pi} \int_{\sigma}^{r_0} \sqrt{y^2 - \sigma^2} \, \frac{d\epsilon(y)}{dy} \, dy \qquad (5A.11)$$

Differentiating both sides of Eq. (5A.11) with respect to σ^2

$$-\frac{\Delta n(\sigma)}{2} = +\frac{\lambda_0}{2\pi} \int_{\sigma}^{r_0} \frac{d\epsilon(y)/dy}{\sqrt{y^2 - \sigma^2}} \, dy \qquad (5A.12)$$

or if σ is set equal to r,

$$\Delta n(r) = -\frac{\lambda_0}{\pi} \int_{r}^{r_0} \frac{d\epsilon(y)/dy}{\sqrt{y^2 - r^2}} \, dy \qquad (5A.13)$$

This equation can be used to evaluate interferograms of axisymmetric fields. It relates the measured fringe shift, really its derivative, which is a function of y to the desired index of refraction which is a function of the radial position r. In practice the measured $\epsilon(y)$ can be represented by best-fit least-square polynomials. This representation and the required differentiation and integration can be easily done on a high-speed computer to yield $n(r)$ and from this the density or temperature distribution.

An alternative method for evaluating interferograms of axisymmetric fields [A3] is useful when a discontinuity of density is present as through a shock wave.

REFERENCES

A1. Hildebrand, F. B., "Methods of Applied Mathematics," 2nd ed., p. 276, Prentice-Hall, Englewood Cliffs, N.J., 1965.

A2. Kahl, G. D., and Mylin, D. C., Refractive Deviation Errors of Interferograms, *J. Opt. Soc. Amer.*, vol. 55, p. 364, 1965.

A3. Bradley, J. W., Density Determination from Axisymmetric Interferograms, *AIAA J.*, vol. 6, p. 1190, 1968.

6 Spectroscopic temperature determination in high temperature gases

E. PFENDER
Heat Transfer Division, Department of Mechanical Engineering
University of Minnesota, Minneapolis

In this chapter techniques will be discussed which are suitable for temperature measurements at very high temperature levels at which matter is in a gaseous, more or less ionized, state. The term *plasma* has been adopted to describe such gases, and this chapter will survey the most important spectrometric methods that have been suggested for the determination of temperatures in plasmas.

The upper limit of temperatures considered in this survey will be approximately $30 \times 10^3\,\mathrm{K}$ which corresponds to the maximum temperatures experienced in present engineering applications, as, for example, in arc or high frequency gas heaters utilized for chemical or material processing or for space-related applications. The lower temperature limit is in the order of $10^3\,\mathrm{K}$. Such temperatures are of importance for molecular radiation. Spectrometric methods have been developed to determine temperatures in this range and at least one of them will be included in this chapter.

Another limitation in this survey is imposed by the pressure of the high temperature plasma. In general, such plasmas represent multicomponent mixtures (molecules, atoms, ions, electrons) in which numerous chemical reactions may occur. A single temperature concept of this mixture is meaningful only when thermodynamic (TE) or local thermodynamic equilibrium (LTE) prevails. Otherwise the various components may have different temperatures and the possible chemical reactions can no longer be described by an equilibrium temperature. One of the decisive parameters which determines whether or not LTE may be expected is the pressure. Low pressure plasmas of laboratory dimensions frequently show strong deviations from LTE, whereas high temperature plasmas at atmospheric or higher pressures approach a state of LTE. Since the majority of the spectrometric temperature measurement techniques are based on the existence of LTE, the discussion will be essentially restricted to these situations.

Spectrometric methods offer two important advantages compared with methods which use probes or thermocouples for sensing the temperature. Since radiation emitted by the plasma is used for the temperature evaluation, the diagnostic tool, which may be in a remote location relative to the plasma, does not have any influence on the plasma, i.e., it does not alter the quantity to be measured. Furthermore, this method provides an excellent spatial resolution of the measured temperature field.

In general, gaseous radiation sources do not have a uniform temperature distribution. In many cases of practical interest, however, such sources display rotational symmetry. In such cases, conversion of the observed side-on radiation intensities into local intensities is a straightforward procedure.

The first section of this chapter (Sec. 6.1) will be devoted to a discussion of the important laws and concepts in gaseous radiation. Since most of the spectrometric methods to be discussed are based on the existence of TE or LTE in the plasma, Sec. 6.2 will review the conditions which are favorable for this situation. In the next section (Sec. 6.3) the various types of radiation emitted by a plasma and crucial for the development of spectroscopic methods are considered. Finally, in Sec. 6.4, the most important spectroscopic methods that have been adopted for diagnostic purposes in high temperature gases will be discussed. References to specific research papers will not be included in this review because they may be found in texts, proceedings of symposia, and/or special review articles [1-17].

This survey covers only the methods which are well known today in high temperature spectroscopy; it does not include spectrometers and the associated hardware.

6.1 BASIC CONCEPTS OF GASEOUS RADIATION

When electromagnetic radiation passes through a medium with varying index of refraction, the wavelength λ as well as the propagation velocity c of the wave vary with the index of refraction. The frequency $\nu = c/\lambda$, however, does not depend on the particular properties of the medium and it will, therefore, be used as a wave parameter in this review, although the index of refraction in gases is usually close to unity.

6.1.1 Intensity and Density of Gaseous Radiation

The monochromatic radiation intensity I_ν is defined as the amount of radiant energy $\Delta\chi$ which passes per unit time Δt through an area element Δa which is perpendicular to the direction of a radiation pencil of solid angle $\Delta\omega$. This definition includes only radiation which is emitted in a frequency interval from ν to $\nu + \Delta\nu$. Writing this definition as a formula yields

$$I_\nu = \frac{\Delta\chi}{\Delta t \, \Delta a \, \Delta\omega \, \Delta\nu} \left(\frac{\text{joule}}{\text{sterad cm}^2} \right) \tag{6.1}$$

The unit solid angle is measured in steradians (abbreviated: sterad). The total radiation intensity I is obtained by integrating over the entire spectrum

$$I = \int_0^\infty I_\nu \, d\nu \left(\frac{\text{watt}}{\text{sterad cm}^2}\right) \tag{6.2}$$

The dimension of monochromatic and total radiation intensity is, of course, not the same.

The monochromatic radiation density u_ν is defined as the amount of monochromatic radiant energy $\Delta\chi$ contained in a volume element ΔV

$$u_\nu = \frac{\Delta\chi}{\Delta V \, \Delta\nu} \left(\frac{\text{joule } s}{\text{cm}^3}\right) \tag{6.3}$$

This radiation density is connected with the radiation intensity through the relation

$$u_\nu = \frac{1}{c} \int_\omega I_\nu \, d\omega \tag{6.4}$$

or for isotropic radiation

$$u_\nu = \frac{4\pi}{c} I_\nu$$

The total radiation density is then

$$u = \int_0^\infty u_\nu \, d\nu \left(\frac{\text{joule}}{\text{cm}^3}\right) \tag{6.5}$$

6.1.2 Emission and Absorption in Gases

The monochromatic radiant energy $\Delta\chi$ which is emitted by a volume element ΔV into a solid angle $\Delta\omega$ per unit time Δt is termed the emission coefficient ϵ_ν.

$$\epsilon_\nu = \frac{\Delta\chi}{\Delta\nu \, \Delta V \, \Delta\omega \, \Delta t} \left(\frac{\text{joule}}{\text{sterad cm}^3}\right) \tag{6.6}$$

As it stands here, ϵ_ν shall describe spontaneous emission only. The total radiation flux stemming from the volume element ΔV may be expressed as

$$\phi = \Delta V \int_0^\infty \int_\omega \epsilon_\nu \, d\nu \, d\omega \ (\text{watt}) \tag{6.7}$$

For isotropic radiation

$$\phi = 4\pi \, \Delta V \int_0^\infty \epsilon_\nu \, d\nu$$

Intensity Loss by Absorption

If monochromatic radiation of intensity I_ν passes in normal direction through an absorbing gaseous slab of thickness dl an attenuation of the initial radiation intensity occurs, according to

$$dI_\nu = -\kappa_\nu' I_\nu \, dl \tag{6.8}$$

where κ_ν' (cm^{-1}) represents the monochromatic absorption coefficient. This absorption coefficient is in general a function of the state and the properties of the gas, of the radiation frequency, and of the direction in which the radiation propagates.

Besides the usual volume absorption coefficient, the mass absorption coefficient $\kappa_{\nu m}' = (1/\rho)\kappa_\nu'$ is sometimes used where ρ represents the mass density of the gas.

If $I_{\nu,0}$ is the radiation intensity entering a gaseous layer of finite thickness L, integration of Eq. (6.8) yields

$$I_\nu = I_{\nu,0} \exp\left(-\int_0^L \kappa_\nu' dl\right)$$
$$= I_{\nu,0} \exp(-\tau_\nu) \tag{6.9}$$

where $\tau_\nu = \int_0^L \kappa_\nu' \, dl$ represents the optical depth of the layer. For the special case of a homogeneous gas with uniform temperature distribution the optical depth becomes $\tau_\nu = \kappa_\nu' L$.

An analogous derivation for the attenuation of radiation by scattering yields

$$I_\nu = I_{\nu,0} \exp\left(-\int_0^L \gamma_\nu dl\right) \tag{6.10}$$

with γ_ν as the monochromatic scattering coefficient which like the absorption coefficient is a function of the state and the properties of the gas, of the radiation frequency, and sometimes also of the direction in which the radiation travels.

The similarity of Eqs. (6.9) and (6.10) permits a description of radiation attenuation due both to absorption and scattering with $\beta_\nu = \kappa'_\nu + \gamma_\nu$ as the monochromatic extinction coefficient.

6.1.3 Radiation in Perfect Thermodynamic Equilibrium

The conditions under which an emitting-absorbing gas is in a state of thermodynamic equilibrium (TE) or local thermodynamic equilibrium (LTE) will be discussed in the next section (Sec. 6.2). In this section a radiation field in a gaseous medium will be considered which is in equilibrium with its boundaries (cavity radiation). Such a radiation field is isotropic with a blackbody radiation intensity B_ν corresponding to an equilibrium temperature T. In such a system Kirchhoff's law may be applied which connects emission and absorption coefficient with the intensity of the radiation field, namely,

$$\frac{\epsilon'_\nu}{\kappa'_\nu} = B_\nu \tag{6.11}$$

where ϵ'_ν is a modified emission coefficient which includes the contribution of induced or forced emission. This emission coefficient is always larger than that for spontaneous emission, as can be readily seen from the Einstein relation

$$\epsilon'_\nu = \frac{\epsilon_\nu}{1 - e^{-h\nu/kT}} \tag{6.12}$$

In this equation h is Planck's and k is Boltzmann's constant. T is the equilibrium temperature. The blackbody radiation intensity B_ν may be expressed by the Planck function

$$B_\nu = \frac{2h\nu^3}{c^2} \frac{1}{e^{h\nu/kT} - 1} \tag{6.13}$$

and the radiation density of blackbody radiation assumes the value

$$u_\nu = \frac{8\pi h\nu^3}{c^3} \frac{1}{e^{h\nu/kT} - 1} \tag{6.14}$$

By adding the emission term ϵ'_ν to Eq. (6.8) and retaining the assumptions under which Eq. (6.8) has been derived, the radiation transport equation for an emitting-absorbing medium becomes

$$\frac{dI_\nu}{dl} = \epsilon'_\nu - \kappa'_\nu I_\nu = \kappa'_\nu(B_\nu - I_\nu) \tag{6.15}$$

Integration of Eq. (6.15) for a layer of finite thickness L with the boundary condition $I_\nu = 0$ for $L = 0$ yields

$$I_\nu = B_\nu(1 - e^{-\kappa'_\nu L}) \qquad (6.16)$$

From this relation follows for an optically thin medium ($\kappa'_\nu L \ll 1$)

$$I_\nu = \kappa'_\nu L B_\nu \qquad (6.17)$$

By applying Kirchhoff's law, Eq. (6.11), then Eq. (6.17) becomes

$$I_\nu = \epsilon'_\nu L \qquad (6.18)$$

which states that the radiation intensity of an optically thin layer increases proportionally with the layer thickness. Without the contribution of induced emission Eq. (6.18) reads

$$I_\nu = \epsilon_\nu L \qquad (6.19)$$

For large optical depths ($\kappa'_\nu L \gg 1$) follows from Eq. (6.16)

$$I_\nu = B_\nu$$

i.e., the highest possible radiation intensity that a high temperature gas in TE is able to emit is the blackbody radiation of its own equilibrium temperature.

6.1.4 Temperature Definitions

High temperature, high density gases are able to emit spectra in which continuous radiation predominates. The various radiation mechanisms which contribute to these spectra will be discussed in Sec. 6.3. The nature of a predominately continuous spectrum suggests a comparison with the spectrum of a blackbody radiator. Depending on the basis of such a comparison, different temperature definitions of gaseous radiation sources have been suggested.

Color Temperature

The observed intensity distribution of a gaseous radiator is matched on a relative scale with that of a blackbody radiator over a frequency interval from ν_1 to ν_2. The temperature of the blackbody radiator is then identified with the color temperature of the gaseous radiator in the specified frequency interval. By shrinking the frequency interval to a single value $\bar{\nu}$ the slopes $(dI_\nu/d\nu)_{\bar{\nu}}$ of the two distributions have to be matched resulting in a definition of the color temperature for a single frequency $\bar{\nu}$.

Radiation Temperature ("Black Temperature")

The same comparison as in Sec. 6.1.4, under "Color temperature," based on absolute intensities, results in the definition of the radiation temperature or "black temperature" of the radiation source in the specified frequency interval. Again, shrinking of the frequency interval to a single frequency $\bar{\nu}$ and matching of the two absolute intensity distributions at this frequency determines the radiation temperature of the gaseous radiation at $\bar{\nu}$.

Effective Temperature

A comparison of the gaseous radiator with a blackbody radiator based on the total emissive power leads to the definition of an effective temperature.

The temperature definitions, in use in the literature, are based on descriptions of the measurements. No specific physical meaning can be attached to them. The color temperature, for example, may be higher or lower than the corresponding LTE gas temperature; the effective and radiation temperatures are always lower than the LTE temperature of the gaseous radiator.

6.1.5 Einstein's Transition Probabilities

The knowledge of the energies which are involved in an emission or absorption process is not sufficient to predict the intensity of the emitted or absorbed radiation. In addition we need to know the number of radiation processes per time and volume unit. The latter is given by the number density of particles participating in the radiation process; the rate of the radiation process is governed by the transition probabilities. Transition probabilities for spontaneous and induced emission, as well as those for absorption, will be discussed along with the relationships among them.

Spontaneous Emission

An excited atom in a higher quantum state s may return to a lower energy state t by giving off a photon of energy $h\nu$ according to Bohr's frequency relation

$$\chi_s - \chi_t = h\nu \tag{6.20}$$

This equation refers to an energy jump within a neutral atom. A similar jump may occur in an r-times ionized atom as long as $r \leqslant Z - 1$, where Z is the number of protons in the nucleus. Equation (6.20) may be rewritten for this more general case as

$$\chi_{r,s} - \chi_{r,t} = h\nu \tag{6.20a}$$

With $n_{r,s}$ r-times ionized atoms per cm^3 in an excited state s, the number of quantum transitions per sec and cm^3 becomes

$$A^{r,s}_{r,t} n_{r,s} \tag{6.21}$$

$A^{r,s}_{r,t}$ is the transition probability for spontaneous transition in an r-times ionized atom from a higher energy state s to a lower state t.

Induced or Forced Emission

An ionized gas imbedded in a radiation field of density u_ν may also radiate by induced emission. In the elementary process of induced emission, a photon of frequency ν interacting with an excited ion in a higher quantum state will force this ion to emit a photon of the same frequency and in the same direction as the oncoming photon. The number of such transitions per sec and cm^3 is given by

$$B^{r,s}_{r,t} n_{r,s} u_\nu \tag{6.22}$$

Note that the dimension of the transition probability for induced emission $B^{r,s}_{r,t}$ is cm^3/joule s^2 whereas the dimension for the transition probability of spontaneous emission $A^{r,t}_{r,s}$ is s^{-1}.

Absorption

In the case of absorption, transitions occur within an r-times ionized atom from a lower quantum state t to a higher quantum state s. If $n_{r,t}$ is the number density of r-times ionized atoms in the quantum state t, the number of absorption processes for photons of frequency ν per sec and cm^3 is given by

$$B^{r,t}_{r,s} n_{r,t} u_\nu \tag{6.23}$$

with $B^{r,t}_{r,s}$ as the transition probability for an r-times ionized atom from a lower quantum state t to a higher quantum state s.

If an ionized gas is in a state of perfect thermodynamic equilibrium, a balance of the radiation process must exist, in other words, the number of processes $s \rightarrow t$ per sec and cm^3 must be equal to the number of opposite processes $t \rightarrow s$. From Eqs. (6.21), (6.22), and (6.23) follows for this balance

$$A^{r,s}_{r,t} n_{r,s} + B^{r,s}_{r,t} n_{r,s} u_\nu = B^{r,t}_{r,s} n_{r,t} u_\nu \tag{6.24}$$

In addition TE requires that the excited states follow a Boltzmann distribution

$$\frac{n_{r,s}}{n_{r,t}} = \frac{g_{r,s}}{g_{r,t}} \exp - \frac{\chi_{r,s} - \chi_{r,t}}{kT} = \frac{g_{r,s}}{g_{r,t}} \exp \left(- \frac{h\nu}{kT} \right) \tag{6.25}$$

The symbols $g_{r,s}$ and $g_{r,t}$ represent the statistical weights of the energy states s and t, respectively.

Replacing $n_{r,s}/n_{r,t}$ in Eq. (6.24) with Eq. (6.25) and solving for u_ν yields

$$u_\nu = \frac{A_{r,t}^{r,s}}{(g_{r,t}/g_{r,s})B_{r,s}^{r,t}e^{h\nu/kT} - B_{r,t}^{r,s}} \tag{6.26}$$

Since TE requires also that the radiation density is that of blackbody radiation, a comparison of Eqs. (6.14) and (6.26) establishes the following relationships between the transition probabilities:

$$B_{r,t}^{r,s}g_{r,s} = B_{r,s}^{r,t}g_{r,t}$$

$$A_{r,t}^{r,s} = B_{r,t}^{r,s}\frac{8\pi h\nu^3}{c^3} \tag{6.27}$$

If, for example, $A_{r,t}^{r,s}$ is known, the other two transition probabilities can be calculated from Eq. (6.27).

6.1.6 Radiation from a Nonhomogeneous Layer
(Abel Inversion)

The radiation transport equation (see Eq. [6.15]) may be integrated in closed form for a homogeneous emitting-absorbing layer of uniform temperature. Equation (6.15) may still be applied for radiation emitted from a nonhomogeneous source of varying temperature, but the integration of this equation is now much more complex because κ_ν' as well as B_ν are functions of the temperature and, therefore, of the position in the source.

An optically thin, rotationally symmetric plasma source is frequently of interest since it is obtained in the laboratory as well as in numerous applications. From such a source the side-on radiation intensity of a spectral line $I_L(x)$ includes contributions from layers having different emission coefficients $\epsilon_L(r)$. (See Fig. 6.1.)

$$I_L(x) = 2\int_0^y \epsilon_L(r)\,dy$$

$$= 2\int_{r=x}^R \epsilon_L(r)\frac{r\,dr}{(r^2 - x^2)^{1/2}} \tag{6.28}$$

The solution of the Abel integral equation (Eq. [6.28]) is given by (cf. Appendix to Ch. 5)

$$\epsilon_L(r) = -\frac{1}{\pi}\int_r^R \frac{dI_L/dx}{(x^2 - r^2)^{1/2}}\,dx \tag{6.29}$$

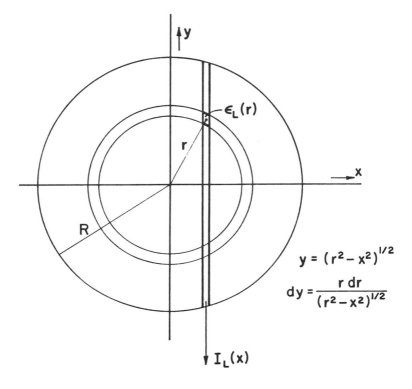

$$\text{OBSERVED SIDE-ON INTENSITY}: I_L(x) = 2\int_0^{(R^2-x^2)^{1/2}}\epsilon_L(r)\,dy = 2\int_{r=x}^R \epsilon_L(r)\frac{r\,dr}{(r^2-x^2)^{1/2}}$$

$$\text{ABEL INVERSION}: \epsilon_L(r) = -\frac{1}{\pi}\int_r^R \frac{I_L'(x)}{(x^2-r^2)^{1/2}}\,dx$$

FIG. 6.1 Abel inversion for a rotationally symmetric, optically thin source.

The experimentally determined distribution $I_L(x)$ may be approximated by polynomials in order to solve for $\epsilon_L(r)$. Frequently, Eq. (6.28) is solved numerically by using a finite difference approach adaptable to a digital computer.

6.2 THERMODYNAMIC STATE OF A HIGH TEMPERATURE PLASMA

Since most of the spectrometric methods to be discussed in Sec. 6.4 are based on the assumption of TE or LTE, the main requirements imposed by these concepts will be included in this survey. A more comprehensive treatment of this subject may be found in the literature [1, 3-7]. As already mentioned, emphasis will be on high temperature plasmas which do not contain molecular species, i.e., the temperature level is either high enough so that molecules are completely dissociated or the plasma is generated from an atomic gas.

6.2.1 High Temperature Plasmas in Perfect Thermodynamic Equilibrium (TE)

Thermodynamic equilibrium prevails in a uniform, homogeneous plasma volume if kinetic and chemical equilibria as well as every conceivable plasma property are unambiguous functions of the temperature. The temperature, in turn, is the same for all plasma constituents and their possible reactions. More specifically, the following conditions must be met:

a. The velocity distribution functions for particles of every species r which exist in the plasma, including the electrons, follow a Maxwell-Boltzmann distribution

$$f(v_r) = \frac{4v_r^2}{\sqrt{\pi}(2kT/m_r)^{3/2}} \exp\left(-\frac{m_r v_r^2}{2kT}\right) \tag{6.30}$$

where v_r is the velocity of particles of species r, m_r is their mass, and T is their temperature, which is the same for every species r, and which is, in particular, identical to the plasma temperature.
b. The population density of the excited states of every species r follows a Boltzmann distribution

$$n_{r,s} = n_r \frac{g_{r,s}}{Z_r} \exp\left(-\frac{\chi_{r,s}}{kT}\right) \tag{6.31}$$

where n_r is the total number density of ions of species r, Z_r is their partition function, and $\chi_{r,s}$ is the energy of the sth quantum state. The excitation temperature T which appears explicitly in the exponential term and implicitly in the partition function Z_r is identical to the plasma temperature.
c. The particle densities (neutrals, electrons, ions) are described by the Saha-Eggert equation which may be considered as a mass action law

$$\frac{n_{r+1}n_e}{n_r} = \frac{2Z_{r+1}}{Z_r} \frac{(2\pi m_e kT)^{3/2}}{h^3} \exp\left(-\frac{\chi_{r+1}}{kT}\right) \tag{6.32}$$

where χ_{r+1} represents the energy which is required to produce an $(r+1)$-times ionized atom from an r-times ionized atom. The ionization temperature T in this equation is identical to the plasma temperature.

d. The electromagnetic radiation field is that of blackbody radiation of the intensity B_ν as described by the Planck function

$$B_\nu = \frac{2h\nu^3}{c^2} \frac{1}{e^{h\nu/kT} - 1} \qquad (6.33)$$

The temperature of this blackbody radiation is again identical to the plasma temperature.

In order to generate a plasma which follows this ideal model as described by Eqs. (6.30) to (6.33), the plasma would have to dwell in a hypothetical cavity whose walls are kept at the plasma temperature or the plasma volume would have to be so large that the central part of this volume, in which TE prevails, would not sense the plasma boundaries. In this way the plasma would be penetrated by a blackbody radiation of its own temperature. An actual plasma will, of course, deviate from these ideal conditions. The observed plasma radiation, for example, will be much less than the blackbody radiation because most plasmas are optically thin over a wide wavelength range. Therefore, the radiation temperature as defined in Sec. 6.1.4, under "Radiation temperature," of a gaseous radiator deviates appreciably from the kinetic temperature of the plasma constituents or the already-mentioned excitation and ionization temperatures. In addition to radiation losses, plasmas suffer irreversible energy losses by conduction, convection, and diffusion, which also disturb the thermodynamic equilibrium. Thus, laboratory plasmas as well as some of the natural plasmas cannot be in a perfect TE state. In the following sections, deviations from TE and their significance will be discussed.

6.2.2 The Concept of Local Thermal Equilibrium (LTE)

This concept will be discussed for optically thin plasmas, a situation which is frequently approached by laboratory plasmas. In contrast to a complete TE situation, LTE in optically thin plasmas does not require a radiation field which corresponds to the blackbody radiation intensity of the respective LTE temperature. It does require, however, that collision processes and not radiative processes govern transitions and reactions in the plasma and that there is a microreversibility among the collision processes. In other words a detailed equilibrium of each collision process with its reverse process is necessary. Steady state solutions of the respective collision rate equations will then yield the same energy distribution pertaining to a system in complete thermal equilibrium with the exception of the rarefied radiation field. LTE further requires that local gradients of the plasma properties (temperature, density, heat conductivity, etc.) are sufficiently small so that a given particle which diffuses from one location to another in the plasma finds sufficient time to equilibrate, i.e., the diffusion time should be of the same order of magnitude as or larger than the equilibration time. From the equilibration time and the particle velocities an equilibration length may be derived which is smaller in regions of small plasma property gradients (e.g., in the center of an electric arc). Therefore, with regard to spatial variations LTE is more probable in such regions. Heavy particle diffusion and resonance radiation from the center of a nonuniform plasma source help to reduce the effective equilibrium distance in the outskirts of the source.

In the following a systematic discussion of the important assumptions for LTE will be undertaken based on laboratory-generated plasmas. In practice, the electric arc appears as a simple and convenient method to generate a high temperature, high density plasma.

Kinetic Equilibrium

It may be safely assumed that each species (electron gas, ion gas, neutral gas) in a dense, high temperature plasma will assume a Maxwelliam distribution. However, the temperatures defined by these Maxwellian distributions may be different from species to species. Such a situation which leads to a two-temperature description will be discussed for an arc plasma.

The electric energy fed into an arc is dissipated in the following way. The electrons, according to their high mobility, pick up energy from the electric field which they partially transfer by collisions to the heavy plasma constituents. Because of this continuous energy flux from the electrons to the heavy particles, there must be a temperature gradient between these two species, so that $T_e >$ T_a; T_e is the electron temperature and T_a the temperature of the heavy species, assuming that ion and neutral gas temperatures are the same.

In the two-fluid model of a plasma, defined in this manner, two distinct temperatures T_e and T_a may exist. The degree to which T_e and T_a deviate from each other will depend on the thermal coupling between the two species. The difference between these two temperatures can be expressed by the following relation [1]:

$$\frac{T_e - T_a}{T_e} = \frac{m_a}{8m_e} \frac{(\lambda_e eE)^2}{(3/2kT_e)^2} \tag{6.34}$$

where m_a is the mass of the heavy plasma constituents, λ_e the mean free path length of the electrons, and E the electric field intensity. Since the mass ratio $m_a/8m_e$ is already about 230 for hydrogen, the amount of (directed) energy $(\lambda_e eE)$ which the electrons pick up along one mean free path length has to be very small compared with the average thermal (random) energy $3/2kT_e$ of the electrons. Low field intensities, high pressures $[\lambda_e \sim (1/p)]$ and high temperature levels are favorable for a kinetic equilibrium among the plasma constituents. At low pressures, for example, appreciable deviations from kinetic equilibrium may occur. Figure 6.2 shows in a semischematic diagram how electron and gas temperatures separate in an electric arc with decreasing pressure. For an atmospheric argon high intensity arc with $E = 13V/cm$, $\lambda_e = 3 \times 10^{-4}cm$,

TEMPERATURE (°K)

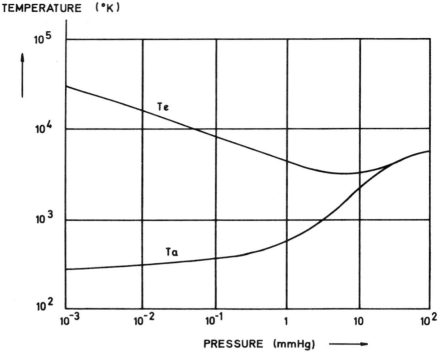

FIG. 6.2 Electron and gas kinetic temperatures in an arc-generated plasma.

$m_a/m_e = 7 \times 10^4$ and $T_e = 30 \times 10^3 \mathrm{K}$, the deviation between T_e and T_a is only 2% [1].

Excitation Equilibrium

In order to determine the excitation equilibrium, every conceivable process which may lead to excitation or de-excitation has to be considered. This discussion is restricted to the most prominent mechanisms which are collisional and radiative excitation and de-excitation.

Excitation	De-excitation
Electron collision	Collision of the second kind
Photoabsorption	Photoemission

For the case of TE, microreversibilities have to exist for all processes, i.e., in the above scheme, excitation by electron collisions will be balanced by the reverse process, namely, collisions of the second kind. Also, excitation by the photoabsorption process will be balanced by photoemission processes which include spontaneous and induced emission. Furthermore, the population of excited states is given by a Boltzmann distribution (see Eq. [6.31]). The microreversibility for the radiative processes holds only if the radiation field in the plasma reaches the intensity B_ν of blackbody radiation. However, actual plasmas are frequently optically thin over most of the spectral range, so that the situation for excitation equilibrium seems to be hopeless. Fortunately, if collisional processes dominate, photoabsorption and photoemission processes do not have to balance. Only the sums on the left-hand side and the right-hand side of the scheme above have to be equal. Since the contribution of the photoprocesses to the number of excited atoms is almost negligible when collisional processes dominate, the excitation process is still close to LTE.

Ionization Equilibrium

For the ionization equilibrium again only the most prominent mechanisms which lead to ionization and recombination will be considered.

Ionization	Recombination
Electron collision	Three-body recombination
Photoabsorption	Photorecombination

In a perfect thermodynamic equilibrium state with cavity radiation, a microreversibility among the collisional and radiative processes would exist and the particle densities would be described by the Saha-Eggert equation. Without cavity radiation, the number of photoionizations is almost negligible requiring instead of the microreversibility, a total balance of all processes involved. Photorecombinations, especially at lower electron densities, are not negligible. The frequency of the three remaining elementary processes is a function only of the electron density leading, for $n_e = 7 \times 10^{15}$ cm^{-3}, to the same order-of-magnitude frequency of these elementary processes. The result is an appreciable deviation between actual and predicted values (from Eq. [6.32]) of the electron densities. Only for values $n_e > 7 \times 10^{15}$ cm^{-3} does the Saha-Eggert equation predict correct values. For smaller electron densities the Corona formula has to be used, which considers ionization by electron impact and photorecombination only.

$$\frac{n_e}{n_0} = \frac{\sqrt{27}}{16\alpha^3} \frac{\varsigma_n}{n} \frac{\chi_{i,H}^2}{\chi_i^3} \frac{kT}{g} \exp\left(-\frac{\chi_i}{kT}\right) \qquad (6.35)$$

In this equation α is Sommerfeld's fine structure constant, ς_n the number of valence electrons, n the principal quantum number of the valence shell, $\chi_{i,H}$ the ionization energy of hydrogen, and g a constant with a value between 1.4 and 4.

The particle concentrations in low intensity arcs at atmospheric pressure, e.g., have to be calculated with this formula. Significant deviation of the electron density predicted by the Saha-Eggert equation from the true electron density may also occur in the fringes of high intensity arcs and plasma jets.

In summary it has been found that LTE exists in a steady state optically thin plasma when the following conditions are simultaneously fulfilled:

a. The different species which form the plasma have a Maxwellian distribution.
b. Electric field effects are small enough, and the pressure and the temperature are sufficiently high so that $T_e = T_a$.
c. Collisions are the dominating mechanism for excitation (Boltzmann distribution) and ionization (Saha-Eggert equation).
d. Spatial variations of the plasma properties are sufficiently small.

Besides the conditions for the two extreme cases, namely LTE (based on Saha-Eggert ionization equilibrium) and Corona equilibrium, conditions in the regions between these two limiting cases are also of interest. In this range, three-body recombination as well as radiative recombination and de-excitation are significant. A number of theories have been advanced for ionization equilibrium over the entire range of radiative-collisional elementary processes. In particular, detailed calculations of optically thin and optically thick hydrogen plasmas have been reported. Some results of these calculations follow for the optically thin case.

If α is the combined collisional-radiative recombination coefficient and S the corresponding ionization coefficient, rate equations may be established which describe the effective rate of population and depopulation. The rate of population of the ground state is described by

$$\left(\frac{dn_{0,0}}{dt}\right)_{\text{pop}} = \alpha n_e n_1 \qquad (6.36)$$

In this relation, $n_{0,0}$ represents the number of neutral hydrogen atoms in the ground state and n_e and n_1 are the electron and ion densities, respectively. The rate of depopulation of the ground state is given by

$$\left(\frac{dn_{0,0}}{dt}\right)_{\text{depop}} = -S n_e n_{0,0} \qquad (6.37)$$

Under steady state conditions

$$\left(\frac{dn_{0,0}}{dt}\right)_{\text{pop}} + \left(\frac{dn_{0,0}}{dt}\right)_{\text{depop}} = 0 \qquad (6.38)$$

or
$$\frac{S}{\alpha} = \frac{n_1}{n_{0,0}} = \frac{n_e}{n_{0,0}} \qquad (6.39)$$

Figure 6.3 shows a state diagram for values of S/α as a function of the electron density with pressure and electron temperatures as parameters. At high electron densities ($\geqslant 10^{24}$ m^{-3}) pairs of LTE and non-LTE curves plotted for the same electron temperature merge. At low electron densities ($\leqslant 10^{21}$ m^{-3}) the non-LTE curves merge into curves valid for Corono equilibrium. The divergence of the non-LTE curves from the LTE curves at lower pressures and/or lower electron temperatures and densities shows how large the deviation from LTE may become in such parameter ranges. Taking values for 1 atm it can be seen that LTE is closely approached for electron temperatures in the interval $14{,}000 < T_e < 28{,}000$K.

Deviations of this kind from LTE may be found, for example, in plasma regions adjacent to walls where the electron density drops appreciably and in all types of low density plasmas of laboratory dimensions. Well-known examples of the latter are: the positive column of glow discharges and the plasma generated in fluorescent lamps.

6.3 RADIATION EMITTED BY A HIGH TEMPERATURE GAS

In this section the different types of radiation which may be emitted by a plasma will be discussed with emphasis on the emission from plasmas which contain only atoms as the neutral component. The pressure in the plasma shall be sufficiently high (> 0.1 atm) so that LTE becomes feasible.

An analysis of the total emitted radiation of a plasma reveals a number of different radiation mechanisms. Their relative importance is a function of temperature and pressure and, in magnetized plasmas, also of the magnetic field. From a theoretical point of view this radiation is useful for diagnostic purposes if its origin and temperature dependence is known. The experiment requires, in addition, that there is sufficient intensity of the spectrally resolved radiation and that this radiation is sensitive to temperature changes.

6.3.1 Line Radiation (Bound–Bound Transitions)

Excited neutral atoms or ions may return to the ground state in one or several steps. Because of the discrete nature of bound energy states the emitted radiation appears as spectral lines according to Eqs. (6.20) and (6.20a). By combining Eqs. (6.20a) and (6.21) the line emission coefficient for an optically thin homogeneous plasma is found to be

$$\epsilon_L = \frac{1}{4\pi} A_{r,\,t}^{r;\,s} n_{r,\,s} h\nu \qquad (6.40)$$

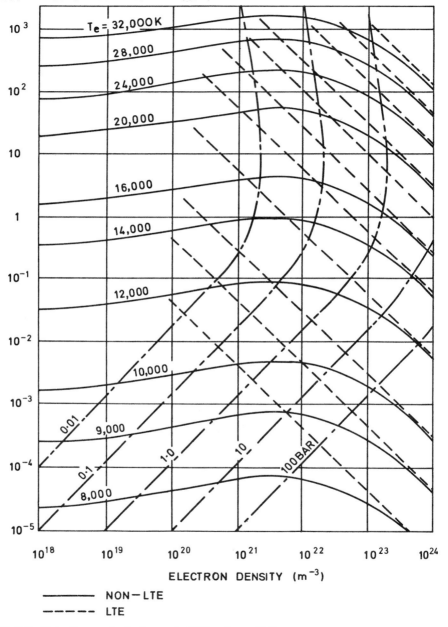

RATIO OF IONIZATION AND
RECOMBINATION COEFFICIENT (S/a)

ELECTRON DENSITY (m^{-3})

NON−LTE
−−−−− LTE

FIG. 6.3 State diagram for hydrogen in LTE and non-LTE.

The factor $1/4\pi$ represents the unit solid angle. Therefore, the dimension of the line emission coefficient is watt/sterad cm^3. The line emission coefficient is already integrated over the natural width Δx of a spectral line

$$\epsilon_L = \int_{\Delta x} \epsilon_\nu \, d\nu$$

where ϵ_ν is the monochromatic emission coefficient (see Eq. [6.6]). For the following considerations, it will be assumed that LTE prevails in the plasma, i.e., the number density of excited atoms or ions may be expressed by a Boltzmann distribution (see Eq. [6.31]). Replacing $n_{r,s}$ in Eq. (6.40) by this Boltzmann distribution yields

$$\epsilon_L = \frac{1}{4\pi} A_{r,t}^{r,s} n_r \frac{g_{r,s}}{Z_r} \exp\left(-\frac{\chi_{r,s}}{kT}\right) h\nu \qquad (6.41)$$

In this equation the temperature appears explicitly only in the exponential term, but n_r also depends strongly on the temperature whereas the partition function $Z_r = \sum_s g_{r,s} \exp(-\chi_{r,s}/kT)$ is a rather weak function of the temperature. For constant pressure, $\epsilon_L(T)$ assumes a maximum at a certain temperature T^* because of the decreasing tendency of n_r owing to the depletion of particles of species r by increasing ionization to species $r + 1$ and the effect of the perfect gas law ($n_e + \sum_r n_r = p/kT$, where p is the total pressure). As an example, Fig. 6.4 illustrates the emission coefficients of two argon lines ($r = 0$ and $r = 1$) on a relative scale. The neutral line will, according to this figure, show up in the spectrum only in the temperature interval $10^4 \, \text{K} < T < 3.10^4 \, \text{K}$. As the temperature increases, lines corresponding to higher ionization stages will appear in the spectrum and those of lower ionization stages will vanish entirely.

6.3.2 Recombination Radiation (Free–Bound Transitions)

In the process of radiative recombination a free electron is captured by a positive ion into a certain bound energy state and the excess energy is converted into radiation according to the relation

$$\frac{m_e v_e^2}{2} + \chi_{r,s} = h\nu$$

In this equation, m_e represents the electron mass, v_e the electron velocity, and $\chi_{r,s}$ the energy level s of an r-times ionized atom to which the electron is trapped. Since the captured free electrons possess a continuous kinetic energy spectrum according to a Maxwellian distribution, the emitted radiation will also

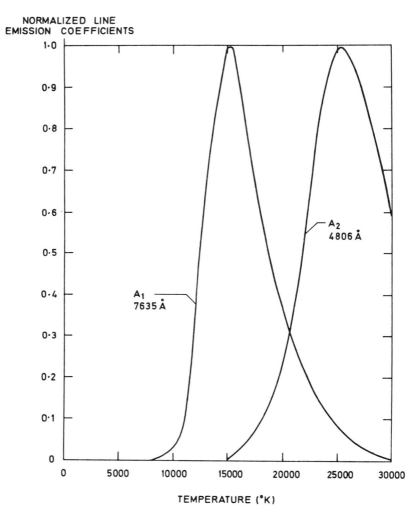

FIG. 6.4 Relative emission coefficients of two argon lines.

be continuously distributed, but with a threshold value (or series limit) λ_{max} due to the trapping of electrons with near-zero velocity where $\lambda_{max} = hc/\chi_{r,s}$.

Recombinations may occur into all possible energy levels $\chi_{r,s}$ so that the number of continuous spectra will coincide with the number s of energy states of these particular ions of species r. The entire free-bound continuum will then consist of a superposition of all continuous spectra emitted from the different species r which are present in the plasma. An exact calculation of the total free-bound continuum over the entire wavelength range is only possible for hydrogen [2]. Since the rate with which radiative recombination processes occur is proportional to the electron as well as the ion density, the emitted free-bound radiation intensity is proportional to the product of both. In the following section the contribution of free-bound radiation to the total continuum will be discussed.

6.3.3 Bremsstrahlung (Free–Free Transitions)

Free electrons in a plasma may lose kinetic energy in the Coulomb field of positive ions and this energy is readily converted into radiation. Since the initial as well as the final states of the electrons are free states in which the electrons may assume arbitrary energies within the Maxwellian distribution, the emitted radiation is of the continuum type. Radiation in a frequency interval between ν and $\nu + d\nu$ will be emitted by free electrons which have a kinetic energy $m_e v_e^2 / 2 \geqslant h\nu$, i.e., the spectrum of the emitted radiation depends on the kinetic energy or temperature of the electrons. In the elementary process of bremsstrahlung, one electron and one ion are involved. Therefore, the intensity of this radiation is expected to be proportional to the product of electron and ion density. The actual emission coefficient of a free-free continuum is given by

$$\epsilon_\nu = \overline{C(Z' + s)^2} \frac{n_e n_1}{(kT)^{1/2}} \exp\left(-\frac{h\nu}{kT}\right) \tag{6.42}$$

where

$$C = \frac{32\pi^2 e^6}{3\sqrt{3}c^3 (2\pi m_e)^{3/2}}$$

where e designates the elementary charge, n_e and n_1 the electron and ion densities, and $Z'e$ is the ion charge. The number s accounts for the fact that fast electrons may penetrate some of the outer electron shells of ions with higher nuclear charge ($s = 0$ for hydrogen ions). Therefore, such electrons will be exposed to a higher positive charge than the ionic charge $Z'e$.

Considering the total radiation continuum consisting of free-free and free-bound radiation, the emission coefficient turns out to be independent of the frequency for $\nu \leqslant \nu_g$

$$\epsilon_\nu = \overline{C(Z' + s)^2} \frac{n_e n_1}{\sqrt{kT}} \tag{6.43}$$

The energy $h\nu_g$ is taken between the ionization level and the energy level corresponding to the limiting frequency ν_g. Since Eq. (6.43) includes only free-bound radiation of closely neighboring energy levels (quasi-continuum), which are usually found in the vicinity of the ionization level, the validity of Eq. (6.43) breaks down for $\nu > \nu_g$. Free-bound radiation stemming from trapping of electrons to lower energy levels which are farther separated from each other leads to pronounced series limits.

For a plasma in LTE the product $n_e n_1$ may be expressed by the Saha-Eggert equation which assumes the following form in a singly ionized gas (Eq. [6.32] for $r = 0$)

$$\frac{n_e n_1}{n_0} = \frac{2Z_1}{Z_0} \frac{(2\pi m_e kT)^{3/2}}{h^3} \exp\left(-\frac{\chi_1}{kT}\right) \tag{6.44}$$

where n_0 designates the number density of neutral atoms, Z_1 and Z_0 the partition functions for ions and neutral atoms respectively, and χ_1 the ionization energy. With Eq. (6.44) the emission coefficient for the combined continuum radiation may be written as

$$\epsilon_\nu = \frac{64\pi^2 \overline{(Z' + s)^2} e^6}{3\sqrt{3}h^3 c^3} \frac{Z_1}{Z_0} n_0 kT \exp\left(-\frac{\chi_1}{kT}\right) \tag{6.45}$$

In this equation $n_0 kT$ may be replaced by the total pressure p provided that the degree of ionization is small ($\xi < 10\%$). For this particular situation the continuum emission coefficient is directly proportional to the pressure.

6.3.4 Blackbody Radiation

The emission coefficients described in Secs. 6.3.1–6.3.3 are all based on the assumption that the plasma is optically thin, or, in other words, that there is no appreciable absorption of radiation in the plasma itself. This assumption may fail for line as well as for continuum radiation. Very strong absorption, for example, occurs for resonance lines for which the absorption coefficient κ_ν' is so high that a layer thickness L of a fraction of a millimeter is already sufficient for complete absorption. In the immediate neighborhood of such a resonance line, the absorption coefficient may be a factor of 10^8 smaller so that layer thicknesses of 10^6 cm or more are required for complete absorption.

Finkelnburg and Peters [2] calculated the conditions for which the continuous radiation of a laboratory plasma would approach blackbody radiation ($I_\nu \gtrsim 0.9 B_\nu$) in the visible range of the spectrum (5,000 Å). Figure 6.5 shows the result of their calculations for five different gases. By considering only singly ionized species of these gases, all curves merge into a common curve which corresponds to an ionization degree of 100%. Above this common curve a laboratory plasma with a layer thickness of 2 mm or larger would be a cavity radiator at the plasma temperature. Argon, for example, with an ionization potential of 15.8 V, would fall between the curves for helium and hydrogen and become a blackbody radiator for temperatures $T > 2 \times 10^4$ K and pressures $p \geqslant$ 200 atm. Cesium, with the lowest ionization potential, would require a minimum pressure of about 5 atm to become a cavity radiator at 5,000 K.

Conceivably, there are other physical processes in plasmas which also lead to the emission of continuous radiation. Neutral atoms or molecules of certain

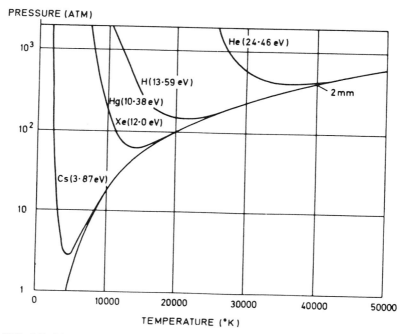

PRESSURE (ATM)

He (24·46 eV)

H(13·59 eV)

Hg(10·38 eV)

2 mm

Xe(12·0 eV)

Cs(3·87eV)

TEMPERATURE (°K)

FIG. 6.5 Blackbody radiation of plasmas for a layer thickness ⩾ 2 mm; the indicated plasmas emit blackbody radiation in the visible range of the spectrum. (From Finkelnburg and Peters, 1957, Ref. [2].)

elements, for example, may have an affinity for electrons. This causes free-free and free-bound radiation by similar mechanisms as described for the interaction of positive ions with electrons. The recombination process corresponds in this case to the formation of negative ions.

Another process which may be responsible for the generation of continuous spectra is the chemical reaction between neutral particles in the plasma. Such a reaction may be considered as a "recombination" process with a corresponding "recombination" continuum. This type of chemical reaction plays an important role in reentry plasmas as well as in plasmas emanating from rocket exhausts. In the latter case the situation may become rather complex because of the numerous combustion products involved.

Finally, in the presence of a magnetic field in the plasma, electrons which are forced into an orbital motion around the magnetic flux lines give rise to a continuous radiation called *cyclotron radiation*. However, the number of collisions which the electrons suffer in the plasma has to be small compared with the number of electron orbits. This requirement is usually expressed by the relation

$$\omega_e \tau_e = \frac{\lambda_e}{r_L} \gg 1 \qquad (6.46)$$

where ω_e is the electron cyclotron frequency, τ_e the average time interval between two electron collisions, λ_e the mean free path length of the electrons, and r_L the average Larmor radius.

Since this review deals essentially with rather dense plasmas and moderate magnetic field intensities, cyclotron radiation assumes little importance in this chapter.

6.3.5 Some Facts About Molecular Radiation

Finally, some aspects of the radiation emitted by molecular gases will be discussed. Since molecular gases are disociated at higher temperatues, appreciably lower temperature levels ($<10^4$ K) in conjunction with pressures in the order of 1 atm will be considered in this paragraph.

In contrast to atomic gases, molecular gases emit rather complex spectra: Excitation of vibrational and rotational energy states alone results in the emission of vibrational-rotational spectra that consist of line sequences in the infrared range of the spectrum. Simultaneous electronic and vibrational-rotational transitions result in band spectra that extend into the visible or even ultraviolet range of the spectrum. Continuous radiation may be superimposed on these bands (recombination continuum, for example). Each single band of a molecular spectrum consists of a number of regularly arranged lines. The intensity distribution of these lines may be used to determine the gas temperature.

The energy eigenvalues of the rotational states of a diatomic molecule within a single band are given by

$$\chi_s = \frac{h^2}{8\pi^2 \theta} J(J + 1) \tag{6.47}$$

disregarding the constant electronic excitation energy which has to be added to Eq. (6.47) where θ represents the moment of inertia of the molecule and J is the rotational quantum number ($J = 0, 1, 2, \ldots$). An approximate expression for the temperature dependence of the line emission coefficients of the rotational lines within a band is obtained by inserting the rotational energy states χ_s into Eq. (6.41).

$$\epsilon_L \sim (2J + 1) \exp\left[-\frac{h^2}{8\pi^2 \theta kT} J(J + 1)\right] h\nu \tag{6.48}$$

In this equation the factor $2J + 1 = g_s$ represents the statistical weight of rotational energy states.

6.4 SPECTROMETRIC METHODS

The discussion of the spectrometric methods shall be subdivided into methods which allow for the determination of the temperature directly and indirect methods which determine the electron density as the unknown and then apply the Saha-Eggert equation for a temperature evaluation.

6.4.1 Direct Methods

Absolute Emission Coefficient of a Spectral Line

For the sake of simplicity this method will be discussed for lines emitted from neutral atoms ($r = 0$). The method may be readily extended to ionic lines ($r > 0$) by calculating the equilibrium composition of the plasma in question. As an example of such a calculation, Fig. 6.6 shows the equilibrium composition of an argon plasma at atmospheric pressure.

The formula for the line emission coefficient (Eq. [6.41]) has been derived assuming an optically thin plasma in which LTE prevails. One of the most accurate methods of determining the temperature field in a plasma is based on an absolute measurement of ϵ_L. In order to solve for the unknown temperature in Eq. (6.41), the temperature dependence of n_0 and Z_0 must be known. As discussed in conjunction with Eq. (6.41) the partition function is only a weak function of the temperature and may, in a first approximation, be replaced by a constant. The number density n_0, which is the total number density of neutral atoms regardless of their excitation state, may be found from Dalton's law

$$p = (2n_e + n_0)kT \tag{6.49}$$

where p is the known total pressure in the plasma. The Saha-Eggert equation (see Eq. [6.32] for $r = 0$) provides the necessary relationship between electron density and temperature. Equation (6.49) and the Saha equation may be applied in this simple form only if the plasma is not generated from a gas mixture but rather from a homogeneous gas. The values for ν, χ_s, and g_s are tabulated for a large number of spectral lines. The transition probability, A_t^s, which is accurately known only for a limited number of spectral lines constitutes a problem for this method. With the exception of hydrogen, these transition probabilities are experimental values.

In summary, the temperature may be derived from the following set of equations, assuming that the absolute line emission coefficient of a neutral line has been measured.

$$\epsilon_L = \frac{1}{4\pi} A_t^s n_0 \frac{g_s}{Z_0} \exp\left(-\frac{\chi_s}{kT}\right) h\nu$$

$$p = (2n_e + n_0)kT \tag{6.50}$$

$$\frac{n_e^2}{n_0} = \frac{2Z_1}{Z_0} \left(\frac{2\pi m_e kT}{h^2}\right)^{3/2} \exp\left(-\frac{\chi_1}{kT}\right)$$

PARTICLE DENSITY (m⁻³)

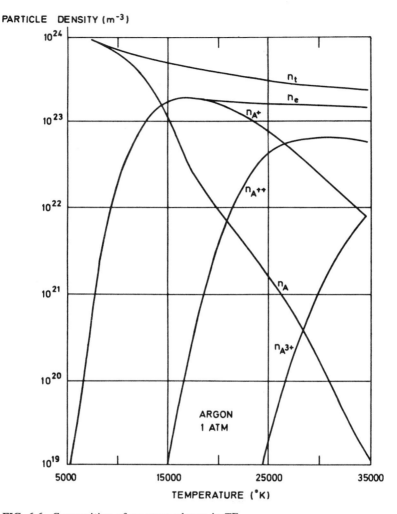

FIG. 6.6 Composition of an argon plasma in TE.

In the last two equations the plasma equation $n_e = n_1$ (quasineutrality) for a singly ionized gas has been introduced. The ionization energy χ_1 which appears in the exponential term of the Saha equation requires a correction frequently, especially if this equation is applied to high pressure plasmas. In such plasmas a lowering of the ionization potential is observed, e.g., the electron continuum appears already at longer wavelengths than anticipated from the theoretical ionization limit. The originally discrete energy levels of the atoms split up in energy bands due to the disturbing influence of neighboring particles. This effect is known as *pressure* or *collision broadening*. By virtue of this broadening effect the upper energy levels of the atoms, which are very close to each other, begin to overlap and form a quasicontinuum. This quasicontinuum, which leads to a reduction of the ionization potential, has been already mentioned in connection with Eq. (6.43). For hydrogen plasmas Unsöld [9] calculated the lowering of the ionization potential to $\Delta\chi_H = 7 \times 10^{-7} \, n_e^{1/3}$. In fact, this relation may be used as an approximation for other elements as well because the arrangement of the upper energy levels of other elements is similar to that of hydrogen. Broadening of hydrogenic energy levels due to the linear Stark effect will be further discussed in Sec. 6.4.2 under "Stark broadening."

For the measurement of absolute intensities the spectrometer and the associated optical components must be calibrated with a radiation standard (tungsten ribbon lamp or carbon arc).

Line Intensity Ratios (Relative Line Emission Coefficients)

For a simplification of the notation we will consider ratios of neutral line emission coefficients. The emission coefficients of two different neutral lines stemming from the same location in a plasma may be written as

$$
\begin{aligned}
\epsilon_{L1} &= \frac{1}{4\pi} (A_t^s)_1 n_0 \frac{g_{s1}}{Z_0} \exp\left(-\frac{\chi_{s1}}{kT}\right) h\nu_1 \\
\epsilon_{L2} &= \frac{1}{4\pi} (A_t^s)_2 n_0 \frac{g_{s2}}{Z_0} \exp\left(-\frac{\chi_{s2}}{kT}\right) h\nu_2
\end{aligned}
\tag{6.51}
$$

The ratio of these two emission coefficients

$$
\frac{\epsilon_{L1}}{\epsilon_{L2}} = \frac{(A_t^s \nu g_s)_1}{(A_t^s \nu g_s)_2} \exp\left(-\frac{\chi_{s1} - \chi_{s2}}{kT}\right)
\tag{6.52}
$$

does not contain the atom density n_0 and the partition function Z_0. This method offers three advantages over the absolute intensity method: (a) a relative measurement of the emission coefficients is sufficient, which is much simpler experimentally than an absolute measurement, (b) the transition probabilities are required on a relative scale only, and (c) the evaluation of the temperature is straightforward. Unfortunately, the accuracy of this method is frequently rather poor because the energy eigenvalues of ions of the same ionization stage are very close to each other. Considering errors of energy eigenvalues only, the relative error of this method is

$$
\frac{\Delta T}{T} = \frac{\Delta(\chi_{s1} - \chi_{s2})}{\chi_{s1} - \chi_{s2}}
\tag{6.53}
$$

An improvement of the accuracy may be achieved by extending this method to more than two lines. Rewriting either one of Eqs. (6.51) with $n_0 h/4\pi Z_0 = C$ results in

$$\log\left(\frac{\epsilon_L}{A_t^s \nu g_s}\right) = -\frac{\chi_s}{k}\frac{1}{T} + \log C \qquad (6.54)$$

or
$$y = -x\,\frac{1}{T} + C' \qquad (6.54a)$$

Plotting $y = \log(\epsilon_L/A_t^s \nu g_s)$ as a function of $x = \chi_s/k$ results in a straight line with slope $-1/T$. Although the application of a large number of lines may reduce the error, the method as such is not very sensitive to the temperature.

The same method may be applied for the rotational lines within a single band of a molecular gas. From a logarithmic plot of $\epsilon_L/(2J + 1)h\nu$ vs. $h^2 J(J + 1)/8\pi^2\theta k$ (see Eq. [6.48]) a straight line with slope of $-1/T$ is obtained. In this way temperatures may be determined in plasmas which contain molecular species, existing only at lower temperature levels. Figure 6.7 shows, as an example, the variation of the equilibrium molecule concentration in a nitrogen plasma as a function of the temperature.

Instead of considering the ratio of two spectral lines, this method may be modified for the ratio of a spectral line with the continuum intensity.

Intensity Distribution of a Spectral Line (Larenz Method)

The relative distribution of the line emission coefficient as a function of the temperature is determined by $n_r(T)$, $Z_r(T)$ and the exponential function in Eq. (6.41). Assuming that the temperature $T = T^*$ is known at a certain location $\rho = \rho^*$ in the plasma source, the ratio of the emission coefficients at an arbitrary location ρ to that at ρ^* is then

$$\frac{\epsilon_L(T)}{\epsilon_L(T^*)} = \frac{n_r(T)Z_r(T)^*}{n_r(T^*)Z_r(T)}\exp\left[-\frac{\chi_{r,s}}{k}\left(\frac{1}{T} - \frac{1}{T^*}\right)\right] \qquad (6.55)$$

Equation (6.55) permits the determination of the entire temperature distribution in the source from the measured relative intensities $\epsilon_L(T)$. A modification of this method, known as the off-axis-peaking or Larenz method, makes use of the fact that the emission coefficient of a spectral line reaches a maximum at a certain temperature. By identifying this temperature with T^*, which can be accurately calculated, a "calibration" temperature within the source is obtained, provided that the maximum temperature in the source $T_m > T^*$. In rotationally symmetric plasmas, as, for example, in wall-stabilized electric arcs, the temperature reaches a maximum in the axis. If this temperature $T_{\text{axis}} > T^*$, the line emission

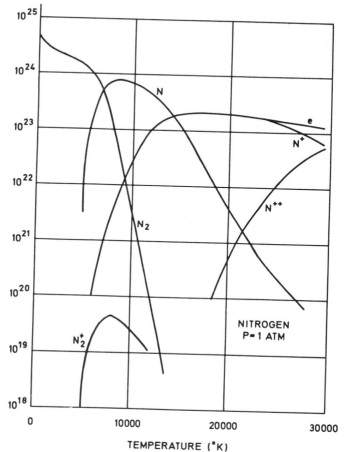

PARTICLE DENSITY (m⁻³)

FIG. 6.7 Composition of a nitrogen plasma in TE.

coefficient will show an off-axis peak. From the calculated relative distribution $(\epsilon_L)_{rel} = f_1(T)$ and the measured relative distribution $(\epsilon_L)_{rel} = f_2(\rho)$, the desired temperature distribution $T(\rho)$ may be derived. This may be done graphically as shown in Fig. 6.8.

Relative Line Emission Coefficients of Different Ionization Stages

A modification of the method discussed in Sec. 6.4.1, under "Line intensity ratios" makes use of the fact that the energy levels of an atomic line and of an ionic line of the same element are much farther separated from each other than the levels of the same ionization stage. This method will be discussed for lines

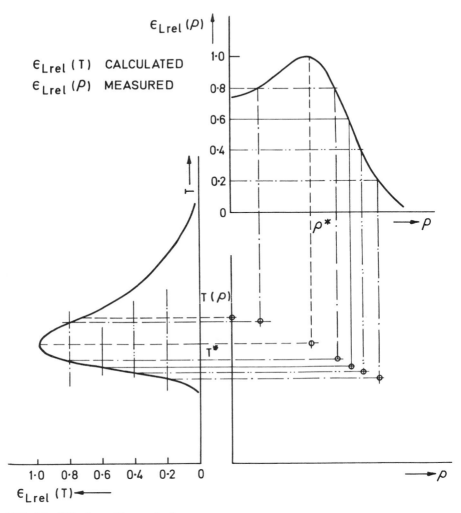

FIG. 6.8 Off-axis peaking method.

stemming from a neutral and a singly ionized atom in a plasma generated from a homogeneous gas. By indicating the neutral component by a subscript 0 and the singly ionized component by a subscript 1, the ratio of the emission coefficients is

$$\frac{\epsilon_{L1}}{\epsilon_{L0}} = \frac{n_1 Z_0}{n_0 Z_1} \frac{(A_t^s g_s \nu)_1}{(A_t^s g_s \nu)_0} \exp\left(-\frac{\chi_{s1} - \chi_{s0}}{kT}\right) \tag{6.56}$$

Since the plasma is assumed to be in LTE, the Saha-Eggert equation (see Eq. [6.32]) may be applied. Combining Eq. (6.32) for $r = 0$ with Eq. (6.56) one finds

$$\frac{\epsilon_{L1}}{\epsilon_{L0}} = \frac{2}{n_e(T)} \frac{(A_t^s g_s \nu)_1}{(A_t^s g_s \nu)_0} \left(\frac{2\pi m_e kT}{h^2}\right)^{3/2} \exp\left(-\frac{\chi_1 + \chi_{s1} - \chi_{s0}}{kT}\right) \qquad (6.57)$$

The electron density $n_e(T)$ follows again from Dalton's law and from the Saha-Eggert equation as

$$n_e(T) = S(T)\left[\sqrt{1 + \frac{p}{kT}\frac{1}{S(T)}} - 1\right] \qquad (6.58)$$

In this equation $S(T)$ is used as an abbreviation for the right-hand side of the Saha-Eggert equation. If the temperature in a plasma is sufficiently high so that ionic and atomic lines appear simultaneously in the spectrum, application of this method is advantageous. A measurement of relative line intensities leads, in this case, to accurate temperatures.

6.4.2 Indirect Methods

Absolute Continuum Emission Coefficient

As discussed in connection with Eq. (6.43) a superposition of free-bound and free-free radiation may lead, in a certain frequency range $\nu \leqslant \nu_g$, to a continuum emission coefficient which is independent of the frequency. This condition holds for free-bound radiation stemming from trapping of electrons to the upper closely spaced energy levels.

Considering a plasma which contains neutral and singly ionized species only ($n_e = n_1$), the frequency independent emission coefficient may be written as (see Eq. [6.43])

$$\epsilon_\nu = C\overline{(Z' + s)^2} \frac{n_e^2}{\sqrt{kT}} \qquad (6.59)$$

In order to determine the electron density from the absolute measurement of ϵ_ν, a value of s is required. For the plasmas discussed in this review, s is always close to unity. For $\sqrt{kT}$ a rough estimate is made which may be, if necessary, improved by iteration using the Saha-Eggert equation.

Assuming LTE in the plasma, the temperature may be derived from the known electron density using the Saha-Eggert equation and Dalton's law. This method results in accurate temperatures because the electron density depends very sensitively on the temperature.

In summary, the desired temperature may be derived from the following set of equations if the specified assumptions hold in a plasma generated from a homogeneous gas

$$\epsilon_\nu = \overline{C(Z' + s)^2} \frac{n_e^2}{\sqrt{kT}} \qquad \nu \leqslant \nu_g$$

$$\frac{n_e^2}{n_0} = \frac{2Z_1}{Z_0} \left(\frac{2\pi m_e kT}{h^2}\right)^{3/2} \exp\left(-\frac{\chi_1}{kT}\right) \qquad (6.60)$$

$$p = (2n_e + n_0)kT$$

The method may be extended directly to plasmas generated from gas mixtures if the number density ratios of the mixture are known. If these ratios are not known an additional measurement is required to determine the ratio of a neutral and an ionic line emission coefficient for two lines belonging to the same element [1].

Stark Broadening

The last method to be discussed in this chapter is suitable for a direct measurement of electron densities in plasmas. It is unique in the sense that it does not require the existence of LTE in the plasma. Measurements based on this method may, therefore, be used for a check of the validity of LTE assumptions.

Broadening of a spectral line may be caused by a number of processes which are usually grouped under the term *pressure or collision* and *Doppler broadening*. The broadening mechanism which is especially of interest for the plasmas considered in this review is the linear Stark effect. Broadening of lines emitted by hydrogenic atoms or ions due to the linear Stark effect occurs as soon as these particles are exposed to strong electric fields. The electric field removes the degeneracy of hydrogenic energy levels, i.e., an originally discrete energy level is now split into a large number of very closely spaced levels (energy band). Electronic transitions between these energy bands result in more or less broadened lines. The half-width ($\Delta\lambda_{1/2}$) of a broadened line depends on the Stark coefficient of an individual element and increases with increasing field strength. The hydrogen H_β line is especially sensitive to the linear Stark effect as shown in Table 6.1 which also demonstrates that the broadening is a strong function of the electron density and a rather weak function of the plasma temperature.

The magnitude of the electric field to which a radiating atom or ion is exposed depends on the electron density. Assuming that $n_1 (= n_e)$ is the number density of singly ionized atoms in a plasma, the average distance z between two neighboring ions is then proportional to $n_1^{-1/3}$. The Coulomb force between two ions separated by a distance z is $f = e^2/z^2 \sim e^2 n_1^{2/3}$. The corresponding electric field strength is $E \sim e n_1^{2/3}$. Because of the random motion of the ions in the

TABLE 6.1 Linear Stark effect

Line	T (K)	n_e (cm^{-3})	$\Delta\lambda_{1/2}$ (A)
H$_\beta$	10^4	10^{14}	.42
4,861 A	10^4	10^{17}	48
	4×10^4	10^{14}	.42
	4×10^4	10^{17}	50
He II	2×10^4	10^{15}	.27
3,203 A	2×10^4	10^{18}	24
A II	2×10^4	10^{15}	.0014
4,806 A	2×10^4	10^{18}	1.4

plasma this field strength assumes a statistical distribution. By introducing a normal field strength $E_0 = 2.61\ en_i^{2/3}$, the actual reduced microfield intensity may be written as $\beta = E/E_0$. Holtzmark (see, for example, [5]) calculated the distribution function $W(\beta)$ of the microfields and the resulting Stark broadened profiles. If $\Delta\lambda$ is the distance from the center of a broadened line, the shape of a line may be characterized by a function $S(\alpha)$ where $\alpha = \Delta\lambda/E_0$. The function $S(\alpha)$ is normalized so that $\int_{-\infty}^{+\infty} S(\alpha)d\alpha = 1$. The original Holtzmark theory only takes positive ions into account and predicts reliable values of line broadening for electron densities $n_e \leqslant 10^{16}$ cm^{-3}. For higher electron densities Griem [3] modified this theory to take into proper account the electrons neglected in the Holtzmark theory.

The procedure to be followed for measuring electron densities with this method in a plasma will be discussed for a specific example, namely, the H$_\beta$ line. The theoretical normalized profile $S(\alpha)$ is available from the literature.

a. Plot the theoretical profile $S(\alpha)$ for $E_0 = 1$ choosing a logarithmic scale for both coordinates.
b. Enter the data points in relative intensity units in the same diagram. The resulting experimental curve contains the unknown value of E_0.
c. Shift the experimental curve to a best fit on the theoretical curve. The shift of the abscissa then yields the desired electron density

$$n_e = \left[\frac{(E_0)_{\text{shift}}}{2.61e} \right]^{3/2} \qquad (6.61)$$

The graphical procedure is shown schematically in Fig. 6.9. According to Eq. (6.61) an error in the measurement of $\Delta\lambda$ causes an error in the electron density which is larger by a factor of 1.5.

As discussed before, if the plasma is in LTE the temperature may be derived by applying the Saha-Eggert equation and Dalton's law. For the chosen example the corresponding set of equations is

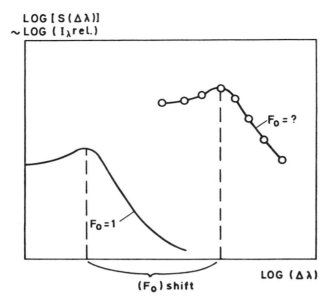

$$n_e = \left(\frac{(F_0)\,\text{shift}}{2\cdot 61e}\right)^{3/2}$$

FROM $F_0 = 2\cdot 61\,en_e^{2/3}$ $\longrightarrow$ $n_e = \left(\dfrac{(F_0)\,\text{shift}}{2\cdot 61e}\right)^{3/2}$

FIG. 6.9 Determination of the electron density from line broadening.

$$n_e = \left[\frac{(E_0)_{\text{shift}}}{2.61e}\right]^{3/2}$$

$$\frac{n_e^2}{n_0} = \frac{2Z_1}{Z_0}\left(\frac{2\pi m_e kT}{h^2}\right)^{3/2}\exp\left(-\frac{\chi_1}{kT}\right) \tag{6.62}$$

$$\chi_1 = 13.59\,eV \qquad 2Z_1/Z_0 \approx 2$$

$$p = (2n_e + n_0)kT$$

Temperatures evaluated from Eq. (6.62) are accurate within a few percentage points even if the error of the measured electron density reaches 20%, because the temperature depends essentially on the logarithm of the electron density only.

6.4.3 Methods for Emitting–Absorbing Gases

The most popular method for determining temperatures in emitting-absorbing gases in a temperature range from approximately 1,000K to 8,000K is the line reversal technique. This method has been successfully applied for temperature measurements in flames, in internal combustion engines, in rocket exhausts, in shocks, and in high temperature expansion flows [7-14]. It may also be applied for high temperature plasmas. In fact, the upper temperature limit may be as high as 50,000K depending on the availability of sufficiently strong, reliable radiation standards for these temperature levels [3]. For temperatures below 1,000K the luminosity of most gases in the visible range of the spectrum becomes too small for accurate measurements. Seeding of the gas with elements or salts of elements with low excitation potentials may appreciably enhance the luminosity. Sodium is frequently used for this purpose.

The line reversal technique considers emission and absorption of the same spectral line within a hot, luminous gas volume. Since line emission and absorption are governed by the populations of the respective energy levels, the temperature obtained from the reversal technique will be the effective electronic excitation temperature of the particular element employed for the measurement. This excitation temperature is the same as the kinetic gas temperature only if the ratio of lower- and upper-state populations for the chosen spectral line is controlled by particle impact; in other words, LTE is required down to the lower energy level of the electronic transition considered. At the same time scattering should be insignificant compared to the other radiation processes (absorption, spontaneous emission, induced emission). The latter requirement may be violated, for example, in flames which contain a large number of solid particles. Other possible deviations from LTE should be carefully checked before this method is applied.

If the method is applicable, reversal temperatures are found by viewing radiation over a certain frequency interval from a comparison source through the hot gas whose temperature is desired. Usually, the chosen frequency interval will cover a strong line (resonance line) emitted by the hot gas or the added seeding material. If the comparison light source (continuous emitter) is relatively weak, the chosen line will appear bright on the background of the spectrum of the comparison source due to net emission. If the radiation of the comparison source is sufficiently strong, the chosen line will appear dark due to net absorption in the hot gas. At the reversal point the intensity of the comparison source and of the hot gas is the same and, therefore, the line is no longer discernible on the background spectrum of the comparison source. The brightness temperature of the comparison source and the blackbody temperature of the gas are now the same for the selected wavelength interval. The "true" reversal method as well as a modification of this technique which makes it more versatile for many applications will be discussed in the following paragraphs.

These reversal methods permit a rather quick determination of the unknown gas temperature if the gas is essentially isothermal. If there are significant temperature gradients within the hot gas the required procedure for evaluating the temperature distribution becomes much more complex than that employed for optically thin gases using the Abel transform. This chapter is restricted to a discussion of the isothermal case. For nonisothermal situations the reader is referred to the literature (see, for example, [15]).

Basic Equations

Line reversal techniques to be discussed in this review are based on Kirchhoff's law which, in turn, requires the existence of TE in the gas. As pointed out in Sec. 6.2.1 this condition is never met rigorously in laboratory plasmas but it holds frequently for resonance lines which require small absorption lengths only. For a hot gas having a uniform temperature T_g Kirchhoff's law may be written for a resonance line as

$$E_\nu(T_g) = \alpha_\nu(T_g) \qquad (6.63)$$

In this equation $E_\nu = I_\nu/B_\nu$ represents the monochromatic emittance and α_ν the monochromatic absorptance of the gas. Both quantities are functions of the gas temperature T_g and the dimensions of the emitting-absorbing gaseous volume. If t_ν is the monochromatic transmittance of a gas (no scattering) the monochromatic absorptance may be expressed by

$$\begin{aligned} \alpha_\nu &= 1 - t_\nu(T_g) \\ \alpha_\nu &= 1 - \exp(-\tau_\nu) \end{aligned} \qquad (6.64)$$

with τ_ν as the optical depth of a gas layer of thickness L. For very large optical depths ($\tau_\nu \gg 1$) the line being viewed appears "black" and the temperature can be directly determined from the measured absolute intensity $B_\nu(T_g)$. The determination of temperatures from optically thin radiation sources ($\tau_\nu \ll 1$) has already been discussed in Secs. 6.4.1 and 6.4.2. The reversal method is applicable between these two extremes, i.e., for intermediate values of the optical depth.

The radiation intensity $I_{\nu g}$ emitted by an emitting-absorbing gas of uniform temperature T_g within a frequency interval from ν to $\nu + \Delta\nu$ is

$$I_{\nu g} = E_\nu(T_g)B_\nu(T_g) \qquad (6.65)$$

By viewing a comparison source through the hot gas in the same frequency interval, a total intensity $I_{\nu t}$ is observed consisting of the radiation emitted by the gas plus a fraction of the radiation from the comparison source which is transmitted through the gas

$$I_{\nu t} = I_{\nu g} + [1 - \alpha_\nu(T_g)]I_{\nu c}$$

Applying Kirchhoff's law to this equation yields

$$I_{\nu t} = I_{\nu g} + [1 - E_\nu(T_g)]I_{\nu c} \qquad (6.66)$$

$I_{\nu c}$ is the radiation intensity of the comparison source in the frequency interval from ν to $\nu + \Delta\nu$. Equations (6.63)–(6.66) constitute the working equations of the reversal technique.

"True" Reversal Technique

When the "true" reversal method is employed the radiation intensity of the comparison source is adjusted until it becomes equal to the total intensity $I_{\nu t}$ viewed through the gas. From Eq. (6.65) and (6.66) follows

$$I_{\nu t} = I_{\nu c}(T_b) = E_\nu(T_g)B_\nu(T_g) + [1 - E_\nu(T_g)]I_{\nu c}(T_b)$$

or
$$I_{\nu c}(T_b) = B_\nu(T_g) \tag{6.67}$$

The known brightness or "black" temperature T_b obtained for the reversal point is then identical with the desired gas temperature T_g, because the brightness temperature T_b is defined as the temperature at which a blackbody would emit the same intensity $I_{\nu c}$ as the comparison source does in the specified frequency interval.

Optical pyrometers are usually calibrated for brightness temperatures. If another radiation standard is employed as a comparison source (e.g., a tungsten ribbon lamp) with a known surface emittance $E_\nu(T_c)$, the comparison intensity becomes $I_{\nu c}(T_c) = E_\nu(T_c) B_\nu(T_c)$ and the desired gas temperature follows from the equation

$$E_\nu(T_c)B_\nu(T_c) = B_\nu(T_g)$$

and Eq. (6.13) as

$$T_g = \frac{h\nu}{k} \left\{ \log\left[1 + \frac{2h\nu^3}{c^2 E_\nu(T_c)B_\nu(T_c)} \right] \right\}^{-1} \tag{6.68}$$

The application of the "true" reversal technique is limited by the brightness temperature range of available comparison sources. An extension beyond this range is provided by a modified reversal technique which will be discussed in the following paragraph.

Modified Reversal Technique

This method is frequently applied when "true" reversal cannot be obtained as, for example, in high temperature flames or in shock wave generated plasmas requiring high speed resolution. The basic equations will apply for this method but an independent measurement is now required of $I_{\nu g}$, $I_{\nu c}$, and $I_{\nu t}$. Equation (6.66) written in terms of these three measured quantities yields

$$I_{vt} = I_{vg} + \left[1 - \frac{I_{vg}}{B_v(T_g)}\right] I_{vc}$$

or
$$B_v(T_g) = \frac{I_{vc} I_{vg}}{I_{vc} + I_{vg} - I_{vt}} \tag{6.69}$$

The right-hand side of this equation contains measured values only; therefore, the desired temperature T_g may be derived from the experimentally determining value of $B_v(T_g)$. Although this method provides a wider range of applicability than the "true" reversal technique, a limitation is imposed by the increasing error as the ratio of gas to comparison source radiation intensity increases. This fact is

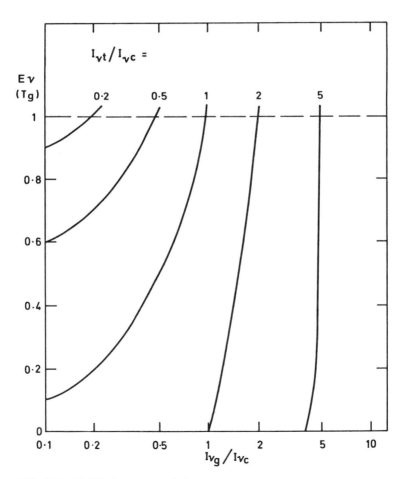

FIG. 6.10 Modified reversal technique.

illustrated in Fig. 6.10 which represents a plot of Eq. (6.69) which may be rewritten in the form

$$E_\nu(T_g) = \frac{I_{\nu g}}{B_\nu(T_g)} = 1 + \frac{I_{\nu g}}{I_{\nu c}} - \frac{I_{\nu t}}{I_{\nu c}}$$

The solid lines in this diagram represent possible operating conditions. For $I_{\nu t}/I_{\nu c} < 1$ the temperature of the comparison source is larger than the unknown gas temperature, whereas for $I_{\nu t}/I_{\nu c} > 1$ the opposite is true. The ratio $I_{\nu t}/I_{\nu c} = 1$ represents "true" reversal.

The steep slope of curves $I_{\nu t}/I_{\nu c} > 1$ indicates that the error of $E_\nu(T_g)$ increases rapidly with increasing values of $I_{\nu t}/I_{\nu c}$ for a given error in the latter and in $I_{\nu g}/I_{\nu c}$, and becomes intolerable for values of $I_{\nu t}/I_{\nu c} > 2$. It is, therefore, desirable to keep $I_{\nu t}/I_{\nu c}$ close to unity or even at values smaller than unity. In addition, the accuracy of measured line reversal temperatures depends on the sensitivity of the line intensity on temperature. The rising portion of the intensity distribution of a spectral line is essentially proportional to $\exp(-\chi_s/kT)$. (See also Fig. 6.4.) By choosing a suitable seeding material (χ_s) for the estimated gas temperature the accuracy of reversal temperatures may be in the order of 1% or even better. Using the sodium-D lines for reversal measurements in a temperature interval from 1,800K to 2,500K, for example, results in an error of approximately 1%. This error is caused mainly by the calibration error of the comparison source.

NOMENCLATURE

$A_{r,t}^{r,s}$	probability for spontaneous transition within an r-times ionized atom from a quantum level s to a lower quantum level t
$B_{r,t}^{r,s}$	transition probability for induced emission
$B_{r,s}^{r,t}$	transition probability for absorption
B_ν	intensity of blackbody radiation
C, C'	constants
c	light velocity (3×10^8 m/sec)
E	electric field strength, emittance
e	electronic charge (1.6×10^{-19} amp sec)
f	Coulomb force
g	statistical weight, constant
h	Planck's constant
I	total radiation intensity
I_ν	monochromatic radiation intensity
$I_{\nu,0}$	initial spectral intensity
J	rotational quantum number
k	Boltzmann constant

L, l	length
m	mass of particles
n	number density of particles, principal quantum number
p	pressure
R	radius of plasma source
r_L	average Larmor radius
r, x, y	coordinates
S	ionization coefficient
$S(T)$	abbreviation for the right-hand side of the Saha-Eggert equation
$S(\alpha)$	normalized function
s	correction for the penetration of the outer electron shells by fast electrons
T	temperature
$t, \Delta t$	time, transmittance
u	total radiation density
u_ν	monochromatic radiation density
$V, \Delta V$	volume
υ	thermal velocity
$W(\beta)$	distribution function of microfields
Z	partition function
$Z'e$	ionic charge
z	average distance between two neighboring ions in a plasma
α	combined collisional-radiative recombination coefficient, Sommerfeld's fine structure constant, $\alpha = \Delta\lambda/E_0$, absorptance
$\beta = E/E_0$	dimensionless parameter
β_ν	monochromatic extinction coefficient
γ_ν	monochromatic scattering coefficient
Δa	area element
Δx	natural width of a spectral line
$\Delta\lambda$	distance from center of a spectral line
$\Delta\omega$	unit solid angle
ϵ_ν	spontaneous spectral emission coefficient
ϵ_ν'	total spectral emission coefficient
ζ_n	number of valence electrons
θ	moment of inertia of a diatomic molecule
κ_ν'	volume absorption coefficient
$\kappa_{\nu m}'$	mass absorption coefficient
λ	wavelength, mean free path length
$\nu, \Delta\nu$	frequency
ξ	degree of ionization
ρ	radial coordinate
τ_ν	optical depth
τ_e	average time interval between two electron collisions
Φ	radiation flux

SPECTROSCOPIC TEMPERATURE DETERMINATION

$\chi, \Delta\chi$	energy
χ_i, χ_1	ionization energy
$\chi_{i,H}$	ionization energy of hydrogen
ω_e	electron cyclotron frequency

Subscripts

a	heavy particle
b	brightness
c	comparison source
e	electron
g	limiting value, gas temperature
L	spectral line
max	maximum value
r	ionization stage, different species
s	energy state
t	energy state, total
ν	monochromatic
0	neutral particles, reference state
1	singly ionized particles
1, 2	different spectral lines (see text)

Superscript

*	reference value

REFERENCES

1. Finkelnburg, W., and Maecker, H., Elektrische Bögen und thermisches Plasma, in S. Flügge (ed.), "Encyclopedia of Physics," vol. 22, Springer-Verlag, Berlin, 1956.
2. Finkelnburg, W., and Peters, T., Kontinuierliche Spektren, Spectroscopy II, in S. Flügge (ed.), "Encyclopedia of Physics," vol. 28, Springer-Verlag, Berlin, 1957.
3. Griem, H. R., "Plasma Spectroscopy," McGraw-Hill, New York, 1964.
4. Huddlestone, R. H., and Leonard, S. L. (eds.), "Plasma Diagnostic Techniques," Academic, New York, 1965.
5. Lochte-Holtgreven, W. (ed.), "Plasma Diagnostics," North-Holland, Amsterdam, 1968.
6. Marr, G. V., "Plasma Spectroscopy," Elsevier, Amsterdam, 1968.
7. Drawin, H. W., and Felenbok, P., "Data for Plasmas in Local Thermodynamic Equilibrium," Gauthier-Villars, Paris, 1965.
8. Dickerman, P. J. (ed.), "Optical Spectrometric Measurements of High Temperatures," University of Chicago Press, Chicago, 1961.
9. Unsöld, A., "Physik der Sternatmosphären," 2nd ed., Springer-Verlag, Berlin, 1955.
10. American Institute of Physics, "Temperature, Its Measurement and Control in Science and Industry," vol. 1, Reinhold, New York, 1941.
11. American Institute of Physics, "Temperature, Its Measurement and Control in Science and Industry," vol. 2, Reinhold, New York, 1955.

E. PFENDER

12. American Institute of Physics, "Temperature, Its Measurement and Control in Science and Industry," vol. 3, pt. 1, Reinhold, New York, 1962.

13. Gaydon, A. G., "Spectroscopy of Flames," Wiley, New York, 1957.

14. Gaydon, A. G., and Wolfhard, H. G., "Flames," 2nd ed., Chapman & Hall, London, 1960.

15. Marrodimeann, R., and Bioteux, H., "Flame Spectroscopy," Wiley, New York, 1965.

16. Weinberg, F. J., "Optics of Flames," Butterworth Scientific Publications, Washington, D.C., 1963.

17. Penner, S. S., "Quantitative Molecular Spectroscopy and Gas Emissivities," ch. 16, Addison-Wesley, Reading, Mass., 1959.

7 Probe measurements in high temperature gases and dense plasmas

JERRY GREY
Calprobe Corporation, New York, N.Y.

The study of high temperature gases and plasmas has received great attention in the past decade. Much of this effort stemmed from the practical problems associated with plasma confinement, the anticipated possibility of controlled thermonuclear reactions, and various forms of electric propulsion devices for space application.

Most of the early plasma research was performed at relatively low pressures. A reasonably comprehensive bibliography of much of the work through the nineteenth century and the first half of the twentieth century appears in [1]. An important recent trend has been the study of higher pressures; i.e., "dense" plasmas characteristic of hyperthermal wind tunnels, reentry material test facilities, and high-temperature chemical processing operations. For the purpose of this section, a *dense plasma* may be defined as a partly or fully ionized gas, in which the relevant mean free path for elastic collisions is much smaller (e.g., at least one order of magnitude) than the characteristic length associated with the hardware (e.g., probe radius).

In the study of dense plasmas, one of the first problems encountered by the investigator is that of measurement: how does one measure the properties of a fluid whose enthalpy and density are such that heat transfer rates to immersed diagnostic probes are measured in terms of kW/cm^2 or whose optical characteristics are such as to render the fluid opaque even to microwaves, or whose gradients are so steep as to invalidate spectroscopic, microwave, or laser-beam diagnostic approaches?

This chapter describes a family of probe-type diagnostic instruments designed specifically for the dense plasma environment, and the included references illustrate their use both in relatively classical problems such as stream mixing and in entirely novel applications such as determination of the degree of nonequilibrium in partly ionized plasmas.

The high temperature, of course, leads to a number of difficult measurement problems; e.g., dense plasmas are characterized by many forms of nonequilibrium radiation, metastable excited electronic states, collisional nonequilibrium such as electron-heavy-particle energy differences, chemical nonequilibrium represented by different dissociation/reassociation and ionization/recombination rates, and the thermodynamic nonequilibria of vibrational and sometimes rotational relaxation. Thus, even when measurements can be made, interpretation is often subject to some question. This uncertainty is illustrated by some of the examples described in the cited references, and often represents one of the principal limitations on correlations between experiment and theory.

Historical development of probe techniques in high-temperature gases began, of course, with Langmuir's classical studies [2] in ionized gases of very low density and therefore low heat flux to the probes. The first example of a relatively sophisticated diagnostic probe utilizing forced cooling methods was the double-sonic-orifice temperature probe developed in the late 1940's and described in some detail, with the proper literature citations, in [3]. Subsequent developments during the past decade have led to the evolution of sophisticated cooled-probe techniques capable of measuring virtually every property of flowing hot gases and plasmas. These techniques are described in detail in subsequent parts of this chapter.

7.1 TECHNIQUES

7.1.1 Calorimetric Probes

The concept of calorimetric enthalpy and heat flux measurements is not new, but the technology required for their useful application in high-temperature, high-density gases and plasmas was not developed until the middle 1950's.

Early calorimetric probes attempted, unsuccessfully, to isolate the calorimeter completely from the exterior cooling jacket required for high heat-flux environments. The first successful approach avoided the problem by utilizing a *tare* measurement [4]. In this probe concept, the energy required to cool the probe without aspiration of a gas sample is subtracted from the energy required to cool the probe with the aspirated gas sample flowing, thereby permitting determination of the gas sample enthalpy.

This probe configuration is shown in Fig. 7.1. Construction of the probe itself is generally of copper, with stainless steel supports. Cooling water from a high-pressure source—up to 55 atm—enters through the mounting block, passes up the front stainless steel support, and through the outermost coolant channel to the probe tip, returning via the inner coolant channel. Sheathed, ungrounded thermocouple junctions are located precisely at the probe coolant channel inlet and outlet.

The central tube carries a steady flow of sample gas from the probe tip past a thermocouple junction located precisely opposite the "water out" thermocouple, and then through a gas sample tube to one or more instruments as described later.

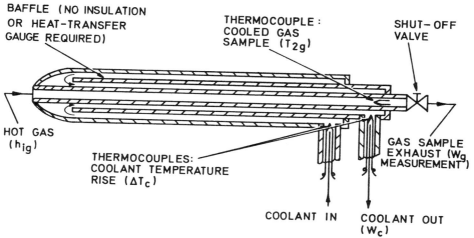

FIG. 7.1 Calorimetric probe used with tare-measurement technique.

In the unique tare measurement, which effectively eliminates errors due to external cooling requirements, a valve in the gas sample line is closed, thus preventing gas from entering the probe, and observations of coolant temperature rise and flow rate are made. The valve is then opened, allowing a gas sample to flow through the probe, and the same measurements are repeated, together with those of the steady gas sample temperature at the probe exit and steady gas sample flow rate. The rate of heat removal from the gas sample is thus given by the difference between the two coolant rates

$$\dot{\omega}_g(h_{1g} - h_{2g}) = (\dot{\omega}_c C_{pc} \Delta T_c)_{\text{flow}} - (\dot{\omega}_c C_{pc} \Delta T_c)_{\text{no flow}}$$

where $\dot{\omega}_g$ = gas sample mass flow rate, $\dot{\omega}_c$ = coolant water mass flow rate, h_{1g} = unknown gas enthalpy at probe entrance, h_{2g} = gas enthalpy at probe exit thermocouple, C_{pc} = coolant specific heat, and ΔT_c = coolant temperature rise = $(T_c)_{\text{out}} - (T_c)_{\text{in}}$.

The technique was found to be quite successful, since the tare measurement not only eliminates the error due to heat transfer from the outer portion of the jacket, but also the error due to radiation heating of the probe. Further, fabrication is comparatively simple compared to double-jacketed or tip-thermocouple designs, and models with outer diameters as small as 1.6 mm have been run successfully in atmospheric pressure arcjet exhausts at temperatures over 13,900°K. Further, the probe may be used to measure impact pressure when the gas sample flow is shut off during the tare measurement (and thereby the velocity or Mach number), and when the gas sample is extracted it may be analyzed for chemical composition. A typical experimental configuration for such determinations is shown in Fig. 7.2.

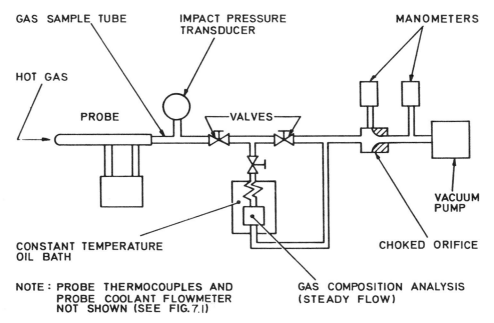

GAS SAMPLE TUBE   IMPACT PRESSURE   MANOMETERS
                         TRANSDUCER

HOT GAS

            PROBE   VALVES

CONSTANT TEMPERATURE   VACUUM
OIL BATH   PUMP

                                                  CHOKED ORIFICE

NOTE : PROBE THERMOCOUPLES AND   GAS COMPOSITION ANALYSIS
       PROBE COOLANT FLOWMETER   (STEADY FLOW)
       NOT SHOWN (SEE FIG. 7.1)

FIG. 7.2 Diagram of instrumentation used with tare measurement calorimetric probe to measure enthalpy, velocity, and gas composition.

The two disadvantages of the tare measurement technique are the necessity for intermittent probe operation, requiring either a steady-state environment or duplication of the test conditions for the flow and no-flow data points; and the selection of a sufficiently small gas sample flow rate so that approximate flow conditions near the probe tip with no gas sample flowing are closely duplicated when the gas sample is being extracted.

The latter condition is readily established in a steady-state environment by simply making the tare measurement, and then taking a series of data at increasingly larger and larger gas sample flow rates. Calculation of gas enthalpy by the above equation for each sample flow rate should give the same value until the sample flow rate becomes large enough to violate the tip-flow duplication requirement, at which point an increasingly large error will be noted. A typical determination of this type, showing the resulting optimum gas sample flow rate to be used, appears in Fig. 7.3. Note that the highest error-free gas sample rate should be used in order to obtain maximum probe sensitivity.

One consideration in any calorimetric probe analysis is the conversion of measured enthalpy to temperature, which requires a knowledge of the thermodynamic state of the unknown gas. In the case of nonreacting gases this is no problem. However, for partly dissociated or ionized media it is necessary either to establish that the unknown gas flow is in equilibrium, so that an equation of state may be used (e.g., the Saha equation) or else to measure independently the electron or dissociated radical concentration. This requirement does not, however, prejudice the enthalpy measurement. A second consideration, as discussed previously, is that these probes measure only stagnation enthalpy, and therefore require either very low subsonic flow ($M \ll 1$) or, in the case of supersonic or high subsonic Mach numbers, a separate determination of Mach number or velocity in order to determine the free-stream temperature T. In the case of the tare measurement probe, this measurement is readily made while the tare measurement is being taken. Furthermore, as in all high-temperature measurement devices, calibration of the calorimetric probe at operating temperatures presents something of a problem, since some sort of calibration standard is required. This was done for the probe of Fig. 7.1 by utilizing an arcjet as a calorimeter; i.e., the total power of the jet issuing from the exit plane of an arcjet nozzle was computed from carefully measured input power and arcjet cooling-jacket power. The calorimetric probe of Fig. 7.1 was then used to survey the nozzle exit plane (about 15 points on the diameter of an axisymmetric 19.1 mm diameter jet), and the resulting enthalpy, density, and velocity distributions were integrated to give the total power in the jet at the nozzle exit-plane. The departure from unity of the ratio of probe-measured power to arcjet-measured power

$$\frac{\int_0^r 2\pi r(\rho V h)\, dr}{EI - \dot{m}_{ca} C_{ca} \Delta T_{ca}}$$

then provided a direct indication of the measurement error. Typical results are shown here in Fig. 7.4.

It is illustrated that the good agreement of this figure was not fortuitous or the result of compensating errors, by the equally good agreement between the

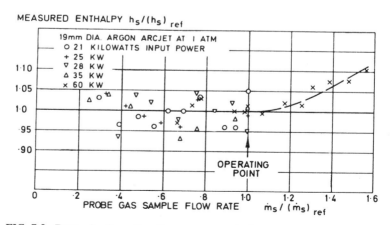

FIG. 7.3 Determination of optimum gas sample flow rate for tare measurement calorimetric probe.

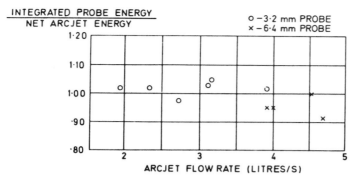

FIG. 7.4 Energy calibration of tare measurement calorimetric probe.

ratio of integrated probe-measured mass flow rate to arcjet input mass flow rate. The measured mass flow rate ratio

$$\frac{\int_0^r 2\pi r(\rho V)\, dr}{\dot{m}_a}$$

is plotted in Fig. 7.5, illustrating the same degree of precision as did the power ratio of Fig. 7.4.

Each experimental point shown in Figs. 7.4 and 7.5 represents a 15-data-point integration, illustrating the excellent resolution of the small-diameter probe. This characteristic, essential for making local measurements in a gas stream having high gradients, is more clearly brought out in Fig. 7.6, which shows a series of temperature profiles at different axial locations in a turbulent subsonic arcjet [5].

An additional feature of the simple single-jacketed probe of Fig. 7.1 is that

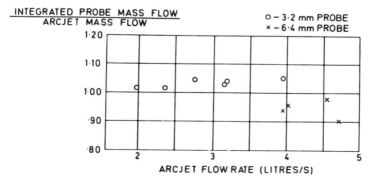

FIG. 7.5 Mass flow calibration of tare measurement calorimetric probe.

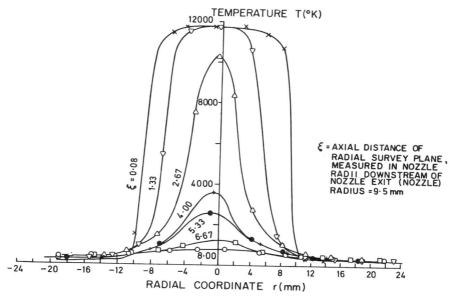

FIG. 7.6 Temperature surveys of axisymmetric subsonic turbulent arcjet exhausts using the tare measurement calorimetric probe of Fig. 7.1.

it may be bent up to 90° in order to remove all support and auxiliary hardware from the hot region. Two commercially available configurations used extensively in high-temperature environments are shown in Fig. 7.7. The 90° probe in this figure has an outer diameter of 3.6 mm, and the 30° probe an outer diameter of 1.9 mm.

A final advantage of the tare measurement probe is the comparatively modest requirement for auxiliary equipment. A 115 l capacity 55 atm pressurized-water source provides both a one hr controlled coolant supply and the necessary flow and pressure instrumentation, and a rack-mounted gas sample analysis system embodying the equipment of Fig. 7.2 can be assembled from standard components.

In applying tare measurement calorimetric probes to environments other than the one-atmosphere argon arcjet described by Grey, Jacobs, and Sherman, an important consideration in determining attainable accuracy levels is the so-called probe sensitivity. This has been defined as

$$\sigma = \frac{(\Delta T_c)_f - (\Delta T_c)_{\text{no flow}}}{(\Delta T_c)_{\text{flow}}}$$

The sensitivity σ, which must be at least 0.05 in order that conventional thermocouples provide adequate accuracy, depends not only on the hot-gas

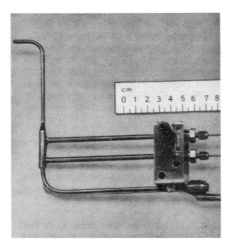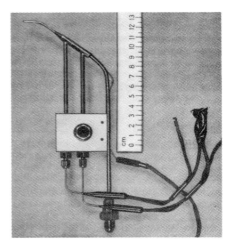

FIG. 7.7 Bent configurations of tare measurement calorimetric probes of Fig. 7.1.

environment but also on the characteristics of the probe itself. An approximate analysis of this dependence has been performed by Grey [6] for a straight probe of the configuration shown in Fig. 7.1. The analysis for the general case of a partly-dissociated, partly-ionized gas as given by Grey requires numerical integration and cannot be expressed in closed form; however, for small degrees of dissociation and/or ionization the sensitivity is given [6] by the approximate expression

$$\frac{\sigma}{\sigma_{ref}} \approx \frac{(\dot{m}/A)_{ref}}{\dot{m}/A} \frac{d_i/d_o}{(d_i/d_o)_{ret}} \left(\frac{d_i}{d_{i_{ref}}}\right)^2 \frac{L_{ref}}{L} \frac{P}{P_{ref}} \frac{\Delta P}{\Delta P_{ref}} \left(\frac{T_{ref}}{T}\right)^{3/2}$$

where $()_{ref}$ indicates any reference set of values. This result was compared with experimental measurements for probes of two sizes for the following set of reference conditions:

P_{ref}	1 atm
σ_{ref}	0.14
T_{ref}	11,900°K
$(\dot{m}/A)_{ref}$	6.4 Kg/m^{2}S
L_{ref}	98.3 mm
$(d_i)_{ref}$	0.91 mm
$(d_o)_{ref}$	3.58 mm
$(\Delta P)_{ref}$	48.5 mm Hg
Gas:	Argon

The result is shown here in Fig. 7.8, which plots measured sensitivity σ against the sensitivity computed from the above equation. The ranges of conditions covered by these tests were

T	6,700 to 13,300°K
ΔP	2.54 to 50.8 mm Hg
$\dot{m}/A$	7 to 70 Kg/m²S
d_i	0.91 to 1.78 mm
d_o	3.58 to 6.4 mm

Despite the apparently unexplainable wide scatter of four of the data points, the first-order sensitivity formulation given above appears adequate, at least for estimation of the behavior of the tare measurement probe of Fig. 7.1 under different operating conditions.

This simple probe concept has been extremely successful in making rather sophisticated measurements at relatively high temperatures over a wide range of flow and environmental conditions (e.g., [5], [7-13] etc.). Its principal advantages are simplicity, capability for withstanding extremely high heat flux conditions in very small-diameter configurations (1.6 mm OD), and mechanical ruggedness. A typical set of heat-flux design conditions is shown in Fig. 7.9, and this probe design has demonstrated the capability of continuous immersion in

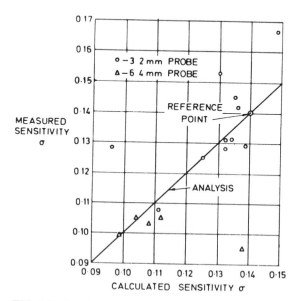

FIG. 7.8 Correlation of tare measurement calorimetric probe sensitivity analysis with experimental results.

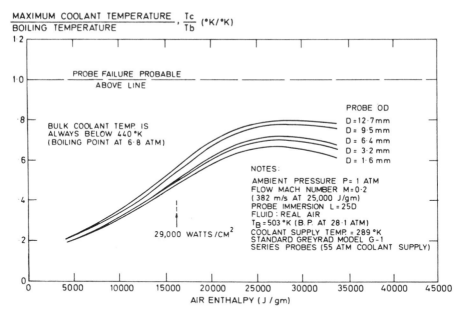

MAXIMUM COOLANT TEMPERATURE , $\frac{Tc}{Tb}$ (°K/°K)
BOILING TEMPERATURE

PROBE FAILURE PROBABLE
ABOVE LINE

PROBE OD

D = 12·7 mm
D = 9·5 mm
D = 6·4 mm
D = 3·2 mm
D = 1·6 mm

BULK COOLANT TEMP. IS
ALWAYS BELOW 440 °K
(BOILING POINT AT 6·8 ATM)

NOTES:

AMBIENT PRESSURE P = 1 ATM
FLOW MACH NUMBER M = 0·2
(382 m/s AT 25,000 J/gm)
PROBE IMMERSION L = 25D
FLUID : REAL AIR
T_B = 503 °K (B. P. AT 28·1 ATM)
COOLANT SUPPLY TEMP. = 289 °K
STANDARD GREYRAD MODEL G-1
SERIES PROBES (55 ATM COOLANT SUPPLY)

29,000 WATTS/CM^2

AIR ENTHALPY (J / gm)

FIG. 7.9 Typical heat transfer capability of simple tare measurement probe.

high-pressure (100 atm) arc-heated wind tunnels at probe wall-surface heat transfer rates exceeding 8.2 Kw/cm^2.

Principal disadvantages of the simple tare measurement probe are: (a) the need to ensure sufficiently repeatable flow duplication at the tip between the tare and actual calorimetric measurements, as discussed above, (b) the decrease in sensitivity which occurs as the energy extracted from the gas sample flow becomes small relative to the total energy absorbed by the probe coolant, e.g., at low pressures, and (c) the restriction to stagnation-point measurements in super-sonic flows, imposed by both the blunt tip (needed for high heat flux capability) and the tare measurement requirement.

The first of these problems is readily accounted for by the calibration technique discussed earlier. The second and third disadvantages, however, repre-sent true limitations on the capability of the tare measurement probe technique, and therefore, should test conditions be such as to introduce problems of low sensitivity or probe bow-shock interactions, different probe designs are required, as will be discussed later. Parameters governing probe sensitivity are summarized in [6]. Comparisons of blunt and shock-swallowing probes are made in [14] and [15], demonstrating the inconclusive nature of the current state of knowledge on probe shock effects ([14] indicates a major difference; [15] demonstrates none at all).

The real problem of reduced sensitivity at lower test pressures, however, needs to be approached by returning to the old idea of completely isolating the calorimetric portion of the probe from the cooling jackets. This concept, now practical in probe dimensions as small as 4.8 mm OD, is shown in Fig. 7.10 (other possible configurations being suggested in [16]). Thermal isolation of the calorimeter, whose coolant flow may now be decreased to obtain any desired temperature rise, is achieved by (a) gas or vacuum space between the calorimetric (inner) jacket and exterior cooling jacket, (b) silvering of the opposing jacket surfaces, and (c) regulation of flow passage dimensions so that with the prescribed coolant temperature rise in each jacket, there is no temperature difference between opposite points on the inner and outer jackets, thus eliminating interjacket heat flux.

The double-probe configuration of Fig. 7.10 has been quite successful in reducing the error due to low sensitivity of the original tare measurement concept described in [4]. One interim method which proved unsuccessful because of unknown interjacket heat transfer was the "split-flow" probe, described in [17-20].

A logical extension of the blunt double-probe configuration shown in Fig. 7.10 is the sharp-inlet design of Fig. 7.11. Operating on the same isolated-calorimeter principle as the blunt double-probe, this shock-swallowing probe concept provides two additional advantageous features: (a) ingestion of a much higher gas sample mass flow, particularly at hypersonic Mach numbers, thus improving sensitivity considerably [14], and (b) provision of mass-flux measurement capability. Because of the large material concentration at the sharp edge, however, this probe cannot meet the heat-flux capability of either the simple tare measurement probe (Fig. 7.1), or the blunt double-probe configuration of Fig. 7.10. Furthermore, the fabrication problems associated with this geometry have limited minimum probe diameters to the order of one inch. However, for large

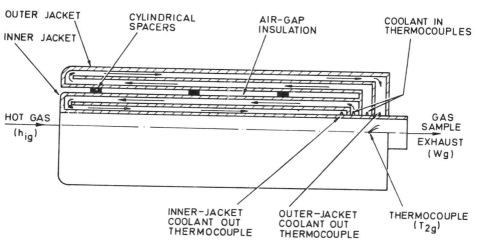

FIG. 7.10 Double probe (fully isolated calorimeter) for stagnation-point measurements.

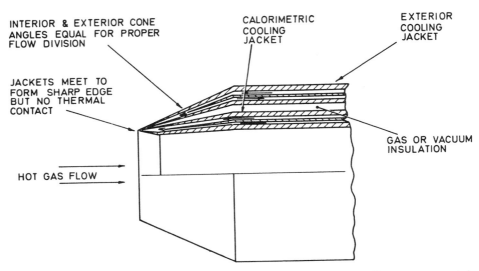

FIG. 7.11 Shock-swallowing double probe for enthalpy and mass flux measurement in supersonic/hypersonic flows.

very low density (hypersonic) applications, the sharp-inlet probe has proven to be quite useful (for example, [14, 15, 21, 22]).

The concept of utilizing gaseous rather than liquid coolants with the above-described calorimetric probe configurations makes them extremely useful for measurement of enthalpy in condensible fluids; e.g., iodine vapor, lithium or other metal vapors, or in environments where the presence of water might be hazardous; e.g., nuclear rocket plenums or exhausts. Other gas-cooled probe applications are described later in these notes. One possible application of probe gas cooling has appeared in the literature in a number of instances [23–25] but without much practical development: the transpiration-cooled probe. To date, no truly effective transpiration-cooled probes have evolved.

One serious problem with any probe designed for extreme temperature measurements is that of calibration. The only methods utilized to date are comparison methods, rather than true calibrations, but nevertheless these provide a basis for rather high confidence levels in these probes. Arcjet energy balances have demonstrated accuracies (standard deviation) within 2%, as described earlier [3, 4]; comparisons with spectroscopic methods in regions where these comparisons were possible have demonstrated agreement within the same level [8], as illustrated in Figs. 7.12 and 7.13, and comparisons with electrostatic-probe electron temperature measurements in equilibrium gases have shown better than 2% agreement [26], as will be discussed in detail later.

7.1.2 Cooled Electrostatic Probe

Although electrostatic probes have been applied to the measurement of electron and ion temperatures and densities for decades [1, 2], their utilization for

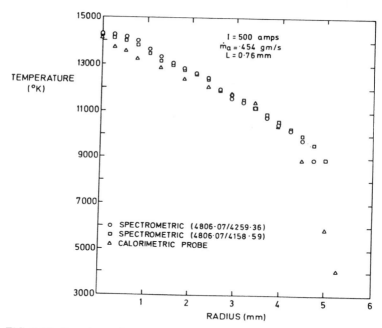

FIG. 7.12 Experimental comparison of calorimetric and spectrometric methods of temperature determination.

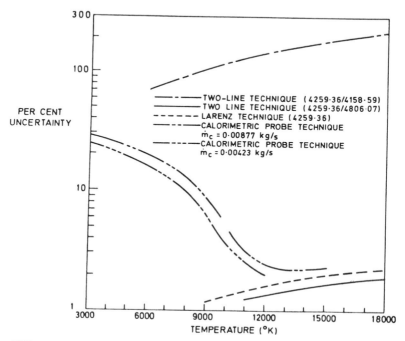

FIG. 7.13 Experimental uncertainty of plasma jet temperature-measurement techniques.

so-called dense plasmas has been quite limited. Despite an enormous growth in the recent literature on cooled or diffusion controlled electrostatic probe theory, only a handful of experimental papers have appeared.

In principle, however, the use of a few relatively nonrestrictive assumptions makes possible the measurement of electron and ion characteristics in dense plasmas with surprisingly high accuracy, within the specific conditions required by the necessary assumptions. One excellent example of this simplified approach is indicated in [26], in which rather sophisticated nonequilibrium measurements in argon plasmas were achieved, without requiring theoretical treatment of any greater complexity than Langmuir's original concept. A detailed presentation of the analytical background leading to these measurements of electron temperature and density appears in [27], and an illustration of the simple theory that predicts the experimentally observed difference between electron and heavy-particle temperatures is described in [28]. Subsequent applications of the cooled-probe technique have been utilized in measuring electron, positive ion, and negative ion densities in rocket exhaust plasmas.

The fundamental difficulty in the application of cooled-probe techniques to electrostatic-probe measurements is the analysis of the effect of a cooled probe surface on observed current-voltage traces, in order that proper interpretation of the data be made possible.

Because the detailed theoretical probe problem is still largely a matter of conjecture, a somewhat simplified approach has been used, principally for the purpose of establishing direct comparisons with experimental data. The approximate theory does, however, deal with all the problems encountered by the application of a cooled probe, i.e., the presence of temperature and velocity boundary layers, distortion of charged and neutral particle densities in the region near the probe, and the mechanisms of ionization and recombination in the probe neighborhood. The sole purpose of this analysis is, of course, to relate the information obtained from the probe to the conditions in the undisturbed plasma. The approximate analysis is directed at a determination of the freestream electron temperature and, if possible, the freestream electron density, based on characteristics that can be measured by a cooled probe.

In order to demonstrate that these measurements are valid, it is also of interest to determine the ratio of saturated electron current (high positive probe voltage) to saturated ion current (high negative probe voltage) and the floating potential (voltage at which electron and ion currents are equal). Since the electron temperature is the critical parameter to be measured, it is essential that knowledge of the probe's effect on this quantity be established. The saturation current ratio and the floating potential can be directly measured experimentally and therefore provide a satisfactory check on the assumptions used in the analysis once the electron temperature has been determined.

We begin by separating the overall problem into four regions as shown in Fig. 7.14: a sheath, in which the electric field effects are dominant; a transition region, in which electric field effects may still exist but in which the effects of

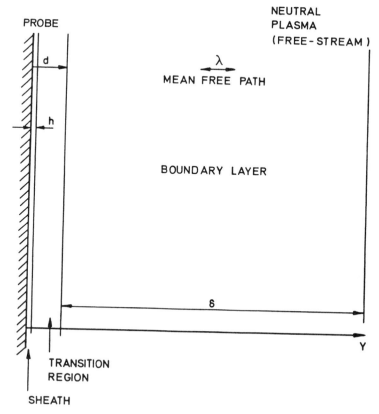

FIG. 7.14 Schematic diagram of four regions near a cooled electrostatic probe.

collisions become important; a boundary layer in which there are no field effects; and finally the freestream conditions, which extend to infinity. Note that we may characterize the probe by a flat plate as in Fig. 7.14, since for the proposed environment the probe dimension will be thousands of times greater than a mean free path or the sheath thickness.

Establishing the relative dimensions of the three bounded regions in Fig. 7.14, we first assume that the sheath thickness is of the order of the Debye length. Using Bohm's criterion [29] for the transition region thickness d, that the field beyond d is smaller than the variation in thermal energy over an electron-ion mean free path, we find (for the typical set of environmental conditions in Table 7.1) that d is of the order of 20 Debye lengths but still less than the order of a mean free path [30–32]. It is thus concluded that all electric-field effects are confined to a region within one pertinent mean free path from the probe. Further discussion of this conclusion appears later.

TABLE 7.1 Typical conditions for cooled-electrostatic probe evaluation

Properties	Speeds, cm/sec	Lengths, cm
$T_\infty = 12{,}000°K\ (21{,}500°R)$	$C_e = 7 \times 10^7$	$L/2 = 10^{-1}$
$T_W = 300°K\ (540°R)$	$C_A = 2.5 \times 10^5$	$\delta = 1.0 \times 10^{-2}$
$P = 1$ atm	$U_\infty = 3 \times 10^4$	$\lambda_{e-A} = 1.43 \times 10^{-3}$
$N_{A\infty} = 6 \times 10^{17}$ cm^{-3}		$\lambda_{e-I} = 1.5 \times 10^{-4}$
$N_{AW} = 2.7 \times 10^{19}$ cm^{-3}		$h = 2.87 \times 10^{-6}$
$N_{e\infty} = 6.3 \times 10^{16}$ cm^{-3}		
$\alpha_\infty = 0.105$		

Consider the specific case of a positive cooled probe of dimension $L \gg \lambda$ immersed in a subsonic plasma flow, the electrons of which are in a Maxwellian distribution around a temperature no lower than the heavy-particle temperature. The foregoing arguments permit us to assume that (a) the boundary-layer thickness δ is much greater than that of the sheath ($\delta \gg h$), and (b) electrons are captured by the probe when they reach the sheath boundary. Furthermore, the heavy-particle temperature in the boundary layer is assumed to decrease from its freestream value to the cooled-probe wall temperature in accordance with usual boundary-layer behavior. Finally, we assume for the moment that although the electrons may lose energy to the ions or atoms by elastic collisions, they do not suffer recombination, i.e., the boundary layer is chemically frozen.

The fraction of energy lost by an electron to a heavy particle in a single elastic collision may be written as

$$\frac{\overline{\Delta E}}{E_0} = \overline{F(\chi)}\,\frac{4m_e m_H}{(m_e + m_H)^2}$$

where $\overline{F(\chi)}$ is the average steric factor over all possible deflection angles χ and has the value $1/2$.

Thus, for $m_e \ll m_H$

$$\frac{\overline{\Delta E}}{E_0} \approx 2\left(\frac{m_e}{m_H}\right)$$

For argon

$$\frac{\overline{\Delta E}}{E_0} \approx \frac{1}{36{,}600}$$

so that after n collisions, where $n \ll 36{,}600$

$$\frac{E_n}{E_0} \approx \left[1 - \left(\frac{n}{36{,}600}\right)\right]$$

We may now approximate the number of heavy-particle collisions which an electron will experience in traversing the boundary layer by

$$n = (6)^{1/2} \left(\frac{\delta}{\overline{\lambda}}\right) (\text{Re}_L)^{1/2}$$

where the factor $(6)^{1/2}$ accounts for the three-dimensional movement of the electron, δ is the Schlichting stagnation-point boundary-layer thickness [33], $\overline{\lambda}$ is the mean value of the pertinent mean free path, and $(\text{Re}_L)^{1/2}$ (Reynolds number based on probe radius L) accounts for the effective increase in δ due to the potential flow around the stagnation point.

Values of n for various temperatures in the 1 atm subsonic argon plasma flow under consideration are shown in Table 7.2, using actual Ramsauer cross-sectional data for argon (σ_{e-A} and σ_{e-I} based on electron energy).

It is clear that in the medium temperature range the electrons suffer so few collisions across the boundary layer (~ 0.010 cm thick) that they lose very little energy: e.g., when $T = 10,500°$K (19,000°R), $n = 370$, and the final electron energy is $\sim 99\% E_0$. Clearly in this range it may be assumed that the electron temperature change across the probe boundary layer will be negligible. This conclusion is in excellent agreement with the work of Talbot [34], Kumura and Kanzawa [35], and Pytte and Williams [36]. Calculations were also performed to determine the effects of variable heavy-particle temperature profiles, electron and atom densities, and mean free paths through the boundary layer, which showed essentially no difference from the results in Table 7.2.

At higher temperatures, however, the reduction in ion-electron mean free path causes n to increase, so that at $13,500°$K (24,000°R) the electrons suffer an energy loss of about 5%. Although this effect levels off when the plasma becomes fully ionized (i.e., over 15,000°K or 27,000°R), electron temperature corrections will then be of the order of 10 or 20% and clearly represent a limitation of the cooled-probe technique.

It has been established that under the proper plasma conditions, the electron

TABLE 7.2 Calculation of number of electron collisions n in 0.01 cm boundary layer on 0.095 cm (radius) probe in argon at 1 atm

T_∞, °K	T_∞, °R	α_∞	$N_{e\infty}$, cm^{-3}	$N_{A\infty}$, cm^{-3}	λ_∞, cm	n
13,500	24,300	0.300	1.5×10^{17}	4.5×10^{17}	6.7×10^{-5}	2,000
12,000	21,500	0.100	6.3×10^{16}	6.0×10^{17}	1.5×10^{-4}	840
10,500	18,900	0.035	2.5×10^{16}	7.1×10^{17}	4.6×10^{-4}	370
8,500	15,300	0.003	2.4×10^{14}	8.0×10^{17}	1.0×10^{-4}	175
7,200	13,000	10^{-4}	1.0×10^{14}	1.0×10^{18}	1.0×10^{-3}	175
6,100	11,000	10^{-5}	1.3×10^{13}	1.3×10^{18}	1.0×10^{-3}	175
5,200	9,400	10^{-6}	1.5×10^{12}	1.5×10^{18}	1.0×10^{-3}	175
1,000	1,800	10^{-13}	10^6	7.2×10^{18}	7.3×10^{-4}	220
300	540	10^{-13}	10^6	2.7×10^{19}	1.8×10^{-4}	1,000

temperature in the freestream plasma will be preserved throughout the boundary layer, so that $T_e \approx T_{e\infty}$ and $T_I \approx T_w$, and it is assumed that as before, $(h/L)^2 \ll 1$, $y \geqslant \lambda \geqslant h$, and $\delta \ll L$. One further assumption is required in order to utilize these expressions to evaluate the performance of the cooled electrostatic probe. Since it has already been shown that most of the probe potential drop has occurred at $y = \lambda$, the effective potential φ_λ is approximated by $\varphi_\lambda \approx V - V$ plasma. If it is now assumed that V plasma $\approx$ constant for slightly negative applied voltages V, the classical Langmuir electron-current expression can be differentiated:

$$J_e \approx \frac{1}{4} N_{e\lambda} \overline{C}_{e\lambda} \exp\left(-\frac{e\varphi_\lambda}{kT_e}\right)$$

replacing $d\varphi_\lambda$ by dV, obtaining the familiar result:

$$\frac{d(\log_e J_e)}{dV} = -\frac{e}{kT_{e\lambda}} \approx -\frac{e}{kT_{e\infty}}$$

This result could be subject to some doubt on two counts: first the plasma potential may change appreciably in the range $-\varphi_F < V < O$, and second, the conclusion $T_{e\lambda} \approx T_{e\infty}$ is valid only in the neighborhood of $(V = V$ plasma). Experimental data have shown, however, as will be discussed later, that the indicated proportionality of $\log J_e$ to V is valid over a sufficiently wide range of applied negative voltages to be useful in determining the electron temperature.

It can also be concluded that if this expression is used to calculate the electron temperature $T_{e\infty}$ from measured current-voltage characteristics the comparable first-order (Langmuir) expressions for saturation current ratio and floating potential may be experimentally evaluated by direct measurement. It is, of course, first necessary to validate the all-important electron temperature determination by some independent calibration measurement, which is performed as described later.

It must be carefully noted that the preceding estimates are not presented in the guise of a theory of the operation of an electrostatic probe in a dense plasma, but rather as an identification of the reasonable degree of approximation that may be employed to gain useful experimental information on plasma characteristics in a specific regime. The basis for application of this analysis is discussed at length in [26].

In one final problem, that of estimating the freestream electron density by cooled-probe measurements, it must be determined in detail what the effect is of the cooled boundary layer on electron densities inside the boundary layer, i.e., between the freestream and a location one mean free path from the probe surface. This involves the solution of the ambipolar diffusion equation in the neighborhood of the probe, which has been performed by Brundin and Talbot [37]. They assumed steady, one-dimensional, weakly ionized, constant-pressure flow in a stagnation-point boundary layer in which the plasma was quasi-neutral (outside the sheath), with frozen electron temperature $(T_{e\infty} = T_e = T_{e\lambda})$, equal ion and atom temperatures $(T_I = T_A)$ and frozen chemistry (no recombination). Their analysis allowed for variable heavy-particle temperature, density, viscosity, diffusion coefficient, and thermal conductivity within the boundary layer. A similar analysis for full constant-pressure equilibrium chemistry [27] produced similar results for N_e/N_A, except that large differences were observed near the wall.

The very useful result of this numerical calculation (Fig. 7.15), i.e., that $N_{e\infty} \approx N_e$, appears to be fortuitous rather than general and cannot be applied indiscriminately. However, the breadth of application of this result can be determined by measuring the random electron current $(J_e)_r$ at the plasma potential and calculating $N_{e\lambda}$ for known $T_{e\lambda} \approx T_{e\infty}$. Then, since $N_{e\infty}$ is known from electron Saha equilibrium about $T_{e\infty}$, a direct comparison may be made. Conversely, we may assume, based on the result of Fig. 7.15, that $N_{e\infty} \approx N_{e\lambda}$ and deduce $T_{e\infty}$ from measurements of $(J_e)_r$ under the previous approximations $T_{e\lambda} \approx T_{e\infty}$ and $(J_e) \approx (J_e)_r$. These values of $T_{e\infty}$ may then be compared with values obtained by other measurements, as will be discussed later.

One further consideration that deserves some comment is the assumption implicit in the foregoing analysis, that electron-ion recombination was negligible

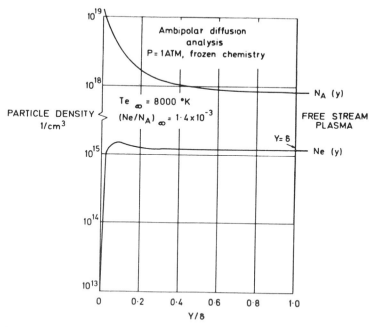

FIG. 7.15 Argon neutral and electron particle density boundary-layer profiles.

in the probe boundary layer. This has been checked [27] by the detailed consideration of all possible recombination mechanisms, i.e., radiative, dielectronic, dissociative, and three-body. It was concluded that the characteristic recombination times for a 1 atm argon plasma were considerably greater than the electron or ion transit times in the boundary layer for all temperatures below 11,000°K (20,000°R), but that the two times become comparable at about 12,000°K (21,500°R). This contribution becomes increasingly more important up to about 15,000°K (27,000°R), where the recombination time is about an order of magnitude smaller than the transit time. Above 15,000°K (27,000°R), the plasma is virtually fully ionized, and the probe boundary layer may be assumed to be in chemical equilibrium. Thus, the assumption of frozen chemistry is valid in the same range as is the earlier approximation that $T_{e\infty} \approx T_{e\lambda}$ and thereby does not further restrict the useful regime of the cooled-probe technique.

The primary purpose of the experiments was to evaluate the cooled-probe method for measuring electron temperature and if possible, electron density, in a dense plasma. A secondary purpose was to examine the validity of the rough analytical approach in the dense plasma regime by comparing predicted and measured values of saturation current ratio and floating potential.

Tests were run by inserting the cooled calorimetric-electrostatic probe of Fig. 7.16 into the 1 cm diameter exhaust of the laminar atmospheric-pressure argon arcjet and making measurements at different axial locations on the jet centerline. The probe consisted of a 1.9 mm diameter calorimetric probe [4] insulated with a zirconium oxide/boron nitride composition over all but a small known area (1.6 mm diameter) on the probe tip. The measurements were of total enthalpy, velocity, and chemical composition (to verify that the argon was undiluted by

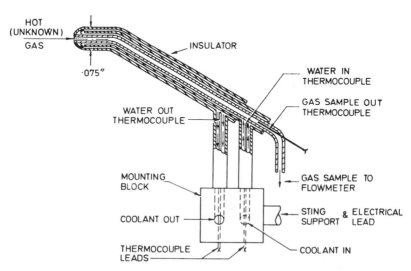

FIG. 7.16 Calorimetric-electrostatic probe.

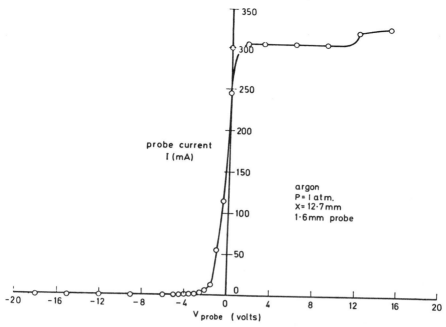

FIG. 7.17 Steady-state electrostatic probe characteristic.

traces of air in the jet environment), as measured by the calorimetric portion of the probe [4], as well as the current-voltage trace of the electrostatic portion. The latter measurement was performed when gas flow through the probe was shut off so that the probe gas sample tube mouth was at stagnation conditions, and the effective collecting area therefore did not include the heavily cooled interior of the gas sample tube. Both dc and sweep techniques were used, i.e., using a dc battery at different voltages to power the probe and reading the resulting direct current output on a Keithley electrometer, or using an x-y oscilloscope both to drive the probe (x axis) and to display the current output (y axis). Oscilloscope sweep frequencies used were 60 Hz and revealed excellent reproducibility of the probe current-voltage trace in laminar arcjet flows.

Typical dc test data of the electrostatic probe appear in Fig. 7.17, and Fig. 7.18 amplifies the negative portion of Fig. 7.4. These figures clearly show the expected ion-current saturation, as predicted for the high values of L/λ and L/h, and indicate the degree of precision obtainable in the measurement of saturation current ratio $(J_e/J_I)_{sat}$ and floating potential φ_F for comparison with first-order theory.

Determination of the electron temperature was performed by the previously discussed classical method as illustrated in Fig. 7.19, which by its appreciable range of linearity clearly illustrates the validity of the assumption that for the range of conditions tested, the probe electric field effects do not penetrate very far into the boundary layer. Note also the clear definition of the plasma potential.

A final measurement extracted from the cooled electrostatic probe data was the electron current at the plasma potential ($\varphi_\lambda = 0$). Because the saturation ion current is so low, we can assume that at $\varphi_\lambda = 0$, the total current is that due to the electrons and will be the random electron current:

$$J_e(\varphi_\lambda = O) = \frac{1}{4} N_{e\lambda} \bar{C}_{e\lambda}$$

Taking $T_{e\lambda} \approx T_{e\infty}$, therefore, we may now evaluate the range of validity of the results of Fig. 7.15, i.e., the observation that for one particular numerical condition, $N_{e\lambda} \approx N_{e\infty}$. The measurement of J_e at $\varphi_\lambda = 0$ is thus used, with the assumption that $N_{e\lambda} = N_{e\infty}$, to determine $T_{e\infty}$ and the result is compared with $T_{e\infty}$ as determined by two other methods. Agreement within $\pm 2\%$ on values of $T_{e\infty}$ implies that the assumption $N_{e\lambda} \approx N_{e\infty}$ is valid to within about $\pm 10\%$.

The principal experimental problem was the evaluation of electron temperature measurements by the cooled electrostatic probe. This was done by utilizing as a calibration measurement the enthalpy measured by the calorimetric portion of the probe, which was then used to determine the average gas temperature (very close to the heavy-particle temperature T_H for the low degrees of ioniza-

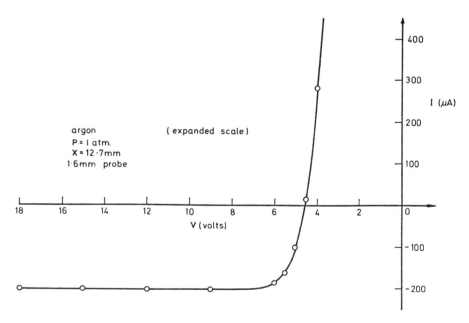

FIG. 7.18 Steady-state electrostatic probe characteristic (expanded scale).

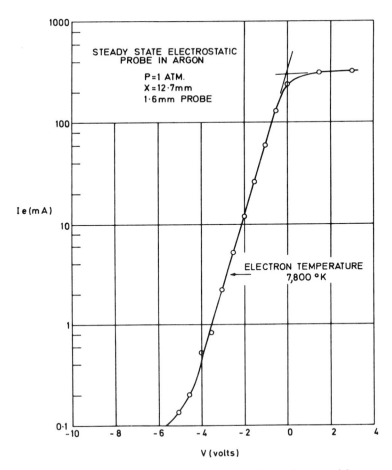

FIG. 7.19 Logarithm of electron current vs. applied probe potential.

tion of these tests). Electron temperatures were determined by the classical method shown in Fig. 7.19, from current-voltage characteristics of the electrostatic portion of the probe. Also, the random electric current (at the plasma potential) was used to calculate electron temperature in order to evaluate the postulate that $N_{e\lambda} = N_{e\infty}$ as was discussed previously.

The foregoing measurements were made at a number of axial stations along the centerline of a laminar argon jet as mentioned earlier, the jet cooling thereby providing a suitable temperature range. Results of the axial survey at 1 atm pressure are shown in Fig. 7.20, and the detailed comparison between T_H, $T_{e\infty}$ from Fig. 7.19, and $T_{e\infty}$ from random-current measurements (assuming $N_{e\lambda} = N_{e\infty}$) is shown in Table 7.3. Agreement at the higher temperatures between $T_{H\infty}$ and $T_{e\infty}$ by both methods is quite good, indicating an accuracy level for the

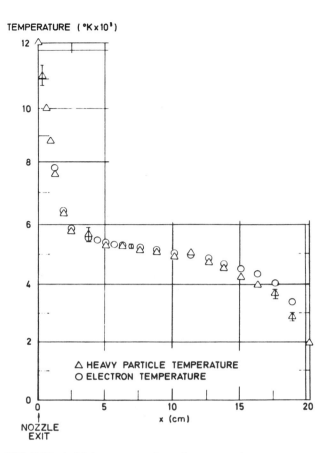

TEMPERATURE (°K x 10³)

12

10

8

6

4

2

0

△ HEAVY PARTICLE TEMPERATURE
○ ELECTRON TEMPERATURE

0   5   10   15   20

x (cm)

NOZZLE
EXIT

FIG. 7.20 Axial temperature decay in argon at 1 atm.

TABLE 7.3 Calibration of electron temperature and density measurements in argon plasma at 1 atm

Location X, cm	Electron temperature T_e Electrostatic probe, °K		Heavy particle temperature T_H Calorimetric probe, °K
	In I_e vs. V	Random current at $\varphi = 0$ (assuming $N_e \approx N_{e\infty}$)	
1.27	7,840	7,840	7,840
2.54	5,890	5,890	5,780
3.81	5,550	5,560	5,660
5.08	5,390	5,360	5,290

360

electrostatic probe which is at least as good as that of the calorimetric probe, i.e., better than 3% for $T_{e\infty}$, and thus of the order of 10% for $N_{e\infty}$.

Finally, the probe data used to prepare Fig. 7.20 and Table 7.3 were applied as a check on the saturation current ratio and floating potential estimates of the first-order (Langmuir) theory. Results are presented in Table 7.4 and indicate that the first-order approximations are reasonably valid, with no greater than about 10% error within the region studied. However, note that there is a consistent variation between experiment and theory in both of these properties, indicating the likelihood of a systematic contribution by the higher-order terms in both expressions.

Having established the validity of the cooled-probe method for electron temperature measurements, a series of tests was run at reduced pressure (1/2 atm) in order to detect any departure from equilibrium by simultaneous measurement of heavy-particle temperatures (calorimetric probe) and electron temperatures (electrostatic probe). Results are shown in Fig. 7.21, clearly indicating a divergence in the two temperatures starting at about 5000°K (9000°R) and increasing rapidly at lower temperatures, a trend predicted by Jacobs and Grey [28]. Applying the electron temperature "freezing" criterion established in [28] to these 1/2 atm tests, the first recognizable difference between electron and heavy-particle temperatures should appear at about 5500°K (9900°R), which is in reasonable agreement with the experimental observation of Fig. 7.21.

A final observation, in confirmation of those reported by Demetriades and Doughman [38, 39], was that when the electrostatic probe was driven by 60 Hz ac an excellent semiqualitative indication of turbulence was observed. That is, in laminar flows, the current-voltage characteristic of the probe, as observed on an oscilloscope screen, was reproducible. However, as soon as the probe was moved to a transition region in the subsonic jet, the current-voltage trace was slightly different on each sweep, giving the appearance on a retentive oscilloscope screen of a line-broadening effect. This broadening was observed to increase qualitatively with intensity of turbulence and is most likely to be caused by fluctuations in local electron density.

For the general conditions appropriate to these experiments, e.g., for dense

TABLE 7.4 Saturation current ratio and floating potential in argon plasma at 1 atm

T_e, °K	$(J_e/J_i)_{sat}$		$\varphi_F - \varphi_{plasma}$, V	
	Experiment	Theory	Experiment[a]	Theory
7,840	1450	1380	− 5.0	− 4.72
5,890	1300	1200	− 4.0	− 3.70
5,550	1200	1170	− 3.5	− 3.45
5,390	1100	1150	− 3.0	− 3.25
5,220	1000	1130	− 2.6	− 2.95

[a] ±1/2 V uncertainty in plasma potential.

TEMPERATURE (°K x 10³)

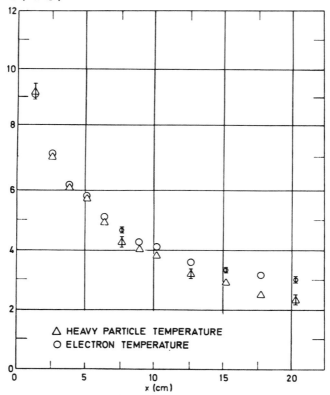

FIG. 7.21 Axial temperature decay in argon at 380 mm Hg.

arcjet environments, it was concluded that many of the factors that result in highly complicated plasma behavior near the probe become relatively insignificant, permitting the use of surprisingly simple probe trace analyses. Although these simple approximations are certainly limited in their applicability as compared with the multitude of elaborate theoretical studies prevalent in the literature, they appear to be reasonably valid over a surprisingly wide range of practical conditions.

7.1.3 Radiation Probe

Measurement of total thermal radiation from arcjets and other difficult environments has often been complicated, or even precluded, by the cool absorbing gas layer which generally surrounds the hot-gas region of interest. At plasma temperatures and pressures for which enthalpy or other probe measurements are useful, the radiated energy can often constitute a significant fraction, and its determination can be quite important. A simple water-cooled collimator (Fig. 7.22) is thus used to penetrate the cool outer regions (which can often absorb as much as 90% of the total energy radiated from a dense, partly ionized gas), and observe the hot core flow directly. This technique is described in detail in [40].

7.1.4 Cyclic Probe

An approach to the measurement of high gas temperatures using an intermittently cooled thermocouple was developed in the early sixties, but it has not seen extensive application. It nevertheless represents an attractive technique for the determination of temperatures in gases whose pressure-enthalpy product is too low for proper sensitivity of the simple tare measurement calorimetric probe, and yet at temperatures too high for thermocouples or other conventional immersion instruments.

The cyclic probe [41] consists of a cooled tube through which a thermocouple junction is alternately injected out into the hot gas stream and retracted inside the cooled tube (see Fig. 7.23). The thermocouple output is thus an oscillating voltage, which alternately increases to some never-attained high asymptotic value while in the hot gas, whose temperature is often well above the thermocouple's melting point, and decreases to a never-attained low asymptotic value while in the cooled tube. The average voltage, as is shown in [41], is a direct function of the cyclic frequency, the ratio of dwell times inside the tube and exposed to the hot gas, and the gas temperature. Since the frequency and dwell-time ratio are known, the temperature can be determined. Calibration of this device may be achieved by simply varying the frequency which, incidentally, can also be adjusted to provide maximum sensitivity without probe failure.

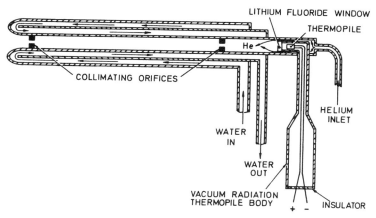

FIG. 7.22 Cooled probe used for radiation measurements inside optically inaccessible plasma regions.

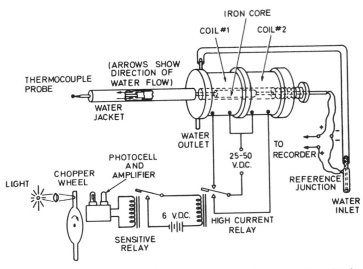

FIG. 7.23 Intermittent (cyclic) thermocouple probe and cycling circuit.

7.1.5 Transient Pressure Probe

This is a unique cooled-probe technique developed for a difficult measurement in an extremely high heat-flux environment. In the measurement of transient detonation wave characteristics inside rocket thrust chambers which are experiencing high-frequency combustion instability, it is necessary to mount a very high-frequency response transducer within the chamber (not at the wall). The only suitable transducer element (in terms of frequency response) is a piezo-electric crystal; unfortunately, however, crystals are subject to severe temperature drift, and therefore require highly effective cooling in the expected 8.2 Kw/cm^2 environment.

The solution, shown in Fig. 7.24, was to employ a cooled probe with a commercial 5.6 mm diameter crystal transducer mounted in its tip (natural frequency approximately 400,000 Hz). The transducer diaphragm was splash-cooled by a ring of tiny re-entrant jets directed at different angles so as to completely cover the transducer face. An ablative coating provided short-term protection during passage of the detonation wave (in case the resulting coolant-jet interruption would have caused damage to the transducer). This design provides effective cooling of the transducer, with no sacrifice in frequency response, and full protection against drift-producing temperature changes.

7.1.6 Heat Flux Probe

As indicated in [3], one method for deducing gas enthalpy in the high-density environment of interest is the measurement of heat transfer rate. In fact, the local heat flux itself is often an extremely important measurement for engineering design purposes. Unfortunately, however, most heat-flux gauges distort the very heat flux pattern they are attempting to measure. That is, unless the heat flux to, say, a cooled body is not distorted by the installation of a gauge, the gauge will measure the wrong heat transfer rate, even though the gauge itself may be calibrated to very high accuracy.

One recent development of considerable interest is the use of an old principle in a new form to measure high heat transfer rates without distorting the heat flux to be measured. Illustrated in Fig. 7.25, this concept utilizes the one-dimensional Fourier heat conduction equation to measure heat flux. The gauge itself consists of three thin, flat wafers selected from the many pairs of thermocouple materials. For example, the outer wafers could be copper and the inner one constantan. Electrical lead wires, insulated and sealed in copper tubing for ruggedness, are brought out of the back of the gauge. If the front of the gauge is now exposed to a high-temperature region and the back is at a lower temperature (e.g., by water cooling), the heat flux through the gauge will produce a temperature difference across the center wafer. Since the wafer thickness, which has been measured to the nearest 0.00013 mm, is much smaller than its diameter, the heat flux is very nearly one-dimensional in nature. The measured temperature drop ΔT, the known wafer thickness Δx, and the mean thermal conductivity $\bar{k}$ of the center wafer (known exactly because the maximum and minimum temperatures of the wafer are known), may then be used in the simple integrated form of the one-dimensional Fourier equation to calculate the heat flux q/A:

$$\frac{q}{A} = -\bar{k}\,\frac{\Delta T}{\Delta x}$$

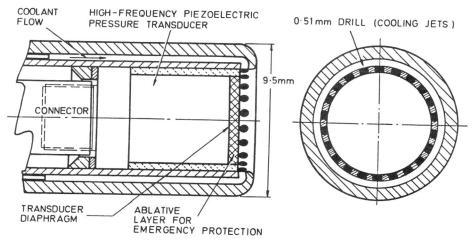

FIG. 7.24 Water-cooled high-frequency transient pressure probe.

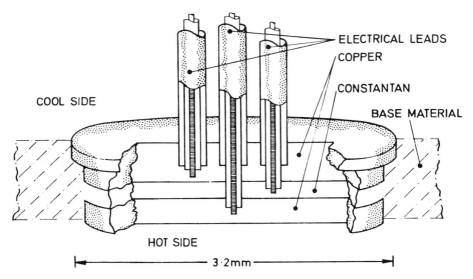

FIG. 7.25 View of cooled thin-wafer heat flux gauge showing its construction and operating principle.

This basic principle, of course, is not at all new in the field of heat flux measurement. However, this heat transfer gauge possesses several unique features which provide major improvements in accuracy over prior applications of the principle. First, relative thicknesses of the three wafers can be varied to match exactly the conductance of the base material in which the gauge is mounted. Thus, lateral heat conduction (in directions other than along the gauge axis) is minimized. For example, the copper-constantan gauge can match the conductance of any base material whose mean thermal conductivity falls between those of copper and constantan. Other base materials can be matched by using different thermocouple materials; e.g., chromel-alumel, tungsten-tungsten/rhenium, platinum-platinum/rhodium, etc.

Second, by using relatively fine electrical leads (0.25 mm maximum), and by using appropriate insulation material and thicknesses so that the overall conductance of the wire installation is identical to that of the center wafer, it is possible to reduce the distortion from true one-dimensional heat flux to very small values.

Third, the "hot-side" surface of the gauge can be coated or plated with the same material and finish as the base material, so that its radiation absorption characteristics will be identical with those of the base material. Thus, the gauge installation produces no effect on base material heat transfer due to radiation or convection in addition to conduction.

This thin-wafer heat flux gauge can be used in a stagnation-point probe (Fig. 7.26) for heat transfer rates as high as 5 Kw/cm^2 or incorporated into a cooled-skin structure such as a hyperthermal wind-tunnel model (Fig. 7.27). In either application, its unique combination of high heat flux capability, ruggedness

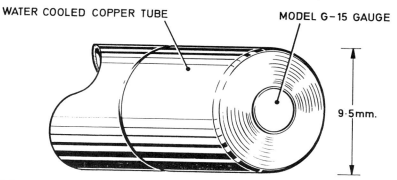

FIG. 7.26 Wafer-type heat flux gauge installed in cooled stagnation-point heat flux probe suitable for measurement up to 5 kW/cm².

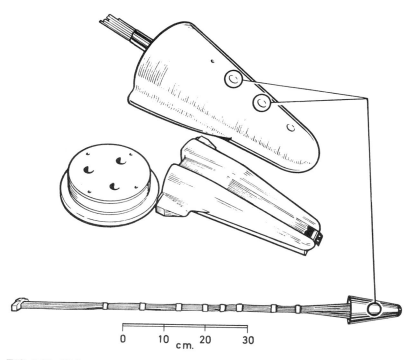

FIG. 7.27 Wafer-type heat flux gauge installed in a water cooled hypersonic wind-tunnel model. Heat transfer rates to this 9° half-angle blunted cone model range from 24 to 350 W/cm².

of construction, small diameter and thickness, and most important, virtually exact simulation of the one-dimensional heat transfer equation, make it one of the most useful techniques for heat flux measurement in any range.

7.2 CONCLUSION

In the somewhat more familiar temperature range (1000–2200K) the utilization of water or gas-cooled probe techniques (calorimetric, heat flux, etc.) for measurement and control of gas temperatures offers several advantages over the more conventional thermocouples and resistance thermometers generally utilized in this range. Most important is their life expectancy: because a cooled probe operates at maximum surface temperatures well below those at which strength and corrosion characteristics begin to deteriorate (e.g., 370–480K), a cooled calorimetric or heat flux probe will far outlast even a high temperature thermocouple, which must, by its very nature, always operate at the gas temperature to be measured. Further, for control purposes, both calorimetric and heat-flux probes sense the product of density and enthalpy (or density and temperature, for nondissociated gases), and therefore lend themselves to utilization as fundamental input-energy control-function sensors for virtually any energy-producing system (e.g., jet engines, furnaces, arcjets or other plasma generators, etc.). Moreover, cooled probes can be used even for the measurement of condensible flows by simply using a gaseous coolant (e.g., hydrogen or helium) to maintain the probe surface temperature above the boiling point of the fluid to be measured.

Finally, probe techniques lend themselves to many areas not suitable for optical methods (spectroscopic, microwave, laser, electron-beam, etc.). Disturbance of the flow to be measured, a definite disadvantage of any probe technique, is minimized by the recent advances achieved in probe miniaturization. On the plus side, they have the capability for accurate local measurement of steady-state or transient characteristics in opaque or otherwise optically inaccessible gases having high property gradients (e.g., 20,000°K/cm). When combined with the relatively low cost and complexity associated with auxiliary equipment, and the ruggedness, simplicity, and low cost of the probes themselves, this often tends to make probe measurements more palatable than the optical methods.

NOMENCLATURE

C_{ca}	arc jet coolant specific heat
C_e	speed of electron
C_{pc}	specific heat at constant pressure of coolant
d	distance from wall to edge of transition region
d_i	inside diameter of probe
d_o	outside diameter of probe
E	free stream electron energy
E_0	energy of heavy particle

EI	electrical power input to arc jet
e	elementary charge
$F(\chi)$	average steric factor
h	enthalpy, thickness of sheath region
h_{1g}	unknown gas enthalpy at probe entrance
h_{2g}	gas enthalpy at probe exit
J_e	electron current
J_I	ion current
k	Boltzmann's constant
$\bar{k}$	mean thermal conductivity
L	probe radius
$\dot{m}_a$	input mass flow rate of arc jet
$\dot{m}_{ca}$	mass flow rate of arc jet coolant
m_e	mass of electron
m_H	mass of heavy particle
$\dot{m}/A$	mass flux (kg/s $\cdot$ m^2)
N_A	atom density
N_e	electron density
n	number of collisions
q/A	heat flux (W/m^2)
Re_L	Reynolds number based on probe radius
r	radial direction
T_A	temperature of atoms
T_e	temperature of electrons
T_I	temperature of ions
$(T_c)_{\text{in}}$	coolant temperature entering probe
$(T_c)_{\text{out}}$	coolant temperature leaving probe
V	velocity
$\dot{w}_c$	coolant water mass flow rate
$\dot{w}_g$	gas sample mass flow rate

Greek Symbols

ΔE	energy lost by an electron to a heavy particle
ΔT	temperature drop across a heat flux probe
ΔT_c	coolant temperature rise
ΔT_{ca}	arc jet coolant temperature rise
$(\Delta T_c)_{\text{flow}}$	coolant temperature change with gas sample flow
$(\Delta T_c)_{\text{no flow}}$	coolant temperature change with no gas sample flow
ΔX	thickness of heat flux probe wafer
δ	boundary layer thickness
$\bar{\lambda}$	mean free path
ρ	density
σ	probe sensitivity

σ_{e-A}	collision cross section for an electron atom collision
σ_{e-I}	collision cross section for an electron ion collision
φ_F	floating potential of probe
φ_λ	effective potential of probe
χ	deflection angle

Subscripts

r	random
λ	condition one mean free path from probe
∞	free stream condition

REFERENCES

1. Finkelnburg, W., and Maecker, H., Electric Arcs and Thermal Plasma, "Handbuch der Physik," vol. 22, Springer-Verlag, Heidelberg, Germany.
2. Langmuir, I., The Pressure Effect and Other Phenomena in Gaseous Discharges, *J. Franklin Institute*, vol. 196, pp. 751–762, 1923.
3. Grey, J., Thermodynamic Methods of High-Temperature Measurement, *ISA Transactions* vol. 4, pp. 102–115, April, 1965.
4. Grey, J., Jacobs, P. F., and Sherman, M. P., Calorimetric Probe for the Measurement of Extremely High Temperatures, *Rev. Sci. Instr.*, vol. 33, pp. 738–741, July, 1962.
5. Grey, J. and Jacobs, P. F., Experiments on Turbulent Mixing in a Partially Ionized Gas, *AIAA J.*, vol. 2, pp. 433–438, March, 1964.
6. Grey, J. Sensitivity Analysis for the Calorimetric Probe, *Rev. Sci. Inst.*, vol. 34, pp. 857–859, Aug. 1963.
7. Grey, J., Sherman, M. P., Williams, P. M., and Fradkin, D. B., Laminar Arcjet Mixing and Heat Transfer: Theory and Experiments, *AIAA J.*, vol. 4, pp. 986–993, June, 1966.
8. Incropera, F. P., and Leppert, G., Investigation of Arc Jet Temperature-Measurement Techniques, *ISA Transactions*, vol. 6, pp. 35–41, Jan. 1967.
9. Au, G. G., and Sprengel, U., Kalorimetrische Messungen von Ortlichen Temperaturen and Geschwindigkeiten in einem Stickstoff-Plasmastrahl, *Zeitschrift fur Flugwissenschaften*, vol. 14, pp. 188–194, April, 1966.
10. Williams, P. M. and Grey, J. Simulation of Gaseous Core Nuclear Rocket Characteristics Using Cold and Arc Heated Flows, *NASA Contractor Report No. CR-690*, June, 1967.
11. Grey, J. Cooled Probe Diagnostics of Dense Plasma Mixing and Heat Transfer Processes, *Am. Inst. Chem. Eng. Preprint 9C*, Nov. 26–30, 1967.
12. Massier, P. F., Back, L. H., and Roschke, E. J., Heat Transfer and Laminar Boundary Layer Distributions to an Internal Subsonic Gas Stream at Temperatures up to 13,900°R, *NASA Contract NAS-7-100, Jet Prop. Lab.*, 1968.
13. O'Connor, T. J., Comfort, E. G., and Cass, L. A., Turbulent Mixing of an Axisymmetric Jet of Partially Dissociated Nitrogen with Ambient Air, *AIAA J.*, vol. 4, pp. 2026–2032, Nov. 1966.
14. Anderson, L. A., and Sheldahl, R. E., Flow-Swallowing Enthalpy Probes in Low-Density Plasma Streams, *AIAA Paper No. 68-390*, April 8–10, 1968.
15. Folck, J. L., and Heck, R. R., Operational Experiences and Preliminary Results of Total Enthalpy Probe Measurements in the AFFDL 50-Megawatt Electrogasdynamics Facility, *USAF Report No. FDM-TM-68-2*, April, 1968.

16. Huber, F. J. A., Probes for Measuring Mass Flux, Stagnation Point Heating and Total Enthalpy of High-Temperature Hypersonic Flows, *AIAA Preprint No. 66-750*, Sept. 21-23, 1966.

17. Halbach, C. R. and Freeman, L., The Enthalpy Sensor—A High Gas Temperature Measuring Probe, *Report MR 20, 331, The Marquardt Corporation*, June, 1965.

18. Haas, F. C., and Vassallo, F. A., Measurement of Stagnation Enthalpy in a High Energy Gas Stream, *Chem. Eng. Prog. Symposium Series 41, vo. 59, AIChE*, 1963.

19. Malliaris, A. C. et al., Optical and Radar Observables of Ablative Materials, *AFML TR 66-331, Pt. 1, Air Force Systems Command*, Oct. 1966.

20. Grey, J. Enthalpy Probes for Arc Plasmas—First Status Review, *Report to Committee E-21, ASTM*, April 12, 1966.

21. Esker, D. W., A Probe for Total-Enthalpy Measurements in Arcjet Exhausts, *AIAA J.*, vol. 5, pp. 1504-1506, Aug. 1967.

22. Grey, J. Enthalpy Probes for Arc Plasmas—Third Status Review, *Report to Committee E-21, ASTM*, May 22, 1968.

23. Haas, F. C., An Evaporating Film Calorimetric Enthalpy Probe, *Report No. AD-1651-Y-1, Cornell Aeronautical Lab.*, Feb. 1963.

24. Cheng, D. Y., and Blackshear, P. L., Factors Influencing the Performance of a Fast-Response, Transpiration-Cooled, High-Temperature Probe, *AIAA Paper No. 65-359*, July 26-29, 1965.

25. Schneider, P. J., and Maurer, R. E., Coolant Starvation in a Transpiration-Cooled Hemispherical Shell, *J. Spacecraft and Rockets*, vol. 5, pp. 751-752, June, 1968.

26. Grey, J. and Jacobs, P. F., Cooled Electrostatic Probe, *AIAA J.*, vol. 5, pp. 84-90, Jan. 1967.

27. Jacobs, P. F., and Grey, J., Electron-heavy Particle Nonequilibrium in a Dense Argon Plasma, *Report No. ARL 66-0143, OAR, USAF*, July, 1966.

28. Jacobs, P. F., and Grey, J., A Criterion for Electron-heavy–Particle Nonequilibrium in a Partly Ionized Gas, *AIAA Preprint 66-192*, March 2-4, 1966.

29. Bohm, D., The Characteristics of Electrical Discharges in Magnetic Fields, McGraw-Hill, New York, 1949.

30. Cohen, I. M., Asymptotic Theory of Spherical Electrostatic Probes in a Slightly-Ionized, Collision-Dominated Gas, *Phys. Fluids*, vol. 6, pp. 1492-1499, 1963.

31. Waymouth, J. F., Perturbation of a Plasma by a Probe, *Phys. Fluids*, vol. 7, pp. 1843-1854, 1964.

32. Schlichting, H., Boundary Layer Theory, Pergamon Press, New York, 1955.

33. Hall, L. S., Probes and Magnetic Pumping in Plasma, *Lawrence Radiation Lab., UCRL 6535*, July 1961.

34. Talbot, L., Theory of the Stagnation-Point Langmuir Probe, *Phys. Fluids*, vol. 3, pp. 289-297, 1960.

35. Kumura, I., and Kanzawa, A., Experiments on Heat Transfer to Wires in a Partially Ionized Argon Plasma, *AIAA J.*, vol. 3, pp. 476-481, 1965.

36. Pytte, A., and Williams, A. R., On Electrical Conduction in a Nonuniform Helium Plasma, *USAF Aeronautical Research Lab. Report 63-166*, 1963.

37. Brundin, C. L., and Talbot, L., The Application of Langmuir Probes to Flowing Ionized Gases, *AGARD Report 478*, Sept., 1964.

38. Demetriades, A., and Doughman, E. L., Langmuir-probe–Measurement Method for Turbulent Plasmas, *Phys. Fluids*, vol. 8, pp. 1001-1002, 1965.

39. Demetriades, A., and Doughman, E. L., Langmuir Probe Diagnosis of Turbulent Plasmas, *AIAA J.*, vol. 4, pp. 451-459, 1966.

40. Grey, J., Sherman, M. P., and Jacobs, P. F., A Collimated Total Radiation Probe for Arcjet Measurements, *IEEE Transactions on Nuclear Science*, vol. NS-11, pp. 176-186, Jan. 1964.

41. Raezer, S. D., and Olsen, H. L., Temperature–Its Measurement and Control in Science and Industry, vol. 3, pt. 2, paper no. 83, Reinhold, New York, 1962.

BIBLIOGRAPHY

Blackshear, P. J., Jr., Sonic Flow Orifice Temperature Probe for High Gas Temperature Measurements, *NACA TN 2167*, Sept., 1950.

Blackshear, P. L., and Dorman, F. D., Heat-Sensing Probe and Process (Diluent Probe), U.S. Patent No. 3,296,865, Jan. 10, 1967.

Blythe, P. A., Nonequilibrium Flow Through a Nozzle, *J. Fluid Mech,* vol. 17, p. 126, 1963.

Boatright, W. V., Sebacher, D. I., and Guy, R. W., Review of Testing Techniques and Flow Calibration Results for Hypersonic Arc Tunnels, *AIAA Paper No. 68-379,* April 8–10, 1968.

Bryon, S., and Spongberg, R. M., Gasdynamic Instrumentation of High Enthalpy Flows, *IEEE Transactions on Nuclear Science,* vol. NS-11, pp. 381–387, 1964.

Carden, W. H., Heat Transfer in Nonequilibrium Dissociated Hypersonic Flow with Surface Catalysis and Second-Order Effects, *AIAA J.,* vol. 4, no. 10, pp. 1704–1711, 1966.

Christensen, D., and Buhler, R. D., Arcjet Tunnel Development and Calibration for Parabolic Reentry Simulation, *Final Summary Report IFR011-1872, Plasmadyne Corp.,* June 1961.

Cookson, T. S., Dunham, P. G., and Kilham, J. K., Stagnation Point Heat Flow Meter, *J. Sci. Inst.,* vol. 42, pp. 260–262, April, 1965.

Cordero, J., Diederich, F. W., and Hurwicz, H. Aerothermodynamic Test Techniques for Reentry Structures and Materials, *Aerospace Eng.,* vol. 22, no. 1, pp. 166–191, Jan., 1963.

Crites, R. C., and Cysz, P., Inlet and Test Section Diagnostics Using a Miniature Mass Flow Probe in Hypersonic Impulse Tunnel, *AIAA Paper No. 68-398,* April 8–10, 1968.

Edmonson, R. B., Thompson, W. R., and Hines, A. L., Thermodynamic Temperature Probe, *American Rocket Society Preprint No. 1431-60,* December 5–8, 1960.

Fay, J. A., and Kemp, N. H., Theory of Stagnation Point Heat Transfer in a Partially Ionized Diatomic Gas, presented at *IAS Annual Meeting, New York,* Jan. 21–23, 1963.

Fay, J. A., and Riddell, F. R., Theory of Stagnation-Point Heat Transfer in Dissociated Air, *J. Aero/Space Sciences,* vol. 25, no. 2, Feb., 1958.

Fingerson, L. M., Research on the Development and Evaluation of a Two-Sensor Enthalpy Probe, *Report No. ARL 64-161, USAF,* Oct., 1964.

Fontijn, A., Rosner, D. E., and Kurzius, S. C., Chemical Scavenger Probe Studies of Atom and Excited Molecule Reactivity in Active Nitrogen from a Supersonic Stream, *Can. J. Chem.,* vol. 42, pp. 2440–2450, 1964.

Freeman, M. P., "Temperature–Its Measurement and Control in Science and Industry," vol. 3, part 2, Plasma Jet Diagnosis Utilizing the Ablating Probe, Reinhold, New York, 1962.

Fruchtman, I., Temperature Measurement of Hot Gas Streams, *AIAA J.,* vol. 1, no. 8, pp. 1909–1910, August, 1963.

Grey, J., Enthalpy Probes for Arc Plasmas–Second Status Review, *Prelim. Report to ASTM, Toronto,* May 3, 1967.

Incropera, F. P., Temperature Measurement and Internal Flow Heat Transfer Analysis for Partially Ionized Argon, *Tech. Report No. SU 247-11, Dept. of Mech. Eng., Stanford Univ.,* Aug. 1966.

Johnson, D. H., Nonequilibrium Electron Temperature Measurements in a Supersonic Arc Jet Using a Cooled Langmuir Probe, *26th Supersonic Tunnel Assoc. Meeting, Ames Research Center,* May 16–18, 1967.

Kilburg, R. F., A High Response Probe for Measurement of Total Temperature and Total Pressure Profiles Through a Turbulent Boundary Layer with High Heat Transfer in Supersonic Flow, *AIAA Paper No. 68-374,* April 8–10, 1968.

Krause, L. N., Buchele, D. R., and Warshawsky, I., Measurement Technique for Hypersonic Propulsion, *NASA TM X-52299*, May 16-18, 1967.

Krause, L. N., Glawe, G. E., and Johnson, R. C., "Temperature–Its Measurement and Control in Science and Industry," vol. 3, part 2, paper no. 54, Reinhold, New York, 1962.

Kubanek, G. R., and Gauvin, W. H., Plasma Jet Research Facility for Solids–Gas Heat Transfer Studies, *Tech. Report No. 466, Pulp and Paper Research Inst. of Canada*, 1966.

Lam, S. H., A General Theory for the Flow of Weakly Ionized Gases, *AIAA J.*, vol. 2, pp. 256-262, Feb., 1964.

McCroskey, W. J., Density and Velocity Measurements in High Speed Flows, *AIAA Paper No. 68-392*, April 8-10, 1968.

Massey, H. S. W., and Burhop, E. H. S. "Electronic and Ionic Impact Phenomena," Oxford Univ. Press, 1951.

Meyer, N., Investigation of Greyrad Calorimetric Probe, BSE Thesis, Coll. of Eng., Univ. of Cincinnati, 1966.

Moore, D. W., Jr., A Pneumatic Method for Measuring High-Temperature Gases, *Aeronaut. Eng. Rev.*, vol. 7, no. 5, pp. 30-34, May 1948.

Petschek, H., and Bryon, S., Approach to Equilibrium Ionization Behind Strong Shock Waves in Argon, *Ann. Phys.*, vol. 1, p. 270, 1957.

Potter, J. L., Arney, G. D., Kinslow, M., and Carden, W. H., Gasdynamic Diagnosis of High-Speed Flows Expanded from Plasma States, *IEEE Transactions on Nuclear Science*, vol. NS-11, pp. 145-157, Jan., 1964.

Rosner, D. E., Similitude Treatment of Hypersonic Stagnation Heat Transfer, *Am. Rocket Society J.*, vol. 29, no. 2, pp. 215-216, Feb. 1959.

Rosner, D. E., On the Effects of Diffusion and Chemical Reaction in Convective Heat Transfer, *Report No. TM-13, AeroChem Research Lab.*, June 8, 1959.

Rosner, D. E., Application of Heat Flux Potentials to the Calculation of Convective Heat Transfer in Chemically Reacting Gases, *Report No. TP-20, AeroChem Research Lab*, Dec. 14, 1960.

Rosner, D. E., Catalytic Probes for the Determination of Atom Concentrations in High Speed Gas Streams, *ARS J.*, vol. 32, pp. 1065-1073, July, 1962.

Rosner, D. E., Diffusion and Chemical Surface Catalysis in a Low-Temperature Plasmajet, *J. Heat Transfer*, pp. 386-394, Nov. 1962.

Rosner, D. E., Sensitivity of a Downstream Langmuir Probe to Rocket Motor Chamber Conditions, *Report No. TP-109, AeroChem Research Lab*, Jan. 1965.

Rosner, D. E., Fontijn, A., and Kurzius, S. C., Chemical Scavenger Probes in Nonequilibrium Gasdynamics, *AIAA J.*, vol. 2, p. 779, 1964.

Sherman, M. P., and Grey, J., Calculation of Transport Properties for Mixtures of Helium and Partly-Ionized Argon, *Princeton Univ. Aeronautical Eng. Lab. Report No. 673*, Dec. 1963.

Sherman, M. P., and Grey, J., Interactions Between a Partly-Ionized Laminar Subsonic Jet and a Cool Stagnant Gas, *Princeton Univ. Aeronautical Eng. Lab. Report No. 707*, Sept. 1964.

Simmons, F. S., and Glawe, G. E., Theory and Design of a Pneumatic Temperature Probe and Experimental Results Obtained in a High-Temperature Gas Stream, *NACA TN 3893*, Jan. 1957.

Softley, E. J., Use of a Pulse Heated Fine Wire Probe for the Measurement of Total Temperature in Shock Driven Facilities, *AIAA Paper No. 68-393*, April 8-10, 1968.

Sprengel, U., Kalorimetrische Messunger von Ortlichen Temperaturen in einem Stickstoff-Plasmastrahl, *Raumfahrtforschung und -Technik, Beilage zur Atompraxis*, Jan. 1966.

Staats, G. E., McGregor, W. K., and Frolich, J. P., Magnetogasynamic Experiments Conducted in a Supersonic Plasma Arc Tunnel, *AEDC TR-67-166*, Feb. 1968.

Su, C. H., and Lam, S. H., Continuous Theory of Spherical Electrostatic Probes, *Phys. Fluids*, vol. 6, pp. 1479–1491, Oct. 1963.

Thirty Kilowatt Plasmajet Rocket Development—Third Year Development Program, *NASA CR-54079*, 2 July, 1964.

Vassallo, F. A., Miniature Enthalpy Probes for High-Temperature Gas Streams, *Report No. ARL 66-0115, USAF*, June 1966.

Vassallo, F. A., A Fast Acting Miniature Enthalpy Probe, *AIAA Paper No. 68-391*, April 8–10, 1968.

Wethern, R. J., Method of Analyzing Laminar Air Arc-Tunnel Heat Transfer Data, *AIAA J.*, vol. 1, no. 7, pp. 1665–1666, July 1963.

Wildhack, W. A., A Versatile Pneumatic Instrument Based on Critical Flow, *Rev. Sci. Inst.*, vol. 21, no. 1, pp. 25–30, Jan. 1950.

8 Transient experimental techniques for surface heat flux rates

C. J. SCOTT
University of Minnesota, Minneapolis

In order to calculate the thermal performance of bodies undergoing temperature changes it is necessary to accurately know the boundary conditions. An energy balance equates the energy increase within a body to the heat added at the boundaries so that knowledge of the surface heat flux (q_w) is of importance. The specification of q_w is generally complicated by the fact that a convective or radiative heat flux is a function of the surface temperature.

Suppose a surface is gaining heat Q_w by a combination of convective, conductive, and radiative processes, i.e., $Q_w = Q_{conv} + Q_{cond} + Q_{rad}$. Q_w is related to the wall temperature by Newton's law of cooling $d^2 Q_w = h(T_f - T_w) dA_w d\tau$ where h is the heat transfer coefficient, T_f is the ambient fluid temperature, A_w is the surface area, and τ is the time. At the wall of a solid the convective velocity is zero, while the radiative transmissivity is close to zero, such that within the body the energy transport is by conduction only. Fourier's hypothesis governs the internal heat conduction and is written

$$d^2 Q_w = -k \left. \frac{\partial T}{\partial n} \right]_w dA_w d\tau$$

Continuity of heat flow at the surface yields

$$h(T_f - T_w) = -k \left. \frac{\partial T}{\partial n} \right]_w$$

The quantity h is sometimes referred to as the specific surface conductance while the value $1/h = r_s$ is called the specific surface thermal resistance. For combined exchange processes the sum of the individual resistances is the thermal resistance of the overall heat transfer process.

In a transient heating process the heat gained by a body is determined by the resistance to the flow of heat at the surface as well as the internal resistance

of the body itself. Whenever the internal conduction is of importance, a dimensionless conduction Nusselt number, given by $\mathrm{Nu_{cond}} = L/k_w = \mathrm{Bi}$ (usually
referred to as the Biot number), enters as a parameter. The Biot number may be
thought of as the ratio of internal r_i to surface r_s specific resistances since

$$\mathrm{Bi} = \frac{hL}{k_w} = \frac{L/k_w}{1/h} = \frac{r_i}{r_s} \tag{8.1}$$

It is of primary importance to estimate the value of the Biot number in a
given heat flux transducer design, in order to justify any assumptions used to
simplify the design. One limiting case is represented by $\mathrm{Bi} = 0$ which implies
zero internal resistance. This condition is approached in practice by minimizing
the calorimeter thickness L and maximizing the calorimeter thermal conductivity
k_w. The other extreme, $\mathrm{Bi} \to \infty$, implies a zero surface resistance. Since the
purpose of this chapter is to discuss the possibilities for experimentally determining the surface resistance, the limiting case of zero surface resistance will not be
considered at length. Of course, exact solutions exist for the case of transient,
one-dimensional heat conduction in bodies of basic geometric shape [1]. Graphs
of solutions that assume constant material density, specific heat, and thermal
conductivity are available [2] and may be used for reference when examining the
limiting conditions.

8.1 NEGLIGIBLE INTERNAL RESISTANCE
$(r_i = 0)$

When a "thin" body is constructed of material possessing a "large" thermal
conductivity, its internal resistance may be ignored such that the overall heat
transfer process is controlled by the surface resistance. No temperature differences can exist under these conditions—the body is isothermal and $T = T(\tau) =
T_w(\tau)$. Consider a body having an initial temperature T_i $(\tau = 0)$ that is suddenly
exposed to a fluid at constant temperature T_f (or the fluid temperature undergoes a step change at $\tau = 0$ from T_i to T_f). The specific rate of energy transfer
from the surroundings for $\tau > 0$ is governed by

$$q_w = \frac{Q_w}{A_w} = \frac{d^2 Q_w}{dA_w d\tau} = \rho_w V_w C_{pw} \frac{dT_w}{d\tau} + q_L \tag{8.2}$$

where ρ_w, V_w, C_{pw}, and A_w are the density, volume, specific heat capacity, and
area of the wall material, and q_L is the rate of heat gain/loss from the
calorimeter due to internal heat sources or sinks, internal conduction, and
backside conduction, convection and radiation.

In order to discuss a variety of calorimeter applications the concept of an
ideal calorimeter is introduced. An ideal calorimeter is constructed of an infinite
thermal conductivity material. In addition, the parameters ρ_w, C_{pw}, V_w, and A_w
are temperature independent and the loss term, q_L, is zero.

8.1.1 Convection Boundary Condition

In convective heating, the basic energy balance is

$$q_w = \rho_w C_{pw} V_w \frac{dT_w}{d\tau} = h(T_f - T_w) \tag{8.3}$$

Equation (8.3) illustrates a commonly used transient method for determining heat flux rates by experimentally measuring the temperature–time slope. Normally the wall temperature is measured as a function of time and the data differentiated, an inherently inaccurate technique. The heat transfer coefficient h is determined using the right side of Eq. (8.3). Since the convective heat flux is transmitted by wall heat conduction, an energy balance at the wall surface yields $h(T_f - T_w) = - k(\partial/\partial x)\,[T(0, \tau)] -$ or, after normalizing with $\mathrm{Bi} = hL/k$, $\bar{T} = (T - T_i)/(T_f - T_i)$, $\mathrm{Fo} = \alpha\tau/L^2$,

$$\mathrm{Bi}\,[1 - \bar{T}(0, \mathrm{Fo})] = - \frac{\partial}{\partial(x/L)}\,[\bar{T}(0, \mathrm{Fo})] \tag{8.4}$$

If h is not a function of temperature, Eq. (8.3) may be integrated to yield

$$\frac{T - T_f}{T_i - T_f} = e - \left(\frac{hA_w}{\rho_w C_{pw} V_w}\right)\tau = \exp - \frac{\tau}{\theta} \tag{8.5}$$

The constant $(\rho_w C_{pw} V_w/hA_w)$ has the units of time and may be considered as a time constant θ. Once the time constant is obtained, the heat flux may be obtained from Eq. (8.3). For a uniform slab, $(V_w/A_w) = L$, the spatially uniform temperature of the object changes exponentially with time. Figure 8.1 illustrates theoretical convective heating temperature–time curves of an ideal calorimeter based on an $L = 3.18$ mm thick copper slab suddenly exposed to a fluid temperature of $T_f = 3610\mathrm{K}$ and several values of h. Note the difficulty in determining the different slopes $(dT_w/d\tau)$ for large values of h. Since copper melts at 1360K, the times for $T_w = 1360\mathrm{K}$ are maximum operating times for the calorimeter. Copper calorimeters actually deviate from these curves at temperature above 700K because of the variations of C_{pw} with T.

The heat transfer coefficient is obtained explicitly by taking the natural logarithm of both sides of Eq. (8.5):

$$\left(\frac{h}{\rho_w C_{pw} L}\right)\tau = \frac{\tau}{\theta} = \ln\left(\frac{T - T_f}{T_i - T_f}\right) \tag{8.6}$$

It is convenient to plot $(T - T_f)/(T_i - T_f)$ vs. τ on semilog paper. The slope of this curve (a straight line) is the reciprocal time constant θ^{-1} and it is possible to use arbitrary initial conditions. Writing Eq. (8.6) at two separate locations, i.e., (T_{w1}, τ_1), (T_{w2}, τ_2), yields

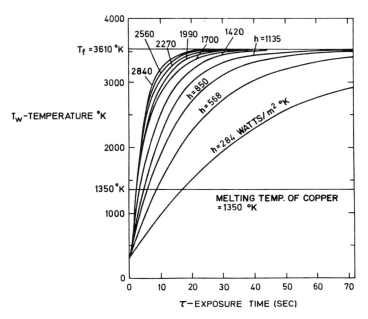

FIG. 8.1 Theoretical convective heat transfer performance of an ideal calorimeter based on a 3.18 mm copper slug.

$$\theta = (\tau_2 - \tau_1) \ln\left(\frac{Tw_1 - T_f}{Tw_2 - T_f}\right) \tag{8.7}$$

8.1.2 Radiation

If convection is not present, or the convective component is small, the following equation is applicable:

$$q_w = \frac{d^2 Q_w}{d\tau dA_w}$$

$$= F_{w-s} \epsilon_{w-s} \sigma(T_s^4 - T_w^4) = \rho_w C_{pw} L \frac{dT_w}{d\tau} = -k \frac{\partial}{\partial x} [T(0,\tau)] \tag{8.8}$$

Consider again an ideal calorimeter that has no heat losses, temperature independent parameters ρ_w, C_{pw}, t_w, and an emissivity ϵ_w that is independent of temperature. Assume that the radiation source is a conical cavity whose emissivity is unity (blackbody). If the calorimeter is aligned with the mouth of the cavity, centered and normal to the axis of the cone, the shape factor F_{w-s} is unity, and the interchange factor ϵ_{w-s} is the product of the emissivity of the calorimeter surface and the emissivity of the conical source. According to Eq. (8.8) the experimental determination of q_w requires records of the rate of change of calorimeter temperature. Normalizing Eq. (8.8) yields a characteristic radiation equation, with $\bar{T} = T/T_0$

$$M(\bar{T}_s^4 - \bar{T}_w^4) = -\frac{\partial}{\partial(x/L)} [\bar{T}(0, \mathrm{Fo})]_w \tag{8.9}$$

The parameter $M = \sigma F_{w-s}\epsilon_{w-s}T_0^3L/k$ is analogous to the Biot number for the convection boundary condition. Figure 8.2 presents a typical surface response curve of a plate of thickness L with insulated back face after sudden exposure to a constant temperature radiation source. For constant properties Eq. (8.8) integrates to, [4]

$$\left(\frac{2\sigma\epsilon_w T_s^3}{\rho_w C_{pw} L}\right)\tau = \tan^{-1}\left(\frac{T}{T_s}\right) - \tan^{-1}\left(\frac{T_i}{T_s}\right)$$
$$+ \frac{1}{2}\ln\left[\frac{1 + (T/T_s)}{1 - (T/T_s)}\right] - \frac{1}{2}\ln\left[\frac{1 + (T_i/T_s)}{1 - (T/T_s)}\right] \tag{8.10}$$

The reciprocal of the quantity in the brackets on the left side of Eq. (8.10) may be considered as a radiation time constant and used to determine ϵ_w. Figure 8.3 presents theoretical radiant heating temperature–time curves for $t_w = 3.18$ mm

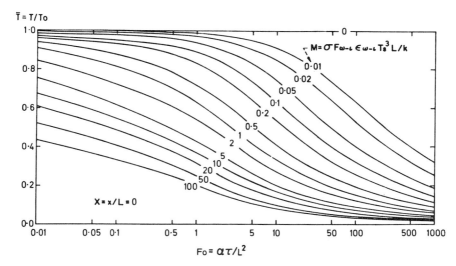

FIG. 8.2 Temperature response of a plate $0 \leqslant x \leqslant L$ with insulated face $X = L$ after sudden exposure to a constant-temperature radiation heat source T_s at $x = 0$.

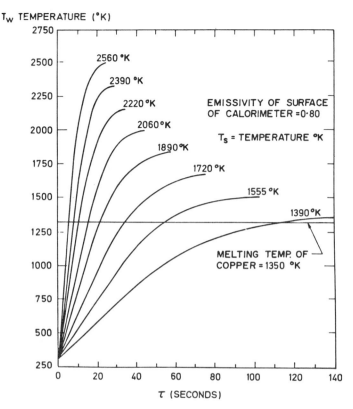

FIG. 8.3 Theoretical thermal history of an ideal calorimeter exposed to a black body heat source based on a 3.18 mm copper slug.

thick copper calorimeter with an emissivity $\epsilon_w = 0.8$ viewing several radiation source temperatures T_s.

8.1.3 Convection with Variable Fluid Temperature

For several special cases involving variable fluid temperatures, Eq. (8.3) may be rewritten

$$\frac{dT_w}{d\tau} + \frac{1}{\theta}\, T_w = \frac{1}{\theta}\, T_f(\tau) \tag{8.11}$$

where T_f is now considered to be a function of time τ.

Integrating Eq. (8.11) yields

$$\theta = \frac{\int_0^\tau T_f \, d\tau - \int_0^\tau T_w \, d\tau}{T_w(\tau) - T_i} \tag{8.12}$$

The integrals in Eq. (8.12) may be evaluated numerically from the measured traces of $T_f(\tau)$ and $T_w(\tau)$; otherwise the solution of Eq. (8.11) is given in terms of an integrating factor, [3]:

$$T_w = \exp\left(-\int \frac{d\tau}{\theta}\right)\left[\int \frac{T_f}{\theta} \exp\left(+\int \frac{d\tau}{\theta}\right) dr\right] + \text{const} \exp\left(-\int \frac{d\tau}{\theta}\right) \tag{8.13}$$

If the time constant $\theta = \rho_w C_{pw} L/h$ is constant and the initial wall temperature $T_w(0) = T_i$, general solution of Eq. (8.11) is

$$T_w = \exp\left(-\frac{\tau}{\theta}\right)\left[\frac{1}{\theta} \int_0^\tau T_f \exp\left(+\frac{\tau}{\theta}\right) d\tau + T_{wi}\right] \tag{8.14}$$

When the fluid temperature varies linearly such that

$$T_f = T_{f_0} + \frac{dT_f}{d\tau} \tau \tag{8.15}$$

and Eq. (8.14) integrates [5] to

$$T_w = T_{f_0} - \theta \frac{dT_f}{d\tau}\left[1 - \exp\left(-\frac{\tau}{\theta}\right)\right] \tag{8.16}$$

the thermal response is exponential and the methods for the explicit determination of θ given for Case A apply. From Eq. (8.16) it is clear that the wall temperature always lags behind the fluid temperature. After the initial transient $(e^{-\tau}/\theta \rightarrow 0)$ the lag becomes $\theta(dT_f/d\tau)$, and this feature may be used to determine θ. Figure 8.4 illustrates the response predicted by Eq. (8.16).

If the fluid temperature is oscillated sinusoidally with a frequency ω and an amplitude ΔT_f such that $T_f = \bar{T}_f + \Delta T_f \sin \omega\tau$, Eq. (8.14) integrates [6] to give, with $\rho_w C_{pw} V_w/hA_w = \theta$

$$\frac{T_w - \bar{T}_f + \dfrac{\Delta T_f}{\sqrt{1 + \omega^2\theta^2}} \{\sin [\tan^{-1}(\omega\theta) - \omega\tau]\}}{T_{w_i} - \bar{T}_f + \dfrac{\Delta T_f}{\sqrt{1 + \omega^2\theta^2}} [\sin \tan^{-1}(\omega\theta)]} = \exp - \frac{\tau}{\theta} \tag{8.17}$$

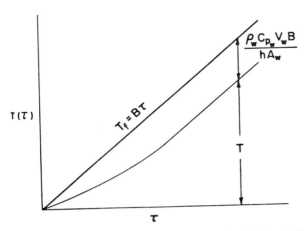

FIG. 8.4 Time temperature history of a plate $0 \leqslant x \leqslant L$ subjected to a ramp fluid temperature.

For large times $e^{-\tau/\theta} \to 0$ and we obtain the steady periodic solution

$$T_w = \bar{T}_f + \frac{\Delta T_f}{\sqrt{1 + \omega^2 \theta^2}} \left[\sin \left(\omega \tau - \tan^{-1} \omega \theta \right) \right] \tag{8.18}$$

As the gas temperature is oscillated the wall temperature lags the gas temperature by an angle $\tan^{-1} (\omega\theta)$ (see Fig. 8.5). This phase lag angle is the product of the forcing frequency ω and the time constant θ. The heat transfer coefficient is obtained by measuring the phase lag from simultaneous measure-

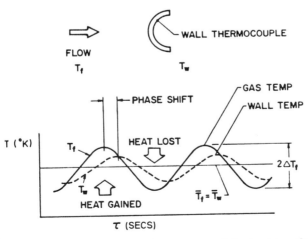

FIG. 8.5 Illustration of the measurement of the heat transfer coefficient by means of sinusoidal temperature oscillations.

ments of the gas temperature and wall temperature. The experiments are conducted with small temperature oscillations ($\pm 6°<$), generating essentially isothermal conditions. The method avoids the inherently inaccurate determination of a temperature–time slope, avoids the requirements of inducing a sharp step change in the gas temperature, and yields data with nearly uniform surface temperatures.

8.2 NEGLIGIBLE SURFACE RESISTANCE ($r_s = 0$)

A class of problems exist in which the surface thermal resistance is negligible and the internal thermal resistance dominates the problem. If the ambient temperature of the surrounding fluid is $T_f(\tau)$ we take the temperature of the surface of the object to be $T_f(\tau)$ for $\tau > 0$.

In the case of $T_f = $ constant, $T_i = T_i(x)$, consider an infinite slab of thickness t_w with the initial temperature distribution $T_i(x)$. The temperature of the surface $T(0, \tau)$ is suddenly changed to $T_f = $ constant for all $\tau > 0$. The solution must satisfy the system

$$\frac{\partial^2 T}{\partial x^2} = \frac{\rho_w C_{pw}}{k_w} \frac{\partial T}{\partial \tau} \tag{8.19}$$

$$T = T_i(x) \quad \text{at} \quad \tau = 0 \quad 0 \leqslant x \leqslant T_w$$

$$T(0, \tau) = T_f = \text{const} \quad \text{at} \quad x = 0 \quad \tau > 0 \tag{8.20}$$

$$\frac{\partial T}{\partial x}(L, T) = 0 \quad \text{at} \quad x = L \quad \tau > 0$$

For the special case of a uniform initial temperature $T_i(x) = T_i$ the general solution of the system (Eqs. [8.19] and [8.20] is presented in [1] and plotted in Fig. 8.6, which illustrates the use of Schneider's time–temperature charts.

8.3 EXACT SOLUTIONS AND THE TIME–TEMPERATURE CHARTS (FINITE INTERNAL AND SURFACE RESISTANCE)

The most realistic heat flux measurements require consideration of both internal and surface resistances. In this discussion we consider the ambient fluid temperature T_f to be uniform, as is the heat transfer coefficient. Consider the convective heating of a large plate of uniform thickness L which again is initially at a uniform temperature T_i. The plate is suddenly exposed to a fluid temperature T_f for $\tau > 0$ while the back face at $x = L$ is insulated. The general solution is, [7]:

$$\frac{T - T_f}{T_i - T_f} = 4 \sum_{n=1}^{\infty} \left(\frac{\sin M_n}{2M_n + \sin 2M_n} \right) \exp\left(-M_n^2 \text{Fo}\right) \cos 2M_n \left(1 - \frac{x}{L}\right) \tag{8.21}$$

$(T-T_i)/(T_f-T_i)$

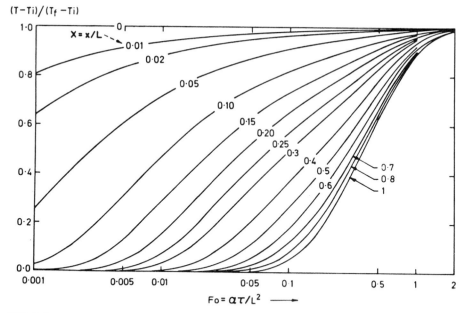

FIG. 8.6 Temperature response of a plate $0 \leqslant x \leqslant L$ with insulated back face $x = L$ after sudden change in external fluid temperature from T_i when $\tau < 0$ T_F when $\tau \geqslant 0$.

where M_n are the eigenvalues given by the characteristic equation

$$M_n \tan M_n = \text{Bi} \qquad (8.22)$$

Note that the solution is a function of two parameters, Fo and Bi, rather than just one as in the case of negligible surface resistance.

Schneider [7] presents a large number (120) of time-temperature charts covering a variety of body shapes for a wide range of Fourier and Biot numbers. These charts are convenient for thermal engineers who need to deduce heat flux rates from transient temperature response measurements while not having sufficient computational experience or direct use of a computer.

Figures 8.7 and 8.8 present the plotted results of Eqs. (8.21) and (8.22) for the locations $x/L = 0$, 1.0, and these curves apply to the case of large Biot and Fourier numbers. To use the curves, one can plot the experimental temperature difference ratio vs. the Fourier number and examine the theoretical Biot number curve which best fits the data. One must remember that to obtain surface heat flux values in this way, accurate values of the thermal conductivity, density, heat capacity, and thickness must be available.

The temperature difference that exists across the thickness of the surface may be obtained by a superposition of Figs. 8.7 and 8.8, and in this way an

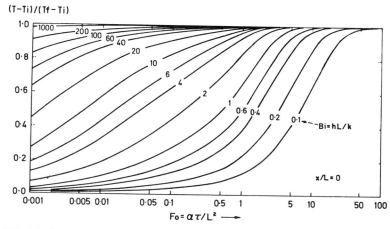

FIG. 8.7 Temperature response of a plate $0 \leqslant x \leqslant L$ with insulated back face $x = L$ after sudden exposure to a uniform convective environment at $x = 0$.

estimate of the errors involved in assuming negligible internal resistance are readily made.

8.4 RAPID RESPONSE MEASUREMENTS

The development of heat transfer gauges has always been directed to some particular application. Earlier work focused on determination of cylinder wall temperature variations in reciprocating engines and the surface temperature history in gun barrels subjected to continuous firing. Later it was necessary to

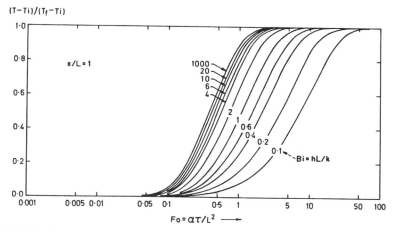

FIG. 8.8 Temperature response of a plate $0 \leqslant x \leqslant L$ with insulated back face $x = L$ after sudden exposure to a uniform convective environment at $x = 0$.

develop instrumentation to determine heat flux rates to aerodynamic models in shock tubes and shock tunnels with 10 microsecond test duration. An early form of a shock wave detector consisted of a thin film mounted flush in a shock tube wall. The gauge provided a very fast electrical pulse for trigger purposes to indicate the passage of the shock wave.

Two basic techniques have evolved for such measurement techniques [8]: thin-film surface thermometry and thick-film calorimetry. The first method records instantaneous surface temperature from which instantaneous heat flux rates are deduced using classical heat conduction theory. In the second method, the gauge absorbs the total heat input to the surface, and the instantaneous heat flux rate is determined by the time rate of change of temperature of the gauge.

It is clear that most heat flux gauges are based on approximations. It is necessary for the experimentalist to arm himself with a variety of exact solutions; i.e., those tabulated in [1]. In dealing with the heat conduction equation, the property of sensor thermal diffusivity ($\alpha_g = k_g/\rho_g C_{pg}$) is fundamental. Recall [9] that for metals the thermal diffusivity is about 400 times that of thermal (and electrical) insulators. The thermal diffusion depth ($\delta_g = \alpha_g \tau$) is a measure of the extreme depth to which a surface heat flux has penetrated in time $\sqrt{\tau}$. Therefore, thermal diffusion depths of metals are 20 times those of insulators.

8.4.1 Thin-Film Gauges

When a film thickness t_g is much less than the thermal diffusion depth ($t_g \ll \delta_g$), temperature gradients in the film may be neglected (see Fig. 8.9 for notation). The film senses the instantaneous surface temperature of the substrate (subscripts) but there exists a response lag relative to zero film thickness because of the small but finite gauge heat capacity. As an example of this response lag, consider a suddenly applied heat flux rate varying inversely with the square root of time. With zero gauge thickness the substrate temperature jumps to a new constant value. For a finite gauge thickness, the substrate surface temperature attains 94% of the ideal temperature change in a time $\tau = 100\,\tau_r$ where

$$\tau_r = \frac{\rho_g^2 C_{pg}^2 t_g^2}{\rho_s C_{ps} k_s} \tag{8.23}$$

and τ_r is the characteristic time of the gauge-substrate combination.

In most applications the heat flux rate into the gauge is one-dimensional. The required relation between the surface heat flux rate and the measured gauge temperature is obtained by solving the one-dimensional, constant-property, heat conduction equation in a semi-infinite slab:

$$\frac{\partial T}{\partial \tau} = \frac{k_s}{\rho_s C_{ps}} \left(\frac{\partial^2 T}{\partial x^2} \right)_{\tau > 0} x > 0 \tag{8.24}$$

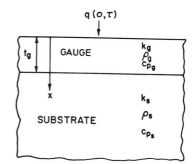

$$q(o,\tau)$$

| GAUGE | t_g | k_g ρ_g c_{p_g} |

| SUBSTRATE | x | k_s ρ_s c_{p_s} |

FIG. 8.9 Gage notation for thick and thin films.

where $T(x, \tau)$ is the difference between the instantaneous and initial temperatures. The boundary conditions are $q(0, \tau) = - k(\partial T/\partial x) (0, \tau)$; $T(x, 0) = 0$. The general solution [10]

$$T(x,\tau) = \frac{1}{\sqrt{\pi \rho_s C_{ps} k_s}} \int_0^\tau \frac{q(0,t)}{\sqrt{\tau - t}} \exp - \frac{x^2}{(4k_s/\rho_s C_{ps})(\tau - t)} \, dt \quad (8.25)$$

is obtained by use of the convolution integral of the Laplace transform. At the substrate surface $x = 0$, the relations between the wall temperature $T(0, \tau)$ and the impressed heat flux $\phi(\tau)$ are

$$T(0,\tau) = \frac{1}{\sqrt{\pi \rho_s C_{ps} k_s}} \int_0^\tau \frac{q(0,t)}{\sqrt{\tau - t}} \, dt \quad (8.26)$$

$$q(0,\tau) = \frac{\sqrt{\rho_s C_{ps} k_s}}{\sqrt{\pi}} \int_0^\tau \frac{dT(0,t)}{dt} \frac{1}{\sqrt{\tau - t}} \, dt \quad (8.27)$$

Equation (8.27) is not suited for numerical evaluation since it involves derivatives of temperature integrals. Equation (8.27) can also be written

$$q(0,\tau) = \sqrt{\frac{\rho_s C_{ps} k_s}{\pi}} \left[\frac{T(\tau)}{\sqrt{\tau}} + \frac{1}{2} \int_0^\tau \frac{T(\tau) - T(t)}{(\tau - t)^{3/2}} \, dt \right] \quad (8.28)$$

Numerical difficulties encountered in evaluating the integral as $t \to \tau$ are the subject of several recent papers. The integrand evaluated at τ takes the form 0/0, and application of L'Hôpital's rule yields an infinite integrand at τ. It is necessary to evaluate the terms enclosed in the square brackets by numerical procedures.

For constant heat flux $q(0, \tau) = q_0 = $ const, Eqs. (8.26) and (8.27) integrate to

$$q(x,\tau) = q_0 \, \text{erfc}\left(\frac{x}{2\sqrt{(ks/\rho_s C_{ps})\tau}}\right) \qquad (8.29)$$

$$T(0,\tau) = 2 \, \frac{q_0}{\sqrt{\pi}} \, \frac{\sqrt{\tau}}{\sqrt{\rho_s C_{ps} k_s}} \qquad (8.30)$$

For a heat flux gauge several possibilities exist for determining the gauge temperature. For example, either a thermocouple or a resistance thermometer may be used for measuring the time-temperature trace. The greater sensitivity of the resistance thermometer gives it an advantage over the thermocouple. For example, sensitivities of resistance thermometers are as great as 1 mv/°C while the sensitivities of conventional thermocouples range from 0.005-0.05 mv/°C. Evaporated thermocouples require two overlapping films, and ceramic films (thermistors) are sometimes considered (vs. metallic films) because their temperature coefficient of resistance (dR/dT) is much larger, but is constant only over a narrow range. Other techniques [10] employ a variable reluctance gauge (a copper mass used to vary reluctance in a magnetic circuit), or a dielectric material as a temperature sensor. When a dielectric material such as barium titanate is heated, an electric charge is generated that is proportional to the temperature change. If the guage is connected to a resistor, the circuit is directly proportional to the time rate of change of gauge temperature.

Some normal operating conditions of thin-film thermometers are [10]:

Film material = platinum evaporated or painted on pyrex substrate

Film thickness = 0.025 μ

Film response time = 10^{-7} sec

Film current = 10^{-2} amp

Film resistance = 100 ± 25 ohms

Film dimensions = 6.35 × 0.635 mm

Film allowable temperature change = 1/4 to 280K

Film $\rho_g C_{pg} k_g$ = 1520 ± 5% W/m²K (Pyrex)

If the temperature changes are not too large ($\Delta T < 30$K), the resistance of the film varies linearly with temperature $R_g = R_{g0} (1 + \alpha T)$ or

$$T = \frac{\Delta R}{R_0 \alpha} \qquad (8.31)$$

The resistance change ΔR is related to the voltage drop across the film voltage $\Delta V(\tau)$ and the constant film current, I_0, by the relation $\Delta R = \Delta V/I_0$ so that the operating equation for constant heat flux [11] is

$$q_0 = \left(\frac{\sqrt{\pi \rho_s C_{ps} k_s}}{2\alpha R_0 I_0}\right)\frac{\Delta V}{\sqrt{\tau}} \tag{8.32}$$

Gauge calibration consists of determining the quantity $\sqrt{\rho_s C_{ps} k_s}/\alpha$ at room temperature plus variations of this parameter with temperature. The bulk material properties cannot be relied upon.

An electrical pulse resulting from the discharge of a charged capacitor across the metal film provides a simple means of generating a known, constant heat flux if the time constant of the discharge circuit is large compared to the calibration test time. A severe limitation of Eq. (8.32) arises from the required knowledge of the gauge area ($q_s = Q_s/A_s$). Early experience showed calibrations accurate to ± 15%, primarily due to errors in film dimensions. Recent accuracies are about ± 5%. The thin-film boundaries are often irregular and difficult to define. Skinner [12] devised a technique for determining $\sqrt{\rho_s C_{ps} k_s}$ that eliminates the necessity of a film area measurement. If the film is calibrated electrically in a thermally conducting fluid, i.e., distilled water, the heat generated will diffuse into both the substrate and the fluid (f) according to the relation

$$T(\tau) = \frac{2}{\sqrt{\pi}} q_0 \frac{\sqrt{\tau}}{\sqrt{\rho_s C_{ps} k_s} + p_f C_{pf} k_f} \tag{8.33}$$

The film dimensions are removed from the problem if the identical calibration is made in both a fluid with known thermal properties and in air, where all of the heat is absorbed by the substrate material.

A platinum film is electrically active and subject to short circuiting when submerged in an ionized or conducting medium. Electrical insulation is accomplished by depositing a thin coating (i.e., magnesium fluoride) over the gauge. This coating—usually several microinches thick—introduces a negligible thermal lag in most circumstances.

For constant impressed heat flux, the error in temperature or heat transfer rate due to neglecting the gauge heat capacity is proportional to the ratio of heat stored in the gauge to that stored in the substrate such that:

$$\frac{\Delta q}{q} = \frac{\Delta T}{T} \cong \frac{\sqrt{\pi}}{2} \sqrt{\frac{\rho_s C_{ps}}{k_s \tau}} \frac{\rho_g C_{pg}}{\rho_s C_{ps}} \tag{8.34}$$

Effect of Temperature Variation on Thermal Properties

The established data reduction techniques are based on constant gauge properties restricted to small temperature changes (50°C). Recent investigations have been made on the effect of temperature-dependent thermal properties on the classical solution for heat conduction. The variable property diffusion equation

$$\rho C_p \frac{\partial T}{\partial \tau} = \frac{\partial}{\partial x}\left(k \frac{\partial T}{\partial x}\right) \tag{8.35}$$

is solved using the integral–conductivity function $\phi(T) = \int_0^T k_s dT$.

For common substrate materials the temperature variation of the thermal conductivity dominates while the temperature effect on thermal diffusivity is small. Two cases were considered [10], assuming $k = A + B \log_{10} T$. The zeroth-order solutions are:

For $q_0 = $ const: $T(\ ) = \dfrac{2q_0\sqrt{\tau}}{\sqrt{\pi \rho_s C_{ps} k_s}}$ \hfill (8.36)

For $q_0 = \dfrac{D}{\sqrt{\tau}}$: $\quad T(\ ) = D \sqrt{\dfrac{\pi k_s}{\rho_s C_{ps}}}$ \hfill (8.37)

where $\quad (\ ) = 1 + \dfrac{B}{k_{s_0}}\left[\dfrac{T_0}{T}\left(1 + \dfrac{T}{T_0}\right)\log_{10}\left(1 + \dfrac{T}{T_0}\right) - 0.434\right]$

$$\tag{8.38}$$

The zeroth order effect of temperature on the thermal properties is to increase the surface temperature.

Advantage of thin-film thermometers [10] are:

1. rapid response,
2. sensitivity for measuring low levels of heat transfer, and
3. output can be numerically evaluated to accurately determine the instantaneous heat flux rate.

Limitations of thin-film thermometers [10] are:

1. numerical evaluation of heat flux is laborious (requires temperature history be read, tabulated, and programmed),
2. measurement accuracy is fixed by calibration of substrate properties $\rho_s C_{ps} k_s$,
3. sensor is electrically active and therefore cannot be used in a conducting media, and
4. thin-film is sensitive to erosion by foreign particles.

8.4.2 Thick-Film Gauges

In the case of thick film, the heat received by the gauge is largely stored within the gauge while only a negligible portion is transferred to the substrate. Thin-film sensors are not successful under conditions of very high heat flux rates and/or long test durations giving excessive temperature rises. For thick films the gauge thickness t_g is comparable to or much larger than the thermal diffusion depth δ_g.

In reality the gauge is a calorimeter in which the governing relation

$$q = q_s + \frac{d}{d\tau}\left[\rho_g C_{pg} t_g \bar{T}(\tau)\right] \qquad \bar{T}(\tau) = \frac{1}{t_g}\int_0^{t_g} T(x,\tau)\, dx \qquad (8.39)$$

where q_s is the heat transferred to the substrate materials and for constant heat flux is found to be

$$q_s = q_0 \left(\frac{2a}{a+1}\right) \sum_{n=1}^{\infty} \text{erfc}\left[\frac{(2n+1)t_g}{2\sqrt{k_g \tau / \rho_g C_{pg}}}\right] \qquad (8.40)$$

and

$$a = \sqrt{\frac{\rho_s C_{ps} k_s}{\rho_g C_{pg} k_g}} \qquad (8.41)$$

With $t_g = \delta_g$ the thick film may be thin by usual standards.

Using platinum and a test duration of 100 μsec, δ_g is 0.0013 mm. Typical thicknesses for shock tube operation are 0.025 mm of platinum for test times of 5×10^{-5} sec.

Many of the techniques used for thin-film gauges apply directly to thick-film sensors, including calibration and temperature sensing techniques. One method of measuring $\bar{T}$ is to use the film as a resistance thermometer. For thick-film calibration, an electrical scheme similar to that developed for thin-film gauges is sometimes used; i.e., a pulse of electrical current is passed through the gauge causing Joule heating of the resistor. Because the resistance of the gauge is quite low (10^{-3} ohms) electrical currents of 500–700 amp are required. In operation 5 to 10 amp are required to develop sufficient voltage differences across the resistance element so that resistance change measurements can be made.

The instantaneous heat flux is proportional to the local slope of $\bar{T}$ vs. τ. Although the data reduction is much simpler for the thick rather than the thin film, the measurement of slopes is inherently inaccurate.

Advantages of the thick film resistor [10] include:

1. Rapid response—response time may be defined as the time for the rear surface temperature–time derivative to equal $d\bar{T}/d\tau$,
2. large thermal capacity—can sense larger q without surface melting at higher temperatures, and
3. insensitive to erosion due to large thickness.

Its limiations include:

1. Smallest heat flux rate is 110 W/cm^2 or 2 orders of magnitude greater than thin-film thermometer,

2. calibration is necessary to determine film properties,
3. large thickness → small resistance → large electrical current passing through film, and
4. it is always difficult to infer instantaneous heat flux rates from data that must be time-differentiated.

The use of thick-film calorimeter gauges simultaneously with thin-film resistance thermometers at intermediate heat flux rates gives a possibility for experimental comparison of the two techniques. In the ranges where both types of gauges have been used good agreement was found [9].

8.4.3 Sweep Operation

The discussion of the preceding apply most generally to a step input of heat flux such as obtained by the use of a shutter, a removable probe cover, or rapid insertion of a probe into a high energy gas flow. In the sweep technique [13], the probe is traversed across a gas flow in a continuous motion; thus a zero dwell time is considered. During sweep operation, the calorimeter response must be sufficient to give a true $dT/d\tau$ in order that a transverse heat flux profile may be obtained. It has been found [13] that for many cases of a small probe sweeping across a large jet flow, adequate response is obtained when $.5 \ (t^2/\alpha) = \tau_r < 0.10P$ where P is the sweep period. If this response time requirement is met, the $dT/d\tau \ (\sim dt/dP)$ is proportional to the local flux at any time during the sweep. Calorimeters of this type have a potential application for measurement of heat flux levels to 11,000 W/cm^2 with test times ranging for 10–300 msec.

Some typical thermocouple traces are plotted in Fig. 8.10 for a series of nozzle temperature profiles. The heat flux rate profiles developed from these profiles are also shown in the lower half of the figure.

8.5 EFFECT OF NONUNIFORM SURFACE TEMPERATURE

Transient measurements obtained from calorimeters are often in gross error due to improper simulation of the test surface. The calorimeter should not disturb the thermal conditions that occur when it is not present. Proper simulation of the contour, smoothness, surface emissivity, and time constant $\rho_m V_m C_{pm}/A_m$ are necessary or deviations of the surface temperature history may occur. For example, if a slab calorimeter $V_m/A_m = t$ has an improper time constant, a nonuniform wall temperature distribution will be set up. An insulated plug-type calorimeter sometimes generates a step in the wall temperature. For laminar flat plate flow, this effect alters the to-be-measured heat flux.

If a plate exposed to a steady flow has a temperature T_f up to $x = x_0$ and a temperature T_w downstream of x_0, the ratio of heat transfer coefficient (c) with

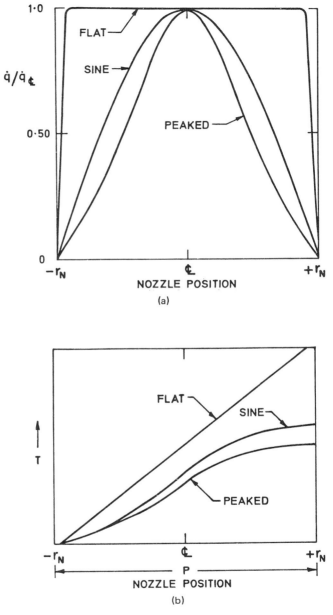

FIG. 8.10 Illustration of sweep technique. (*a*) Typical nozzle heat flux profiles. (*b*) Typical thermocouple traces.

step wall temperature (h) to the heat transfer coefficient with constant wall temperature h_{iso} is [5]

$$\frac{h}{h_{iso}} = \left[1 - \left(\frac{x_0}{x}\right)^{3/4}\right]^{-1/3} \tag{8.42}$$

8.6   THE QUASISTEADY ASSUMPTION

It is often interesting to compare the heat flux results of experiments using transient techniques with analytically predicted values. In general the theoretical analyses apply to the steady state case only, and any comparison is only approximate. The comparison is facilitated if the quasisteady assumption is adopted which supposes that the thermal layer passes through a succession of instantaneous steady states. Truly there is a difference between the actual instantaneous heat transfer and the quasisteady value. Sparrow and Gregg [14] examined this problem for the case of laminar flow over a flat plate and obtained the approximate result for Pr = 0.72:

$$\frac{q_{inst}}{q_{qs}} = 1 + \frac{x}{U_\infty}\left[2.39\,\frac{dT_w/d\tau}{(T_w - T_f)_{qs}} - 0.801\,\frac{d^2 T_w/d\tau^2}{(T_w - T_f)_{qs}}\,\frac{x}{U_\infty} + \cdots\right]$$

In most forced-convection situations the term in the brackets is small compared to the first term. For example, if $T_w - T_f = 55.5K$, $dT_w/d\tau = 55.5K/s$, $x = 0.0305$ m, and $U_\infty = 30.5$ m/s, then $q_{inst}/q_{qs} \simeq 1.02$. In nearly all situations the small values of x/U_∞ assure quasisteady heat transfer.

NOMENCLATURE

Bi   Biot number = hL/k
C_{pw} specific heat of body material, J/Kg°K
c constant
d total derivative
e constant = 2.71828
Fo   Fourier number = $\alpha\tau/t_w^2$
F_{w-s} radiation surface configuration factor
h convective unit surface conductance, W/m²K
k_w thermal conductivity of body material, W/mK
L body thickness, m
Q total heat input = qA, W
q rate of heat input, W/cm²
T temperature, K
$\bar{T}$ dimensionless temperature

t thickness, m
x depth, m
∞ infinitely large

Greek

α thermal diffusivity of body material $= k/\rho C_p$, cm^2/s
δ diffusion depth, m
∂ partial derivative
ϵ_{w-s} emissivity interchange factor
π constant $= 3.14159$
ρ density of body material, kg/m^3
σ Stefan-Boltzmann radiation constant $= 17.3 \times 10^{-10}$, W/m^{2}K^4
τ time, s
ω frequency, cycles/sec

Subscripts

f ambient fluid
max   maximum
0   initial ($\tau = 0$)
w wall

REFERENCES

1. Carslaw, H. S., and Jaeger, J. C., "Conduction of Heat in Solids," Clarendon, Oxford, 1947.
2. Schneider, P. J., "Temperature Response Charts," Wiley, New York, 1963.
3. Kudryavtsev, Y. V., "Unsteady State Heat Transfer," English translation, Scripta Technica, Iliffe, London, 1966.
4. Brookley, C. E., Measurement of Heat Flux in Solid Propellant Rocketry, presented at *Inst. Soc. of America Conf, 18th*, Chicago, Sept. 1963.
5. Eckert, E. R. G., and Drake, R. M., "Heat and Mass Transfer," 2nd Ed., McGraw-Hill, New York, 1959.
6. Bell, J. C., and Katz, E. R., A Method of Measuring Heat Transfer Using Cyclic Temperature Variation, *Heat Transfer and Fluid Mech. Inst.*, 1949.
7. Schneider, P. J., "Conduction Heat Transfer," Addison-Wesley, Reading, Mass., Sept. 1957.
8. Hall, J. G., and Hertzbert, A., Recent Advances in Transient Surface Temperature Thermometry, *Jet Prop. Jour.*, vol. 28, no. 11, pp. 719–723, Nov. 1958.
9. Rose, P. H., Development of the Calorimeter Heat Transfer Gauge for Use in Shock Tubes, *AVCO Research Rep. 17*, Feb. 1958.
10. Vidal, R. J., Transient Surface Temperature Measurements, *Symposium: Measurement in Unsteady Flow, ASME Hydraulic Div. Conf., New York*, May 1962.
11. Bogdan, L., High Temperature Thin-Film Resistance Thermometers for Heat Transfer Measurement, *Cornell Aeronautical Lab. Rep. No. HM-1510-Y-5*, Feb. 1963.

12. Skinner, G. T., Calibration of Thin-Film Backing Materials, *ARS Jour.*, vol. 31, no. 5, May 1961.
13. Starner, K. E., Use of Thin-Skinned Calorimeters for High Heat Flux Jet Measurements, *Aerospace Corp. Rep. TR-0158 (3240-10)-4*, July 1967.
14. Sparrow, E. M., and Gregg, J. L., Prandtl Number Effects on Unsteady Forced Convection Heat Transfer, *NACA TN 4311*, June 1958.

9 Analogies to heat transfer processes

E. R. G. ECKERT
School of Mechanical and Aerospace Engineering
University of Minnesota, Minneapolis

9.1 CONDUCTION HEAT TRANSFER

Fourier's equation describing steady heat conduction in a constant property medium has mathematically the form of Laplace's equation in the absence of internal heat sources or of Poisson's equation in the presence of internal heat sources. These equations also describe various other physical processes which consequently can be used to set up analogs. The flow of electricity according to Ohm's law is described by Laplace's equation and provides the possibility of electric analogs. These are constructed using a conducting paper or an electrolytic trough; the latter device has the advantage that it can easily be adjusted to the solution of rotationally symmetric problems or to general three-dimensional situations.

In fluid mechanics, Laplace's equation describes the flow of an inviscid fluid. Flow of a viscous fluid between two parallel plates at a small Reynolds number generates the same streamlines as flow of an inviscid fluid and can, therefore, also be used as an analog (Hele-Shaw flow). Ludwig Prandtl was the first one to point out the fact that Laplace's or Poisson's equation describes the shape of a thin membrane and that this can again be used for a very useful analog to flow or heat conduction processes.

A hydraulic analog can be constructed to an unsteady heat conduction process in which the thermal capacity is replaced by the capacity of stand pipes and the conductive resistance by capillary tubes which interconnect the stand pipes. The liquid level in the stand pipes is then an analog to the temperature in the heat conduction process. This analog can easily be modified to study the effect of change of phase as it occurs, for example, in a freezing or melting process [1].

The usefulness of these analogs has recently diminished because of the availability of digital electronic computers. Such a computer can actually be considered as an analog which has the advantage that all conductive situations can be studied, including complicating features like variation of the properties involved. For this reason, the analogies mentioned above will not be discussed in any detail and the reader is referred to the relevant literature.

9.2 RADIATIVE HEAT TRANSFER

Electric networks have been proposed to describe radiative heat transfer in engineering systems. In these analogs Ohm's resistances are interconnected in parallel and in series and various electric potentials are applied to the external node points. Such an analog is very useful for demonstration because engineers are often more acquainted with electric networks than with radiative exchange processes. To solve radiative energy processes, again today one prefers the digital electronic computer, especially since the Monte Carlo method offers a procedure by which, in principle, any radiative exchange can be calculated.

Mechanical and optical analogs have been developed to obtain shape or angle factors needed in the analysis of radiative exchange. These are useful for intricate geometries for which the analysis is often extremely involved and an example is shown in Fig. 9.1. The photo is obtained by the optical analogy [2]. The ratio of the areas of the dark shadow to the illuminated disk is the shape factor

FIG. 9.1 Optical determination of the angle factor of a coil. (From Eckert, 1955, Ref. [2].)

describing the radiative energy exchange between a hot coil and its cold surroundings, assuming the coil surfaces to be black, diffuse radiators. Similar analogs can also be conceived which describe the energy exchange between specularly reflecting surfaces.

9.3 CONVECTIVE HEAT TRANSFER

9.3.1 Mass Transfer Analogy for a Constant Property Fluid

Basic Considerations

The flow of a constant property fluid is described by the continuity equation and the Navier-Stokes equations. These will be presented here for a Cartesian coordinate system in tensor notation and with dimensionless parameters. The dimensioned parameters will be denoted by a dash in this section. Accordingly, the three coordinates will be indicated by x'_1, x'_2, x'_3, respectively, or generally by x'_i or x'_j with i or j having values from 1 to 3. In the same way, the three velocity components will be denoted by v'_i or v'_j. Length dimensions will be made nondimensional, dividing them by a reference length L'_0; velocities are made dimensionless by an arbitrarily chosen reference velocity v'_0. The time τ' is made dimensionless by a reference time L'_0/v'_0 and pressures p' are made dimensionless with a reference pressure $\rho' v_0'^2$ with ρ' indicating the fluid density

$$x_i = \frac{x'_i}{L'_0} \qquad v_i = \frac{v'_i}{v'_0} \qquad \tau = \frac{\tau'}{L'_0/v'_0} \qquad p = \frac{p'}{\rho' v_0'^2} \tag{9.1}$$

With this notation, the continuity and momentum equations take on the following forms:

$$\frac{\partial}{\partial x_i}(v_i) = 0 \tag{9.2}$$

$$\frac{Dv_i}{d\tau} = -\frac{\partial p}{\partial x_i} + \frac{1}{Re_0}\frac{\partial}{\partial x_j}\left[\left(1 + \frac{\epsilon}{\nu}\right)\left(\frac{\partial v_i}{\partial x_j} + \frac{\partial v_j}{\partial x_i}\right)\right] \tag{9.3}$$

In reading these equations, it has to be recalled that any term in which a subscript i or j is repeated stands for the sum of three terms with the subscript replaced alternatively by 1, 2, and 3. The operator $D/d\tau$ stands for the substantial derivative towards time τ. The index i in Eq. (9.3) indicates that this equation represents three equations in which the index i is again alternatively replaced by 1, 2, 3. The Reynolds number Re_0 is based on the reference parameters ($Re_0 = v'_0 L'_0/\nu$, with ν indicating the kinematic viscosity), ϵ denotes the turbulent diffusivity of momentum. The following relation can be derived from similarity considerations or from experiments.

$$\frac{\epsilon}{\nu} = f(\text{Re}_0, x_i) \tag{9.4}$$

It may, in addition, be influenced by the turbulence characteristics of the flow entering the system under consideration or by the roughness of the surfaces involved. Dimensional analysis, which will not be pursued here, leads to the result that for objects of similar shape any dimensionless parameter characterizing the flow is a function of Reynolds number and of dimensionless location only. This holds, for instance, for friction factors, drag coefficients, dimensionless pressure, velocity distribution, and so on.

It should also be pointed out that for a constant property fluid, the flow process is completely described by Eqs. (9.2) and (9.3) and by the corresponding boundary conditions. This means, for instance, that the flow process is completely independent of the fact whether heat or mass transfer is present or not. An experimentally determined flow field in an isothermal one-component fluid, therefore, remains unaltered when heat or mass transfer is superimposed on the flow process.

We will now consider a heat transfer process which is superimposed on the velocity field and described by the following equation written again in dimensionless parameters:

$$\frac{Dt}{d\tau} = \frac{1}{\text{Re}_0 \text{Pr}} \frac{\partial}{\partial x_i} \left[\left(1 + \frac{\epsilon}{\nu} \frac{\text{Pr}}{\text{Pr}_t} \right) \frac{\partial t}{\partial x_i} \right] \tag{9.5}$$

The parameter Pr indicates the Prandtl number of the fluid ($\text{Pr} = \nu/\alpha$, where α denotes the thermal diffusivity of the fluid) and the turbulent Prandtl number Pr_t is, in a similar way, defined as the ratio of the turbulent diffusivity ϵ for momentum to the turbulent diffusivity ϵ_H for heat, $\text{Pr}_t = \epsilon/\epsilon_H$. A dimensionless temperature, t stands for the ratio of a temperature difference Δt divided by a reference temperature difference Δt_0 where both temperature differences are counted from an arbitrary reference temperature. For instance, Δt_0 can denote the difference between the surface temperature of an object exposed to the flow to the fluid upstream temperature. The turbulent Prandtl number is given by a relation

$$\text{Pr}_t = f(\text{Re}_0, \text{Pr}, x_i) \tag{9.6}$$

This follows again from similarity considerations and from experimental evidence. In addition, it is again influenced by the upstream turbulence and by surface roughness. From Eqs. (9.2) to (9.6) it follows that local Nusselt numbers as dimensionless expressions for the local heat transfer coefficients will be described by a functional relationship of the following form:

$$Nu = f(Re_0, Pr, x_i) \tag{9.7}$$

We consider now a mass transfer process in a two-component single-phase medium superimposed on the velocity field. This process is described by the equation

$$\frac{Dw}{d\tau} = \frac{1}{Re_0 Sc} \frac{\partial}{\partial x_i} \left[\left(1 + \frac{\epsilon}{\nu} \frac{Sc}{Sc_t} \right) \frac{\partial w}{\partial x_i} \right] \tag{9.8}$$

in which w, a dimensionless mass fraction of one of the components, stands again for the ratio of a mass fraction difference to a reference mass fraction difference defined in the same way as the dimensionless temperature t in Eq. (9.5). $Sc = \nu/D$ is the Schmidt number with D denoting the binary mass diffusion coefficient and $Sc_t = \epsilon/\epsilon_M$ is the turbulent Schmidt number with ϵ_M indicating the turbulent diffusivity for mass. Negligible variation of properties is assumed.

Dimensional analysis requires that local Sherwood numbers, defined as the product of the mass transfer coefficient h_M and a refernce length L_0 divided by the mass diffusion coefficient D, can be expressed in the following way:

$$Sh = f(Re_0, Sc, x_i) \tag{9.9}$$

The fact that Eqs. (9.5) and (9.8) are of the same form, however, allows an additional important conclusion to be drawn. For this purpose it is useful to realize that all available experimental evidence points to the fact that

$$Sc_t = Pr_t \tag{9.10}$$

or, in other words, that Eq. (9.6) describes the turbulent Schmidt number Sc_t as well as the turbulent Prandtl number Pr_t. With this in mind, Eq. (9.8) can be changed to Eq. (9.5) by replacing w with t and Sc with Pr. From this it follows that for any geometrically similar configuration and for similar boundary conditions an equation of the form of Eq. (9.9) can be obtained from the corresponding Eq. (9.7) if one replaces in that equation the Nusselt number by the Sherwood number and the Prandtl number by the Schmidt number.

Dimensional analysis can also be performed on the equations which describe natural convection heat transfer of a fluid with slightly varying properties, on the one hand, and natural convection mass transfer in an isothermal two-component fluid where the properties of the two components differ little. The result is again that a simple relation exists between the Sherwood number describing the mass transfer process and the Nusselt number describing the heat transfer process. The local Sherwood number Sh can be presented as a function of a Grashof number Gr_{0M} based on reference parameters, a Schmidt number Sc, and dimensionless coordinates x_i fixing the location under consideration on the surface of the object at which mass transfer occurs:

$$Sh = f(Gr_{0M} Sc, x_i) \quad \text{or} \quad Sh = f(Ra_{0M}, Sc, x_i) \tag{9.11}$$

More recently, the Rayleigh number $Ra_{0M} = Gr_{0M} Sc$, is used instead of the Grashof number, because the Sherwood number, and also transition to turbulence, depends much less on the Schmidt number when it is expressed by a Rayleigh number, than when it is expressed by a Grashof number. The analogy between heat and mass transfer leads again to the result that, for any fixed geometry and for similar boundary conditions, an equation describing heat transfer can be obtained from Eq. (9.11) when the Sherwood number is replaced by the Nusselt number, the Grashof number or Rayleigh number for mass transfer by a Grashof number or Rayleigh number, respectively, for heat transfer; and the Schmidt number by the Prandtl number. This is indicated in the following lines:

$$
\begin{array}{ccc}
\text{Nu} & \text{Gr}_0 & \text{Pr} \\
\cancel{Sh} = f(\cancel{Gr}_{0M} \cancel{Sc}, x_i) & \text{or} & \cancel{Sh} = f(\cancel{Ra}_{0M} \cancel{Sc}, x_i)
\end{array}
$$

$$
\begin{array}{ccc}
& \text{Nu} & \text{Ra}_0 & \text{Pr}
\end{array}
\tag{9.12}
$$

(For the definition of the dimensionless parameters, see the nomenclature at the end of this chapter.)

In the derivation of the mass transfer analogy, one can also consider Eqs. (9.2), (9.3), (9.5), and (9.8) as describing turbulent heat and mass transfer processes, even when the turbulent contributions expressed by the turbulent diffusivity ϵ are disregarded. In this case the equations describe in principle the time-varying turbulent velocity, temperature, and composition fields when the proper boundary conditions are applied. Our knowledge today is insufficient to solve these equations; however, the similarity considerations can be based on them and the conclusions are the same as presented in the preceding paragraphs. They express the content of the mass transfer analogy. Its value lies in the fact that any heat transfer relation provides also a mass transfer relation for a similar problem but the similarity in the boundary conditions has to be carefully considered. A mass transfer process, for instance, frequently causes a mass release or mass absorption on the solid surfaces involved. This creates a finite velocity normal to the wall surfaces and if these velocities are sufficiently large to effect the mass transfer process, then the corresponding heat transfer relation must be for a process with similar velocities at the surface.

The mass transfer analogy can also be utilized in reverse to obtain heat transfer information through mass transfer experiments because there are situations in which a mass transfer process can be set up with cleaner boundary conditions and can be studied more easily and more accurately than the corresponding heat transfer process. Examples of such experiments are discussed in paragraphs that follow.

Mass transfer experiments have been performed in gases as well as in liquids using absorption, condensation, or evaporation of a gaseous component at a surface. The analogy has also been used in liquids with solution or absorption of a substance at the surface. In the latter case, use of an electrolyte offers special advantages and the corresponding electrochemical method has been used to a considerable degree.

9.3.2 Mass Transfer Analogy in Gases

The first one to make use of the mass transfer analogy in 1921 to study heat transfer was Thoma. The results of his study are contained in a book the title of which translates to "High Performance Steam Boilers" [3]. Thoma used models of tube banks made out of filter paper and soaked with phosphoric acid. An air-ammonia mixture was directed through the tube bank. The ammonia was absorbed so readily on the surface of the tubes that its concentration there was negligible. The amount of ammonia transferred to the surface was determined by titration of the acid in the filter paper. Thoma obtained, in this way, heat transfer coefficients which were, in the main, well verified by later heat transfer measurements.

The same method was used on an enlarged scale by Lohrisch [4] in a wind tunnel shown in Fig. 9.2. He verified essentially the results obtained by Thoma and determined also local heat transfer coefficients in the way that the filter paper covering the tube surface was subdivided into a number of longitudinal strips. Figure 9.3 presents results obtained in this way. Both Thoma and Lohrisch also used mass transfer experiments for flow visualization. In this case the model surfaces were soaked with chloric acid and the developing ammonia chloride clouds were photographed, as in Figs. 9.4 and 9.5. The Schmidt number for diffusion of ammonia into air at $0°C$ has a value $Sc = 0.634$. Experiments performed at this temperature result, therefore, in Nusselt numbers at a Prandtl number 0.634. It is found that the Nusselt number is approximately proportional to $Pr^{1/3}$. With this relation, Nusselt numbers obtained by the mass transfer analogy can be generalized over a moderate range of Prandtl numbers.

The study of laminar or turbulent mass diffusion between two different gas streams offers a convenient analogy to the exploration of thermal diffusion of a gaseous plume with the surrounding gas. Such a study was reported by Malhotra and Cermak [5]. It considered the turbulent diffusion of mass from a point source into a turbulent boundary layer formed on a plane impermeable surface. Ammonia gas with a specific gravity of 0.6 relative to air was supplied through a small hole in the 2-meter-by-4-meter test section forming the lower wall of a wind tunnel. The air-ammonia mixture at various locations in the plume downstream of the hole was collected by a small sampling probe; the ammonia was

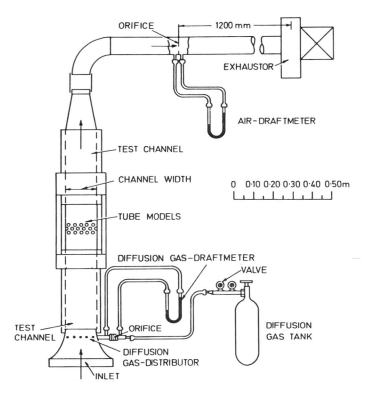

FIG. 9.2 Test setup for mass transfer analog using ammonia absorption. (From Lohrisch, 1929, Ref. [4].)

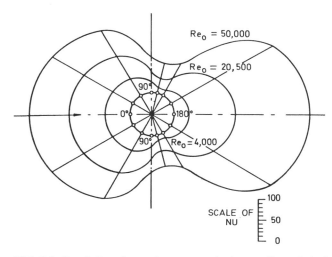

FIG. 9.3 Local Nusselt numbers on a tube in crossflow of air determined by mass transfer analogy. (From Lohrisch, 1929, Ref. [4].)

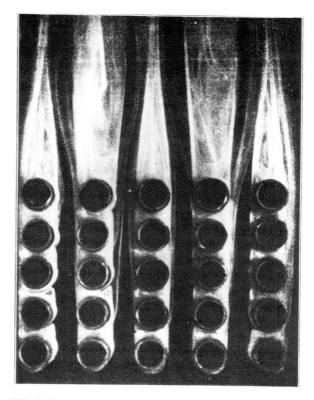

FIG. 9.4 Visualization of mass transfer in flow of air through a bundle of aligned tubes. (From Lohrisch, 1929, Ref. [4].)

absorbed by hydrochloric acid, and measured by a colorimeter. The turbulent boundary layer was either isothermal or heated from the horizontal test section to create a stratified unstable flow. The three-dimensional concentration field obtained in the measurements can again be converted to an analogous temperature field created by turbulent mixing of two streams of the same gas but with different temperatures.

This method has been used with advantage to determine the effectiveness of film cooling, a method used extensively in high temperature gas turbines for structural elements which are exposed to the high temperature gas stream. Cooling air is ejected through slots or rows of holes to protect the downstream portion of the structure.

The film cooling effectiveness is a dimensionless expression describing the temperature which an adiabatic wall assumes under the influence of a hot main stream and a "coolant" gas. This temperature is also useful in calculating the heat transfer on a nonadiabatic surface protected by film cooling. In the mass

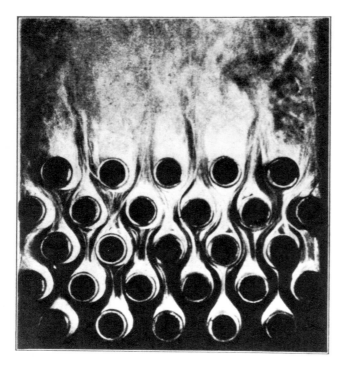

FIG. 9.5   Visualization of mass transfer in flow of air through a
bundle of staggered tubes. (From Lohrisch, 1929, Ref. [4].)

transfer analogy the coolant gas is of a different nature but at the same
temperature as the gas in the mainstream, and the concentration at the wall
surface is analogous to the adiabatic wall temperature. Most heat transfer experiments have used air as the mainstream fluid and as coolant. Air with a tracer gas
as coolant then provides the analogy to a constant property heat transfer
situation. Helium, carbon dioxide, and freon have been used as tracer gases and
the concentration of these gases in the air stream is measured conveniently with
a gas chromatograph. The Schmidt numbers of these gases are, respectively, 0.2,
1, and 1.4. Samples are removed through small holes in the wall to obtain
concentrations at the wall surface.

The analogy can, on the other hand, also be used to investigate the effect of
large temperature differences on the film cooling effectiveness. In the mass
transfer analogy, the density of the coolant gas has then to be considerably
different from the density of the mainstream gas. This analogy is discussed in
more detail at the end of this section. An investigation into film cooling of a
plane surface by ejection of a coolant gas through a row of holes or a porous
slot into the turbulant boundary layer covering the surface has been carried out
by Pedersen [6].

Evaporation of water vapor in measured quantities from surfaces covered by a water film into an air stream have also been used as a mass transfer analogy. Hilpert [7] studied, in this way, combined heat and mass transfer in free convection from a vertical plate. Powell and Griffiths [8] used the same method to obtain local heat transfer coefficients on a flat plate with forced flow over its surface. Figure 9.6 shows the experimental arrangement for these studies. The rate of evaporation from the strips of linen covering the surface of the plate was determined in two ways: directly, by measuring the amount of water fed through the individual supply tubes, and indirectly, by measuring the electric energy needed to keep the surface temperature uniform and at the same value for a wet and dry surface. The Schmidt number for diffusion of water vapor into air at 8°C is 0.615. Experiments at this temperature lead to Nusselt numbers for fluids with a Prandtl number of the same magnitude.

Evaporation of naphthalene has frequently been used in mass transfer analogy studies. The method appears to have been introduced by Jakob and Kezios in 1953 and has been used in a variety of experimental investigations since that time [9-15]. The naphthalene used is usually of crystal grade, chemical symbol $C_{10}H_8$, molecular weight 128.2, melting point between 79 and 80°C, and 0.001% residue after ignition. In [10], the method was used to obtain information on local heat transfer coefficients along the surface of a cylinder exposed to an axial flow of air.

Figure 9.7 presents a sketch of the cylinder. The cylinder was hollow, air flowing through the interior as well as along the outside; in this way starting effects were minimized. The cylinder model was manufactured from steel and had a thin layer of naphthalene along its outer surface. The naphthalene was cast

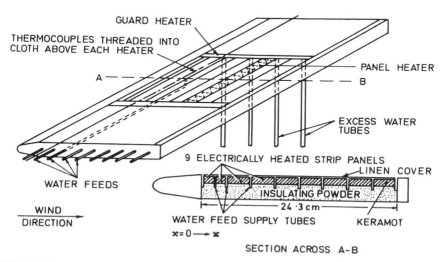

FIG. 9.6 Mass transfer analog to measure heat transfer on an isothermal flat plate by evaporation of water vapor. (From Powell and Griffiths, 1935, Ref. [8].)

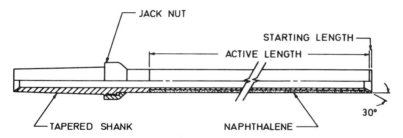

FIG. 9.7 Mass transfer analog using naphthalene sublimation from the surface to a hollow cylinder exposed to air flow parallel to the cylinder axis. (From Christian and Kezios, 1957, Ref. [10].)

onto the steel model in a mold made of dental plaster consisting of two parts to facilitate the removal of the model. The surface of the mold was coated with epoxy resin to make it impermeable to molten naphthalene and was covered with a thin layer of silicone grease to prevent adhesion of the model. The model and the mold were preheated by steam passed through the interior of the metal tube; the molten naphthalene was poured at a temperature of approximately 100°C into the vertical mold, and then cooled by gradually pouring water into the interior of the cylinder, raising the water level slowly so that the cooling proceeded from the lower end upward. In this way a dense layer of naphthalene could be obtained which had an outer diameter of 25 mm, with a tolerance of ± 0.1 mm. The model was exposed to an air stream in a wind tunnel and after completion of a run, removed; the change in diameter was measured by a profilometer graduated to 2.5 μ. In this way the local evaporation rate could be determined. To avoid a mass loss by evaporation, the measurements with the profilometer were made in an air-tight enclosure in which the air was saturated with naphthalene vapor. The cylinder radius changed during an experiment by up to 40 μ.

The experiments were evaluated in the following way: The local evaporation rate was obtained by the profilometer measurements; the concentration of naphthalene at the cylinder surface was obtained from the partial pressure of the naphthalene gas, through the ideal gas equation in which the partial pressure was obtained from the surface temperature, using equilibrium data listed in Table 9.1. The surface temperature itself was not measured and it was assumed that it was within the accuracy of the measurements equal to the temperature in the air stream. The density of the solid naphthalene which enters the weight loss calculation was obtained by weighing a small naphthalene cylinder that was cast in the same way as the model.

The mass transfer analogy proved especially advantageous in the study of heat transfer at the surface of a rotating disk, because the analogy avoided temperature and heat flux measurements which are difficult to perform on a rotating part. Reference [13] reports on these measurements and Fig. 9.8 is a sketch of the test equipment. The rotating horizontal disk, with 250 mm diameter has on its upper surface a layer of naphthalene which was cast onto the surface in a molten condition. After cooling in air, the ring a, which was necessary for casting, was removed and the upper surface of the naphthalene layer was machined on a lathe. The disk was rotated for 6 to 20 min at speeds up to 10,000 rpm and the mass loss was measured by weighing. The average mass transfer coefficient h_M was then obtained with the following relation:

$$h_M = \frac{mR_v T_s}{p_{vs}A} \qquad (9.13)$$

in which m denotes the mass loss per unit time, R_v the gas constant of the naphthalene vapor, T_s the surface temperature of the naphthalene, p_{vs} the vapor pressure of naphthalene at the surface temperature, and A the surface area. The difference between the surface temperature and the measured temperature in the air surrounding the disk was again neglected. It was estimated that the temperature depression of the surface due to evaporation was approximately 0.05°C and the temperature change due to the recovery effect was approximately 0.5°C. Properties in Table 9.1 were again used for the evaluation. The accuracy in the determination of the mass transfer coefficient was estimated to be approximately 6%. Nusselt numbers were obtained through the mass transfer analogy.

Another situation for which the mass transfer analogy proved useful is laminar or turbulent heat transfer in forced convection on a flat plate with

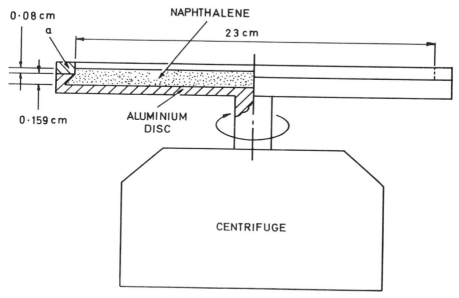

FIG. 9.8 Mass transfer analog to determine heat transfer characteristics of a disk rotating in air using naphthalene evaporation. (From Kreith, Taylor, and Chong, 1959, Ref. [13].)

TABLE 9.1 Physical properties of naphthalene vapor at 1 atm pressure

t		ν		ρ_{so}	p_v
(°F)	(°C)	(cm²/s)	Sc	(g/cm³)	(b*)
70	21.1	546	2.38	1.139	0.0111
75	18.3	556	2.39	1.127	0.0148
80	15.5	565	2.40	1.114	0.0194
85	12.8	575	2.40	1.101	0.0254
90	10.0	584	2.41	1.088	0.0331
95	7.2	533	2.41	1.075	0.0429

Source. Reference [10].

Note. ν = kinematic viscosity; ρ_{so} density of solid naphthalene; p_v = vapor pressure of solid naphthalene; * = 1 b = 10^5 N/m².

spanwise strips which are alternately unheated and heated to a uniform temperature [14, 15]. The stepwise temperature variation connected with this situation is practically impossible to obtain in a heat transfer experiment because of heat conduction within the plate, but it presents no difficulty at all in a mass transfer study. The experimental procedures used in these studies and in [9, 11, 12] were essentially the same as discussed above.

Mass transfer studies using naphthalene evaporation can also be made to simulate conditions as they occur in ablation cooling. For this purpose, the rate of evaporation is increased by operation of the wind tunnel into which the model is placed at elevated temperatures. Such studies were performed some years ago at the Research Laboratory of General Mills, Inc., and Fig. 9.9 shows the plaster cast of a blunt object of naphthalene which has been exposed for some time to a high temperature air stream. The ablation pattern shown resembles closely the pattern on a reentering satellite or spaceship with protection by a plastic ablating material.

The mass transfer analogy using naphthalene sublimation has recently been developed to high accuracy [16]. It has been applied to a study of local heat transfer in a one-row plane fin and tube heat exchanger. Figure 9.10 presents a pictorial view and a plane view of such an exchanger. The geometry investigated is described by the following parameters: h/D = 0.193, S/D = 2.5, and L/D = 2.16. A heat exchanger passage was simulated in the analogy by two naphthalene plates with circular disks made of Delrin used to model the tubes. The naphthalene plates were cast in a mold with accurately manufactured, polished, and lapped surfaces. They were removed from the mold by hammer blows without using any lubricant. Room air was sucked through the assembly as shown in Fig. 9.10. After a proper testing time the setup was disassembled and the local sublimation of naphthalene was measured by a precision dial gauge. The total ablation rate obtained by an integration of these local values agreed with an independent determination by weighing within 1% or 2%. Mass transfer

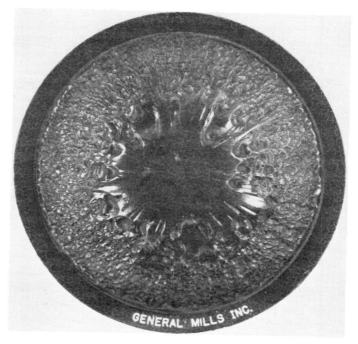

FIG. 9.9 Plaster cast of a blunt model of naphthalene exposed to a supersonic air stream (experiments at General Mills, Inc.).

coefficients were determined in the way that the local ablation rate per unit area was divided by the difference of the surface concentration and the bulk concentration in the respective flow cross section. The surface concentration was obtained through the temperature of the air entering the setup. Special care was taken to obtain an accurate value of this temperature because the vapor pressure of

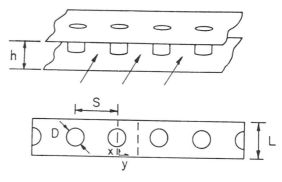

FIG. 9.10 A one-row plane fin and tube heat exchanger.

naphthalene is very sensitive to temperature with a variation of about 10%/°C at room temperature. Local Sherwood numbers $Sh = h_M D_h/D$ were obtained from the mass transfer coefficients with h_M denoting the mass transfer coefficient, D_h the hydraulic diameter, and D the mass diffusion coefficient. The hydraulic diameter is defined as $D_h = 4A_c L/A$, where A_c is the minimum flow area, L the streamwise length (Fig. 9.10), and A the transfer surface area. The Reynolds number based on the hydraulic diameter was varied from 150 to 1270. Figure 9.11 is an example of the results obtained. It presents local Sherwood numbers as a function of the two dimensionless coordinates (Fig. 9.10) at a Reynolds number of 648. It is interesting to observe the sharp peaks in the Sherwood number caused by a vortex forming at the corner between the disks and the plates. In the wake behind the disks the mass transfer is almost zero. The Sherwood numbers can be converted to Nusselt numbers by the analogy relation, Eq. (9.12). As far as we know, the mass transfer analogy provided for the first time the opportunity to obtain detailed information on a two-dimensional field of heat transfer coefficients in a geometry as complex as a fin and tube heat exchanger.

Electrochemical Method

Mass transfer studies which simulate heat transfer in liquids can be done very effectively by the electrochemical method. This method was introduced for the measurement of average mass transfer coefficients by Wagner in 1949 [17], and by Wilke, Tobias, and Eisenberg in 1954 [18]. Grassmann, Ibl, and Trüb [19] extended it to the measurement of local mass transfer coefficients in 1961 and Hanratty and associates used it for turbulence measurements in the immediate neighborhood of a surface in 1962.

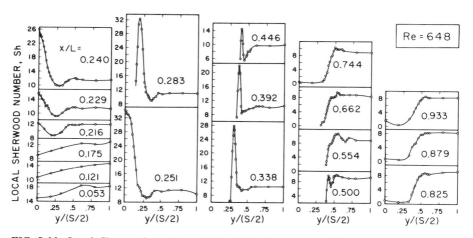

FIG. 9.11 Local Sherwood numbers obtained in the test assembly presented in Fig. 9.10. (From Saboya and Sparrow, 1974, Ref. [16].)

The liquid used in such experiments is an electrolyte. When two electrodes are inserted into it, the negatively charged ions move to the anode, the positively charged ions to the cathode. At a sufficiently large potential difference between the two electrodes, the ions react there releasing or binding one or more electrons. The movement of the ions toward the electrodes is caused by diffusion, by convection if the electrolyte is in motion, and by ion movement in the electric field. The number $\dot{n}$ of moles of ions moving toward an electrode per unit time and area is then described by the following equation:

$$\dot{n} = v_y c - D\frac{dc}{dy} - cB_i \frac{dE}{dy} \qquad (9.14)$$

in which v_y indicates the convective velocity of the electrolyte toward the electrode, c the mole concentration of the ions under consideration, D the mass diffusion coefficient of the ions in the electrolyte, B_i the ion mobility, E the electric potential, and y the direction normal to the electrode surface.

In a normal mass transfer process, an electric field is absent and the last term in the above equation has, therefore, to be made vanishingly small in order to use the ion transport as an analog to a normal mass transfer process. This can be done in the following way: A second strong electrolyte—the so-called "supporting" electrolyte—is added to the liquid, selected so that its ions do not react at the electrodes at the potential difference used in the experiment. If such an electrolyte were to be present alone, then the electric current between the electrodes, which flows when the electric circuit is closed, diminishes rapidly toward zero because layers of ions accumulating near the electrodes create strong space charges. The gradients of the electric field are then concentrated in these thin layers with a thickness of order 10^{-9} m, whereas in the main body of the electrolyte the electric potential is constant. This situation is practically maintained when the supporting electrolyte is added to the active electrolyte simulating the mass transfer process. In this way the last term in Eq. (9.14) can be disregarded and a transport of ions of the active electrolyte occurs by diffusion and convection only. The number $\dot{n}$ of moles can easily be obtained by a measurement of the electric current I flowing into or out of the electrode with the equation

$$I = \dot{n}zF \qquad (9.15)$$

in which z denotes the valence of the ion and F Faraday's constant, which equals $[0.965 \times 10^8 \ (\text{As/kg aeq})]$.

In the use of the electrochemical method to study a mass transfer process, and through the analogy also a heat transfer process, one of the electrodes is the model on which mass transfer or heat transfer, respectively, occurs. The other electrode can be the wall of the flow channel or of the container. Provision should be made that the second electrode is large in area compared to the model so that the resistance for mass transfer is large on the model surface alone. Under normal conditions, the concentration field is concentrated in a boundary layer surrounding the model, whereas concentration differences are minimized in the main body of the electrolyte by convection, which can either be enforced or created by the concentration differences themselves (natural convection).

Figure 9.12 is a sketch of an arrangement to measure local heat transfer coefficients around the periphery of a cylinder exposed to the flow of an electrolyte normal to its axis. The model is in this case the anode, and the channel wall the cathode. A thin strip on the cylinder surface is electrically insulated so that the current into this strip can be measured separately. From the measurement of this current, a local mass transfer coefficient h_M is obtained through the equation

$$\dot{n} = h_M(c_\infty - c_s) \tag{9.16}$$

The ion mole concentration c_∞ is readily measured in the bulk of the fluid. The difficulty of measuring the ion concentration c_s at the model surface is circumvented by adjustment of the electric potential difference ΔE between the two electrodes in such a way that the saturation current is obtained. This is indicated in a diagram as shown in Figure 9.13 by the fact that the current I becomes independent of the magnitude of the electric potential ΔE (by the horizontal portion of the current-potential curve). In this case, the ion concentration c_s at the model surface is zero.

An electrolyte which has frequently been used in this method is the system ferrocyanide/ferricyanide $[K_3Fe(CN)_6/K_4Fe(CN)_6]$. The reactions of the ions on the electrodes are described by the following equation:

$$Fe(CN)_6^{-3} + e^- \rightleftarrows Fe(CN)_6^{-4} \tag{9.17}$$

The reaction goes from left to right at the cathode and from right to left at the anode. In [19] a 0.025 N solution in water was used for the two chemicals and sodium hydroxide [NaOH 2 N] was added as the supporting electrolyte to equalize the electric potential in the main body of the fluid.

Figure 9.14 presents the results obtained in [19]. The ordinate can be

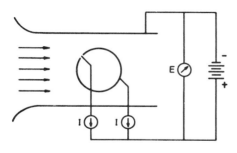

FIG. 9.12 Sketch of the electrochemical method applied to study local heat transfer on a circular cylinder exposed to a transverse liquid flow.

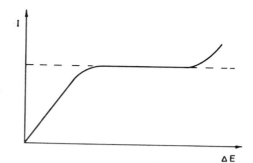

FIG. 9.13 Determination of the saturation current in the electrochemical method.

interpreted either as the Sherwood number or Nusselt number. In the first case, the Schmidt number is Sc = 2,170; in the second case, the Prandtl number has this value. The ordinate, interpreted as $Nu/Pr^{1/3}$, should approximately describe heat transfer in a high Prandtl number fluid. It is of interest to observe the dip in the two numbers close to the stagnation point as such a dip has also been observed in other mass transfer measurements. The cylinder used in these measurements was manufactured of nickel with a diameter of 50 mm and the insulated nickel strip in the cylinder wall used for the local mass transfer measurements had 0.5 mm width and was separated from the rest of the surface by a Teflon strip 0.05 mm thick. This demonstrates one advantage of this

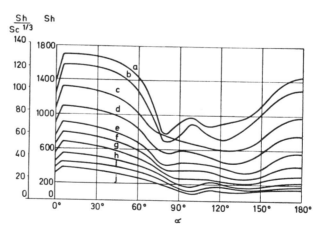

(a) Re =11 730, (b) Re =10 200, (c) Re =7000, (d) Re =4820, (e) Re = 3080, (f) Re =2220, (g) Re =1485, (h) Re = 1020, (i) Re = 656, (j) Re =458.

FIG. 9.14 Local Sherwood numbers around the periphery of a cylinder exposed to a transverse liquid flow measured by the electrochemical method. (From Grassmann, Ibl, and Trüb, 1961, Ref. [19].)

method as very local measurements can be made because sensitive instruments exist to measure an electric current.

The accuracy of the mass transfer measurements is claimed to be $\pm 2\%$–3% and timewise fluctuations can also be readily recorded. Some precautions have to be taken, on the other hand, to obtain accurate results as the electrolyte disintegrates when exposed to light and the concentration of the electrolyte has, therefore, to be checked continuously and the model surface frequently cleaned; otherwise, a layer with high electrical resistance is deposited on it. The test setup was manufactured of nickel and rubber-coated steel.

In [20] copper sulfate ($CuSO_4$) was used with sulfuric acid (H_2SO_4) as the strong electrolyte to study free convection on a horizontal cylinder and on a sphere. Similar studies are also reported in [21] and [22].

The electrochemical method was also used for measurements of wall shear and of turbulence parameters in boundary layers [23], [24]. The reading of a small electrode (of the order of 1 mm^2) in an electrically neutral wall can be interpreted as the velocity gradient at the surface. The time mean signal then determines the time-averaged velocity gradient, while the time variation of the signal describes instantaneous velocity gradients. A probe in a larger electrode surface reads the local mass flux as has been discussed in the preceding paragraph. It can also be used to obtain the limiting law of the turbulent diffusivity at the wall surface. In [24], it was found that the turbulent diffusivity is proportional to the fourth power of the wall distance. In these turbulence measurements, the test setup and the models were manufactured out of Plexiglas with platinum foil serving as electrodes.

The electrochemical method has been discussed in a recent extensive survey article by Mizushina [25]. It includes the theory and the experimental technique of the method and results obtained for a wide variety of situations including free convection on vertical and horizontal plates, horizontal cylinders, and spheres, as well as laminar and turbulent forced convection in tube flow under developed and entry condition, in crossflow around a cylinder, for rotating and vibrating bodies, falling films, packed beds, and agitated vessels. It also discusses the time response and fluctuations of measured mass transfer coefficients. The Mizushina article contains an extensive bibliography.

The capability of the electrochemical method is also demonstrated in a recent publication [26] which considers local natural convection mass transfer measurements, and contains a detailed discussion of the experimental procedure. Copper ions establish the transfer and sulfuric acid is used as supporting electrolyte. The Schmidt number of the fluid has, therefore, a value of approximately 2000. The corresponding relation for Nu is, therefore, useful for large Prandtl number fluids. Instability of the flow creates vortices in the form of rolls with their axes parallel to the flow direction on the upward facing surface of an inclined plate when a critical Rayleigh number is exceeded, the value of which depends on the inclination angle. The spacing between these vortices is quite small (of order 1 mm). They create a periodic variation of the local mass transfer

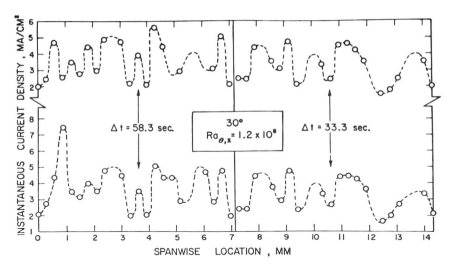

FIG. 9.15 Variation of local and instantaneous mass transfer rates along a horizontal line on a flat plate inclined under an angle of 30 degrees in natural convection. (From Lloyd, Sparrow, and Eckert, 1972, Ref. [26].)

across the span of the plate. These variations could well be detected by small mass transfer probes inserted in the plate surface with a thickness of 0.25 mm and a probe spacing of 0.05 mm. An example of the measurements is shown in Fig. 9.15. The ordinate lists the instantaneous mass transfer rate. The plate was inclined under an angle of 30 degrees against the vertical and the Rayleigh number of the experiments is listed in the figure. The time lapse Δt between two instantaneous distributions is also indicated.

Electrochemiluminescence

A variation of the electrochemical method [27-30] can be used for flow visualization and for an optical measurement of mass fluxes. To the active electrolyte and the supporting electrolyte is added a chemiluminescent substance which produces a glow at the anode. In this way, the flow close to the anode surface becomes visible and flow separation, for instance, can be detected very clearly. The intensity of the glow is proportional to the mass transfer rate so that optical measurements can also be used for quantitative determination of mass fluxes.

In [27] hydrogen peroxide (H_2O_2) is used as the active electrolyte and potassium chloride (KCl) as supporting electrolyte. Luminol (Eastman Kodak Luminol = 5-amino-2,3-dihydro-1,4 phthalazinedione) is the chemiluminescent substance. Finally, potassium hydroxide (KOH) is added to adjust the pH value of the solution since Luminol only exhibits chemiluminescence in an alkaline solution of pH larger than approximately 8. Table 9.2 gives the composition of the solution as it was used in [27].

TABLE 9.2 Composition of a chemiluminescent solution

Substance	Solution (g/1)	Remarks
H_2O	–	Solvent should be distilled
KCl	74.56	Supporting electrolyte
KOH	1.68	Adjust pH; luminol exhibits chemiluminescence in alkaline solution only
H_2O_2 [a]	0.0976	Oxidizing agent, concentration strongly affects light intensity
Luminol	0.15	Chemiluminescent substance soluble in alkaline solutions only

Source. Reference [27].
[a]Corresponds to 0.3 ml of H_2O_2 (30% solution).

The test setup and the model were again made of Plexiglas and the electrodes of platinum. The intensity of the chemiluminescence was measured with a photomultiplier. Best conditions for luminescence were obtained at a potential difference of 0.41 V between anode and electrolyte. The relation between the measured intensity and the mass flux was obtained by calibration, using a flat plate and the stagnation region of a cylinder in crossflow for which the mass flux was known from laminar boundary layer analysis.

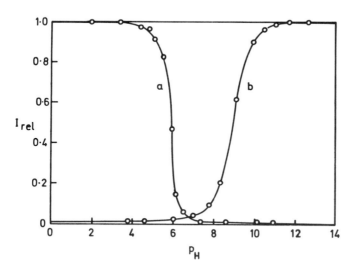

FIG. 9.16 Relative intensity I_{rel} of fluorescence as a function of the pH value for two chemical substances. (*a*) 2-Naphthochinolin. (*b*) Natrium-2-naphthol-6-sulfonate. (From Hiby, Braun, and Eickel, 1967, Ref. [31].)

Related to the above method is a study [31] in which fluorescence was used to determine the thickness of the layer on the surface of a liquid film into which a gas penetrates. Accurate measurements could be obtained regardless of the fact that this layer was only 0.1 mm thick. Figure 9.16 shows that the light intensity I of the two substances a and b, emitted when exposed to ultraviolet light, changes quite suddenly at a certain pH value. Substance b was used to measure the thickness of the alkaline absorption layer. The method also served for flow visualization and demonstrated clearly the Marangoni effect. This name describes local flows which arise in liquids when the absorbed substance creates local variations of surface tension.

9.3.3   Mass Transfer Analogy for a Variable Property Fluid

The mass transfer analogy has been used to a much smaller degree for a study of heat transfer in a variable property fluid. An early paper on this subject appears to be [32]. Local ablation rates and temperatures were measured on a flat plate with a naphthalene coating exposed to subsonic and supersonic air flow and the results, together with a turbulent boundary layer analysis by Deissler, were used to obtain information on local friction coefficients. Another contribution was discussed in Sec. 9.3.2 [6].

In the following lines, the basic considerations will be presented for a mass transfer analogy in a variable property fluid, for steady flow with a velocity small enough so that energy dissipation can be neglected (or for a gas with a Prandtl number equal to 1). For this purpose we consider two cases: Case 1, in which a one-component fluid undergoes a heat transfer process at large temperature differences and Case 2, in which a two-component fluid mixture undergoes a mass transfer process at large concentration differences but for the isothermal condition. The conservation equations are in this case written in the dimensioned parameters. They express for Case 1 conservation of mass, of momentum, and of energy, respectively.

$$\frac{\partial}{\partial x_i} (\rho v_i) = 0 \tag{9.18}$$

$$\rho v_i \frac{\partial v_j}{\partial x_i} = -\frac{\partial p}{\partial x_j} + \frac{\partial}{\partial x_i} \left[(\mu + \rho \epsilon) \left(\frac{\partial v_i}{\partial x_j} + \frac{\partial v_j}{\partial x_i} \right) - (\mu' + \rho \epsilon') \frac{\partial v_k}{\partial x_k} \delta_{ij} \right] \tag{9.19}$$

$$\rho v_i \frac{\partial i}{\partial x_i} = \frac{\partial}{\partial x_i} \left[\left(\frac{\mu}{\text{Pr}} + \rho \frac{\epsilon}{\text{Pr}_t} \right) \frac{\partial i}{\partial x_i} \right] \tag{9.20}$$

The fluid is assumed Newtonian and the equations are again written in tensor notation. The symbol i indicates the fluid enthalpy and δ_{ij} denotes the Kronneker delta with the definition $\delta_{ij} = 1$ for $i = j$ and $\delta_{ij} = 0$ for $i \neq j$. The symbol μ' denotes the viscosity connected with the normal stress and ϵ is a corresponding turbulent viscosity (see [33]).

A mass transfer process according to Case 2 is described by the system of Eqs. (9.18), (9.19), and (9.21):

$$\rho v_i \frac{\partial w}{\partial x_i} = \frac{\partial}{\partial x_i}\left[\left(\frac{\mu}{Sc} + \rho \frac{\epsilon}{Sc_t}\right)\frac{\partial w}{\partial x_i}\right] \tag{9.21}$$

where Eq. (9.21) expresses conservation of mass of one of the components, the mass fraction of which is denoted by w.

Similarity exists between the two cases when the variation of the properties ρ and μ as a function of temperature in Case 1 is similar to the variation of the same properties with mass fraction w in Case 2. In addition, the dependence of Pr on temperature must be similar to the dependence of Sc on mass fraction. The turbulent parameters Pr_t and Sc_t are equal to each other as was pointed out before. The similarity of the properties has to exist only in the parameter range which occurs in a specific problem, a condition which is found to be well approximated in various situations. For gases, the Prandtl and Schmidt numbers can usually be considered constant. The dependence of the density on temperature and on mass fraction respectively can be expressed according to the ideal gas laws through the following equations:

$$\rho = \frac{pM}{R(t + T_0)} \tag{9.22}$$

$$\rho = \frac{p}{RT} \frac{M_1 M_2 /(M_2 - M_1)}{w + M_1 /(M_2 - M_1)} \tag{9.23}$$

It can be observed that the similarity requirement is fulfilled. For the viscosity, this is not exactly the case; however, the variation is often expressible by a linear relation in certain ranges of temperature or mass fraction respectively. In this case, then, the similarity requirement is again fulfilled and Sherwood numbers obtained for a mass transfer situation can be readily interpreted as Nusselt numbers for a heat transfer situation with similar boundary conditions, by replacement of the Sherwood number in the dimensionless expressions by the Nusselt number and the Schmidt number by the Prandtl number.

NOMENCLATURE

A	area
c	mole concentration
D	mass diffusion coefficient
E	electric potential

F	Faraday's constant
g	gravitational acceleration
h	heat transfer coefficient
h_M	mass transfer coefficient
I	electric current per unit area
i	enthalpy
k	thermal conductivity
L_0	reference length
M	molecular weight
m	mass flow per unit area and time
$\dot{n}$	mole flow per unit area and time
p	pressure
R	gas constant
T	absolute temperature
t	temperature
v_0	reference velocity
v_i, v_j	velocity components
w	mass fraction
x_i, x_j	Cartesian coordinates
z	valence
α	thermal diffusivity
β	thermal expansion coefficient
ϵ	turbulent diffusivity of momentum
ϵ_H	turbulent diffusivity of heat
ϵ_M	turbulent diffusivity of mass
μ	viscosity
ν	kinematic viscosity
ρ	density
τ	time
$\mathrm{Re}_0 = \rho v_0 L_0/\mu$	Reynolds number
$\mathrm{Pr} = \nu/\alpha$	Prandtl number
$\mathrm{Pr}_t = \epsilon/\epsilon_H$	turbulent Prandtl number
$\mathrm{Nu} = hL_0/k$	Nusselt number
$\mathrm{Sc} = \nu/D$	Schmidt number
$\mathrm{Sc}_t = \epsilon/\epsilon_M$	turbulent Schmidt number
$\mathrm{Sh} = h_m L_0/D$	Sherwood number
$\mathrm{Gr}_0 = [g\beta L_0^3(t_\infty - t_s)]/\nu^2$	Grashof number for heat transfer
$\mathrm{Gr}_{0M} = [gL_0(\rho_\infty - \rho_s)]/\nu^2\rho$	Grashof number for mass transfer
$\mathrm{Ra}_0 = [g\beta L_0^3(t_\infty - t_s)]/\nu\alpha$	Rayleigh number for heat transfer
$\mathrm{Ra}_{0M} = [gL_0^3(\rho_\infty - \rho_s)]/\nu\alpha\rho$	Rayleigh number for heat transfer

Subscripts

0	reference state
H	heat
M	mass

s surface

v vapor

∞ in mainstream

REFERENCES

1. Gershun, T., "A Study of Conductive Heat Transfer with Change of Phase-Mathematical and Analogue Solutions," Ph.D. thesis, University of Minnesota, Minneapolis, 1955.
2. Eckert, E., *Z. Ver. deut. Ingr.*, vol. 79, p. 1495, 1955.
3. Thoma, H., "Hochleistungskessel," Julius Springer, Berlin, 1921.
4. Lohrisch, W., *Forschung. Geb. Ingenieur.*, no. 322, 1929.
5. Malhotra, R. C., and Cermak, J. E., Mass Diffusion in Neutral and Unstably Stratified Boundary Layer Flow, *Int. J. Heat Mass Transfer*, vol. 7, pp. 169–186, 1964.
6. Pedersen, D. R., "Effect of Density Ratio on Film Cooling Effectiveness for Injection through a Row of Holes and for a Porous Slot," Ph.D. thesis, University of Minnesota, Minneapolis, March 1972.
7. Hilpert, R., *VDI-Forschungsheft*, no. 355, 1932.
8. Powell, R. W., and Griffiths, E., *Trans. Inst. Chem. Engs. (Brit.)*, vol. 13, p. 175, 1935.
9. Sogin, H. H., and Jakob, M., Heat and Mass Transfer from Slender Cylinders to Air Streams in Axisymmetrical Flow, *Heat Transfer Fluid Mech. Inst.*, Preprint of papers, p. 5, Stanford Univ. Press, Stanford, Calif., 1953.
10. Christian, W. J., and Kezios, S. P., Experimental Investigation of Mass Transfer by Sublimation from Sharp-Edged Cylinders in Axisymmetric Flow with Laminar Boundary Layer, *1957 Heat Transfer Fluid Mech. Inst.*, p. 359, Stanford Univ. Press, Stanford, Calif., 1957.
11. Ko, Shao-Yen, and Sogin, H. H., Laminar Mass and Heat Transfer from Ellipsoidal Surfaces of Fineness Ratio 4 in Axisymmetric Flow, *Trans. ASME*, vol. 80, p. 387, 1958.
12. Sogin, H. H., Sublimation from Disks to Air Streams Flowing Normal to Their Surface, *Trans. ASME*, vol. 80, p. 593, 1958.
13. Kreith, K., Taylor, J. H., and Chong, J. P., Heat and Mass Transfer from a Rotating Disk, *J. Heat Transfer*, vol. 81, p. 95, 1959.
14. Sogin, H. H., Laminar Transfer from Isothermal Spanwise Strips on a Flat Plate, *J. Heat Transfer*, vol. 82, p. 53, 1960.
15. Sogin, H. H., and Goldstein, R. J., Turbulent Transfer from Isothermal Spanwise Strips on a Flat Plate, *Proc. 1961-62 Heat Trasnfer Conf. ASME*, p. 447, 1963.
16. Saboya, F. E. M., and Sparrow, E. M., Local and Average Transfer Coefficients for One-Row Plate Fin and Tube Heat Exchanger Configurations, *J. Heat Transfer*, vol. 96, p. 265, 1974.
17. Wagner, C., Theorie und Experiment in der elektrochemischen Verfahrenstechnik, *Chem. Ing. Tech.*, vol. 32, p. 1, 1960.
18. Wilke, C. R., Tobias, C. W., and Eisenberg, M., Free-Convection Mass Transfer at Vertical Plates, *Chem. Eng. Progr.*, vol. 49, p. 663, 1953.
19. Grassmann, P., Ibl, N., and Trüb, J., Elektrochemische Messung von Stoffübergangszahlen, *Chem. Ing. Tech.*, vol. 33, p. 529, 1961.
20. Schutz, G., Untersuchung des Stoffaustausch-Anlaufgebietes in einem Rohr bei vollausgebildeter hydrodynamischer Strömung mit einer elektrochemischen Methode, *Int. J. Heat Mass Transfer*, vol. 7, p. 1077, 1964.
21. Kozdoba, L. A., and Zagoruiko, V. A., Solution of Heat and Mass Transfer Problems by the Electrical Analogy Method, *Int. Chem. Eng.*, vol. 7, p. 202, 1967.
22. Newman, J., The Effect of Migration in Laminar Diffusion Layers, *Int. J. Heat Mass Transfer*, vol. 10, p. 983, 1967.

23. Hanratty, T. J., Study of Turbulence Close to a Solid Wall, *Phys. Fluids, Suppl.*, S126, 1967.

24. Son, J. S., and Hanratty, T. J., Limiting Relation for the Eddy Diffusivity Close to a Wall. *A.I.Ch.E. J.*, vol. 13, p. 689, 1967.

25. Mizushina, T., The Electrochemical Method in Transport Phenomena, in T. F. Irvine and J. P. Hartnett (eds.), "Advances in Heat Transfer," vol. 7, Academic, New York, 1971.

26. Lloyd, J. R., Sparrow, E. M., and Eckert, E. R. G., Local Natural Convection Mass Transfer Measurements, *J. Electrochem. Soc.*, vol. 119, pp. 702–707, 1972.

27. Colello, R. G., and Springer, G. S., Mass-Transfer Measurements with the Technique of Electrochemiluminescence, *Int. J. Heat Mass Transfer*, vol. 9, p. 1391, 1966.

28. Springer, G. S., Use of Electrochemiluminescence in the Measurement of Mass-Transfer Rates, *Rev. Sci. Instrum.*, vol. 35, p. 1277, 1964.

29. Howland, B., Springer, G. S., and Hill, M. G., Use of Electrochemiluminescence in Visualizing Separated Flows, *J. Fluid Mech.*, vol. 24, p. 697, 1966.

30. Luikov, A. V., Shulman, Z. P., and Puris, B. I., Mass Transfer of a Cylinder in Forced Non-Newtonian Liquid Flow, Preprints, *All-Union Heat Mass Transfer Conf.*, 3rd, Minsk, 1968.

31. Hiby, J. W., Braun, D., and Eickel, K. H., Eine Fluoreszenzmethode zur Untersuchung des Stoffübergangs bei der Gasabsorption im Rieselfilm, *Chem. Ing. Tech.*, vol. 39, p. 297, 1967.

32. Sherwood, T. K., and Träss, O., Sublimation Mass Transfer through Compressible Boundary Layers on a Flat Plate, *J. Heat Transfer*, vol. 82, p. 313, 1960.

33. Eckert, E. R. G., and Drake, R. M., Jr., Analysis of Heat and Mass Transfer, pp. 258–270, McGraw-Hill, New York, 1972.

10 Thermal radiation measurements

D. K. EDWARDS
University of California, Los Angeles

Man cannot yet calculate from "first principles" the radiation received from the sun. But he can measure without great difficulty the heat radiation received within a few percentage points. One cannot predict with any confidence from first principles the infrared reflectance of sandblasted aluminum. But one can measure the reflectance rather easily to within a few percentage points. With knowledge gained from such measurements engineers can predict and can arrange to use or provide for radiant energy transfers. Measurements together with a careful assessment of their probable accuracy are essential to the applied worker.

Thermal radiation measurements fall into two main categories: Measurement of the radiant energy received from some source and measurement of the radiation characteristics of materials, often expressed as a ratio of power reflected, absorbed, or transmitted to power incident. Some aspects of problems in both categories will be reviewed briefly in what follows, but the entire field of thermal radiation measurements is too broad to be reviewed in a short work. A worker in the field must have knowledge of applications, radiative transfer theory, the physics of interactions between radiation and matter, optics, and, by no means least important, hardware.

In what follows there is first a review of the concept of radiant intensity which is basic to the subject of radiation heat transfer and measurement techniques (Sec. 10.1). Included in this first section is a brief treatment of coherency and polarization, which is usually neglected in discussions of thermal radiation. While such phenomena are absent at thermodynamic equilibrium, they arise naturally as soon as the departures from strict equilibrium necessary for net energy transfer occur, and these phenomena often play a role in radiation instruments. There is then a short review of radiation characteristics which affect instrument performance and which themselves are often the object of a measurement (Sec. 10.2). Brief reviews of the fields of radiometry (Sec. 10.3) and radiation property measurement (Secs. 10.4 and 10.5) are then given. In these reviews the reader may note a recurring theme, an admonition to formulate the measured quantity and to consider at least the first-order departures from the elementary concept forming the basis for the measurement technique.

10.1 THERMAL RADIATION

10.1.1 Blackbody Radiation

Thermal agitations of particles with charge give rise to electromagnetic radiation within matter in the gaseous, liquid, or solid state. Bose-Einstein statistics [1] indicate that the number of photon wave packets per unit volume with energies between ϵ' and $\epsilon' + d\epsilon'$ is, at thermodynamic equilibrium,

$$\frac{dN}{d\epsilon'} = \frac{8\pi\epsilon'^2/h^3c^3}{e^{\epsilon'/kT} - 1} \tag{10.1}$$

where k is Boltzmann's constant (1.38054×10^{-23} joule/K) and c is the velocity of the radiation (2.997925×10^8 m/sec in vacuum). At equilibrium the directions of propagation are uniformly distributed over a sphere of 4π steradians. The number of photons crossing a small area dA within a small pencil of solid angle $d\Omega$ around a direction θ, ϕ (see Fig. 10.1) is then

$$dn = \frac{dN}{d\epsilon'} d\epsilon' \frac{d\Omega}{4\pi} c \, dA \cos\theta \tag{10.2}$$

Each photon carries energy ϵ' which is related to wavenumber ν or wavelength λ by Einstein's photoelectric law

$$\epsilon' = hc\nu = \frac{hc}{\lambda} \qquad hc \, d\nu = |d\epsilon'| = \left(\frac{hc}{\lambda^2}\right) d\lambda \tag{10.3}$$

where h is Planck's constant (6.6256×10^{-34} joule-sec). The power per unit projected area per unit solid angle per unit spectral band width is then

$$I_b(\lambda, T) = \frac{\epsilon' \, dn}{\cos\theta \, dA \, d\Omega \, d\lambda} = \frac{2hc^2 \lambda^{-5}}{e^{hc/\lambda kT} - 1} \tag{10.4a}$$

$$I_b(\nu, T) = \frac{\epsilon' \, dn}{\cos\theta \, dA \, d\Omega \, d\nu} = \frac{2hc^2\nu^3}{e^{hc\nu/kT} - 1} \tag{10.4b}$$

This quantity is referred to in engineering literature as the Planck blackbody spectral radiant intensity [2].

A particularly useful relation is the fraction of Planckian radiation having wavelengths between 0 and λ

$$f_b(\lambda, T) = f_b(\lambda T) = \frac{\int_0^\lambda I_b(\lambda', T)\, d\lambda'}{\int_0^\infty I_b(\lambda', T)\, d\lambda'} \tag{10.5}$$

That the external Planck fraction f_b is a function of only the product λT may be seen readily by making a change of variable from v or λ to hcv/kT or $k\lambda T/hc$ and noting that

$$\int_0^\infty \frac{x^3\, dx}{e^{ax} - 1} = \frac{\pi^4}{15a^4} \tag{10.6}$$

For engineering purposes the dimensional variable λT used by Dunkle [3] usually serves best. (The reader may consult Chapter 2 of [4] for another view.) Czerny and Walther [5] term f_b the fractional function of the first kind. The notation $f_b(\lambda T)$ makes it possible to write the blackbody intensity as

$$I_b(\lambda, T) = \frac{1}{\pi}\, \sigma T^5 f_b'(\lambda T) \tag{10.7}$$

where the prime denotes differentiation with respect to the argument λT, and the quantity σ is the Stefan-Boltzmann constant

$$\sigma = \pi \left(\frac{2hc^2}{15}\right)\left(\frac{\pi k}{hc}\right)^4 \tag{10.8}$$

Laboratory blackbody or hohlraum sources are used to approach Planckian radiation. Such sources are often a heated tube with a dark specularly reflecting wall and a grooved dark back end. Sometimes the front of the tube is covered with a plate having a small orifice. A number of commercial models are available (e.g., Table 3-3 of [4]). A good source should be made with a thick wall of highly conducting material and be heated so as to minimize temperature gradients. Heating by means of condensing vapor in a jacket gives good uniformity in temperature. Errors due to a slightly cold front end may be partially mitigated by not viewing that portion of the cavity directly and giving it a low emittance. Of course, the temperature of the high emissivity zone viewed must be determined in an accurate manner.

10.1.2 Radiant Intensity

A geometrical optics viewpoint is most often taken, because of its practicality and simplicity. The photons are imagined to travel in straight lines along paths from one area of concern to another, the path lengths being much greater than the photon wavelength being considered. The optical elements themselves may have films or particles small compared to wavelength or features regular even in terms of wavelengths, but the behavior of these elements is described in terms of radiation characteristics. In addition, the state of the matter in the optical elements is taken to be that characterized by a temperature T describing an equilibrium condition. Heat conduction (through collisions) is often sufficiently large to make such a characterization reasonable.

From this viewpoint, the quantity most often desired by a worker interested in calculating radiation heat transfer rates is the spectral radiant intensity I, the power per unit projected area, per unit solid angle, per unit band width. If this quantity is known, power transfer rates can easily be found by integration (usually numerical integration). One integration is made over the spectrum of photon wavelengths (or wavenumbers or energies), and a second is made over the hemisphere over an area. Denote the first by a total transfer operator operating upon a spectral intensity I

$$\overline{\mathfrak{F}}\,[I(\theta,\phi,\lambda)] = \int_0^\infty I(\theta,\phi,\lambda)\,d\lambda \tag{10.9}$$

Denote the second integration by a hemispherical transfer operator operating upon I

$$\overline{\mathfrak{K}}[I(\theta,\phi,\lambda)] = \int_0^{2\pi}\int_0^{\pi/2} I(\theta,\phi,\lambda)\sin\theta\cos\theta\,d\theta\,d\phi \tag{10.10}$$

where θ and ϕ are polar and azimuthal angles as shown in Fig. 10.1. The radiant power per unit area leaving one side of a plane is then

$$q_R^+ = \overline{\mathfrak{F}}\,\overline{\mathfrak{K}}[I^+(\theta,\phi,\lambda)]\,\text{W/m}^2 \tag{10.11}$$

The order of the two operators may be reversed, of course. The operators operating on the blackbody intensity give the Stefan-Boltzmann relation

$$\overline{\mathfrak{F}}\overline{\mathfrak{K}}\,[I_b(\lambda,T)] = \sigma T^4 \tag{10.12}$$

A superscript $+$ or $-$ is used to distinguish whether or not the photons are traveling away from or toward the surface for which the θ, ϕ angles are defined. The quantity q_R^+ is often termed the *radiosity* (symbol J or B) and q_R^- the *irradiation* (symbol G or H). The net radiation heat flux across a control surface is then

$$q_R = q^+ - q^- \tag{10.13}$$

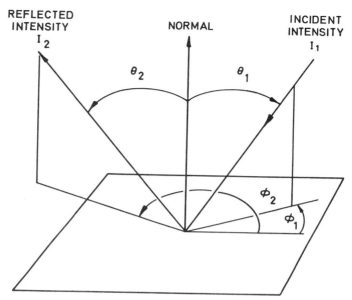

FIG. 10.1 Polar and azimuthal angles of incidence and emergence.

Two such control surfaces may be taken in the vicinity of an interface between two media. Spalding [6] has named these the S surface just inside the considered medium and the L surface just within the second one. Since the volume between the S and L surfaces is arbitrarily small, the net flux q_R across one surface is the same as that across the other. The q^+ crossing the S surface is referred to as the emissive power when the irradiation q^- at the S surface is zero, since it is the result of nonexternally induced emission within the second medium. For a strongly absorbing second medium it is possible to consider an M surface somewhat below the L surface where the net radiative flux is not distinguished from the conductive flux. For quasi-steady state conditions the conductive heat flux crossing the M surface will be the same as the sum of the radiation and conduction, if any, crossing the L surface (and the S surface, when no mass transfers occur).

10.1.3 Coherence and Polarization

In the classical limit, the stream of photon wavepackets forms a continuous propagating electromagnetic wave whose intensity I is proportional to the square of the magnitude of the electric field vector e, which oscillates perpendicular to the direction of propagation. The electric vector e can be resolved into a set of mutually perpendicular components, e_p and e_s. Let those components (and the square root of the constant of proportionality giving I) be represented by the real parts of the following complex numbers [7]:

$$z_p = M_p \exp \{i [2\pi\nu(ct - S) + \delta_p]\} \qquad (10.14a)$$

$$z_s = M_s \exp \{i [2\pi\nu(ct - S) + \delta_s]\} \qquad (10.14b)$$

where ν is the wavenumber, t time, S path length, δ the instantaneous phase angle, and i the unit imaginary number, $i = \sqrt{-1}$. The radiant intensity I is the sum of the intensities of the components

$$I = I_p + I_s \qquad (10.15)$$

where $\qquad\qquad I_p = \langle z_p z_p^* \rangle \qquad I_s = \langle z_s z_s^* \rangle \qquad (10.16)$

The asterisk denotes the complex conjugate, and the angular brackets denote a time and wavelength average over a short time period around time t, and a narrow wavelength band around wavelength λ. The brackets are a tacit admission that the experimenter can never measure an instantaneous power flux or an infinitely narrow spectral band, since a detector responds only to a finite amount of energy.

Planckian radiation has a random character; there is no fixed relation between δ_p and δ_s, and I_p equals I_s. Such radiation is said to be "incoherent" in phase, "unpolarized," or "natural." However, reflection of Planckian radiation from a plane surface at an angle from the normal, or transmission of Planckian radiation through an optical element, may alter the ratio of M_p to M_s and/or cause a fixed shift in δ_p or δ_s for a particular set of s and p directions. The fact that such modifications occur means that the spectral radiant intensity alone is not sufficient to describe the nature of the radiant flux and the way that it interacts with matter.

Suppose one considers, say, transmission of a beam of radiant intensity I through a retardation plate, causing a shift in δ_s of Δ_s and a shift in δ_p of Δ_p, followed by reflection from a glossy black paint at an angle of $55°$ or so, an angle near the Brewster angle. One finds that, in order to predict the amount of radiation absorbed by the black paint, one must know not only $\langle z_p z_p^* \rangle$ and $\langle z_s z_s^* \rangle$ but also the real and imaginary parts of $\langle 2z_p z_s^* \rangle$. These latter quantities are the third and fourth Stokes coefficients [8] I_u and I_v. One must know a four-component vector then in order to describe properly the radiant intensity

$$[I] = \begin{bmatrix} I_p \\ I_s \\ I_u \\ I_v \end{bmatrix} = \begin{bmatrix} z_p z_p^* \\ z_s z_s^* \\ \Re(2z_p z_s^*) \\ \Im(2z_p z_s^*) \end{bmatrix} \qquad (10.17)$$

The intensity I is the sum of the first two components

$$I = [E]^t \cdot [I] \qquad [E]^t = [1 \ 1 \ 0 \ 0] \qquad (10.18)$$

and the blackbody intensity vector has only two nonzero components

$$[I_b] = [x_b]I_b \qquad [x_b] = \begin{bmatrix} 1/2 \\ 1/2 \\ 0 \\ 0 \end{bmatrix} \qquad (10.19)$$

Reference [7] shows also how a transformation from one set of p_1, s_1 directions to a second set p_2, s_2 rotated by angle β from the first. Resolving the z_{p1} and z_{s1} vectors into new components and applying Eq. (10.17) shows that the transformation can be represented compactly by matrix multiplication

$$\begin{bmatrix} I_{p2} \\ I_{s2} \\ I_{u2} \\ I_{v2} \end{bmatrix} = \begin{bmatrix} \cos^2\beta & \sin^2\beta & \frac{1}{2}\sin 2\beta & 0 \\ \sin^2\beta & \cos^2\beta & -\frac{1}{2}\sin 2\beta & 0 \\ -\sin 2\beta & \sin 2\beta & \cos 2\beta & 0 \\ 0 & 0 & 0 & 1 \end{bmatrix} \cdot \begin{bmatrix} I_{p1} \\ I_{s1} \\ I_{u1} \\ I_{v1} \end{bmatrix} \qquad (10.20)$$

$$[I_2] = [D_{1-2}] \cdot [I_1] \qquad (10.20a)$$

Polarization arises incidentally or inadvertently in many radiation instruments. Reflection from a mirror at an angle from the normal, or transmission through a prism when the beam enters or emerges at an off-normal angle, introduces such polarization. One might insert other polarizing elements deliberately into such an instrument to counteract the incidental polarization or to cause complete polarization so that the measurement will be more precisely defined. Analysis of the effect of such elements can be made using the radiant intensity vector just introduced and radiation characteristics defined below.

10.2 THERMAL RADIATION CHARACTERISTICS

10.2.1 Specular Surface of an Isotropic Material

An isotropic material with a plane surface reflects specularly, that is, the radiant intensity incident at θ_1, ϕ_1 causes a reflected beam at $\theta_2 = \theta_1$ and $\phi_2 = \phi_1 + \pi$. The magnitudes of the reflected components parallel and perpendicular to the plane of incidence are changed and the phases are changed so that

$$
\begin{bmatrix} I_{p2} \\ I_{s2} \\ I_{u2} \\ I_{v2} \end{bmatrix} = \begin{bmatrix} \rho_{pp} & 0 & 0 & 0 \\ 0 & \rho_{ss} & 0 & 0 \\ 0 & 0 & \rho_{uu} & \rho_{uv} \\ 0 & 0 & \rho_{vu} & \rho_{vv} \end{bmatrix} \begin{bmatrix} I_{p1} \\ I_{s1} \\ I_{u1} \\ I_{v1} \end{bmatrix} \tag{10.21}
$$

where the reflectivity matrix elements are given by the Fresnel relations (see also Sec. 8.5 of [4]):

$$
\rho_{ss} = \frac{(\cos\theta - a)^2 + b^2}{(\cos\theta + a)^2 + b^2} \tag{10.22a}
$$

$$
\rho_{pp} = \frac{[(n^2 - k^2)\cos\theta - a]^2 + [2nk\cos\theta - b]^2}{[(n^2 - k^2)\cos\theta + a]^2 + [2nk\cos\theta + b]^2} \tag{10.22b}
$$

$$
\rho_{uu} = \rho_{vv} = \sqrt{\rho_{ss}\rho_{pp}} \cos\delta \tag{10.22c}
$$

$$
\rho_{vu} = -\rho_{uv} = \sqrt{\rho_{ss}\rho_{pp}} \sin\delta \tag{10.22d}
$$

$$
\delta = 2\pi + (\Delta_p - \Delta_s) \tag{10.22e}
$$

$$
\tan\Delta_p = \frac{2\cos\theta\,[(n^2 - k^2)b - (2nk)a]}{(n^2 + k^2)^2 \cos^2\theta - (a^2 + b^2)} \tag{10.22f}
$$

$$
\tan\Delta_s = \frac{2b\cos\theta}{\cos^2\theta - a^2 - b^2} \tag{10.22g}
$$

$$
a^2 = \frac{1}{2}\,[\sqrt{(n^2 - k^2 - \sin^2\theta)^2 + 4n^2 k^2} + (n^2 - k^2) - \sin^2\theta]
$$
$$
\tag{10.22h}
$$

$$
b^2 = \frac{1}{2}\,[\sqrt{(n^2 - k^2 - \sin^2\theta)^2 + 4n^2 k^2} - (n^2 - k^2) + \sin^2\theta]
$$
$$
\tag{10.22i}
$$

The above relations are written for a medium of $n = 1$, $k = 0$ (a vacuum) above a material of refractive index n and absorptive index k. The subscript p denotes a direction parallel to the plane of incidence, and s designates one perpendicular to that plane, fixed by the surface normal and incident ray.

For dielectrics k is much smaller than n, and for metals k is generally greater than n, but when the Hagen-Ruebens reflection law holds at long wavelengths, k equals n. Most low density gases have n very close to unity and k zero away from absorption bands, but the refractive index of air, 1.000293, is sufficiently different from that of helium, 1.000036, to make a noticeable shift in calibration when helium is used to purge a prism monochromator.

Figure 10.2 shows how the specular reflectance elements ρ_{pp} and ρ_{ss} vary with angle from the normal. Note that at an angle of $\theta_B = \sin^{-1} n$ reflection is strongly polarizing. One can note also that the reflectivity of poor reflectors like

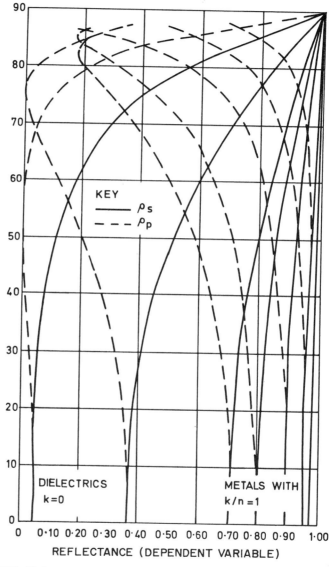

ANGLE OF INCIDENCE, DEGREES
(INDEPENDENT VARIABLE)

KEY
—— ρ_s
--- ρ_p

DIELECTRICS
k = 0

METALS WITH
k/n = 1

REFLECTANCE (DEPENDENT VARIABLE)

FIG. 10.2 Reflection components for a plane interface between isotropic media. (From Edwards and Bevans, 1965, Ref. [7].)

433

glossy paint resins rises to high values at large angles from the normal. For metals the ρ_{pp} reflectance drops at high angles from the normal and then increases to unity just at completely grazing incidence.

The directional reflectivity of a material irradiated by (partly) polarized radiation is $[\rho_v]$ where

$$[\rho_v]^t = [E]^t \cdot [\rho_M] \tag{10.23}$$

For an isotropic material

$$[\rho_v]^t = [\rho_{pp} \ \rho_{ss} \ 0 \ 0] \tag{10.23a}$$

The superscript t denotes a row matrix, that is, the transpose of the column matrix. The quantity $[\rho_M]$ is the matrix in Eq. (10.20), and $[E]^t$ was defined in Eq. (10.18). The intensity reflected is

$$I_2 = [\rho_v]^t [I_1] \tag{10.24}$$

If I_1 is incoherent, we can write

$$I_2 = [\rho_v]^t [x_b] I_1 = \rho(\theta_1, \phi_1) I_1 \tag{10.25}$$

so that the directional reflectivity $\rho(\theta, \phi)$ is

$$\rho(\theta, \phi) = [\rho_v]^t [x_b] \tag{10.26}$$

For the isotropic solid

$$\rho(\theta) = \frac{1}{2} [\rho_{pp} + \rho_{ss}] \tag{10.26a}$$

The directional absorptivity of a material sufficiently thick to be opaque is, by conservation of energy,

$$\alpha(\theta, \phi) = 1 - \rho(\theta, \phi) \tag{10.27}$$

If the irradiation is polarized, we must use the directional absorptivity vector

$$[\alpha_v]^t = [E]^t - [\rho_v]^t \tag{10.28}$$

The heat flux absorbed is then

$$q_{ABS} = \bar{\jmath} \, \bar{\mathfrak{K}} \{ [\alpha_v]^t \cdot [I_S^-] \} \tag{10.29}$$

At thermodynamic equilibrium the principle of detailed balancing [1] requires among other things that photons absorbed from a certain direction and of a certain energy be re-emitted back to their source. This aspect of the principle of detailed balancing is called Kirchhoff's law, which has several forms:

$$[\epsilon_v]^t = [\alpha_v]^t \tag{10.30}$$

$$\epsilon(\theta, \phi, \lambda, T) = [\epsilon_v]^t \cdot [x_b] = [\alpha_v]^t \cdot [x_b] = \alpha(\theta, \phi, \lambda, T) \tag{10.30a}$$

$$\bar{\mathfrak{J}}[\epsilon(\theta, \phi, \lambda, T)I_b(\lambda, T)] = \bar{\mathfrak{J}}[\alpha(\theta, \phi, \lambda, T)I_b(\lambda, T)]$$

$$\epsilon_T(\theta, \phi, T)\left(\frac{\sigma T^4}{\pi}\right) = \alpha_T(\theta, \phi, T, T)\left(\frac{\sigma T^4}{\pi}\right) \tag{10.30b}$$

$$\epsilon_T(\theta, \phi, T) = \alpha_T(\theta, \phi, T, T)$$

$$\bar{\mathcal{K}}\bar{\mathfrak{J}}[\epsilon(\theta, \phi, \lambda, T)I_b(\lambda, T)] = \bar{\mathcal{K}}\bar{\mathfrak{J}}[\alpha(\theta, \phi, \lambda, T)I_b(\lambda, T)]$$

$$\epsilon_{TH}(T) = \alpha_{TH}(T, T) \tag{10.30c}$$

The quantities with subscript T are the total emissivity and absorptivity. Those with subscript H are the hemispherical emissivity and absorptivity. The double subscript TH indicates total hemispherical emissivity and absorptivity. These are defined as indicated above Eqs. (10.30b) and (10.30c). The total absorptivity is

$$\alpha_T(\theta, \phi, T, T_s) = \mathfrak{J}_{T_s}[\alpha(\theta, \phi, \lambda, T)] \tag{10.31}$$

where the operator is the total property operator for blackbody radiation with source temperature T_s.

$$\mathfrak{J}_{T_s}[\alpha(\theta, \phi, \lambda, T)] = \int_0^\infty \alpha(\theta, \phi, \lambda, T)f_b'(\lambda T_s)\, d(\lambda T_s) \tag{10.32}$$

The hemispherical absorptivity is

$$\alpha_H(\lambda, T) = \mathcal{K}[\alpha(\theta, \phi, \lambda, T)] \tag{10.33}$$

where the hemispherical property operator is

$$\mathcal{K}[\alpha(\theta, \phi, \lambda, T)] = \frac{1}{\pi}\int_0^{2\pi}\int_0^{\pi/2} \alpha(\theta, \phi, \lambda, T)\sin\theta\,\cos\theta\,d\theta\,d\phi \tag{10.34}$$

In this convenient notation the total hemispherical absorptivity is then

$$\alpha_{TH}(T, T_s) = \mathfrak{K}\mathfrak{I}_{T_s}\{[\alpha_v]^t \cdot [x_b]\} \tag{10.35}$$

In many engineering problems involving radiation transfer within an enclosure a second kind of total average is desired. This total net quantity may be defined by

$$\epsilon_{T_n}(T, T_s) = \frac{\epsilon_T(T)\sigma T^4 - \alpha_T(T, T_s)\sigma T_s^4}{\sigma T^4 - \sigma T_s^4} \tag{10.36}$$

Often the limit of $\epsilon_{T_n}(T, T_s)$ as T_s approaches T is desired. In our operator notation this internal total emissivity is $\epsilon_{T_i}(T)$

$$\epsilon_{T_i}(\theta, \phi, T) = \mathfrak{I}_{T_i}[\alpha(\theta, \phi, \lambda, T)] \tag{10.37}$$

where $$\mathfrak{I}_{T_i}[\alpha(\theta, \phi, \lambda, T)] = \int_0^\infty \alpha(\theta, \phi, \lambda, T)f_i'(\lambda T) \, d(\lambda T) \tag{10.38}$$

where, from Eqs. (10.5) and (10.7), the internal fraction f_i is given by

$$f_i'(\lambda T) = \frac{5}{4} f_b'(\lambda T) + \frac{1}{4} (\lambda T)f_b''(\lambda T) \tag{10.39}$$

As in Eq. (10.7) the prime represents differentiation with respect to the argument. As before this internal total operator may be combined with the hemispherical operator to obtain an internal total hemispherical emissivity. The quantity in Eq. (10.30c) can be referred to as the *external total hemispherical emissivity*. The adjectives *internal* and *external* are used because the internal characteristic is usually appropriate for internal surfaces in a space vehicle, and the external characteristic is appropriate for external surfaces. Confusion of the total net emittance with the total external emittance has been referred to as the *nongray error* [9]. The internal fraction f_i is tabulated in [5]. There it is called the *fractional function* of the second kind (symbol B^*).

10.2.2 Specular Surfaces of Anisotropic Materials

In general it is possible that all 16 terms in the four-by-four reflectance matrix in Eq. (10.21) could be nonzero. Equations (10.23) and (10.26) to (10.30) would still hold. In general it is also possible to have a thin isothermal slab which transmits. In this case a transmittance matrix similar to the reflectance matrix exists to describe the transmitted intensity. Equation (10.28) is then replaced with

$$[E]^t = [\alpha_v]^t + [\rho_v]^t + [\tau_v]^t \tag{10.40}$$

10.2.3 Diffuse Surfaces

If the surface is rough or if the material below the surface contains scattering particles, the surface is said to be *imperfectly* diffuse. In this case the reflected radiant intensity in any direction θ_2, ϕ_2, not necessarily the specular direction but not excluding it, is given by the bidirectional reflectance matrix

$$d[I_2] = \frac{1}{\pi} [\rho_{BM}] [I_1] \cos \theta_1 \, d\Omega_1 \qquad (10.41)$$

The bidirectional reflectance (scalar) is the value that results when the irradiation is incoherent. In this case the reflected intensity is

$$dI_2 = \frac{1}{\pi} \rho_B(\theta_1, \phi_1, \theta_2, \phi_2, \lambda, T) I_1 \cos \theta_1 \, d\Omega \qquad (10.42)$$

where $$\rho_B = [E]^t [\rho_{BM}] [x_b] \qquad (10.43)$$

Apparatuses to measure ρ_B are called *bidirectional reflectometers* or *goniospectrophotometers*. Early apparatus [10] was used for total measurements in illumination engineering.

The principle of detailed balancing requires that there be no net exchange between two radiators via the surface when equilibrium exists. This requirement is expressed by

$$[\rho_{BM}(\theta_1, \phi_1, \theta_2, \phi_2, \lambda, T)] = [\rho_{BM}(\theta_2, \phi_1, \theta_1, \phi_2, \lambda, T)] \qquad (10.44)$$

$$\rho_B(\theta_1, \phi_1, \theta_2, \phi_2, \lambda, T) = \rho_B(\theta_2, \phi_1, \theta_1, \phi_2, \lambda, T) \qquad (10.44a)$$

This requirement is often called the *Helmholtz reciprocity principle* [11] and is, unfortunately, not well understood by some workers. This principle has had a profound influence upon the design of directional reflectometers and emittance inspection devices as will be shown.

If it is desired to know all the radiation reflected regardless of its directional distribution, one integrates the bidirectional reflectance over the hemisphere

$$[\rho_v(\theta_1, \phi_1, \lambda, T)]^t = \mathcal{K}_2 \{[E]^t \cdot [\rho_{BM}(\theta_1, \phi_1, \theta_2, \phi_2, \lambda, T)]\} \qquad (10.45)$$

For incoherent irradiation the directional reflectance is

$$\rho(\theta_1, \phi_1, \lambda, T) = \mathcal{K}_2 \{[E]^t \cdot [\rho_{BM}] \cdot [x_b]\} \qquad (10.46)$$

As a result of the Helmholtz reciprocity, Eq. (10.46) can just as well be written

$$\rho(\theta_2, \phi_2, \lambda, T) = \mathfrak{K}_1\left[\rho_B(\theta_1, \phi_1, \theta_2, \phi_2, \lambda, T)\right] \qquad (10.46a)$$

But the experimental arrangement implied is quite different. For the case described by Eq. (10.46), a beam of radiation is brought onto the specimen, and all radiation reflected is detected. For the case described by Eq. (10.46a) the specimen is perfectly diffusely irradiated, and the reflected intensity in direction θ_2, ϕ_2 is compared to the intensity of the irradiation. Because of the completely different experimental arrangements the quantity in Eq. (10.46) has been called the *angular-hemispherical* or *directional-hemispherical reflectance* and that in Eq. (10.46a) called the *hemispherical-angular* or *hemispherical-directional reflectance* [12]. The quantities are, of course, equal except for errors in measurement or for departures from thermodynamic equilibrium populations. While the redundance in the double adjective has the desirable effect of calling the attention of the reader to the measurement technique, it has the undesirable side effect of confusing some as to whether the quantity obtained by subtraction from unity (for an opaque surface) is the directional or hemispherical absorptance (it is the directional absorptance).

Total and hemispherical characteristics of imperfectly diffuse specimens are defined as was previously done for the isotropic solid. One simply applies the total and hemispherical operators to the directional characteristics. Note that the word *characteristic* is used in preference to *property* when a complex "surface system" is involved. Note also the use of the *ance* ending in preference to the *ivity* ending for such systems. Finally one should keep in mind that all the characteristics are defined for the nearly isothermal system in the near-equilibrium state. Transmittances are defined only for thin slabs in which lateral photon movement is negligible.

10.3   RADIOMETRY

10.3.1   Detectors, Signals, and Noise

Two broad classes of detectors are in common usage, photoelectric and thermal. The first broad class depends upon a photon-electron interaction. In the photomultiplier vacuum tube, a photon causes emission of an electron which is accelerated by an electric field to collide with a second emitter and so release a cascade of electrons which are detected. In a barrier layer photovoltaic cell a photon causes an electron to be elevated from the semiconductor's valence band into a conduction band. When the electrical charge carrier diffuses across the barrier, power is delivered to an external circuit. In a photoconductor the creation of a hole-electron pair in a solid or an ion-electron pair in a gas allows current to flow under an imposed field caused by an external bias. In film or in the eye photochemical reactions occur due to photon-electron interactions.

These photon-electron detectors tend to be spectrally selective to a high degree. At photon energies below that necessary to overcome the surface work function of a photoemitter or to enable an electron to cross the band gap in a semiconductor, the chances of a detectable photon-electron interaction are quite small. At photon energies above this threshold value, the excess energy causes no increase in detector signal, so that the relative sensitivity of the detector tends to decrease directly proportional to wavelength, as λ becomes shorter. On the other hand these detectors have a high sensitivity and a fast response time (1–1000 μs). Photomultipliers can be used from the ultraviolet to approximately 1 μm; a room temperature lead-sulfide photoconductor can be used to approximately 3 μm; cooled photoconductors such as doped germanium can be used as far as 100 μm in the infrared. Chapter 11 of Ref. [4] shows typical detector characteristics.

A thermal detector is one in which the photons are merely absorbed to create a heat effect, an increase in the thermal motions of the constituents of matter. The heat effect then has to be detected in some manner. In the thermistor the effect is an increase in electrical resistance; in the thermocouple the effect is generation of thermoelectric power; in the Golay cell the effect is a pressure rise and expansion in a gas causing the deflection of a small mirror which in turn sweeps a light beam over a photocell. These thermal detectors are relatively insensitive and slow in response (0.1 to 10 sec time constant). Their advantage is that they can be made to be linear and relatively nonselective over a wide spectral range.

Table 10.1 lists some characteristics which must be considered in selecting a detector. It may be seen that noise equivalent power, that is, the radiant power just sufficient to give a signal of the same order as the noise, is just one of many considerations. Spectral, directional, and spatial selectivity, stability, ruggedness, and many other considerations must be given weight.

10.3.2 The Ideal Radiometer

From the foregoing it is evident that the energy received between time t_1 and t_2, wavelengths λ_1 and λ_2, within a solid angle defined by some θ, ϕ limits, and within an area element specified by some x–y limits is

$$Q = \int_{t_1}^{t_2} \int_{\lambda_1}^{\lambda_2} \int_{\phi_1}^{\phi_2} \int_{\theta_1}^{\theta_2} \int_{y_1}^{y_2} \int_{x_1}^{x_2} [\alpha_v]^t \cdot [\mathbf{I}] \cos\theta \, dx \, dy \, d\theta \, d\phi \, d\lambda \, dt \quad (10.47)$$

It would be ideal to have a radiometer which would respond to this power only, which could have y_1 and y_2 varied arbitrarily as a function of x; ϕ_1 and θ_2 similarly varied arbitrarily as a function of θ, the values of λ_1, λ_2, t_1, and t_2 chosen at will, and $[\alpha_v]^t$ variable at will to find the nature of $[\mathbf{I}]$. In actuality we obtain a signal $V - V_0$ (subject to uncertainty $\pm \Delta V$). The quantity V_0 results when a chopper or shutter is viewed and V results when the source is viewed:

TABLE 10.1 Characteristics influencing detector selection

Characteristic	Comments
Noise equivalent power	This quantity is the radiant power sufficient to give a signal of the same order as the noise.
Time response	Photon detectors are fast. Thermal detectors are slow.
Spectral selectivity	Photo detectors are spectrally selective. Thermal detectors can be quite nonselective.
Directional selectivity	Exposing the edges of a photoconductor cell or using a diffusing cover can improve omnidirectional characteristics.
Spatial selectivity	Photoconductors are more sensitive near their electrodes. Thermocouple detectors are more sensitive near their junctions. Time responses may differ spatially over thermocouple detectors so that response may be quite sensitive to phasing in chopped systems.
Linearity	Photodetectors saturate.
Stability	For example, photomultipliers may suffer a permanent loss of sensitivity by exposure to too large a flux. Plated thermopiles may slowly change over a period of months or years due to diffusion and chemical changes. There may be short-term reversible changes due to, say, sensitivity to temperature level. Thermal detectors are susceptible to zero drift when used in the dc mode.
Ease of operation and maintenance of detector and auxiliary equipment; ruggedness and reliability	Golay cells used to have a very limited useful life, but may now last as long as vacuum thermocouples. Both eventually fatigue and are somewhat delicate.
Size, cost, weight, volume	These factors seldom influence laboratory equipment, but do influence field or flight instruments.

$$V = \int_{-\infty}^{0} \int_{0}^{\infty} \int_{0}^{2\pi} \int_{0}^{\pi/2} \int_{y_1}^{y_2} \int_{x_1}^{x_2} [R]^t \{[I] - [x_b]I_b\} \cos\theta \, dx \, dy \, d\theta \, d\phi \, d\lambda \, dt \tag{10.48}$$

where $[R]$ is a four-component response function, each component of which is a function of time, wavelength, direction, and position on the detector. Its vector nature arises from the fact that it contains a product of 4-by-4 transmittance or reflectance matrices and the four-component detector absorptivity vector. It is impossible to give $[R]$ a sharp step-function characteristic which would give Eq. (10.47). The quantity I_b in Eq. (10.48) is the Planck spectral intensity for the detector temperature.

One obtains a defined measurement (a "clean" measurement) in certain limiting cases. Consider for example the field of view. If the object viewed entirely fills the field of view, and the intensity $[I]$ is constant over the field of view, then the significant instrument characteristic is an integral of $[R]$. On the other hand if the object viewed falls well within the field of view, where $[R]$ is essentially constant, then the significant source characteristic is an integral of [I], and the significant instrument characteristic is the central value of [R]. Similar considerations hold for the other parameters. It is interesting to note, however, that one never knows that [I] is constant over the field of view. One must infer that it is. To aid in drawing such inferences, it is well to be able to vary or sweep the solid angle field and spectral band.

10.3.3 Spectral, Band–Pass, and Total Radiometers

The ideal radiometer doesn't exist, and certainly for many applications a relatively crude characterization of the irradiation is adequate. In those cases where spectral measurements are necessary, a spectral radiometer is used. In this case the instrument is essentially a spectrometer with an optics system for viewing the source. In the infrared, first surface aluminized mirrors are usually used in the optics train, and an infrared transmitting prism such as LiF, NaCl, KBr, or CsBr or an aluminized blazed grating is used to disperse the radiation, so that the spectrometer entrance slit image coincides with the exit slit image only for a certain wavelength. With equal exit and entrance slits the spectral slit function (transmittance) is triangular in shape.

Since a grating has higher orders of shorter wavelengths superposed on the first order, either a fore-prism or reflection or transmission filters must be used to eliminate unwanted short-wavelength radiation. Such filters may be selective absorbers such as black polyethylene, selective scatterers such as aluminized rough glass or powders embedded in polyethylene, or selective reflectors such as reststrahlen crystals. Photon detectors are desirable for spectral measurements because of their sensitivity and fast response time, but thermocouples or thermistors are often used to obtain a wide spectral range without the need for detector changes. Table 19-3 of [4] gives a summary of some commercially available spectrometers. Recently Fourier transform interferometers and rapid scan spectrometers have come onto the market [13].

In those cases where a wide band or a total measurement is desired, a thermal detector is almost invariably used [14]. Filters may be inserted in the optics intentionally to limit the spectral response, or the filter may be in the optics by virtue of a requirement for a window or by the convenience introduced through use of a lens. Figures (7-24) to (7-27) of [4] show the spectral transmittance of some of the older commercially available filters. Due to advances in thin-film technology there is now quite a variety of reflection and transmission filters.

10.3.4 Radiometers with Thermal Detectors

The thermal radiometer consists of a housing to provide an isothermal environment for the detector, the dectector itself, and field or filter optics. In a directional radiometer a field stop, or a field-stopped mirror or lens, limits the detector to a view in a relatively small solid angle. In a hemispherical radiometer the detector is given a view of the entire hemisphere. In this latter case the radiometer case may take the degenerate form of a jet of air over a mirror as in the case of the aspirated hemispherical radiometer used in meteorology.

Of concern in the design and use of total radiometers is the thermal resistance from the detector to the case R and the thermal capacity of the detector C. The RC-product is the time constant of the instrument; the value of R fixes the dc sensitivity, and the nature of R determines the conditions of use. If R is small, the time constant is fast but the dc signal is also small. If C is large, the time constant is large, but the detector may be rugged and inexpensive. The thermal resistance is the net effect of conduction, convection, and radiation from the detector to the housing. A conduction-coupled radiometer is one with a large thermal conductance from the detector to the housing so that convection and radiation transfer has a negligible effect in the detector heat balance. Such a radiometer is desirable because it is not too sensitive to atmospheric pressure, radiometer temperature level, and radiometer orientation with respect to gravity. A heat balance on the detector can be formulated using the principles set forth in any good heat transfer text [15]. A fin type of analysis is often applicable.

In addition to radiation coupling, factors causing calibration shift with temperature level are a nonlinear thermocouple or thermistor response. With thermistors it is possible to arrange for compensation either by using two different ones in series or by arranging for compensation in the external bridge circuit. In the first case two thermistors having opposite signs for the second derivative of the resistivity vs. temperature are combined to give a second derivative of zero in the total resistance vs. temperature. In the latter case the bridge resistors R_c are chosen with a resistivity temperature coefficient such that the quantity $(R_e + R_d)^{-1}$ (dR_d/dt) is independent of temperature, where R_d is the thermistor resistance.

Radiometers with thermal detectors are notorious for zero drift. For this reason an automatic internal chopper is often used, and two identical detectors coupled in the same way to the case, one exposed and one unexposed, are often employed. One can cite Coblentz's cryptic observation [16] regarding zero drift, "Moll's surface thermopile made of thin sheets of copper-constantan had the defect that the 'cold junctions' were soldered to relatively heavy posts." When working with small values of $Id\Omega$, frequent zero readings should be taken using a highly reflecting shutter at the radiometer temperature. A chopper, i.e., a rotating internal shutter, if used, should be highly reflecting and located on the source side of the main field stop. Because of the term I_b in Eq. (10.48), the detector housing temperature must be well defined and accurately known, and the specular shutter and chopper should be viewed at an angle so that the detector does not look itself in the eye, i.e., does not view a mirror image of itself.

10.3.5 Calibration of Radiometers

Calibration of radiometers is usually accomplished by viewing a Planckian cavity source, as mentioned in the text following Eq. (10.8). Secondary standard lamps which are calibrated against other standard lamps calibrated with a cavity source are also used [17]. Heated gases in which the temperature is obtained spectro-scopically and for which the absorption coefficient has been measured have also been used to calibrate spectral radiometers. Radiometer detectors equivalent to "microcalorimeters" have been calibrated by "absolute" methods, for example, by replacing the radiant heating with electrical heating as in the Callendar and Angstrom radiometers or by knowing the heat of fusion of melting ice, the heat capacity of a silver disk, etc. [18, 19].

10.4 SURFACE RADIATION MEASUREMENTS

10.4.1 The Surface–System Concept

Radiation is of course not emitted or absorbed by a surface but by a volume. In some cases this volume is a thin layer along a surface of a strongly absorbing medium, that is, the volume between the Spalding S surface and M surface already mentioned in the discussion following Eq. (10.13). Such a layer is referred to as an *opaque surface system*, or for short, an *opaque surface*. In other cases the material of concern is a thin slab, such as a thin pane of glass, or a thin plastic film. Such a slab is termed a *transmitting surface system,* or for short, a *transmitting surface*. When such "surfaces" (surface systems) have their energy level populations characterized by a temperature T corresponding to a thermo-dynamic equilibrium population distribution, we can speak of surface thermal radiation characteristics [20] as reviewed in Sec. 10.1. Such characteristics facilitate engineering heat transfer accountings.

10.4.2 Types of Measurements

As noted earlier there is an extensive array of characteristics which can be measured. There are two characteristics, reflectance and transmittance, to which can be applied the four direction adjectives (bidirectional, specular, directional, and hemispherical) and the three spectrum adjectives (spectral, internal total, and external total) for a subtotal of 24 possible combinations. In addition there is the spectral emittance or absorptance (which are equal), the external total emittance and absorptance (not equal), and the internal total emittance for which the adjectives *directional* and *hemispherical* apply, for a subtotal of 8 more. The total of 32 by no means exhausts the possibilities, for one sometimes desires to use moments or integrals of the spectral, bidirectional, or directional characteristics over some portion of the spectrum or some portion of the incident or emergent hemispheres. And, of course, a spectral measurement must have the wavelength specified, a bidirectional measurement must have two direc-tions (four angles) specified, a directional measurement one direction (two angles) specified, and one can even measure specular (regular) reflectance or transmittance at off-specular directions, for example, the back direction in the case of a corner reflector.

If polarization enters the problem, one then can multiply some of the first

24 quantities by the 16 elements needed for the four-by-four polarization matrix, and measurements of the last 8 quantities would have to be increased four-fold to obtain the four components of the polarization vector. There is no particular point in trying to catalog all of these quantities. The reader would do better to spend his time rereading the previous sections on thermal radiation, and thermal radiation characteristics (Secs. 10.1 and 10.2).

The engineer must consider what surface characteristics he really needs. For surfaces inside an enclosure the internal total hemispherical emittance is usually adequate. For surfaces outside, say, a space vehicle, the total hemispherical emittance and short wavelength spectral directional absorptance will usually suffice. Sometimes an indication is useful that the surface is, from a power transfer point of view, specular or not specular. The engineer must consider the cost of making the measurements and the cost of making use of the measurements. It should not be overlooked that a quantity can be measured either directly or by making a more primitive measurement and making the appropriate integration. Often, because of instrumental errors, it is desirable to compare two values; one from a direct measurement, and one from integration of the more primitive measurements. For example, total hemispherical emissivity can be measured directly and from integrations of easily made measurements of spectral directional reflectivity and transmissivity.

10.4.3 Bidirectional Characteristics

Bidirectional reflectance or transmittance is measured by irradiating a specimen in a small solid angle with radiation from a source, and viewing the specimen in a small solid angle with a radiometer as shown schematically in Fig. 10.3. Schmidt and Eckert [21], Eckert [22], Münch [23], and others have made total measurements, and Birkebak and Eckert [12] have reported spectral measurements. It is best to introduce chopping on the source side and measure the ac detector signal so that dc sample emission is not detected. After experimenting with monochromatic irradiation [24] and total irradiation schemes, I believe it best to have no optics on the source side, and to use a blackbody source of a well-defined area, so that the solid angle of irradiation is known from simple geometric considerations, and to use a spectrometer detector.

Specular reflection or transmission may occur, and such behavior amounts to a delta-function bidirectional characteristic. It is important for the experimenter to try to establish whether or not the measured quantity appears to have a limit in the mathematical sense as the solid angle of irradiation approaches zero [25]. The fact that both the solid angle of irradiation and that of detection are desired to be small means that the amount of energy for detection is small. These factors make accurate spectral measurements in the infrared very difficult and time consuming. There have been claims made that the Fourier-transform interferometer is a panacea for these difficulties and it is true that the information theory aspects of making bidirectional measurements deserve careful consideration. To

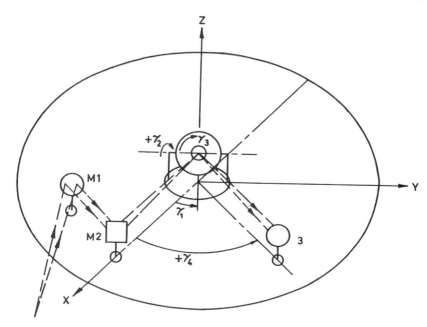

FIG. 10.3 Diagram of bidirectional apparatus (arrows can be reversed for recipro-cal mode). (From Herold and Edwards, 1966, Ref. [24].)

my knowledge the Fourier-transform interferometers have not been subjected to a thoughtful error analysis, but the various dispersion spectrometers have been.

10.4.4 Directional Absorptance and Emittance

Directional absorptance is usually measured indirectly, by making directional reflectance plus transmittance measurements and subtracting from unity, or measuring directional emittance and utilizing Kirchhoff's laws. In the integrating sphere to be described in more detail below, the sum of reflectance and transmittance can be measured simultaneously. Direct measurement of directional absorptance, by coating a detector or calorimeter, is sometimes attempted for the total characteristic but never to my knowledge for the spectral characteristic.

Directional emittance can be measured at moderate temperatures (usually above room temperature) by viewing a heated specimen with a directional radiometer [26–29]. Both total and spectral determinations are made. Spectral measurements usually require temperatures considerably above room temperature, so that spectral directional reflectance and transmittance are usually measured instead for low sample temperatures, to give spectral directional absorptance and, from Kirchhoff's law, spectral directional emittance. In other cases where sample temperature is difficult or inconvenient to measure, reflectance and transmittance measurements are preferable.

10.4.5 Direct and Reciprocal Measurements
of Directional Reflectance

Directional reflectance (or transmittance) can be attempted in a direct manner by irradiating a sample from a particular direction, then attempting to detect all reflected (or transmitted) radiation regardless of its directional distribution. A more successful approach in many instances is to attempt to irradiate the specimen perfectly diffusely and detect the reflected intensity in a small solid angle. This mode of measurement is referred to as a *reciprocal determination,* as it rests on the Helmholtz reciprocity principle, that is, the principle of the detailed balance from thermodynamics.

The practical advantage of the reciprocal mode is that it is much easier to make a large area uniform source than a large area uniform response detector. Measurements can be made single-beam by observing a 100% reference signal when the sphere wall is irradiated (direct mode) or observed (reciprocal mode), a zero when the beam is blocked, and a sample signal when the sample is irradiated out of the view of the detector. In a ratio-recording instrument the chopper is replaced by a nutating mirror which alternately switches the beam from the sample to the reference spot on the wall. It is best to arrange the geometry so that the reference spot on the wall is that spot fixed by the specular ray to or from the sample.

10.4.6 Spectral Directional Reflectance
in the 0.3–2.5 μm Region

A device which works equally well in the direct or reciprocal mode is the integrating sphere [30–34]. Figures 10.4, 10.5, 10.6 and the top of Fig. 10.7 show diagrams of these instruments. The shape factor from any receiver element on a sphere to any source element on a sphere is simply the source element area divided by the total area of the sphere. It follows that, if the sphere wall is perfectly diffuse and if the detector views all parts of the sphere equally well (a diffuse plane detector on the wall or a spherical one in the center), the sphere can act as a 2π-steradian detector for a specimen anywhere within the sphere. In the reciprocal mode the sphere wall can be uniformly irradiated (from a plane perfectly diffuse source on the wall or a spherical source in the center) so that a specimen anywhere within the sphere receives perfectly diffuse irradiation. It is essential in the direct mode that the detector does not see the sample or in the reciprocal mode that the sample does not see the source. In the direct mode, with a directional detector viewing a small area of the sphere wall, for purposes of analysis the detector should be regarded as the viewed area on the wall. Chopping should be done on the source side so as to eliminate sample emission

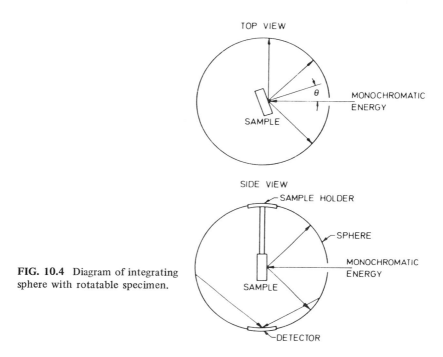

FIG. 10.4 Diagram of integrating sphere with rotatable specimen.

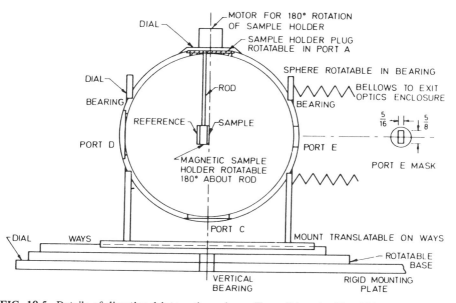

FIG. 10.5 Details of directional integrating sphere. (From Edwards, Gier, Nelson, and Roddick, 1961, Ref. [34].)

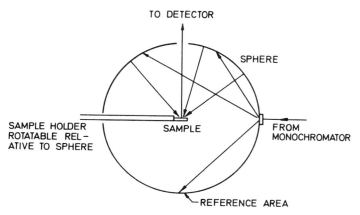

FIG. 10.6 Diagram of Toporets integrating sphere.

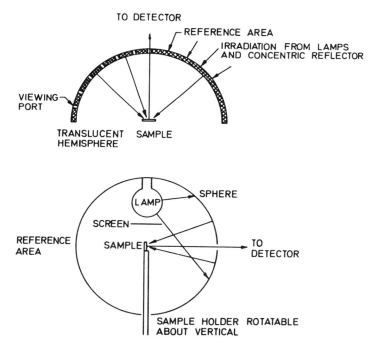

FIG. 10.7 Diagram of McNicholas and Tingwaldt directional reflectometers.

from the measurement, but unless high temperature specimens are used, sample emission is not too consequential in the short wavelengths. Highly reflecting highly diffuse sphere coatings such as magnesium oxide or barium sulfate work well only in the short wavelengths, and the sphere is relatively inefficient and therefore requires an efficient photon detector. Thus the integrating sphere is not usually employed outside the 0.3-to-2.5 μm region.

A reciprocal mode short wavelength reflectometer is the McNicholas milk-glass sphere [10, 30] shown at the top of Fig. 10.7. In this device a cavity is made uniformly radiating by transmission through a diffusing wall. This instrument has been largely supplanted by the integrating sphere.

10.4.7 Directional Reflectance in the 2–20 μm Region

The reciprocal mode heated cavity is made uniformly radiating by thermal emission [35–37]. If, in either the heated cavity or integrating sphere, the sample is on the wall, as it was on most early devices, only one polar angle of view is convenient. Mounting the sample at the end of a rotatable rod within the cavity or sphere permits the polar angle to be varied at will [36, 37]. Figures 10.8 and 10.9 show diagrams of a single beam heated cavity reflectometer with a rotatable sample [37] and Figs. 10.10 and 10.11 show two sample spectra [38] measured with the instrument. Another observation that can be made with the directional heated cavity with rotatable sample is shown schematically in Fig. 10.12. A quick crude measure of "diffuseness" is obtained by rotating the specimen to normal incidence so that specularly reflected radiation is not detected.

Except for an ingenious modification by Nelson, Luedke, and Bevans [39], the heated cavity is used with chopping on the detector side, so that it is subject to sample emission errors. In the Nelson scheme a cylinder split into halves, one heated and one cooled, is rotated above the specimen which is viewed through a

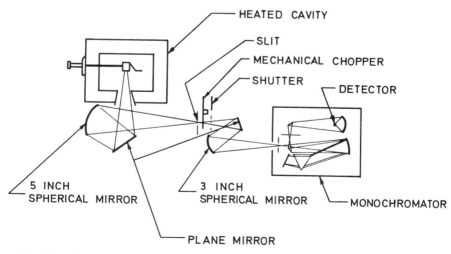

FIG. 10.8 Diagram of heated cavity reflectometer. (From Dunkle, Edwards, Gier, Nelson, and Roddick, 1962, Ref. [37].)

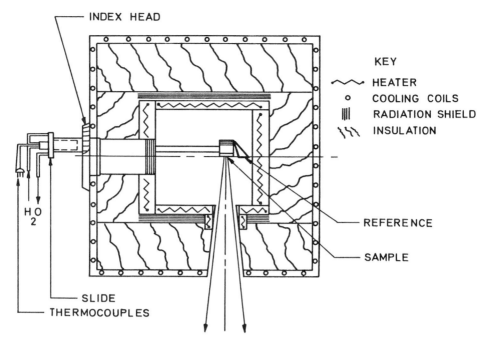

FIG. 10.9 Details of heated cavity reflectometer. (From Dunkle et al., 1962, Ref. [37].)

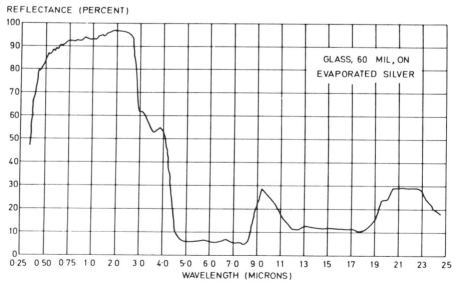

FIG. 10.10 Sample spectrum measured with integrating sphere and heated cavity reflectometers: second surface silver mirror. (From Edwards and Roddick, 1963, Ref. [38].)

REFLECTANCE (PERCENT)

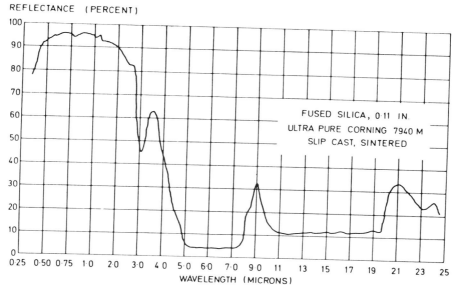

FIG. 10.11 Sample spectrum measured with integrating sphere and heated cavity reflectometers: sintered fused silica. (From Edwards and Roddick, 1963, Ref. [38].)

slot in the cylinder as shown in Fig. 10.13. As it has not proved too convenient to have a large temperature difference between the halves, the device is used only for total measurements. The internal total reflectance is measured, unless a filter such as that indicated in Fig. 10.14 is used to correct the integral weighting factor. Black polyethylene serves well. The device is limited to one angle of view.

The heated cavity works best in the intermediate wavelength range from 2 to 20 μm where there is ample thermal radiation from, say, a 1,000K source and

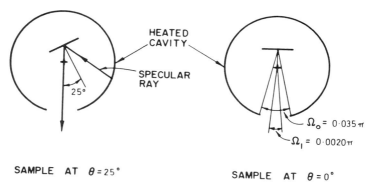

FIG. 10.12 Use of the exit port effect in directional reflectometers to obtain a measure of "diffuseness."

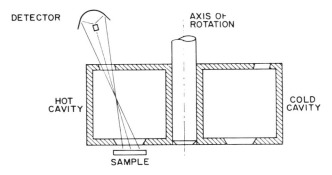

FIG. 10.13 The double-cavity reflectometer.

not too much sample emission error, if the sample temperature and chopper
surrounds temperature are kept within a few degrees of each other.

10.4.8 Directional Reflectance in the Far Infrared

A device which works equally well in the long and short wavelengths (but not as
well as the ones just mentioned in their respective spectral ranges) is the
2π-steradian mirror. Paschen and Coblentz used hemispheres at the turn of this
century and Birkebak, Hartnett, and Eckert [28] and others have likewise used a
hemisphere in the direct mode. Gier, Dunkle, and Bevans [40] used two identical

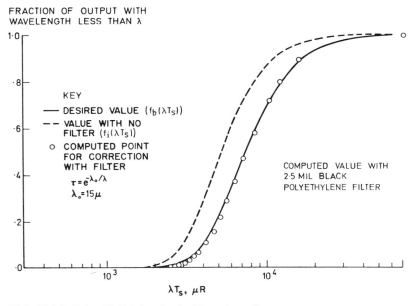

FIG. 10.14 Ogive $f(\lambda T_S)$ for the double-cavity reflectometer.

opposed 2π-steradian paraboloids, and Dunn, Richmond, and Wiebelt [41] used an ellipsoid in the direct mode. In the direct mode, radiation is brought onto the sample directionally through a hole in the mirror or via a reflection from a spot on the mirror collected by the 2π-steradian mirror or mirrors, and somewhat focused on a detector. Errors occur from optical aberrations and more importantly, nonuniformly sensitive detectors.

I have advocated [42, 43] the McNicholas [30] reciprocal mode for the 2π-steradian mirror. Those 2π-steradian mirror instruments used in the reciprocal mode [44–47] have displayed superior performance, particularly when the interreflection error is minimized [46]. A special advantage of the 2π-steradian mirror is that it lends itself to source-side chopping in either mode of operation. For this reason it is preferable to the heated cavity at wavelengths beyond 20 microns; it is more easily used with specimens which are difficult to cool; and it permits measurements with specimen temperatures at either elevated or cryogenic levels.

Figure 10.15 shows a schematic diagram of the apparatus described in [46, 47] and Fig. 10.16 shows a photograph. Visible in the foreground is a surface source which radiates upward to an inclined water cooled roughened and sooted mirror. Above the mirror a mechanical chopper exists to interrupt radiation going from the mirror to the off-axis paraboloid, whose edge is visible at the right of the source. Radiation is reflected from the highly off-axis paraboloid onto the slightly off-axis one [47]. Radiation reflected from this paraboloid is focused somewhat on a specimen holder which is shown turned toward the camera out of its natural position. If one were able to put an infrared eye at the specimen location, one would see only the source via reflection in the mirrors. The sample is therefore diffusely irradiated except for the square observation port, visible near the center of the mirror just above the chopper blade hub. The sample is

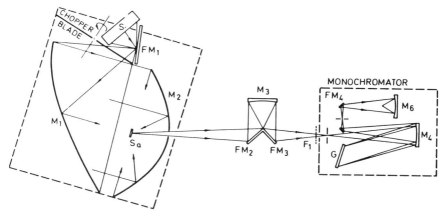

FIG. 10.15 Diagram of reciprocal paraboloid reflectometer. (From Neher and Edwards, 1965, Ref. [46].)

FIG. 10.16 Photograph of reciprocal paraboloid reflectometer. (From Neher and Edwards, 1965, Ref. [46].)

viewed through this port by a grating monochromator, shown with cover removed behind the reflectometer. A vacuum thermocouple with diamond window is used for a detector. (A Golay cell would extend the spectral range.) Figure 10.17 shows some data obtained with the instrument [46].

10.4.9 Hemispherical Characteristics

Total hemispherical emittance is measured rather easily calorimetrically at temperatures above room temperature, particularly when the specimen is a coating which can be applied to a ribbon, cylinder, or sphere. At temperatures significantly below room temperature such factors as lead conduction, the possibility of phase changes, temperature drift, and edge effects cause difficulties, as pointed out by Nelson and Bevans [48]. Geometries tending to be focusing such as concentric spheres or cylinders, are troublesome also. The measurement made is said to be a calorimetric one, because a heat balance is made on a specimen suspended within a blackened and baffled evacuated enclosure.

A radiometric measurement of hemispherical emittance can be made, if a

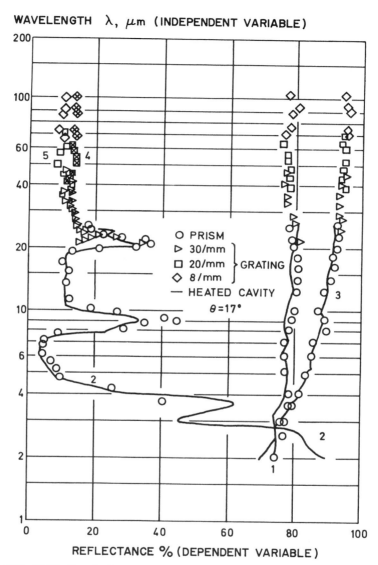

WAVELENGTH λ, μm (INDEPENDENT VARIABLE)

REFLECTANCE % (DEPENDENT VARIABLE)

FIG. 10.17 Sample far infrared spectra measured with reciprocal paraboloid reflectometer. *Curve 1*: Vacuum evaporated gold on wire mesh bonded to mylar; wire spacing, 0.24 mm; wire dia, 0.080 mm. *Curve 2*: Sintered fused silica Corning 7940 M 2.8 mm thick. *Curve 3*: Polished 17-7 PH stainless steel. *Curve 4*: Alundum heater cement; Norton RA1161 3 mm thick. (From Neher and Edwards, 1965, Ref. [46].)

hemispherical radiometer is used with a plane specimen. However, except at high fluxes hemispherical radiometers are rather difficult to use. Eichhorn [49] has pointed out that a directional radiometer can be used for hemispherical emittance when a spherical specimen lies entirely within the radiometer's field of view or, perhaps more conveniently, when a cylindrical specimen has a triangular area masked off within the radiometer's field of view.

In the short wavelength region, hemispherical reflectance can be measured using a single integrating sphere. In this case the sample is neither irradiated by the source nor viewed by the detector, but rather the efficiency of the sphere is observed as the sample is substituted for a portion of the sphere wall, a portion masked from view of the source and detector. Hemispherical transmittance can be measured with a double integrating sphere scheme, one for the source and one for the detector. Again neither the source nor detector should directly view the specimen.

10.4.10 Error Analyses

No measurement is useful without an accompanying estimate of its uncertainty. After the basic principle on which an instrument functions is recognized, the first step is to reject all idealizations made. Common idealizations include some of the following:

The specimen is perfectly diffuse.

The specimen is perfectly specular.

The specimen is in a near-thermodynamic-equilibrium state.

All energy transfer into the specimen is by radiation.

All energy storage within the specimen is negligible or accounted for with a constant specific heat.

The surrounds are isothermal.

The surrounds are diffuse.

Interreflections are negligible or accounted for perfectly.

The irradiation is unidirectional.

The irradiation is incoherent.

The irradiation all has the same chopping phase.

The detector is at steady state.

The detector is linear.

The detector is insensitive to the state of polarization of the irradiation.

The detector is uniformly sensitive independent of direction, image location, and phase.

The optics are isothermal.

The sample and reference optics are identical.

The optics are nonpolarizing.

The optics are nonscattering.

The optics focus perfectly.

The spectral slit width is small.

The source is steady.

The source, optics, and detector are nonselective over the spectral slit region.

Usually a perturbation analysis is adequate, that is, the effect of rejecting an idealization can usually be analyzed without rejecting all of them at once. Errors can then be added with account taken of their sign, and uncertainties with nonskewed distributions can be combined in a root mean square [50]. For example, consider an integrating sphere. Some spheres used for relative measurements have the sample in clear view of the detector. In this case the intensity of the sample I_s in the direction toward the detector is given by (assuming incoherent irradiation)

$$\pi I_S A_S = P \rho_B(\theta, \phi, \theta_S, \phi_S) \tag{10.49}$$

where θ, ϕ is the direction of incidence and θ_S, ϕ_S gives the direction toward the sample, P is the power incident on the sample, and A_S is the irradiated area. We define

$$p_{DS} = \int_{A_D} \frac{\rho_B(\theta, \phi, \theta_S, \phi_S)}{\rho(\theta, \phi)} \left[\frac{\cos \theta_S \cos \theta_D}{\pi r_{DS}^2} \right] dA_D \tag{10.50}$$

so that the quantity $p_{DS}\rho_S P$ is the power reflected directly to the detector with area A_D when sample S is irradiated. The signal put out by a linear detector with response independent of angle of incidence is then

$$V_S = K[p_{DS}\rho_S P + \rho_S(1 - p_{DS})P\eta] \tag{10.51}$$

where η is the sphere efficiency, which for a small detector in a large sphere is

$$\eta = \frac{A_D}{A_W} \frac{\rho_W}{1 - \rho_W} \tag{10.52}$$

The quantity ρ_W is the average reflectance of the sphere wall including the detector, openings, sample, etc. The reference signal is

$$V_R = K[p_{DR}\,\rho_R P + \rho_R(1 - p_{DR})P\eta] \qquad (10.53)$$

so that the ratio is

$$\frac{V_S}{V_R} = \left[\frac{p_{DS} + \eta(1 - p_{DS})}{p_{DR} + \eta(1 - p_{DR})}\right]\frac{\rho_S}{\rho_R} \qquad (10.54)$$

The error in taking $V_S/V_R = \rho_S/\rho_R$ is then

$$\delta\!\left(\frac{\rho_S}{\rho_R}\right) = \frac{(\rho_S/\rho_R) - (V_S/V_R)}{(\rho_S/\rho_R)} = \frac{(p_{DR} - p_{DS})(1 - \eta)}{p_{DR} + (1 - p_{DR})} \qquad (10.55)$$

If the reference is diffuse, p_{DR} may be of the order of 1% or 2%, and η is typically of this order. The quantity p_{DS} may have any value from zero to unity. Large errors are seen to be possible and have in fact been detected, but not always understood [51, 52].

As a second example consider entrance port error in an integrating sphere. The power lost out of the entrance port E is $p_{ES}P$ so that the signal, when the sample is not viewed by the detector, is

$$V_S = K(1 - p_{ES})\rho_S(\theta, \phi)\eta P \qquad (10.56)$$

The reference signal when a portion of the wall viewed by the detector is

$$V_R = K\eta P \qquad (10.57)$$

The ratio is then

$$\frac{V_S}{V_R} = (1 - p_{ES})\rho(\theta, \phi) \qquad (10.58)$$

For a specular sample irradiated somewhat off normal p_{ES} is zero. However, if the sample is slightly diffusing p_{ES} may be appreciable. Some manufacturers have used two entrance ports both just slightly away from the specularly reflected ray. Needless to say, such irrational design is pure folly. It is best to be able to vary the angle of incidence appreciably as in the directional integrating sphere with suspended sample [34].

As a final example, we can consider an easily understood error in a directional heated cavity. In what was otherwise a nice design, Tingwaldt [36] suspended a cooled specimen at the center of a jacketed Pyrex glass sphere which was heated by condensing sulfur. With the center position there is an appreciable interreflection error caused by specular reflection from the sphere wall back to the cold specimen. Glass reflects fairly well in the 9-to-11 and 18-to-22 reststrahlen bands, and anomalous results were in fact observed in the reststrahlen region.

For further details of error analyses of sample emission error in heated cavities; substitution, shadowing (or interreflection) errors in integrating spheres; stray light and polarization errors; the interested reader should see [30, 31, 37, 43, 46, 47, 51, 53, and 54]. Table 10.2 summarizes some of the more significant errors to be considered when using directional reflectometers.

10.5 GAS RADIATION MEASUREMENTS

10.5.1 Types of Measurements

In the case of a gas volume in a state defined by temperature T, that is, in a state of local thermodynamic equilibrium [56], it is desired to know the absorption and scattering mass coefficients k_a and k_s. For an isotropic gas, perhaps with entrained particles, the quantity k_a will not depend upon direction, but the distribution of scattered radiation will depend upon the angle between the scattered ray and original one. The absorbed power in path dS is

$$dI \, d\Omega_1 = -\rho_m k_a \, dSI_1 \, d\Omega_1 \qquad (10.59)$$

and the scattered intensity emerging is

$$dI(\theta_2) = \frac{1}{4\pi} \rho_m k_s(\theta_2 - \theta_1) \, dSI_1 \, d\Omega_1 \qquad (10.60)$$

where ρ_m is the mass density of the absorbing-scattering species. If the gas contains scatterers one has a problem akin to that of measuring the bidirectional reflectance to find $k_s(\theta_2 - \theta_1)$, and the determination of the forward-scattering or backward-scattering coefficient

$$k_f = \frac{1}{2\pi} \int_0^{\pi/2} k_s(\theta_2 - \theta_1) \, d(\theta_2 - \theta_1) \qquad (10.61)$$

$$k_b = \frac{1}{2\pi} \int_{\pi/2}^{\pi} k_s(\theta_2 - \theta_1) \, d(\theta_2 - \theta_1) \qquad (10.62)$$

is akin to the determination of the directional transmittance or reflectance. Further, one has polarization to account for. In the interests of keeping this survey of thermal radiation measurements within the bounds of a short work,

TABLE 10.2 Directional reflectometers and principal errors

Instruments	Typical errors	Refs.	Comments
Integrating sphere	1. Direct irradiation error	[34]	1. Sample should not see detector in a direct mode or source in reciprocal mode. See [51] and [52] for bad examples.
	2. Port error*	[34]	
	3. Directional sensitivity error	[34]	3. Detector should have cosine response in direct mode or wall source should be perfectly diffuse in reciprocal mode.
	4. Polarization*	[34]	
	5. Stray radiation*	[35]	
	6. Nondiffuseness of sphere wall	[34]	
	7. Shadowing	[34]	
	8. Interreflections (substitution)	[34]	
	9. Nonlinearity of detection system*	[34]	
Heated cavity	1. Sample emission	[51]	1. Common to all instruments with no source-side chopping.
	2. Port error*	[37]	
	3. Temperature gradients	[37]	3. Use of reference fin reduces error.
	4. Shadowing by retainer	[37]	
	5. Interreflections	[37]	5. See [36] for a bad example.
	6. Stray radiation*	[55]	
	7. Nonlinearity*	[55]	
	8. Polarization*	[55]	
2π mirror	1. Optical aberrations	[46]	1. Optical quality images may be sacrificed to obtain hemispherical irradiation or detection. See [46].
	2. Directional sensitivity of detector or source	[46]	2. Source should not be viewed at high angles from the normal.
	3. Spatial sensitivity of detector or source (phase and amplitude)	[46]	3. Strongly favors reciprocal mode operation.
	4. Port error* and shadowing	[47]	4. See [46] for a bad example.
	5. Interreflections	[46]	
	6. Variation of mirror reflectances with angle and position.	[46]	
	7. Stray radiation*	[46]	7. See also [55].
	8. Nonlinearity*	[46]	
	9. Polarization*	[46]	

*Common to all instruments.

only a few observations are made in what follows on the problem of making measurements for equilibrium gases containing no scattering particles.

For no scattering particles Eq. (10.59) can be integrated over a path of length L to yield

$$I = I_0 e^{-\int_0^L \rho_m k_a \, dS} \tag{10.63}$$

and for constant pressure and temperature

$$I = I_0 e^{-k_a w} \tag{10.64}$$

where w is the mass path length

$$w = \rho_m L \tag{10.65}$$

From a conceptual point of view the measurement problem is then very simple. One simply views a chopped source through a gas cell length L with a spectrometer to measure $e^{-k_a w}$ or views a cold black target through the hot gas with detector side chopping to measure $(1 - e^{-k_a w})I_b$. However, there are problems with the gas containment system and with the interpretation of the measurement.

The types of measurements which are made are consequently characterized by adjectives describing the gas containment system and the spectral response of the detector. The gas systems fall into the following categories:

Hot window cell

Cold window cell

Nozzle seal cell

Free jet

The spectral nature of the measurement is described in the following ways:

Narrow band measurement

Band absorption measurement

Total measurement

10.5.2 Gas Containment Systems

The hot window cell contains the gas within a container having two windows (or a window and a mirror) heated to the cell temperature, so that the gas and cell are essentially in equilibrium. Such cells have been used by Tingwaldt [57], Penner [58], and others [59, 60]. Troubles arise from lack of suitable windows

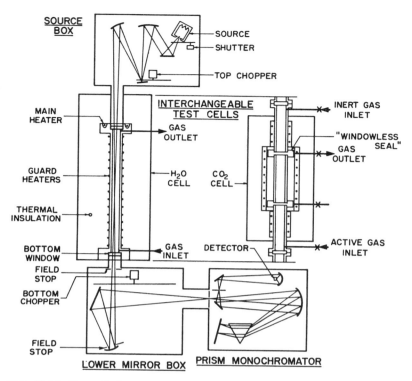

FIG. 10.18 Gas radiation apparatus. (From Edwards et al., 1967, Ref. [65].)

for particular combinations of temperature, pressure, and desired spectral range. There may be chemical attack and gas adsorption on the window, and it has been this writer's experience that polycrystalline windows should be avoided at elevated temperatures. But even with crystal windows high temperatures and long wavelengths are unobtainable.

Cold window cells are cells in which the windows are cooled. Shock tubes are a good example [61, 62]. Tien and Giedt [63] used a heated cell with movable cooled windows. Cold window cells introduce inhomogeneities into the gas temperature and density. Under some circumstances these inhomogeneities are not too serious and, of course, in high temperature shock tube studies they have to be accepted. The cold window cell does lend itself to studies of nonhomogeneous gases [64, 65].

Nozzle seal cells are open flow cells in which absorbing gas meets nonabsorbing gas in the throat of a nozzle containing a slot, through which the gas mixture is withdrawn. Such a cell makes it possible to contain an infrared absorbing emitting gas, that is, a gas with an asymmetric diatomic or polyatomic molecule having infrared vibration-rotation bands, between two layers of an infrared nonabsorbing gas, that is, a nonabsorbing symmetric diatonic or monatomic gas.

Hottel and Mangelsdorf [66], Eckert [67] and other early workers used such cells to make total radiation measurements for H_2O and CO_2. Edwards and coworkers used a nozzle seal cell to make band absorption measurements of CO_2, CH_4, and H_2O [68], and Fig. 10.18 shows the cells used in these studies. The nozzle seal system eliminates some of the problems with windows, but does cause some inhomogeneities in composition at the ends of the cell (and temperature also, unless the seal gases are preheated, which is strongly recommended). In addition some degree of scattering is introduced from the turbulent eddies of differing indices of refraction in the nozzle throats. This problem has not been serious for spectral measurements where the incident energy is tagged by source-side chopping, but it led to significant problems in the older total measurements. Figure 10.19 shows a typical low resolution spectrum used to obtain band absorption.

Jet and burner measurements are those used to obtain high temperature spectra. Schmidt [69] used a nozzle for his early investigation of radiation from water vapor. Recently there have been a number of measurements reported using, in essence, a small rocket engine. For example, Ferriso and Ludwig [70] used a 3.12 cm diameter jet at 1 atm, and they show a shadowgraph of the image of the monochromator exit slit upon the jet. Scattering from the turbulent eddies

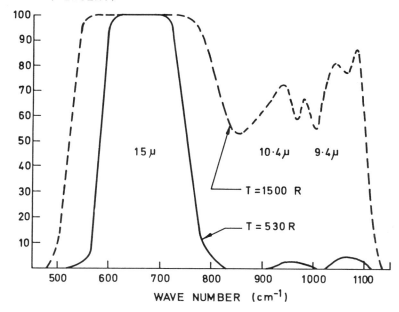

FIG. 10.19 Sample gas radiation spectra: The 9.4-, 10.4-, and 15-μ CO_2 bands. The high temperature curve is 100% CO_2 at 10 atm with a 39-cm path. The room temperature curve is 10% CO_2 in N_2 with a 129-cm path, 10 atm pressure.

and nonhomogeneities on the jet boundaries are evident in the shadowgraph, but just at the beginning of the jet the mixing layer between the jet and surrounding atmosphere is reasonably thin.

10.5.3 Interpretation of Measurements

A major difficulty in gas radiation measurements is that unless the pressure and/or temperature are high the spectral lines of the gas are narrow, on the order of 0.1 wavenumbers (cm^{-1}). A band may have hundreds of thousands of lines in a region of a few hundred wavenumbers in extent, and a given species may have one to ten or more important bands distributed throughout the infrared. As the temperature is increased above room temperature the lines which were important become weaker and new ones become stronger. The fact that the lines are narrow and that there are many of them dictates that the measured quantity usually is not the absorption coefficient k_a but rather an effective or mean absorptivity or emissivity. The mean absorptivity may be written

$$\bar{\alpha}(\nu_0) = \frac{1}{\Delta\nu} \int_{\Delta\nu} (1 - e^{-k_a w}) S(\nu - \nu_0)\, d\nu \qquad (10.66)$$

where $S(\nu - \nu_0)$ is the spectral response function.

In the late 1930s and early 1940s the approach taken was to measure total emissivity and absorptivity vs. temperature, pressure and path-length-pressure-product, as mentioned in the preceding section. Charts [71] reasonably accurate at 1 atm, were developed for the total emissivity of CO_2 and H_2O, and approximate (in some cases very approximate) corrections were developed to account for pressure, composition, and the difference between absorptivity and emissivity [71]. The quantity measured was

$$\alpha_T(w, P, T, T_S) = \int_0^\infty (1 - e^{-k_a w}) \frac{\pi I_b(\nu, T_S)}{\sigma T_S^4}\, d\nu \qquad (10.67)$$

The more recent works cited in the preceding section have measured band absorption

$$A(w, P, T) \equiv \int_{\nu_1}^{\nu_2} (1 - e^{-k_a w})\, d\nu \doteq \int_{\nu_1}^{\nu_2} \bar{\alpha}(\nu_0)\, d\nu_0 \qquad (10.68)$$

and narrow band absorptivity $\bar{\alpha}(\nu_0, w, P, T)$. In order for the measurements of A or $\bar{\alpha}$ to be useful it is necessary to have band model correlations which can represent at least the pressure and mass path dependency of the quantities. It is highly desirable to be able to represent the temperature and wavenumber dependency as well. Models capable of correlating measurements of A are termed *wide band models*, and models capable of correlating $\bar{\alpha}$ are *narrow band models*.

Progress in the ability to make measurements has been accompanied by progress in models to represent the data. Plass [72, 73] has contributed much to the understanding of narrow band representations which were pioneered by Elsasser [74] and Goody [75]. Reardon and Huffaker [76] have used narrow band correlations in predicting liquid fuel rocket engine exhaust radiation. Howard, Burch, and Williams [77] used linear, square-root, and logarithmic relations to fit the w dependency of wide band absorption, and Edwards and Menard [78] developed a wide band model which has been put into a form more convenient for some applications by Tien and Lowder [79]. A somewhat sophisticated band model [80] gives a reasonable representation of the temperature and wavenumber dependencies of $\bar{\alpha}$, and the temperature dependency of A, in addition to approximately predicting the mass path length and pressure dependencies of both quantities.

10.5.4 Factors to Be Considered in Error Analyses

It should be clear from the preceding remarks that some factors to be considered are the following:

Reflection and scattering by windows when transmission is measured and reflection, scattering, and emission by windows when emission is measured.

Scattering by mixing zones in nozzle seals or in free jets.

Inhomogeneities in temperature, pressure, and composition.

Adsorption on windows.

In addition to these experimental problems there are problems in data interpretation or correlation:

Insufficient overlapping of lines when attempting to measure integrated intensity.

Pressure-induced changes in dipole moment.

Improper correlation model.

Insufficiently low pressure when attempting to measure line-width-to-line-spacing parameters.

The measured quantity should be formulated taking into account all the factors which can be treated theoretically. On the first list items 1 and 3 are easily treated, and on the second list items 1 and 4 can be treated theoretically. The remaining items on the list must be at least investigated experimentally.

10.6 CONCLUDING REMARKS

In this short chapter an attempt has been made to survey techniques used to measure radiant energy fluxes and thermal radiation characteristics. The references cited were selected to get the interested reader off to a good start. They are by no means exhaustive. The very old works of Paschen, Ulbricht, Coblentz, and others have not been cited specifically, because this note is not intended to be a history. Nor have many of the very modern ones in symposia and reports been cited, because many of these papers are concerned with the objects or properties being measured and use standard techniques. An attempt has been made to mention most of the works I regard as important, for example [30-37, 39, 45-48]. It should not be inferred that all the works cited are meant to be models. Table 10.2 indicates some cases of improper experimental arrangements.

A convenient notation and nomenclature has been propounded. Key adjectives are the ones describing the directions, namely, bidirectional, directional, and hemispherical and the ones describing the spectrum, namely, spectral, internal total, and external total. The reader should be cautioned that the words total, diffuse, and hemispherical are frequently garbled elsewhere. The reader should also be warned that there are committees and editors who enforce other schemes of notation and nomenclature. In regard to this latter point, it doesn't make a great deal of difference whether one uses or is forced to use emissivity, emittance, or emissance; intensity, radiance, or steradiance; radiosity or emittance. What is important is whether or not one's thinking is clear. One must know, for example, that there is both an internal and an external total emittance characteristic; that the radiant intensity (as used here) of a perfectly diffuse surface doesn't vary with $\cos\theta$ but is constant; that diffuse reflection may be far from perfectly diffuse and may, in fact, be essentially specular as far as computing power transfer is concerned.

In summary, in Sec. 10.1 the concept of radiant intensity, the definition of radiant intensity vector used to describe the state of polarization of the radiant intensity, and relations between the radiant intensity and the radiant flux have been reviewed. In Sec. 10.2 the concept of radiation characteristics to describe reflection, transmission, absorption, and emission from an external accounting viewpoint was presented. Polarization and spectral and directional dependencies were provided for. Section 10.3 considered the problem of measuring radiant intensity and radiant flux, and key features of spectral and total radiometers were briefly stated. Measurement techniques for the radiation surface-system characteristics were reviewed in Sec. 10.4. It was shown that the major problem faced by the experimentalist is contriving to measure directional reflectance, regardless of the bidirectional reflectance characteristic of the specimen. Instruments suitably contrived were shown to be integrating spheres, heated cavities, and 2π-steradian mirrors. The advantages of reciprocal measurement arrangements over direct ones were described and examples of perturbation error analyses were presented. Last was a short review of gas radiation measurements (Sec. 10.5) where ' the concept of band absorption and the use of band models were reviewed. It was shown that the major problem faced by the experimentalist in the gas radiation property measurement area is accounting for discrete line structure in gas spectra.

NOMENCLATURE

a	Fresnel reflection parameter, Eq. (10.22h); also dummy parameter in Eq. (10.6)
A	area, Eqs. (10.2), (10.49), et seq.; also band absorption, Eq. (10.68)
b	Fresnel reflection parameter, Eq. (10.22i)
c	speed of light, Eq. (10.1)
d	differential operator
D	direction matrix, Eq. (10.20a)
E	fractional energy, Eq. (10.18)
f_b	fractional function for blackbody radiation, Eq. (10.5)
f_b'	derivative of f_b
f_i	internal fractional function
f_i'	derivative of f_i, Eq. (10.39)
h	Planck's constant, Eq. (10.1)
i	unit imaginary number $\sqrt{-1}$
I	radiant intensity, Eqs. (10.4), (10.11)
k	Boltzmann constant, Eqs. (10.4a), (10.4b); also absorptive index, Eq. (10.22b) et seq.
k_a	absorption coefficient, Eq. (10.59)
k_b	backward-scattering coefficient, Eq. (10.62)
k_f	forward-scattering coefficient, Eq. (10.61)
k_s	scattering coefficient, Eq. (10.60)
K	calibration constant, Eq. (10.51)
M	quantity proportional to magnitude of electric field, Eq. (10.14)
n	refractive index, Eq. (10.22b); also number of photons, Eq. (10.2)
N	number of photons per unit volume, Eq. (10.1)
p	fraction of imperfectly diffusely reflected radiation, Eq. (10.50)
P	power, Eq. (10.49); also pressure, Eq. (10.67)
q	flux, i.e., power per unit area, Eq. (10.11)
Q	energy, Eq. (10.47)
r	distance along a ray, Eq. (10.50)
R	response function, Eq. (10.48)
S	distance, Eqs. (10.14), (10.59); also spectral slit function, Eq. (10.66)
t	time, Eqs. (10.14), (10.47)
T	temperature, Eq. (10.1) et seq.
V	instrument signal, e.g., voltage, Eq. (10.48)
w	density-path-length product, Eq. (10.65)

x dummy variable, Eq. (10.6)

x_b blackbody fraction, Eq. (10.19)

x_1, x_2 distance, Eq. (10.47)

y_1, y_2 distance, Eq. (10.47)

z quantity proportional to instantaneous electric field strength, Eq. (10.14)

Script Symbols

$\mathcal{X}$ hemispherical property operator, Eq. (10.34)

$\overline{\mathcal{X}}$ hemispherical flux operator, Eq. (10.10)

$\mathcal{I}$ imaginary part, Eq. (10.17)

$\mathcal{R}$ real part, Eq. (10.17)

$\mathcal{J}_{T_s}$ total property operator, Eq. (10.32)

$\mathcal{J}_{T_n}$ total net property operator, Eq. (10.36)

$\mathcal{J}_{T_i}$ internal total property operator, Eq. (10.37)

$\mathcal{J}$ total flux operator, Eq. (10.9)

Greek Symbols

α absorptivity or absorptance, Eq. (10.27)

β angle of rotation of coordinates, Eq. (10.20)

γ experimental angular degree of freedom, Fig. 10.3

δ instantaneous phase difference, Eq. (10.14)

Δ Fresnel reflection quantity, Eq. (10.22f)

ϵ emissivity or emittance, Eq. (10.30)

ϵ' photon energy, Eq. (10.1)

η efficiency, Eq. (10.52)

θ polar angle, Fig. 10.1

λ wavelength, Eq. (10.3) et seq.

μ 10^{-6} meters

ν wavenumber, Eq. (10.3) et seq.

π 3.14159 . . .

ρ reflectivity or reflectance, Eq. (10.21) et seq.

ρ_m mass density, Eq. (10.59)

σ Stefan-Boltzmann constant, Eq. (10.8)

τ transmissivity or transmittance, Eq. (10.40)

ϕ azimuthal angle of incidence, Fig. 10.1

Ω solid angle, Eq. (10.2)

Subscripts

a absorption

b blackbody, backward

B bidirectional

D	detector
E	entrance port
f	forward
H	hemispherical
i	internal
m	mass
M	matrix
n	net
o	reference value or set point
p	parallel to plane of incidence
R	reference, radiant
s	perpendicular to plane of incidence (from *senkrecht*)
S	sample or surface
T	total
u	third Stokes coefficient
v	fourth Stokes coefficient
V	vector or column matrix
w	wall
1	incoming, surface 1, first value
2	outgoing, surface 2, second value

Superscripts

t	transpose of a matrix
$+$	outgoing
$-$	incoming

Special Symbols

[ ]	matrix or vector quantity
{ }	function operated upon

REFERENCES

1. Knuth, E. L., "Introduction to Statistical Thermodynamics," McGraw-Hill, New York, 1966.
   Davidson, N., "Statistical Mechanics," pp. 230–235, McGraw-Hill, New York, 1962.
2. Sparrow, E. M., and Cess, R. D., "Radiation Heat Transfer," Brooks Cole, Belmont, Calif., 1966.
3. Dunkle, R. V., Thermal Radiation Tables and Applications, *Trans. Amer. Soc. Mech. Eng.*, vol. 76, pp. 549–552, 1954.
4. Wolfe, W. L. (ed.), "Handbook of Military Infrared Technology," United States Government Printing Office, Washington, D.C., 1965.
5. Czerny, M., and Walther, A., "Tables of the Fractional Functions for the Planck Distribution Law," Springer-Verlag, Berlin, 1961.

6. Spalding, D. B., Heat Transfer from Chemically Reacting Gases, in W. Ibele (ed.), "Modern Developments in Heat Transfer," pp. 19-64, Academic, New York, 1963.
7. Edwards, D. K., and Bevans, J. T., Effect of Polarization on Spacecraft Radiation Heat Transfer, *AIAA J.,* vol. 3, pp. 1323-1329, 1965.
8. Chandrasekhar, S., "Radiative Transfer," p. 27, Dover, New York, 1960.
9. Edwards, D. K., and Nelson, K. E., Maximum Error (Discrepancy) in Total Emissivity Measurements Due to Non-Grayness of Samples, *ARS J.,* vol. 31, pp. 1021-1022, 1961.
10. McNicholas, H. J., Equipment for Measuring the Reflective and Transmissive Properties of Diffusing Media, *J. Res., Nat. Bur. Stand. (U.S.),* vol. 13, pp. 211-236, 1934.
11. Nicodemus, F. E., Directional Reflectance and Emissivity of an Opaque Surface, *Appl. Opt.,* vol. 4, pp. 767-773, 1965.
12. Birkebak, R. C., and Eckert, E. R. G., Effects of Roughness of Metal Surfaces on Angular Distribution of Monochromatic Reflected Radiation, *J. Heat Transfer,* vol. 87, pp. 85-94, 1965.
13. Dolin, S. A., Kruegle, H. A., and Penzias, G. J., A Rapid Scan Spectrometer That Sweeps Corner Mirrors through the Spectrum, *Appl. Opt.,* vol. 6, pp. 267-274, 1967.
14. Drummond, A. J., Hickey, J. R., Scholes, W. J., and Laue, E. G., Multichannel Radiometer Measurement of Solar Irradiance, *J. Spacecraft Rockets,* vol. 4, pp. 1200-1206, 1967.
15. Eckert, E. R. G., and Drake, R. M., "Analysis of Heat and Mass Transfer," McGraw-Hill, New York, 1972.
16. Coblentz, W. W., Thermopile Construction and Use, in W. E. Forsythe (ed.), "Measurement of Radiant Energy," pp. 191-198, McGraw-Hill, New York, 1937.
17. Stair, R., Schneider, W. E., and Fussell, W. B., The New Tungsten-Filament Lamp Standards of Total Irradiance, *Appl. Opt.,* vol. 6, pp. 101-105, 1967.
18. Guild, J., Investigations in Absolute Radiometry, *Proc. Roy. Soc. (London),* vol. 161, p. 1, 1937.
19. Abbot, C. G., Studying the Sun's Heat on Mountain Peaks in Desert Land, *Ann. Rept. Smithsonian Inst.,* p. 145, 1920. (See also p. 319, 1910; p. 153, 1912.)
20. Dunkle, R. V., Thermal Radiation Characteristics of Surfaces, in J. A. Clark (ed.), "Fundamental Research in Heat Transfer," pp. 1-31, Macmillan, New York, 1963.
21. Schmidt, E., and Eckert, E. R. G., Uber die Richtungs-verteilung der Warmestrahlung von Oberflachen, *Forsch. Geb. Ingenieur.,* vol. 6, p. 175, 1935.
22. Eckert, E. R. G., Messung der Reflexion von Warmestrahlen an Technischen Oberflachen, *Forsch. Geb. Ingenieur.,* vol. 7, pp. 265-270, 1936.
23. Münch, B., "Die Richtungsverteilung bei der Reflexion von Warmestrahlung und ihr Einfluss auf die Warmeubertragung," Mitteilungen a.d. Institut fur Thermodynamik und Verbrennungsmotorenbau, no. 16, Verlag Leeman, Zurich, 1955.
24. Herold, L. M., and Edwards, D. K., Bidirectional Characteristics of Rough, Sintered-Metal, and Wire-Screen Systems, *AIAA J.,* vol. 4, pp. 1802-1810, 1966.
25. Moon, P., and Laurence, J., Construction and Test of a Goniophotometer, *J. Opt. Soc. Amer.,* vol. 31, p. 130, 1941.
26. Jakob, M., "Heat Transfer," vol. II, pp. 87-89, Wiley, New York, 1957.
27. Snyder, N. W., Gier, J. T., and Dunkle, R. V., Total Normal Emissivity Measurements on Aircraft Materials Between 100 and 800 F, *Trans. Amer. Soc. Mech. Eng.,* vol. 77, pp. 1011-1019, 1955.
28. Birkebak, R. C., Hartnett, J. P., and Eckert, E. R. G., Measurement of Radiation Properties of Solid Materials, *Progr. Int. Res. Thermodynamic Transport Prop., ASME,* pp. 563-574, 1962.
29. Richmond, J. C., Harrison, W. N., and Shorten, F. J., An Approach to Thermal Emittance Standards, in J. C. Richmond (ed.), "Measurement of Thermal Radiation Properties of Solids," *NASA* SP-31, pp. 403-423, 1963.

30. McNicholas, H. J., Absolute Methods in Reflectometry, *J. Res., Nat. Bur. Stand. (U.S.)*, vol. 1, pp. 29–73, 1928.

31. Karrer, E., Use of the Ulbricht Sphere in Measuring Reflection and Transmission Factors, *Sci Papers, Nat. Bur. Stand.*, vol. 17, pp. 203–225, 1922.

32. Tingwaldt, C. P., Uber die Messung von Reflexion, Durchlassigkeit und Absorption an Prufkorpern beliebiger Form in der Ulbrichtschen Kugel, *Optik*, vol. 9, pp. 323–332, 1952.

33. Toporets, A. S., Study of Diffuse Reflection from Powders under Diffuse Illumination, *Opt. Spectros.*, vol. 7, pp. 471–473, 1959.

34. Edwards, D. K., Gier, J. T., Nelson, K. E., and Roddick, R. D., Integrating Sphere for Imperfectly Diffuse Samples, *J. Opt. Soc. Amer.*, vol. 51, pp. 1279–1288, 1961.

35. Gier, J. T., Dunkle, R. V., and Bevans, J. T., Measurement of Absolute Spectral Reflectivity from 1.0 to 15 Microns, *J. Opt. Soc. Amer.*, vol. 44, pp. 558–562, 1954.

36. Tingwaldt, C. P., Measurement of Spectral Sensitivity in the Infrared, *Trans Conf. Use Solar Energy*, vol. II, no. 1, sec. A, p. 57, 1955.

37. Dunkle, R. V., Edwards, D. K., Gier, J. T., Nelson, K. E., and Roddick, R. D., Heated Cavity Reflectometer for Angular Reflectance Measurements, *Progr. Int. Res. Thermodynamic Transport Prop.*, ASME, pp. 541–562, 1962.

38. Edwards, D. K., and Roddick, R. D., Spectral and Directional Thermal Radiation Characteristics of Surfaces for Heat Rejection by Radiation, *Prog. Aeronaut. Astronaut.*, vol. 11, pp. 427–446, 1963.

39. Nelson, K. E., Luedke, E. E., and Bevans, J. T., A Device for the Rapid Measurement of Total Emittance (sic), *J. Spacecraft Rockets*, vol. 3, pp. 758–760, 1966.

40. Gier, J. T., Dunkle, R. V., and Bevans, J. T., Final Progress Report Snow Characteristics Project, Institute of Engineering Research, University of California, Berkeley, 1955.

41. Dunn, S. T., Richmond, J. C., and Wiebelt, J. A., Ellipsoidal Mirror Reflectometer, *J. Opt. Soc. Amer.*, vol. 55, p. 604, 1965; abstract of paper TB15 presented at the Spring 1965 meeting of the Opt. Soc. Am. (See also *J. Spacecraft Rockets*, vol. 3, pp. 961–975, 1966.)

42. Edwards, D. K., Thermal Radiation Characteristics of Solids, lecture, *Princeton Univ. Conf. Heat Transfer in Major Technologies*, April 6, 1961.

43. Edwards, D. K., Measurement of Thermal Radiation Characteristics, *1963 Proc., Inst. Environmental Sci., Mt. Prospect, Ill.*, pp. 417–424, 1963.

44. Janssen, J. E., and Torberg, R. H., Measurement of Spectral Reflectance Using an Integrating Hemisphere, in J. C. Richmond (ed.), "Measurement of Thermal Radiation Properties of Solids," *NASA* SP-31, pp. 169–182, 1963.

45. White, J. U., New Method for Measuring Diffuse Reflectance in the Infrared, *J. Opt. Soc. Amer.*, vol. 54, pp. 1332–1337, 1964.

46. Neher, R. T., and Edwards, D. K., Far Infrared Reflectometer for Imperfectly Diffuse Specimens, *Appl. Opt.*, vol. 4, pp. 775–780, 1965.

47. Edwards, D. K., Comments on Reciprocal-Mode 2π-Steradian Reflectometers, *Appl. Opt.*, vol. 5, pp. 175–176, 1966.

48. Nelson, K. E., and Bevans, J. T., Errors of the Calorimetric Method of Total Emittance Measurement, in "Measurement of Thermal Radiation Properties of Solids," *NASA* SP-31, pp. 55–63, 1963.

49. Eichhorn, R., Lecture, *Princeton Univ. Conf., Heat Transfer in Major Technologies*, April 6, 1961.

50. Kline, S. J., and McClintock, F. A., Describing Uncertainties in Single-Sample Experiments, *Mech. Eng.*, vol. 75, pp. 3–8, 1953.

51. Dunkle, R. V., Spectral Reflectance Measurements, in F. J. Clauss (ed.) "Surface Effects on Spacecraft Materials," pp. 117–136, Wiley, New York, 1960.

52. Olson, O. H., and Pontarelli, D. A., Asymmetry of an Integrating Sphere, *Appl. Opt.*, vol. 2, pp. 631–633, 1963.

53. Dunkle, R. V., Ehrenburg, F., and Gier, J. T., Spectral Characteristics of Fabrics from 1 to 23 Microns, *J. Heat Transfer*, vol. 82, pp. 64-70, 1960.

54. Brandenberg, W. M., Focusing Properties of Hemispherical and Ellipsoidal Mirror Reflectometers, *J. Opt. Soc. Amer.*, vol. 54, pp. 1235-1237, 1964.

55. Edwards, D. K., and Bayard de Volo, N., Useful Approximations for the Spectral and Total Emissivity of Smooth Bare Metals, in S. Gratch (ed.), "Advances in Thermophysical Properties at Extreme Temperatures and Pressures," pp. 174-188, American Society of Mechanical Engineers, New York, 1965.

56. Goody, R. M., "Atmospheric Radiation," pp. 29-39, Oxford Clarendon Press, London, 1964.

57. Tingwaldt, C. von, "Die Absorption der Kohlensaure in Gebeit der Bande $\lambda = 2.7\mu$ zwischen $300°$ und $1100°$ absolut ($°$K)," *Phys. Z.*, vol. 35, p. 715, 1934. (For 4.3 μ see Ibid., vol. 39, p. 1, 1938).

58. Penner, S. S., "Quantitative Molecular Spectroscopy and Gas Emissivities," chap. 6, Addison-Wesley, Reading, Mass., 1959.

59. Goldstein, R., Measurements of Infrared Absorption by Water Vapor at Temperatures to $1000°$K, *J. Quant. Spectrosc. Rad. Trans.*, vol. 4, pp. 343-352, 1964.

60. Oppenheim, U. P., and Goldman, A., Spectral Emissivity of Water Vapor at $1200°$K, *Symp. (Int.) Combustion, 10th, Combustion Inst., Pittsburgh*, pp. 185-188, 1965.

61. Sulzmann, K. G. P., High Temperature, Shock Tube CO_2 Transmission Measurements at 4.25μ, *J. Quant. Spectrosc. Rad. Trans.*, vol. 4, pp. 375-413, 1964.

62. Menard, W. A., Thomas, G. M., and Helliwell, T. M., Experimental and Theoretical Study of Molecular, Continuum, and Line Radiation from Planetary Atmospheres, *AIAA J.*, vol. 6, pp. 655-664, 1968.

63. Tien, C. L., and Giedt, W. H., Experimental Determination of Infrared Absorption of High Temperature Gases, in S. Gratch (ed.), "Advances in Thermophysical Properties at Extreme Temperatures and Pressures," pp. 167-173, American Society of Mechanical Engineers, New York, 1965.
    Abu Romia, M. M., and Tien, C. L., Measurements and Correlations of Infrared Radiation of Carbon Monoxide at Elevated Temperatures, *J. Quant. Spectrosc. Rad. Trans.*, vol. 6, pp. 143-167, 1966.

64. Simmons, F. S., An Analytical and Experimental Study of Molecular Radiative Transfer in Nonisothermal Gases, *Symp. (Int.) Combustion, 10th, Combustion Inst., Pittsburgh*, pp. 177-184, 1965.

65. Edwards, D. K., Glassen, L. K., Hauser, W. C., and Tuchscher, J. S., Radiation Heat Transfer in Nonisothermal Nongray Gases, *J. Heat Transfer*, vol. 89, pp. 219-229, 1967.
    Weiner, M. M., and Edwards, D. K., Nonisothermal Gas Radiation in Superposed Vibration-Rotation Bands, *J. Quant. Spectrosc. Rad. Trans.*, vol. 8, pp. 1171-1183, 1968.

66. Hottel, H. C., and Mangelsdorf, H. G., Heat Transmission from Non-Luminous Gases. II. Experimental Study of CO_2 and H_2O, *Trans. A. I. Ch. E.*, vol. 31, p. 517, 1935.

67. Eckert, E. R. G., Messung der Gesamtstrahlung von Wasserdampf und Kohlensaure in Mischung mit nichtstrahlenden Gazen bei Temperaturen bis $1300°$C, *VDI-Forschungsheft.*, vol. 387, pp. 1-20, 1937.

68. Edwards, D. K., Absorption by Infrared Bands of Carbon Dioxide Gas at Elevated Pressures and Temperatures, *J. Opt. Soc. Amer.*, vol. 50, 617-626, 1960. (See also *Appl. Opt.*, vol. 3, pp. 847-852, 1964; vol. 3, pp. 1501-1502, 1964; vol. 4, pp. 715-721, 1965; Studies of Infrared Radiation in Gases, *UCLA Dept. Eng. Rep.* 62-65, January, 1963; *Chem. Eng. Progr. Symp. Series*, vol. 64, pp. 173-180, 1968.

69. Schmidt, E., Messung der Gesamtstrahlung des Wasserdampf bei Temperaturen bis zu $1000°$C, *Forsch. Geb. Ingenieur.*, vol. 3, p. 57, 1932.

70. Ferriso, C. C., and Ludwig, C. B., Spectral Emissivities and Integrated Intensities of the 2.7 μ H_2O Band between 530 and $2200°$K, *J. Quant. Spectrosc. Rad. Trans.*, vol. 4, pp. 215-227, 1964.

71. Hottel, H. C., and Sarofim, A. F., "Radiative Transfer," chap. 6, McGraw-Hill, New York, 1967.
72. Plass, G. N., Useful Representations for Measurements of Spectral Band Absorption, *J. Opt. Soc. Amer.,* vol. 50, pp. 868–875, 1960.
73. Plass, G. N., The Influence of Numerous Low-Intensity Spectral Lines on Band Absorptance; *Appl. Opt.,* vol. 3, pp. 859–866, 1964.
74. Elsasser, W. M., "Heat Transfer by Infrared Radiation in the Atmosphere," Harvard Meteorological Studies, no. 6, Harvard, Cambridge, Mass., 1942.
75. Goody, R. M., A Statistical Model for Water Vapour Absorption, *Quart. J. Roy. Meteor. Soc.,* vol. 78, p. 165, 1952.
76. Reardon, J. E., and Huffaker, R. M., Radiative Heat Transfer Calculations for Saturn Exhaust Plumes, Molecular Radiation and Its Application to Diagnostic Techniques, *NASA* TMX-53711, pp. 184–218, 1967.
77. Howard, J. N., Burch, D. L., and Williams D., Near Infrared Transmission through Synthetic Atmospheres, *Geophysi. Res. Paper no.* 40, Air Force Cambridge Res. Cent., 1953. (See also *J. Opt. Soc. Amer.,* vol. 46, pp. 186, 237, 334, 1956.)
78. Edwards, D. K., and Menard, W. A., Comparison of Models for Correlation of Total Band Absorption, *Appl. Opt.,* vol. 3, pp. 621–625, 1964.
    Edwards, D. K., Studies of Infrared Radiation in Gases, *UCLA Dept. Eng. Rep.* 62–65, January, 1963.
79. Tien, C. L., and Lowder, J. E., A Correlation for Total Band Absorptance of Radiating Gases, *Int. J. Heat Mass Transfer,* vol. 9, pp. 698–701, 1966.
80. Weiner, M. M., and Edwards, D. K., Theoretical Expression of Water Vapor Spectral Emissivity with Allowance for Line Structure, *Int. J. Heat Mass Transfer,* vol. 11, pp. 55–65, January, 1968.

11 Measurement of thermophysical properties

W. LEIDENFROST

School of Mechanical Engineering, Purdue University
Lafayette, Indiana

Rapid technological developments during the last decades have generated an increasing effort to expand our knowledge of properties of materials. This seems to be especially true for those property values needed whenever heat transfer must be evaluated, and whenever the proper design of heat transfer elements is vital, as in nuclear engineering, rocketry, space travel, and in many other areas of our modern technology.

In addition to direct application of the property values in engineering computation and design, there is the need of scientists who try to predict property values by statistical mechanics, and who require basic information in the form of accurate property values to verify and check their models. The theoretical prediction of properties is of great importance when property data are needed at conditions where measurements are as yet impossible.

Experimental work on viscosity, thermal conductivity, and specific heats has now been in progress for well over a century, and a comprehensive and detailed review of the numerous experimental arrangements which have been used for this purpose cannot be given in a single paper. We will only discuss, therefore, some general methods used for the determination of each of the three properties of interest and will illustrate some of the most commonly used apparatus and, in addition, try to demonstrate the approaches necessary to achieve reliable data.

11.1 GENERAL REMARKS

Determination of properties, especially transport properties, is difficult and time consuming and requires painstaking attention to detail, great skill, and much patience. These facts will be experienced by everyone who sets out to measure properties. The reasons for the difficulties can be summarized with Kestin [1] as follows. In order to produce a measurable effect due to any one of the transport properties of a sample, it is necessary to subject it to an irreversible thermo-dynamic process in the course of which it inevitably departs from thermodynamic equilibrium. Thus the state at which the property is measured is only an intermediate one which is averaged over a range of states existing in the apparatus. Consequently every effort must be made to disturb the system as little as possible from an equilibrium state, which implies that the effect to be measured is inherently very small. The smaller the departure from equilibrium the smaller the effect to be measured and the greater the difficulty in determining it precisely. Under actual experimental conditions it is always and unavoidably necessary to work at the threshold of sensitivity with respect to the measured effect, in order to preserve a near-equilibrium state. Actually in all measurements it is desirable to be in a position to extrapolate to a state of equilibrium and to perform experiments over a decreasing range of the effect which is being measured. The lower end of this series is naturally set by the resolution of the detectors used in the investigation.

Everyone who sets out to determine a property should select a method for which an exact theory exists and which can be applied in an instrument with the least number of corrections and possibly only those corrections for which proper account can be made. This point will be discussed in detail later, for each investigation whenever necessary.

11.2 DETERMINATION OF VISCOSITY

As this chapter is related especially to heat transfer, only the viscosity of fluids is of interest and will be discussed. The effects of viscosity on a macroscopic scale manifest themselves in the presence of shear and it is clear that the measurements of viscosity must involve a state of motion. It is, of course, practically impossible to measure local shear stress and it is necessary to base the measurements on some effect of the motion, which must be very slow, i.e., laminar and at very low Reynolds numbers—since it is only for this condition that exact solutions of the governing equations are known.

For example we can measure the pressure drop in a fluid flowing through a tube, or the drag on a body falling through or rotating or oscillating in a stagnant fluid.

11.2.1 The Use of Poiseuille Flow

The use of Poiseuille flow seems at a first glance the simplest arrangment from the theoretical point of view. The laminar flow through a straight tube seems to be stable and the principal solution is elementary. For fully developed parabolic velocity profile the volumetric rate of flow $\dot{V}$ of the incompressible fluid is given by the Hagen-Poiseuille equation

$$\dot{V} = \frac{\pi d^4 \Delta p}{124 \mu L} \tag{11.1}$$

where Δp is the pressure drop over a length L of the tube whose diameter is d. Hence the viscosity μ can be calculated if V and Δp are measured.

Completely stable laminar flow will exist if the Reynolds number, based on the average velocity, satisfies the condition that Re is below some critical value, usually taken as 2300. This condition together with the requirement of a small but still measureable pressure drop, is the cause of the difficulties of the method especially when gases are the test fluids. The pipe diameter must be small and L must be large. d enters into Eq. (11.1) raised to the fourth power and must therefore be determined extremely accurately; this becomes increasingly difficult with decreasing diameter of the bore. Therefore, proper selection of diameter-to-length ratio must be made; this ratio normally can be proper only for a certain value or small range of viscosity. In addition it must be kept in mind that a tube of a given bore can be manufactured precisely only for a certain length.

Operating the capillary viscometer makes it necessary to consider several corrections for reasons that the actual flow departs from that assumed in Eq. (11.1). When a gas flows through the capillary it will expand slightly due to its pressure having decreased by Δp. This means that the volume rate of flow increases along the capillary. The gas is accelerated slightly and requires a larger pressure drop, and a compressibility correction must be introduced. In many instances it is necessary to use very narrow capillaries. The mean free path of a gas molecule may become comparable with the diameter of the tube, and a slip correction may be necessary (i.e., [2]). The formula holds true only for a parabolic velocity profile which develops asymptotically, theoretically for $L = \infty$ from the inlet. The shear stress in the undeveloped flow is larger, and inlet corrections must be applied for which an exact solution is still needed. The most extensive treatment of the influence was done by Schiller [3], also by others (Couette [4] Goldstein [5]). If the capillary is curved, further corrections are needed [6] due to the change in flow pattern and vortices. Coiled capillaries have been used by Vasilesco [7], Hawkins, Solberg, and Potter [8] and others to achieve a uniform temperature control which otherwise could not be done in capillaries of large length, chosen for the sake of obtaining an appreciable pressure drop in fluids of low viscosity. For the same reason flow in narrow annulus was used. Due to the curvature, centrifugal forces are acting on the fluid, secondary motion is introduced, and the velocity profile ceases to be parabolic so Eq. (11.1) is not valid anymore. Since these vortices are unstable the correction proves to be extremely difficult.

Capillary viscometers have been used extensively by many researchers [9-11] and others, in a variety of constructional forms. Consideration of all of these is beyond the scope of this paper and, therefore, only two principal arrangements will be described. The first one, developed by Flynn, Hanks, Lemaire, and Ross [12], is a steady state instrument applying Eq. (11.1) directly, and the layout of the instrument is shown in Fig. 11.1. The capillary a is accommodated in a solid membrane which divides the thermostated pressure vessel b into two compartments. The vessel is pressurized with the aid of a pressure balance c which

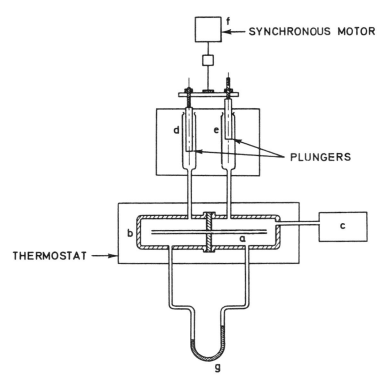

FIG. 11.1 Capillary viscometer. (From Flynn, Hanks, Lemaire, and Ross, 1963, Ref. [12].)

maintains a constant pressure in the downstream chamber of the pressure vessel. The test fluid is made to circulate with the aid of two precision pumps d and e driven by a synchronous motor f through precision gear boxes. In this manner gas is exhausted from one compartment of the pressure vessel and an equal volume is forced into the other. The resulting pressure drop is measured with a differential manometer g. The flow rate $\dot{V}$ is determined from the displacement of the plunger of the pumps and from the speed of the synchronous motor. Another steady flow type instrument utilizes radial flow between disks. Normally, inward flow is used because the flow tends to be more stable. Apparatus of this kind appears to be capable of producing satisfactory results.

A quasi-steady viscometer was first proposed by Rankine [13]. Figure 11.2 shows an instrument of this type as it was designed by Michels and Gibson [14]. The Rankine viscometer is shaped in the form of a loop consisting of the capillary a and a wider tube b, both connected at the lower end to a common branch c. The glass viscometer filled with the gas is placed in the steel vessel d containing mercury and the open end of c is immersed in the mercury. In order to bring the instrument into operational condition the pressure in d is increased and as a result of it mercury rises compressing the gas before it. When the mercury level has reached the junction of the wider tube b to the common branch c, it will rise in b faster than in a due to the higher resistance of the gas flow in the capillary. An initial driving pressure drop is thereby established as shown in the insert of Fig. 11.2. The pressure in the pressure vessel is now maintained constant, and gas is forced through the capillary under the action of Δp falling gradually. The volume forced through is measured by timing the motion of the mercury meniscus rising in the wider tube of well known diameter connected to the lower end of the capillary. This volume must be identical to the one evaluated by integrating Eq. (11.1) over the range of Δp observed.

The Rankine-type viscometer was modified by Ubbelohde [15] and Cannon, Manning, and Bell [16] resulting in an instrument in which the test liquid runs through the capillary due to its own hydrostatic head. The kinematic viscosity is evaluated by observing the time required for the liquid to run out of a reservoir of well known volume attached to the upper end of the capillary.

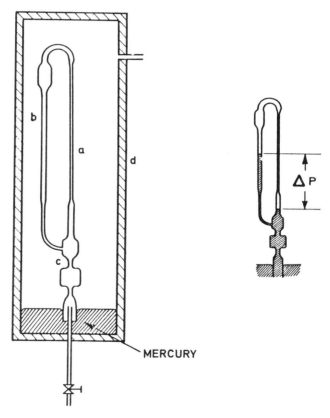

FIG. 11.2 Michels and Gibson's Rankine-type viscometer. (From Michels and Gibson, 1931, Ref. [14].)

11.2.2 The Use of Falling Bodies

Stokes equation is, like the Hagen-Poiseuille equation, an exact solution of the Navier-Stokes equations and relates the drag W on a sphere to the viscosity μ, the radius of the sphere, and the velocity u.

$$W = 3\pi\mu Du \tag{11.2}$$

Equation (11.2) is valid only for unaccelerated motion. This means that u must be the terminal velocity u_t which is reached whenever the drag equals the weight of the sphere. Equation (11.2), in addition, is applicable only for Reynolds numbers of the order of unity.

The requirements of $u = u_t$ and Re $\leqq 1$ can be fulfilled only with very small spheres, or larger but hollow ones of uniform and thin walls. Spheres of these two kinds are almost impossible to fabricate. The permissible radius may be increased by creating an upward force on the sphere by an electrical or a magnetic field—however, those fields may affect the viscosity of fluids to a certain degree. Other variations of the method had been proposed and are in use, i.e., the fall of a sphere through a vertical or inclined tube of only slightly larger diameter. The idea is to ensure laminar friction on the small annulus in the hope that it will constitute almost all of the drag, but in reality wakes and vortices are formed behind the body, causing an increase in drag and departure from the theory of Stokes. For the inclined tube there exists, in addition, the influence of the eccentricity. All these facts cannot be properly accounted for and for this reason at the present time the otherwise very attractive falling body viscometer is not used as a primary instrument, but largely applied to relative measurements.

11.2.3 The Use of Rotating Bodies

The rotating cylinder method is in principle one very accurate way to obtain absolute viscosity measurements, and has been used frequently to establish standard values. The arrangement consists of two concentric cylinders one of which is suspended while the other rotates at a constant speed. For slow speeds the angular deflection of the stationary cylinder of infinite length, produced by the viscous torque, is directly proportional to the viscosity as seen in the equation

$$M = \pi\mu h \, \frac{D^2 d^2}{D^2 - d^2} \, \omega \tag{11.3}$$

where h is the height of either the outside or inside cylinder; M, D, and d their respective radii, and ω the angular velocity. Equation (11.3) is valid only for infinite cylinders. For finite cylinders, corrections must be made for end effects, which can be minimized to a large degree by the use of guard rings. A typical instrumental arrangement is shown schematically in Fig. 11.3. There the outer

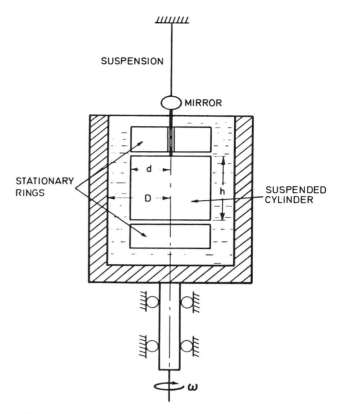

FIG. 11.3 Rotating cylinder viscometer.

cylinder is rotated (see also Gilchrist [17] and Kellström [18]) at a constant rpm while the inner cylinder is stationary, consisting of a middle part guarded at either end by two rings fixed in position. The suspension system might be represented by an elastic wire as shown in Fig. 11.5 and an optical system for measurements of the anular deflection, i.e., the torque. Another quite common arrangement is to suspend the cylinder by a rigid rod held in balance by a known electromagnetic torque. Bearden [19] rotated the inside cylinder to establish very accurate and reliable viscosity values of air. Rotating bodies of other geometries have also been used but with less satisfaction.

The rotating cylinder arrangements have many advantages. First of all the cylinders can be relatively large in size and their precision machining can be done easily; furthermore, presentday technology can keep the angular velocity extremely accurately constant over a long period of time. A disadvantage is that the rotating mechanism, together with the need of using a calibrated wire or electromagnetic coils for torque measurements rule this system out as far as high temperature and high pressure applications are concerned.

11.2.4   The Use of Oscillating Bodies

This method makes use of the observation of the decay of slow torsional oscillations. The use of a torsional pendulum for viscosity measurements was suggested by Coulomb as early as 1784. Maxwell [20] used a pile of disks and since then many other researchers applied different systems but for many years only with limited success. Kestin and Leidenfrost [21] finally succeeded in developing this method for absolute and very accurate viscosity determinations. The achievement was made possible by the development of a precise theory of the instrument initiated first by Kestin and Persen [22] and completed by Newell [23]. The theory is somewhat involved and here we will discuss only its principle. The reader is referred to the works by Kestin and Newell. Figure 11.4 shows the most common arrangement. The disks or spheres axially suspended by an elastic wire are initially deflected slightly and then released. They oscillate with very slow motion in the test fluid surrounding the bodies either in an infinite space or in a narrow gap.

The period of the oscillation T and the damping decrement

$$\Delta = \frac{1}{2\pi} \ln \frac{\alpha_n}{\alpha_{n+1}} \qquad (11.4)$$

are measured. With the vacuum values of T_0 and Δ_0 enough information is available to describe the isochronous harmonically damped motion of the system.

The motion in the fluid due to shear on a disk is described by the Navier-Stokes equation in a cylindrical and symmetrical coordinate system.

$$\frac{\partial u}{\partial t} = \nu \left(\frac{\partial^2 u}{\partial r^2} + \frac{1}{r} \frac{\partial u}{\partial r} - \frac{u}{r^2} \frac{\partial^2 u}{\partial z^2} \right) \qquad (11.5)$$

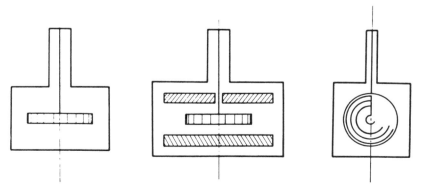

FIG. 11.4   Some oscillating systems for the measurements of viscosity.

under the assumption that only circular motion exists, where u is the tangential velocity, ν the kinematic viscosity and t the time.

Introducing

$$\xi = \frac{r}{\delta} \qquad \eta = \frac{z}{\delta} \qquad \tau = \omega_0^t$$

with

$$\delta = \sqrt{\frac{\mu T_0}{2\pi\rho}} = \sqrt{\frac{\nu}{\omega_0}}$$

δ represents the average distance in the fluid at right angles to the body over which most of the shearing takes place and is referred to as boundary layer thickness or depth of penetration.

The circular frequency in vacuum is $\omega_0 = 2\pi/T_0$. Introducing furthermore $\phi = u/r$, the local angular velocity yields the Navier-Stokes equation in dimensionless form

$$\frac{\partial\phi}{\partial\tau} = \frac{\partial^2\phi}{\partial\xi^2} + \frac{3}{\xi}\frac{\partial\phi}{\partial\xi} + \frac{\partial^2\phi}{\partial\eta^2} \tag{11.6}$$

The motion of the body is described by the ordinary differential equation

$$M_f = I\ddot{\alpha} + I\omega_0^2\alpha \tag{11.7}$$

where I is the moment of inertia of the body and M_f is the moment due to viscous forces depending on the velocity gradient of the solution of the Navier-Stokes equation.

For very slow motion observed at a sufficiently large value of time t, and when the motion can be regarded as isochronous and uniformly damped, the principal solution is given in the form

$$\alpha = A_1 \exp \sigma_1 t + A_2 \exp \sigma_2 t \tag{11.8}$$

where A_1 and A_2 are constants and

$$\sigma_1 ; \sigma_2 = -\frac{\Delta \pm i}{\theta} \qquad \theta = \frac{T}{T_0} \tag{11.9}$$

are two complex conjugate roots (with negative real parts) of a transcendental characteristic equation of the type

$$\sigma^2 + 1 + \sigma^{3/2} F(\rho_i M_k) = 0 \tag{11.10}$$

The function $F(\rho_i M_k)$ depends on the geometrical arrangements and contains one or several similarity parameters m_k. These in turn depend on the viscosity μ and on the density ρ of the fluid.

The determination of the precise form of the characteristic equation constitutes the theoretical problem of the measurements with oscillating bodies. The equations are very complex and can be solved only by iteration processes and high speed computers are necessary. This is the reason why this method of measurement only became successful recently.

For measurement it is of advantage to select in accordance with Fig. 11.4 the systems with infinite spacing much less than boundary layer thickness as only the viscosity enters the characteristic equation. For infinite spacing the density also must be known, in many cases necessitating a separate measurement of this property.

Viscometers using oscillating bodies have been applied very successfully for determination of high pressure data, also for wide ranges of temperature (but this range is limited by the increase of internal damping of the suspension wire with temperature). The apparatus used by Kestin and Leidenfrost is shown in Fig. 11.5. The oscillating system consists of a thin plane parallel and highly polished disk suspended by an elastic wire. A long stem is attached to the bottom of the disk holding a mirror at its lower end. The oscillating system is housed in a high pressure chamber positioning also two stationary plates on either side of the disk. The system is deflected by means of a rod lifted and engaged to the stem and turned with the aid of a magnet. The motion of the system and its decay is observed by optical means through the window in front of the mirror. Kearsley [24] successfully used a hollow sphere filled with the test fluid and oscillating in a vacuum chamber.

11.3 DETERMINATION OF THERMAL CONDUCTIVITY

General discussions of the subject are given in [25-28]. In the viscosity case we only have to consider the measurements of fluids. With respect to thermal conductivity all materials are important to heat transfer, and the measurement of solids, therefore, must be included. Thermal conductivity can be measured in any apparatus which supplies the required boundary condition to a particular solution of the Fourier equation of heat conduction, without heat source or sink in Cartesian coordinates and for k dependent on location.

$$\rho c \frac{dT}{dt} = \frac{\partial}{\partial x}\left(k_x \frac{\partial T}{\partial x}\right) + \frac{\partial}{\partial y}\left(k_y \frac{\partial T}{\partial y}\right) + \frac{\partial}{\partial z}\left(k_z \frac{\partial T}{\partial z}\right) \qquad (11.11)$$

For an isotropic medium, Eq. (11.11) simplifies to

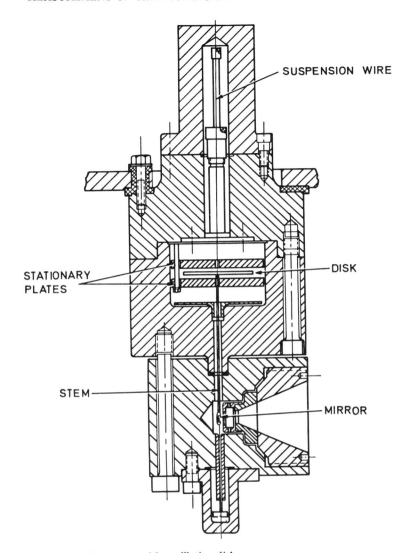

FIG. 11.5 Viscometer with oscillating disk.

$$\frac{\partial T}{\partial t} = a\nabla^2 T \qquad (11.12)$$

This equation normally is applied to a one-dimensional geometry and the thermal diffusivity a can readily be evaluated from the measured temperature distribution T as a function of time t. The thermal conductivity may then be determined with the heat capacity $c\rho$.

In the one-dimensional case of steady state and $k \neq f(T)$ Eq. (11.12) becomes

$$\frac{d^2 T}{dx^2} = 0 \qquad (11.13)$$

which is identical to

$$q_k = -kA \frac{dT}{dx} \quad \text{for} \quad q_k = \text{constant} \qquad (11.13a)$$

and the thermal conductivity can be determined directly.

Practically all measurements of the thermal conductivity are based on Eqs. (11.11) through (11.13a), however, the boundary conditions established in various instruments and the corrections necessary differ from material to material. For this reason it seems to be more convenient to discuss the measurements of fluids and solids seaparately.

11.4 DETERMINATION OF THERMAL CONDUCTIVITY
OF FLUIDS

11.4.1 Steady State Measurements

Fourier's law can be written in a more general form

$$q_k = kB\Delta T \qquad (11.14)$$

where ΔT is the temperature difference observed under steady heat flow through the layer of test fluid of geometry B.

There are two common types of thermal conductivity cells: first, linear heat flow instruments, and second, radial heat flow arrangements. The geometric constant for linear heat flow through a parallel layer of test fluid of area A and thickness L is

$$B = \frac{A}{L}$$

For radial heat flow:

$$B = 2\pi \frac{h}{\ln D/d}$$

for an infinitely long cylindrical annulus of height h and of thickness $D - d$,

$$B = 2\pi \frac{D \cdot d}{D - d}$$

for a spherical arrangement, and

$$B = 2\pi \left(\frac{h}{\ln D/d} + \frac{D \cdot d}{D - d} \right)$$

for a combined arrangement cylinder capped with hemispherical ends. (Diameter ratio chosen so that logarithmic and hyperbolic temperature profiles are practically identical.)

The plane horizontal single layer type apparatus has been used by Schmidt [29], Michels and his coworkers [30, 31, 32], Fritz and Poltz [33], and others [34]. The arrangement is shown principally in Fig. 11.6. The instrument consists of an upper heater plate surrounded by a guard and a lower plate acting as a constant temperature heat sink. The material to be tested lies between the two plates. The guard heater can be eliminated [31] by using a hot plate sandwiched between two cold plates with the test fluid in between.

The radial heat flow method consists of measuring the heat dissipation from a wire or cylinder surrounded by an annulus of test fluid and an outer tube acting as a heat sink. The so-called "hot-wire" apparatus was popularized by Schleiermacher [35]; the arrangement is shown in Fig. 11.7. A fine wire generally made of platinum is axially stretched between the ends of a closed cylinder. The wire serves as heater and also as a resistance thermometer. The heat flow is measured as indicated in the figure by measuring the current and the voltage drop along the measuring section of length h. Coaxial cylinder arrangements have been used extensively by Schmidt and Sellschopp [36], Keyes [37], Ziebland [38], Johannin, Wilson, and Vodar [39], and Thodos and Misic [40]. In most cases guard heaters have been applied in order to assure radial flow only. Figure 11.8 sketches a representative arrangement of this type of apparatus. Riedel [41] used a spherical arrangement. Schmidt and Leidenfrost [42] introduced an instrument, combining cylinder and sphere arrangements, shown in Fig. 11.9.

The various types of apparatus just discussed are in principle of simple construction. Also the formula [Eq. (11.14)] applied to determine the thermal conductivity from the quantities measured is simple. But the reader should not

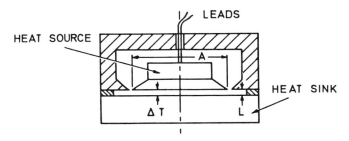

FIG. 11.6 Thermal conductivity apparatus: Linear heat flow.

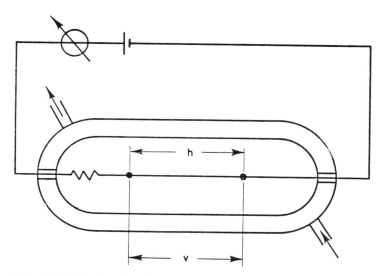

FIG. 11.7 Schleiermacher hot-wire cell.

be misled by these facts, as perfect conditions for which Eq. (11.14) is valid rarely exist and many corrections must be applied. In accordance with the statement made earlier in Sec. 11.1, it is felt of importance to discuss these corrections in detail.

Investigating q_k first we find

$$q_k = (E \pm \Delta E)(1 \pm \Delta I) - q_r - q_c \pm q_L \pm (q_{\text{osc}})_h$$

$$\pm (q_{\text{osc}})_c - (q_{\text{inh}})_h - (q_{\text{inh}})_c - (q_{\text{inh}})_{\text{fluid}} + q_{\text{ch.r}} \pm \cdots \quad (11.15)$$

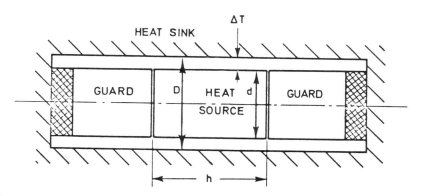

FIG. 11.8 Radial heat flow thermal conductivity cell.

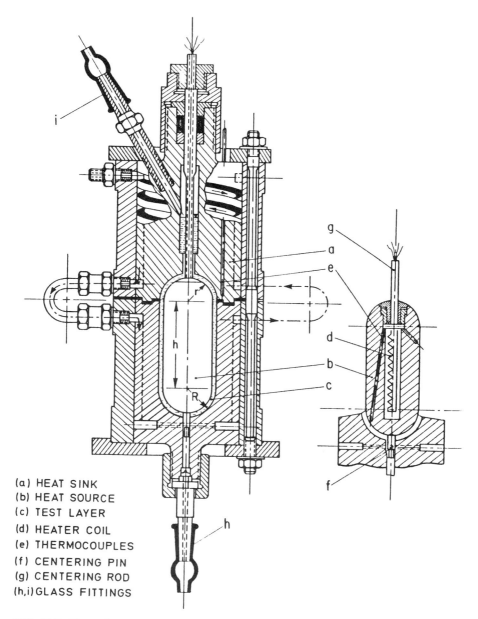

(a) HEAT SINK
(b) HEAT SOURCE
(c) TEST LAYER
(d) HEATER COIL
(e) THERMOCOUPLES
(f) CENTERING PIN
(g) CENTERING ROD
(h,i) GLASS FITTINGS

FIG. 11.9 Thermal conductivity cell.

where the electrical power input (E · I) must be corrected for radiation q_r, free convection q_c, lead-in losses q_L, heat flow due to unsteady state conditions q_{osc} and/or inhomogeneities q_{inh}, and for the possible chemical reaction $q_{ch.r}$ of the fluid with the wetted walls, respectively.

Most of these corrections are very difficult, and some even impossible, to apply. Therefore in order to achieve reliable measurements, it is important to select a measurement technique for which the smallest number of corrections are necessary and yet can be properly applied. Corrections for free convection can be avoided by selecting the plane layer type of instrument where the heat is applied from above. Free convection will be presented in all the other cells discussed but can be made negligibly small by providing conditions for which the Ra number is smaller than its critical value. This can be done in most cases by selecting thin layers of test fluid and by measuring at small ΔT values.

The heat flow due to unsteady state conditions q_{osc} results from temperature fluctuation of the heater or the heat sink and is very difficult to evaluate. The temperature fluctuation in addition influences proper measurements of ΔT. These influences can only be depressed in cells which quickly reach equilibrium, and can be easily maintained at equilibrium long enough to take the readings. The last two types of instruments mentioned above do not use guard heaters and, therefore, measuring conditions are more quickly established. But these instruments with the inner body as heater elements require some kind of centering devices, which normally are also used to enclose wires to the electrical heater coil and temperature sensing elements. Thus a path for heat flow is introduced from the hot to the cold body. This type of heat flow, the so-called "lead-in loss" is also present in most other instruments, even those with guard heaters, since the temperature of the guard and the heater are not necessarily identical at every location. Lead-in losses can be made zero as proved by Leidenfrost [43] for ideal conditions and can be evaluated for real cases to a high degree of certainty according to Leidenfrost and Tree [44].

Another error influencing the measurements is due to inhomogeneities in the temperature field. This field can be disturbed for geometrical reason whenever the geometry of the arrangement deviates from the one described by Eq. (11.14) (i.e., misalignment, eccentricity, surface roughness disturbances at the location where guards join the heater and/or sink and others). All these influences can be properly evaluated as will be demonstrated below.

Disturbances of the temperature field caused by deviation of the surfaces of heater and sink from isothermal conditions cannot be evaluated and every effort must be spent to minimize those effects—also for the reason that they influence the geometric constant B. Deviation from isothermal conditions result from uneven heating or cooling of the heater and/or the sink and from disturbances of the temperature field within those bodies caused by temperature sensing elements placed within the walls. Disturbances also might be introduced by uneven heat flow due to radiant heat transfer between surfaces of nonuniform optical properties.

The geometric constant B cannot be determined with sufficient accuracy by length measurements alone, but it can be determined easily and accurately by measuring the capacitance of the arrangement [43]. This measurement will also account for eccentricity and other inhomogeneities caused by deviation from proper geometry. The geometric constant should be measured electrically for all ranges of temperature and, if possible, pressure, in order to evaluate the change in geometry and of B with temperature and pressure.

The surface temperature of the walls wetted by the test fluid cannot be measured directly, but must be extrapolated from the temperature drop in the wall that is obtained from sensing elements located in the wall itself. This extrapolation is very difficult to achieve, since the sensing element disturbs the temperature field in the respective bodies as already stated above and discussed by Leidenfrost [45]. The more difficult task however is the evaluation of the temperature field and the resulting inhomogeneities and their influence on the geometric constant and the heat flux. Chemical reaction of the test fluid with the walls of the instrument will result in a change of the test substance, in a change of optical properties of walls, and possibly, in the steady or unsteady generation of heat. Since these influences are practically impossible to evaluate, the only way to overcome them is to avoid chemical reaction by using contamination-free materials in the construction of the instrument.

Assuming that all the corrections discussed so far can be properly applied or can be minimized to a large degree by choosing the most feasible arrangement for the measurement, there are still great difficulties which must be overcome.

The necessity to measure small temperature differences represents by itself quite a task which normally can only be achieved by using most reliable sensing elements. Pt-resistance thermometers and selected thermocouples calibrated within the instrument must be used together with sensitive instrumentation shielded and guarded properly. Radiant heat exchange in addition must be accounted for in all cases of nonopaque test fluids. The most simple case of perfectly transparent media cannot be tested easily, because measurements even under a perfect vacuum will not only yield the radiant heat transfer but also the lead-in losses. Those are a function of apparent thermal conductivity and cannot be subtracted to evaluate the radiant heat exchange.

The situation becomes even more complex in case of test substances able to absorb and emit radiation. The theory for simultaneous radiation and conduction was established by Viskanta and Grosh [46] and applied for correction of thermal conductivity measurements by Leidenfrost [43], and later by Poltz [47, 48], Kohler [49], and Schödel [50]. Those calculations were based on the gray gas assumption. Furthermore the temperature distribution within the test fluid was assumed to be linear, due only to conduction and not influenced by radiation. Only recently Leidenfrost [51, 52] expanded the evaluation for non-gray conditions and for the true temperature distribution.

In order to correct the thermal conductivity measurements for radiation losses the conservation of energy equation must be solved. However, in the case of simultaneous radiation and conduction, the energy transfer is governed by a complicated nonlinear integrodifferential equation. Since the radiation energy flux depends on the test cell configuration, the problem was simplified by assuming that the heat transfer takes place between two infinitely large parallel plates. This assumption is permissible for radial heat flow arrangements when the diameter-to-width ratio of the test layer is large. In order to make the problem solvable, additional assumptions were necessary. The walls of the test cell are assumed to be isothermal diffuse absorbers and emitters of thermal radiation, and to have constant radiation properties, except for their variation with wavelength.

Under the two given assumptions the steady state conservation of energy equation reduces to

$$\frac{d}{dx} \left(- k \frac{dT}{dx} + q_r \right) = 0 \tag{11.16}$$

Clearly the total heat transfer is constant, composed only of radiative and conductive contributions.

The computation of the problem is quite involved and can be carried out only with high-speed computers. The results proved that radiant heat transfer in an absorbing and emitting gas can influence the accuracy of thermal conductivity measurements by a larger percentage than is normally claimed for thermal conductivity determinations. The percentage increases with temperature and emissivity of the walls. For example, in the case of steam as a test substance, the error due to radiant heat transfer will be of the order of 35% at critical point conditions and black walls. Evaluating radiant heat transfer by approximate methods can yield results still in error by a considerable amount. However, in many instances the correction cannot be performed because we don't know the optical properties of the test fluids. Therefore, it is of importance to minimize the radiant heat exchange by providing instrument walls with lowest possible emissivity.

The use of fine wires (hot-wire method) and the thin films of test fluid necessary to depress free convection leads to a further complication and deviation from Eq. (11.14).

Whenever the measuring conditions are such that the mean free path of the conducting molecules in the fluid becomes comparable with the size of wire or the gap width between heater and sink there is a temperature discontinuity at the surfaces. This is caused by the imperfect interchange of energy between the fluid molecules and the solid surfaces. In case of viscosity we had to apply under similar conditions a slip correction, which accounted for a velocity of the gas at the surface deviating from zero. In case of thermal conductivity there is a "temperature jump" at the wall. This jump is proportional to the pressure and is a function of the accommodation coefficient, which in turn is dependent on the materials involved. Dickens [53], among others, discussed this influence.

The steady state methods discussed so far will produce, under normal conditions, accurate data whenever the corrections shown can be applied properly or are not necessary. In most instances lead-in losses, free convection, and radiation must be considered; quite often this is very difficult or even impossible. The errors listed will influence all measurements of thermal conductivity and partly also specific heat determination, and are by no means restricted to steady state measurements. Some of the errors might become negligibly small in other measurement techniques but then other influences must be accounted for, e.g., in unsteady measurements the observation of time and sometimes location while the temperature changes are critical.

Unsteady state devices normally operate so fast that free convective motion has no chance to develop. Furthermore radiant heat transfer seems to be only of very small influence and so is the lead-in loss.

11.4.2 Unsteady Heat Flow Methods

Stalhane and Pyk [54] used a thin straight wire acting as a heat line source surrounded by the infinite homogeneous fluid to be measured. They used this arrangement to determine the thermal conductivity of liquids and also of granular materials. It is also often used for measurements on soil or other materials where steady state instruments fail due to possible moisture change during the long time necessary to establish equilibrium conditions.

The constant rate of heat production in the wire will cause a cylindrical temperature field in the fluid. The rise of temperature in the field dependent upon the thermal properties of the fluid will be an indication of the thermal conductivity. Van der Held and Van Druven [55] showed that the temperature T_1 at time t_1 and temperature T_2 at time t_2 are related:

$$T_2 - T_1 = \frac{q}{4\pi k} \ln \frac{t_2}{t_1} \qquad (11.17)$$

The equation is valid after a short initial time depending on the size of wire and the properties of the fluid. For a platinum wire of 10–30 μ diameter the initial time is of 0.1 sec. If the relation between the heat capacity of wire and surrounding liquid is favorable then the initial time may even be shorter than 1/msec. Grassmann, Straumann, Widmer, and Jobst [56] modified the arrangement in order to overcome the painstaking conversion to a logarithmic time scale by recording the increase of the temperature of the wire immediately as a function of the logarithm of time. This was done by using the increase of the temperature of a second wire immersed in a reference liquid with known thermal conductivity as a logarithmic scale. Both line source wires are branches of two Wheatstone bridges and the resulting voltages are recorded on the axes of an X-Y recorder. The recorder will write a straight line after the initial time has elapsed. The slope of this line will be inversely proportional to the thermal conductivity of the liquid to be measured. Deviation of the recorded curve from a straight line indicates the beginning of free convection. This was observed to start at a Ra number of 1800. The instrument so modified is a relative one but still produces data of high precision. If the time interval of one measurement is shorter than 10 sec, the temperature field is not influenced by the walls of the containing vessel if the distance from the wire is larger than 10 mm [56].

We do not favor applying the line method to measure gases; the heat capacity of the gas is small compared to that of the wire (increasing, therefore, the initial time), and free convection might set in too soon. The difficulties were overcome by Briggs, Goldstein and Ibele [57]. The method is not directly applicable for conducting liquids, and provisions must be made to insulate the wire, introducing some of the errors discussed for steady state methods. Allen [58] evaluated sources of errors: finite mass of the line source, temperature variation along the wire, radiation and free convection, and deviation from constant power dissipation in the wire.

Cyclic heat flow for measurements of thermal conductivity of fluids also has been used. The penetration of surface temperature fluctuation from a plane slab into the sample can be evaluated by solving the general conduction equation which for this case has the form

$$\frac{\partial \theta}{\partial t} = a \, \frac{\partial^2 \theta}{\partial x^2} \tag{11.18}$$

where θ is the deviation from the mean temperature at any point at any time and x is the distance measured normal to the surface of the slab.

The penetration of the surface temperature fluctuation into the fluid is given by the well known solution of Eq. (11.18):

$$\theta_{(x, t)} = \theta_0 \exp -x \sqrt{\frac{\omega}{2a}} \, \cos\left(\omega t - x \sqrt{\frac{\omega}{2a}}\right) \tag{11.19}$$

where ω is the angular velocity and θ_0 the amplitude of the signal at $x = 0$. Another relationship for computing thermal diffusivity may be obtained by comparing the amplitude of temperature fluctuation θ_x of the response at distance x with θ_0

$$\frac{\theta_x}{\theta_0} = \exp -x \sqrt{\frac{\omega}{2a}} \tag{11.20}$$

Harrison, Boteler, and Spurlock [59] used this method in a cylindrical arrangement. See also related paper by Tarmy and Bonilla [60].

A new method based upon gas diffusion techniques by Westenburg and Walker [61] has been extended to thermal conductivity measurements by Westenburg and deHaas [62]. A line source of heat is stretched across a uniform laminar gas flow. The thermal conductivity can be calculated from the peak temperature rise at some known distance downstream and the known input heat rate to the wire. The temperature rise at any point downstream from an infinite source can be expressed by:

$$\Delta T = \left(\frac{q}{4uc_p\rho\pi kr}\right)^{1/2} \exp\left(\frac{uc_p\rho(x-r)}{2k}\right) \tag{11.21}$$

where c_p is the specific heat, ρ the density, q the heat flow, u the velocity of the gas, r the distance from the source, and x the distance parallel to flow direction.

11.4.3 Other Methods of Measuring Thermal Conductivity of Fluids

Sengers [63] discussed the transport properties of compressed gases, pointing out that measurements of thermal conductivity are practically impossible to achieve under critical conditions. At those conditions the Gr number and therefore the Ra number become very large and the heat transport is governed by free convection. Under equilibrium conditions $\Delta T = 0$ and Ra has no influence. Measurements under those conditions seem to be possible by optical means.

Sogin and Thompson [64] developed an apparatus to measure transport properties of gases by means of Rayleigh-Jeffreys instability, i.e., by means of free convection. Their system is a linear heat flow cell using a horizontal layer but heated from below. They measure the heat flow under increasing pressure. For all values of Ra $<$ Ra$_{cr}$ this heat flow will remain constant. When increasing pressure causes Ra $>$ Ra$_{cr}$ then free convection will suddenly increase the heat flow.

For a single component or a mixture of ideal gases Ra is given as

$$\text{Ra} = \frac{gL^3 \Delta T c_p p^2}{\mu k R^2 T_m^3} \tag{11.22}$$

where p = pressure
     R = individual gas constant
     T_m = mean temperature

The critical value of Ra has been established by many investigators and is known for the parallel layer arrangement with a high degree of certainty. Therefore, when the measurements under increasing pressure cause a sudden increase in heat flux, Ra$_{cr}$ has been reached and the above relationship, Eq. (11.22), can be solved for viscosity or thermal conductivity when all the other quantities are known.

Since viscosity and thermal conductivity of gases are pressure independent to a certain degree and over small ranges, the change of pressure during the measurements has no influence on the properties. In the liquids' case Ra$_{cr}$ must be established by changing ΔT. This influences the measurements, as viscosity and thermal conductivity are temperature dependent. The measurements carried out so far by the investigators cited above indicate usefulness of this technique.

Another indirect determination of thermal conductivity is the measurement of the Prandtl number, introduced by Eckert and Irvine [65]. Normally the viscosity and the specific heat are known to a better accuracy than the thermal conductivity and the Pr number can be measured with high precision. Therefore the evaluation of k from direct measurement of Pr has advantages. The method of measurement involves the relation describing the temperature which a flat plate assumes in a high velocity steady two-dimensional gas flow and when the plate exchanges heat only by convection. This recovery temperature T_r is usually described by a recovery factor r.

$$r = \frac{T_r - T_\infty}{T_{ad} - T_\infty} \tag{11.23}$$

where T_{ad} is the stagnation or total temperature, T_∞ the static temperature of the flow.

Busemann proved that for certain ranges of Pr

$$r = \sqrt{Pr} \tag{11.24}$$

Thus Pr can be measured by temperature measurements and the method has been used successfully since it was established. Smiley [66] and Lauver [67] describe a shock tube technique and Carey, Carnevale, and Marshall [68] an ultrasonic method. Both apparatuses are useful especially for high temperature work. Leidenfrost [45] described a method to determine thermal conductivity of electrically nonconducting fluids in the absence of temperature and heat flow measurements. He observed time changes of geometry of an annulus filled with the test fluid by means of capacitance measurements. This method applied to a spherical system heated from the outside was analyzed by Leidenfrost and Kennedy [69]. It was proved that the method is an absolute one and the thermal conductivity can be obtained from the lowest capacitance value observed under proper heat input to the system. The new technique, in addition, allows one to determine thermal conductivity quickly and with very small temperature differences within the test fluid. Also discussed in [69] is the procedure to obtain simultaneously the thermal diffusivity.

11.5 MEASUREMENT OF THE THERMAL CONDUCTIVITY OF SOLIDS

Many publications can be found in the series of thermal conductivity conferences; two volumes are [70] and [71]. The thermal conductivity of solids varies from values as low as those known for gases to values several orders of magnitude higher. The high values of thermal conductivity are observed for metals, especially in metals of high electrical conductivity. This fact makes it understandable that quite different approaches must be used to measure each of these materials and for this reason we shall discuss low-conducting and high-conducting solids separately.

11.5.1 Steady State Measurements of Thermal Conductivity of Low–Conducting Solids

Equation (11.14) is principally applied in linear or radial heat flow devices. For linear heat flow, apparatuses like the one shown in Fig. 11.6 are used; similar instruments, with single, or multilayer test substances are also utilized. The layers can be thick because free convection is of no concern. In case of gases and liquids, thermal contact with the surfaces of heat source and sink is always achieved and, therefore, the temperature difference across the test layer is that between surfaces of heater and sink measured by sensing elements placed in the good-conducting walls of these bodies.

For low-conducting media, where contact resistance is of less concern, the same measuring techniques for ΔT can be applied. For somewhat better conducting insulators it becomes more important to achieve good contact between the sample and the metallic surfaces of the instrument. This can be done in many instances by means of unevaporating liquids and it might also be necessary to locate the temperature sensing elements inside the sample. But the good contact with the heater or sink still must be assured—otherwise the temperature field is disturbed and Eq. (11.14) cannot be applied.

Uniform and low contact resistance normally becomes increasingly difficult to achieve with increasing temperature and for better conducting materials. In those cases it is of advantage to provide poor but uniform contact by means of insulating layers sandwiched between heat source and sink and the test sample. For measurements at high temperatures the insulation can be represented by an air space. This avoids or reduces problems of thermal contact even when the specimen is warped or disturbed by expansion resulting from the temperature gradient established in it during the measurements—or when the surfaces cannot initially be made plane.

An instrument of this kind was originally devised by Dickinson for measurements of standards and has been described in detail by Flynn and Didion [72]. The instrument is shown schematically in Fig. 11.10. The apparatus consists of a high temperature furnace, the disk-shaped specimen, and a steam calorimeter used to measure the heat flow through the sample. The specimen is supported slightly above a circular silicon-carbide plate heated by the furnace. The sample receives heat by radiation and conduction through the air gap and looses heat from its upper surface again by radiation and conduction across an air space to the calorimeter. The heat flux through the specimen is determined by the steam calorimeter. This consists of a central metering chamber which is surrounded by a guard chamber operating at the same temperature which prevents edge effects.

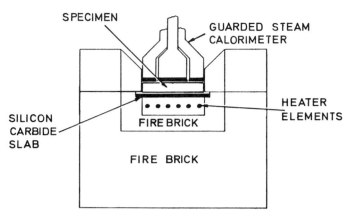

FIG. 11.10 NBS steam calorimeter apparatus for measuring thermal conductivity.

Boiling water or any other suitable fluid has some advantages over circulating liquid calorimeters with regard to temperature stability and uniformity. The temperature drop through the specimen is measured by means of thermocouples imbedded in its surfaces. This will not cause considerable disturbances of the heat flow pattern since the sample size in this apparatus can be rather large.

Some materials are available only in small samples because of the difficulties of obtaining them with adequate uniformity. In small-volume specimens even a small thermocouple may introduce some disturbance and for this reason some investigators prefer to deduce the temperature at the contact surfaces of a specimen by extrapolating temperatures observed in the contacting body. If this is done the difficult problem of the thermal resistance at the contacting interface is encountered.

A so-called "cut-bar apparatus" for measurements of small specimens is shown in Fig. 11.11 which represents an NBS standard instrument [73]. The longitudinal heat flow through the sample is determined from the temperature gradient observed in the reference bars shown in the figure, or it can be measured directly by means of a properly guarded heater located close to the specimen. Radial heat losses are depressed or minimized by guard cylinders surrounding the insulated reference bars and the sample, heated to an identical axial temperature distribution.

Radial heat flow instruments are in most cases of a cylindrical arrangement and spherical only for poor-conducting materials (fibers, powders, or granular materials). A spherical apparatus was described by Black and Glaser [74] and is shown in Fig. 11.12. The instrument consists of an inner sphere and an outer sphere kept at a uniform temperature by means of electrical heaters—or circulating or boiling liquids. The test material is placed between the spheres. The heat flowing through the samples can either be generated electrically in the inner sphere or, as shown in Fig. 11.12, can come from the outer sphere and boil off the liquid contained in the inner sphere. The boiloff rate is an indicator of heat flow and is used to calculate the thermal conductivity. One advantage of this type of apparatus is the simplicity of construction. Also, the sample is completely enclosed and can be kept at a constant condition (i.e., moisture free). However, it has the disadvantage that the test material can be filled into the gap only with a more or less accidental density and in addition, the heat leak along the neck cannot be easily estimated. Cylindrical arrangements use the test material in the form of long cylinders with small inner bores heated either from the inside or the outside. In the latter case the heat flow is normally measured calorimetrically, by observing the change of temperature of a fluid flowing through the inner bore or by measuring the evaporation rate of a liquid at rest.

Axial heat flow in a solid cylinder cannot be avoided completely and it might be advantageous to introduce axial flow resistance by using rings stacked axially on top of each other. The rings close to the ends of the axial arrangement act as guard to the rings located in the central part. The heat flow generated in

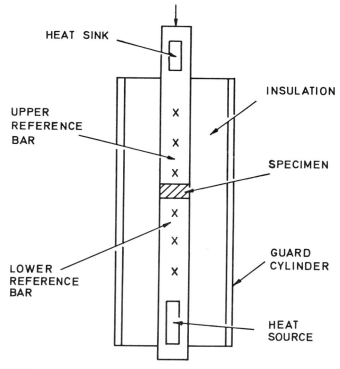

NOTE.

X DENOTES A THERMOCOUPLE POSITION

FIG. 11.11 A schematic drawing of cut-bar apparatus.

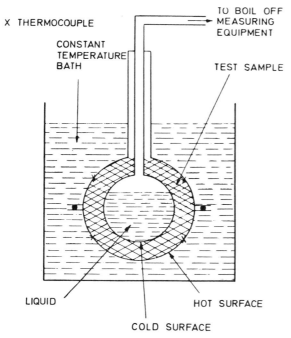

FIG. 11.12 Thermal conductivity cell with spherical test chamber.

this section is measured by observing the voltage drop as indicated in the schematic drawing, Fig. 11.13.

The radial temperature difference is measured by thermocouples (or optically at higher temperature) located in the holes drilled axially at different radii. It is recommended that one of the holes penetrates all the disks so that the longitudinal temperature distribution can be measured.

The size of the holes must be small in order to minimize the disturbance of the temperature field.

11.5.2 Unsteady State Measurements of Thermal Conductivity of Low–Conducting Solids

Cyclic heat flow measurements discussed for liquids are also used for solids and the same holds true for the line source arrangement. A typical instrument of the latter kind is shown in Fig. 11.14. Probes are extensively used for porous media (powder, granular material, and soil). The measurements of k are carried out in the same manner as described above. Details of the system shown in Fig. 11.14 and error analysis are given by Messmer [75].

For measurements of very small samples Parker, Jenkins, Butler, and Abbott [76] have developed an unsteady (so-called "flash" or "heat pulse") method

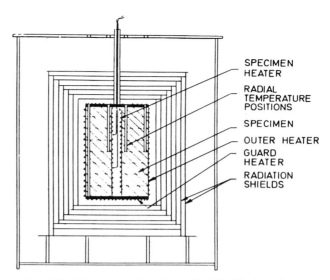

FIG. 11.13 Thermal conductivity cell with disk-shaped sample.

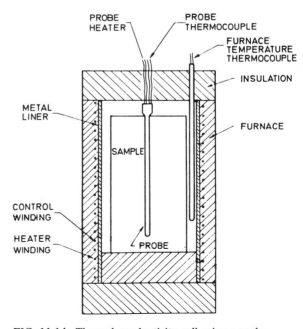

FIG. 11.14 Thermal conductivity cell using a probe.

where a high-intensity short-duration energy pulse is absorbed in the front surface of a thin specimen. The thermal diffusivity is determined by the shape of the temperature versus time curve at the rear surface. Parker solved the general conduction equation (11.12) which yields, under the assumption of temperature-independent properties, the temperature history of the rear surface:

$$\frac{T(L, t)}{T_m} = 1 + 2 \sum_{n=1}^{\infty} (-1)^n \exp - \frac{n^2 \pi^2 at}{L^2} \qquad (11.25)$$

when
$$T(L, t) = 0.5 T_m \qquad t = t_{1/2}$$

the simple relationship for the diffusivity results:

$$a = \frac{1.37 \, L^2}{\pi^2 t_{1/2}} \qquad (11.26)$$

where L is the thickness of the specimen, $T(L, t)$ the temperature of the rear surface and T_m its maximum value. The upper limit of the front surface temperature T_f was also derived by Parker. The result is

$$T_f = \frac{38L}{a^{1/2}} \, T_m \qquad (11.27)$$

which for metals amounts to $10°C$–$20°C$ and for ceramics $50°C$–$100°C$.

Beck [77] has analyzed the method and extended its application to temperature-dependent properties. The heat loss from the sample might become quite high at higher temperatures and can be minimized by using two disks with the rear surfaces facing each other. This technique was applied by Kasper and Zehms [78] but using a periodic flux variation. Another way to minimize the heat loss is to measure in a radial arrangement by heating a cylinder from the outside. Such a method was developed by Nakata [79]. The conduction equation in cylindrical coordinates is

$$\frac{1}{r} \frac{\partial}{\partial r} \left(r \frac{\partial T}{\partial r} \right) + \frac{1}{r^2} \frac{\partial^2 T}{\partial \phi^2} + \frac{\partial^2 T}{\partial z^2} = \frac{1}{a} \frac{\partial T}{\partial t} \qquad (11.28)$$

If a circular cylinder whose axis coincides with the axis of z is heated and the initial and boundary conditions are independent of the coordinates ϕ and z the temperature will be a function of r and t only. Equation (11.28) reduces then to

$$\frac{1}{r} \frac{\partial}{\partial r} \left(r \frac{\partial T}{\partial r} \right) = \frac{1}{a} \frac{\partial T}{\partial t} \qquad (11.29)$$

In this case the heat flow is radial.

For constant heat input to the cylinder surface the temperature as a function of r and t is

$$T(r, t) = A_1(r^2 + 4at) + A_2 \tag{11.30}$$

where A_1 and A_2 are constants. The thermal diffusivity is obtained by measuring the time $\Delta t = t_2 - t_1$ for the temperature at point r_1 to reach that at point r_2. Equation (11.30) gives

$$a = \frac{r_2^2 - r_1^2}{4\Delta T} \tag{11.31}$$

Many of the instruments and steady and unsteady methods discussed so far for solids can be applied for the determination of the thermal conductivity or diffusivity of metals. For this reason we will discuss here only the arrangements used specifically for metals and/or electrical conductors.

11.5.3 Measurements of Thermal Conductivity of Metallic Solids

The NBS metal apparatus for measuring thermal conductivity is shown in Fig. 11.15. The specimen is supported within an outer container and is drilled at each end. The lower cavity encloses an electrical heater while the upper one is cooled. The guard is cooled in a similar way and heated in order to produce an axial temperature distribution similar to that in the sample. The specimen is rather long and has a rather large temperature difference from end to end. The longitudinal temperature distribution is measured by several thermocouples placed with known distance to each other, therefore enabling the calculation of the thermal conductivity at several temperatures.

The methods discussed so far in most instances applied heat indirectly to the specimens. Metals or other reasonably good electrical conductors can be heated directly by an electrical current. Instruments based on the so called "direct electrical heating method" apply the solution of the general conduction equation (11.11) expanded by additional terms.

For cylindrical symetrical geometry the equation is

$$k\left(\frac{\partial^2 T}{\partial x^2} + \frac{\partial^2 T}{\partial r^2} + \frac{1}{r}\frac{\partial T}{\partial r}\right) + \frac{\partial k}{\partial T}\left[\left(\frac{\partial T}{\partial x}\right)^2 + \left(\frac{\partial T}{\partial r}\right)^2\right]$$

$$+ \lambda\left[\left(\frac{\partial E}{\partial x}\right)^2 + \left(\frac{\partial E}{\partial r}\right)^2\right] - \mu\frac{I}{A}\frac{\partial T}{\partial x} = c\rho\frac{\partial T}{\partial t} \tag{11.32}$$

where λ is the electrical conductivity, E the electrical potential, I the electrical

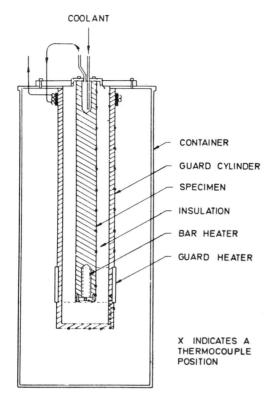

COOLANT

CONTAINER

GUARD CYLINDER

SPECIMEN

INSULATION

BAR HEATER

GUARD HEATER

FIG. 11.15 The NBS metals apparatus for measuring thermal conductivity.

X INDICATES A THERMOCOUPLE POSITION

current, A the cross-sectional area of the sample, and μ the coefficient of Thompson heat.

The influence of the Thompson heat can be minimized by measuring at small temperature gradients. Using small sizes of the sample allows radial gradients to be neglected and Eq. (11.32) becomes for steady state

$$k \frac{\partial^2 T}{\partial x^2} + \frac{\partial k}{\partial T} \left(\frac{\partial T}{\partial x} \right)^2 + \lambda \left(\frac{\partial E}{\partial x} \right)^2 = 0 \qquad (11.33)$$

Measuring at small or zero values of dT/dx or assuming $k \neq f(T)$, the second term vanishes and Eq. (11.33) becomes:

$$k \frac{\partial^2 T}{\partial x^2} + \lambda \left(\frac{\partial E}{\partial x} \right)^2 = 0 \qquad (11.34)$$

Kohlrausch [80] applied Eq. (11.34) in a simple arrangement where the ends of the cylindrical specimen were kept at some constant temperature and radial heat flow was minimized by insulation. With these provisions a paraboliclike axial temperature profile is obtained from which the ratio of thermal and electrical conductivity can be derived. Knowing the electrical conductivity (evaluated by the same measurements) allows one to determine the thermal conductivity.

Neglecting radial heat losses from the sample yields the simple relationship first derived by Kohlrausch

$$\frac{k}{\lambda} = \frac{1}{2} \frac{(E_1 - E_2)^2}{T_2 - T_1}$$
(11.35)

T_2 is the maximum temperature and T_1 is a temperature a certain distance away where the electrical potential is E_1. The parabolic temperature profile will be symmetrical when the two ends of the sample are kept at the same temperature.

Bode [81] heated thin wires, the ends of which were kept at the same temperature, and generated within the sample the same amount of heat which was lost radially by radiation to the walls of a thermostated vacuum chamber. The temperature distribution is then uniform along the wire (only small radial temperature distribution) and the assumption made for the Thompson heat in Eq. (11.33) is fulfilled. Angell [82] established also a practically uniform temperature distribution at the central region of a very long specimen (kept at lower temperatures at the ends). The thermal conductivity from their measurements can be evaluated from

$$k = \frac{1}{4\pi} \frac{\Delta E}{L} \frac{I}{\Delta T}$$
(11.36)

where ΔE is the potential drop observed over a length L in the region of uniform temperature distribution and ΔT is the radial temperature difference at the same location.

A similarly simple relationship can be derived for a hollow tube. Samples of this geometry have been used by Powell and Shofield [83] for the measurements of k and λ for carbon and graphite up to 2400°C. Rudkin, Parker, and Jenkins [84], Krishnan and Jain [85], and others used more or less similar arrangements in vacuum. A detailed description of all the various instrumental setups and the equations derived for the evaluation of the properties would be outside the scope of this chapter and the reader is referred to a survey by Powell, DeWitt, and Nalbantyan [86]. Cezairliyan is conducting unsteady measurements of direct heating by means of capacitance discharge. The work resulted from a study on high-speed thermodynamic measurements by Beckett and Cezairliyan [87].

11.5.4 Some Other Methods of Measuring Thermal Conductivity of Solids

A very simple instrument able to produce data quickly and cheaply is the thermal comparator by Powell [88] which measures relative thermal conductivity. This method consists in principle of heating two metal spheres of equal size to the same temperature and bringing one sphere into contact with, and the other very close to, a colder specimen. If one assumes that the heat exchange by radiation and gaseous conduction from the two spheres is identical, then the sphere brought in contact with the specimen will lose heat in addition by conduction into the specimen across the area of contact. This additional heat flow will result in a more rapid temperature change of this sphere which can be observed by a differential thermocouple placed inside the two spheres.

The radius of the contact with the flat specimen is several orders of magnitude smaller than the sphere and the system can be treated like two semi-infinite large bodies. Under this assumption the system was analyzed by Dahl and Jones [89]. The results of the analysis indicated a very short transient period (involving thermal diffusion) but after this transient period the heat flow is by conduction. The instrument, therefore, is virtually a steady state device. Ginnings [90] analyzed the comparator by including heat transfer by radiation and gaseous conduction from the spheres. This analysis indicated that the sensitivity and accuracy of Powell's comparator are most favorable for measuring materials of moderate conductivity.

Neglecting the transient period, the cooling rate observed is related to the thermal conductivities of sample k_2 and sphere k_1 and the heat capacity of the sphere by

$$\frac{dT}{dt} = \frac{4k_1 k_2 \Delta T_0 r_i}{(k_1 + k_2)\rho c_v} \qquad (11.37)$$

where ΔT_0 is the excess temperature of the sphere over the sample and r_i is the radius of contact. Observing the cooling rate by measuring samples of different k_2, a calibration curve is established. Powell [91] described a direct heating form of the thermal comparator. He gives information about material of the probe and minimum size of specimen. The device records the temperature change at contact location. Kollie and McElroy [92] used the comparator at temperatures up to 400°C and investigated different sphere mountings and tested the reliability of the measured values of thermal conductivity.

Hoch, Nitti, Gottschlich, and Blackburn [93] developed a new method to determine thermal conductivity at high temperatures. A metal cylinder 1 to 2 cm in diameter and of approximately the same height is heated in a high vacuum by means of high-frequency induction to 1000 to 3000°C. At a high current frequency the heating is localized in the skin of the cylindrical surface of the specimen. (Nonmetallic samples can be heated by the same means when surrounded by a metal foil.) If the heat is uniformly generated in cylindrical surface it will be isothermal. The heat flows into the interior by conduction and is dissipated from the circular end-faces of the specimen by radiation. Under steady state conditions the heat conducted to any point on the end surface must equal the heat radiated from that point.

The describing equation was solved by Hoch and coworkers [93] and the thermal conductivity can be obtained from the relationship (valid for $L/D = 0.18$)

$$k = \frac{\epsilon \sigma T^4}{4(dT/dr) \cdot L} \tag{11.38}$$

where T is the temperature in the center of the circular end-face, dT/dr the temperature gradient at the center, D the diameter of the specimen, L its half-height, σ the Stefan-Boltzmann constant, and ϵ the emissivity.

Neuer [94] described a new method for thermal conductivity measurements at high temperatures on poor thermal conductors such as foams. The temperature drop across a thin layer (0.5 mm) is measured on two or more disk-shaped specimens of different thicknesses. The specimen is heated in a vacuum by electron bombardment and the heat flux is determined by measuring the radiation flux from the unheated surface with a total radiation detector. The same equipment can be used for measuring the total normal and angle-dependent emittance.

11.6   MEASUREMENTS OF SPECIFIC HEATS

Specific heat enters heat transfer processes directly due to enthalpy or internal energy changes of the materials involved and indirectly by means of the thermal diffusivity in all unsteady cases.

The methods described for thermal conductivity and diffusivity in the previous sections of this paper obviously can be applied also for the measurements of specific heat when in addition the density of the material is known. The discussion of the measurements of specific heats will refer in some cases to methods already described.

By definition the specific heat is given by

$$c = \frac{1}{m} \frac{dQ}{dT} \tag{11.39}$$

and in principle can be evaluated by measuring the heat (dQ) needed to increase the temperature by dT in a mass m of the specimen.

Since the specific heat is temperature dependent the observation must be carried out with small changes of T; otherwise the so-called "mean specific heat" c_m is measured

$$c_m = \frac{1}{T_1 - T_2} \int_{T_1}^{T_2} c \, dT \tag{11.40}$$

From this relationship c can be obtained only when its temperature dependence is known. Normally it is necessary to observe c_m many times at different values of $T_1 - T_2$.

The change of temperature in the material normally causes a change in volume; this needs additional energy to overcome the intermolecular forces. Measurements under constant pressure, therefore, will yield a higher value than measurements at constant volume:

$$c_p = c_v + \alpha^2 \, \frac{T}{\rho \kappa} \tag{11.41}$$

where α is the volumetric expansion coefficient, ρ the density, and κ the compressibility. For solids and liquids, measurements normally are carried out under constant pressure and c_v is determined by means of Eq. (11.41).

Equation (11.41) also is used for measurements of gases—those in addition can be determined indirectly by means of thermodynamic relationships or experimentally.

11.6.1 Measurements of the Specific Heat of Solids and Liquids

The methods applied differ from each other mainly by different means of measuring the heat causing the temperature change in the specimen of known mass. The sample might be heated directly electrically (in a few cases by a chemical reaction) or indirectly, by observing the effect on a calorimeter body when brought in heat exchange with the sample. The direct heating methods have been used in many varieties and are most common for adiabatic calorimeters. The adiabatic calorimeter consists of a container with the sample or the sample itself heated by an internal heater and surrounded by a shield heated to the same temperature as the calorimeter. Moser [95] heated the shield to undergo a small linear temperature change and heated the calorimeter to follow the temperature of the shield. Schmidt and Leidenfrost [96], West and Ginnings [97], Stansbury et al. [98], and others heated the sample and forced the shield to follow the temperature change. This approach is simpler in respect to determining the heat generated in the sample.

The instrument developed by Schmidt and Leidenfrost is shown in Fig. 11.16. A small heater sphere is surrounded by the test substance (liquids or solid granulates) in a wide spherical annulus. The outside of the calorimeter is surrounded by a spherical shield which in turn is insulated by radiation shields in a vacuum. These investigators solved the temperature distribution in the test sample undergoing a linear temperature change. They proved that after a certain initial time has elapsed the temperature within the sample will increase at the same rate everywhere. Under those conditions the recorded temperature change with time in the sample is directly proportional to the specific heat:

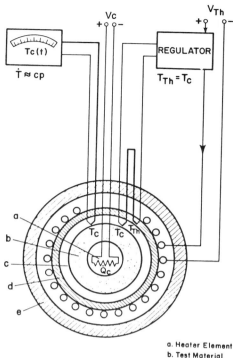

FIG. 11.16 Schematic diagram of calorimeter.

a. Heater Element
b. Test Material
c. Calorimeter Shell
d. Thermostat
e. Insulation

$$c = \frac{1}{m}\left(\frac{q}{\dot{T}} - w_c\right) \tag{11.42}$$

where $\dot{T}$ is the temperature change with time, q the constant heat input, and w_c the heat capacity of the calorimeter body (heater sphere, outer shell, and wires). The specific heat observed will be of high accuracy whenever the rate of temperature change is chosen so that the temperature difference within the sample is small enough for the assumption of c = constant, but high enough to detect the temperature change accurately. The spherical arrangement has some advantages for reasons of eliminating end-losses which are encountered in cylindrical arrangements and providing for the maximum volume-to-surface ratio. The latter minimized the potential heat transfer surface relative to the heat supplied to the specimen.

Schmidt and Leidenfrost used the arrangement to measure the specific heat of nickel up to 500°C and the specific heat of liquids. When the measurements on nickel were carried out to determine the exact temperature of the Curie point, the system was so sensitive that another kind of transition was noticed at about 230°C. The system was also tested for measuring simultaneously the thermal conductivity of poor-conducting fluids by following the transient technique discussed in the previous sections.

A number of problems must be solved in order to attain precision in adiabatic calorimetry. First, it is difficult to achieve complete adiabacy. Resulting from this there will be heat leakages to the surrounding shield. Lead-in losses also are present and so are the effects of nonuniformity and several other influences already discussed in the previous section and by West [99].

A quasi-adiabatic calorimeter was developed by Krischer [100] especially for the measurements of specific heat of poor-conducting solids and for samples of high moisture content. Equally thick layers of the sample are sandwiched together with a foil-type heater in between as shown on the left side of Fig. 11.17. If the samples are large enough and unwanted effects are avoided, the uniformly generated heat will cause, after an initial period of time, a quasi-stationary condition within the sample. The conditions are described by equidistant parabolic isotherms shown in the right-hand side of the figure. The temperature of the heating foil T_f is measured and that of a center location T_c. Observing the time period Δt necessary for T_c to reach T_f allows one to evaluate the specific heat from the equation

$$c = \frac{q}{\rho L} \frac{\Delta t}{T_f - T_c} \tag{11.43}$$

where L is half thickness of the layer. The equation is valid because every point of the sample undergoes the same temperature change during the time Δt. The arrangement obviously can also be used for thermal conductivity.

Adiabatic calorimeters can produce specific heat data with an accuracy

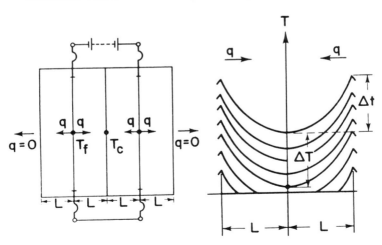

FIG. 11.17 Krischer's instrument for measurements of k and c. (From Krischer, 1954, Ref. [100].)

which equals in certain ranges that of the drop-type calorimeter. In other ranges they have in addition the advantage that the specific heat can be evaluated at any temperature covered in one run and that measurements will detect influences of transitions or other changes which might occur within the sample during the test. In the drop-type calorimeter many drops must be performed in order to establish the specific heat values at regions of transition—or the transition might be similarly frozen during the rapid cooling process in the calorimeter.

The drop-type calorimeter technique heats a sample to a known temperature, usually in a furnace, which is then dropped into a calorimeter where the sample cools rapidly and heat it evolves is measured. This procedure essentially measures the enthalpy of the sample at the higher temperature relative to that at the lower temperature. The heat exchanged from the sample to the calorimeter can be measured by observing the temperature change in a liquid or a solid calorimeter body or by determining isothermally the amount of ice melted, or of liquid evaporated or condensed. Of very high precision is the so-called "ice calorimeter," a very simple, very accurate, best tested, and most commonly applied device.

The calorimeter consists of an inner tube open at the upper end to receive the sample and closed at the bottom. The tube is completely surrounded by a wider vessel. The space between the tube and the vessel is filled with pure, air-free water and some pure mercury at the bottom. Some of the water is frozen. There is only one opening to the outside—by a mercury-filled tube connecting the pool of pure mercury to the outside. The heat evolved by the sample when it cools to the ice point melts a corresponding amount of ice which is very precisely measured by the mass of mercury drawn in (from a weighed outside supply) to compensate for the volume shrinkage. A bath of ice water serves as a heat shield making the heat losses of the calorimeter very small and highly constant. A detailed description of the operation and performance of drop-type calorimeter was given by Douglas [101].

Taylor [102] determined specific heat by a pulse-heating method. The small sample in the form of a wire is heated rapidly by the passage of large currents of the order of several thousand amperes and causing heating rates of 10^3 to 10^5 degrees per second. Taylor lists as advantages: heat losses small, no furnace required, chemical interaction and preferential vaporization of specimen is limited, the upper temperature limit is the melting point. The disadvantages are that the sample must be an electrical conductor and it is difficult to measure the temperature. (Cezairliyan [87] used tubes and measured temperature by optical and fast-recording devices. It appears that he did overcome those difficulties; see also [103].) In Taylor's experiment a dual oscilloscope was used to measure voltage and current simultaneously as a function of time and at discrete time intervals. The specific resistance at each time interval is calculated from the relationship

$$\rho = \frac{A}{L}\frac{E}{I} \qquad (11.44)$$

where A is the cross-sectional area of the wire, L the distance between voltage probes, and E and I the voltage and current, respectively.

The specific resistance as a function of time is then determined. The specific heat at any temperature is given by the expression

$$c_p = \frac{E \cdot I}{m} \frac{d\rho/dT}{d\rho/dt} \tag{11.45}$$

where m is the mass of the sample between the voltage probes and $E \cdot I$ the power dissipated in the region. $d\rho/dT$ is the temperature coefficient of the resistance at T and $d\rho/dt$ the time rate of change of the resistivity at T. The accuracy of the specific-heat data measured by that method agreed well with those observed in drop-type calorimeters operating with samples heated to the same temperature. The adiabatic and drop-type calorimeter can also be used for liquids by providing suitable containers.

Some other measurement techniques for liquids are shown in the following figures. Figure 11.18 represents a differential calorimeter. One of the calorimeter containers is filled with the test sample, the other with a liquid of known specific heat. The specific heat is evaluated from measuring the heat necessary to produce in each calorimeter the same temperature increase, from the masses contained in the calorimeters and the known specific heat. Due to the symmetry of the arrangement it is not necessary to evaluate the heat capacities of the container itself and the heat losses.

Specific heat of liquids quite often is measured by determining the temperature change in a liquid flowing through an insulated tube under constant heat input. A calorimeter of this type is shown in Fig. 11.19 representing a device originally used by Callendar and Barnes [104]. H represents the heater wire

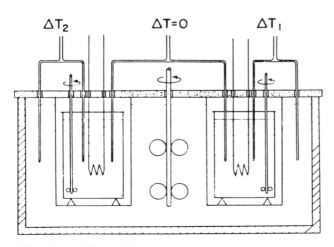

FIG. 11.18 Differential calorimeter.

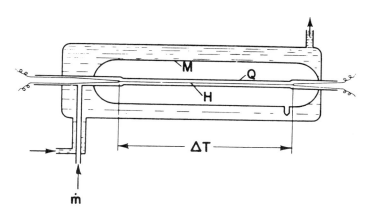

FIG. 11.19 Callendar and Barnes calorimeter [104].

(normally made of platinum) placed in the center of tube Q which is insulated by a vacuum chamber M, kept at a constant temperature by a circulating thermostat fluid. The temperature change is measured by two thermocouples or resistance thermometers located at proper places close to the ends of the tube. The heat losses q_L in the system can be determined by varying the mass flow rate and heat input under constant ΔT

$$c = \frac{1}{\dot{m}}\left(\frac{q}{\Delta T} - q_L\right)$$ (11.46)

Flow calorimeters have been used with success at high pressures and temperatures and also for gases.

Another calorimeter system of interest for reasons of its simplicity and accuracy is that of Schlesinger [105] shown in Fig. 11.20. The calorimeter C is filled completely with the test fluid by means of valves a and b and placed in a thermostat. When equilibrium is established valve b is closed and some mercury is forced to enter the capillary k by slight increase in pressure through a which then is also closed. The liquid is heated H by a small amount and for a short time (ΔQ). If the thermal expansion and specific heat can be assumed constant the displacement of the mercury in the capillary immediately will yield the change of volume even when the liquid is heated only in the neighborhood of the heater H. The liquid next to the walls of C remains practically at the temperature of the surroundings and no heat losses are encountered. The specific heat, therefore, is determined not only quickly but also very accurately.

$$c_p = \frac{\Delta Q}{\Delta V}\Delta v$$ (11.47)

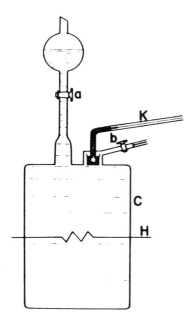

FIG. 11.20 Schlesinger calorimeter [105].

where ΔV is the volume change and Δv the specific volume change of one unit of mass of liquid under a temperature change of 1 degree.

11.6.2 Measurements of the Specific Heat of Gases

The flow-type and the mixing calorimeters can be used for gases as well as heating a gas enclosed in a large vessel by an electrical heater of large surface area. The specific heat of gases furthermore can be determined by means of techniques to observe heating values of gas reaction (i.e., $H_2 + \frac{1}{2}O_2$) but adding the gas of unknown specific heat. Measurements also can be carried out by the differential method and by many other methods.

Of more interest are the indirect methods. The adiabatic change of a gas from a state v_0, p_0, T_0 to another state v_1, p_1, T_1, is given by the thermodynamic relationship for ideal gases:

$$\left(\frac{p_1}{p_0}\right) = \left(\frac{T_1}{T_0}\right)^{\kappa/\kappa - 1} \qquad \left(\frac{p_1}{p_0}\right) = \left(\frac{v_0}{v_1}\right)^{\kappa} \tag{11.48}$$

and for real gases:

$$\left(\frac{\partial T}{\partial p}\right)_s = \left(\frac{\partial T}{\partial p}\right)_v \frac{\kappa - 1}{\kappa} \tag{11.49}$$

The measurements of two of the related quotients yield $\kappa = c_p/c_v$ and with the molar heats

$$C_{p_0} - C_{v_0} = R \tag{11.50}$$

c_p and c_v can be evaluated for ideal gases.

For real gases Eq. (11.50) is not valid and an equation of state must be introduced making Eq. (11.49) quite complex. However, the difficulties can be overcome by special differentiating techniques [106] and with the aid of high-speed computers. Furthermore κ can be computed from the velocity of sound. The following process under constant enthalpy yields the Joule-Thompson coefficient:

$$\mu = \left(\frac{\partial T}{\partial p}\right)_h = \frac{T(\partial v/\partial p)_p - v}{c_p} \tag{11.51}$$

and c_p can be calculated when $T(\partial v/\partial p)_p - v$ can be computed from an equation of state.

NOMENCLATURE

A	area
$A_1; A_2$	constants
a	thermal diffusivity
B	geometric constant
$C_p; C_v$	molar heats
c	specific heat
c_m	mean specific heat
c_p	specific heat at constant pressure
c_v	specific heat at constant volume
D	outer diameter
d	inner diameter
E	electrical potential
$E_1; E_2$	electrical potential at locations 1 and 2
ΔE	potential difference
f	function
h	height section of infinitely long cylinder
I	electrical current; (Eq. [11.7]) moment of inertia
k	thermal conductivity
$k_1; k_2$	thermal conductivity of material 1 and 2
$k_x; k_y; k_z$	thermal conductivity in x, y, and z direction respectively
L	thickness or length
M	viscous torque
M_f	moment due to viscous forces
m	mass

$\dot{m}$	mass flow rate
m_k	similarity parameter
Pr	Prandtl number
p	pressure
$p_0 ; p_1$	pressures at different states
Δp	pressure drop
Q	heat
q	heat flow
q_c	heat flow by convection
q_k	heat flow by conduction
q_L	heat loss
q_{osc}	heat flow due to unsteady state conditions in steady state devices
q_{inh}	errors in heat flow due to inhomogenities in a system
$q_{\mathrm{ch.r}}$	heat generation by chemical reaction
q_r	heat flow by radiation
Ra	Rayleigh number
Re	Reynolds number
R	gas constant
r	radial coordinate, recovery factor
$r_1 ; r_2$	radial location
r_i	radius of a contact area
T	temperature; (Eq. [11.9]) period of oscillation
T_{ad}	stagnation or total temperature
T_c	temperature in the center of a body
T_f	front surface temperature, temperature of a heater foil
$T_1 ; T_2$	temperature at location 1 and 2
T_m	mean temperature (Eq. [11.22]); maximum temperature (Eqs. [11.25] and [11.27])
T_0	period of oscillation in vacuum
T_r	reovery temperature
ΔT	temperature difference
T_∞	static temperature
t	time
$t_1 ; t_2$	time
$t_{\frac{1}{2}}$	time needed to observe half of the maximum temperature rise at rear surface of a sample when heated by a pulse at the front surface
Δt	time interval
u	velocity
u_t	terminal velocity
V	volume
$\dot{V}$	volume flow rate
ΔV	volume change
v	specific volume
$v_0 ; v_1$	specific volume at different states 0, 1

Δv	specific volume change
W	viscous drag
x	coordinate
y	coordinate
z	coordinate

Greek Symbols

α	angular motion, volumetric expansion coefficient (Eq. [11.41])
α_n	nth amplitude of an oscillating system
α_{n+1}	nth + 1 amplitude
$\delta; \Delta$	boundary layer thickness or depth of penetration and damping decrement
ϵ	emissivity
η	dimensionless coordinate
θ	dimensionless frequency (Eq. [11.9]); temperature fluctuation (Eqs. [11.18], [11.19], [11.20])
κ	specific heat ratio
λ	electrical conductivity
μ	dynamic viscosity, coefficient of Thompson heat (Eq. [11.32]), Joule-Thompson coefficient (Eq. [11.51])
ν	kinematic viscosity
ξ	dimensionless coordinate
ρ	density and electrical resistivity
σ	Stefan Boltzmann constant
$\sigma_1; \sigma_2$	complex roots of a characteristic equation
ϕ	local angular velocity
ω	angular velocity
ω_0	circular frequency in vacuum

REFERENCES

1. Kestin, J. On the Direct Determination of the Viscosity of Gases at High Pressures and Temperatures, *Proc. Biennial Gas Dynamic Symp., 2nd,* A. B. Cambel and J. B. Finn (eds.), Northwestern University Press, Evanston, Ill., 1958.
2. Knudsen, M., "Kinetic Theory of Gases," Methuen, London, 1934.
3. Schiller, L., Untersuchung über Laminare und turbulente Strömung, *Forschg. Ing. Wes.,* vol. 6, p. 248, 1922.
4. Couette, M. M., Studies on Viscosity of Liquids, *Ann. Chem. Phys.,* vol. 6, no. 21, pp. 433–509, 1890.
5. Goldstein, S., "Modern Development in Fluid Dynamics," vol. I, Oxford, London, 1938.
6. Dean, W. R., The Streamline Motion of a Fluid in Curved Pipe, *Phil. Mag.,* vol. 7, no. 4, p. 208, 1927 and vol. 5, p. 673, 1928.
7. Vasilesco, V., Récherches Experimentales sur la viscosité des gaz aux températures elévees, *Ann. Phys.,* vol. 20, pp. 137 and 292, 1945.

8. Hawkins, G. A., Solberg, H. L., and Potter, A. A., The Viscosity of Water and Superheated Steam, *Trans. ASME,* vol. 62, p. 677, 1940.
9. Mayinger, F., Messung der Viskosität von Wasser und Wasserdampf bis zu 700°C. und 800 atm., *Int. J. Heat Mass Transfer,* vol. 5, p. 807, 1962.
10. Swindells, J. F., Coe, J. R., and Godfrey, T. B., Absolute Viscosity of Water at 20°C., *J. Res. Nat. Bur. Stand. (U.S.),* vol. 48, p. 1, 1952.
11. Whitelaw, J. H., Viscosity of steam at Supercritical Pressures, *J. Mech. Eng. Sci.,* vol. 2, p. 288, 1960.
12. Flynn, G. P., Hanks, P. R., Lemaire, N. A., and Ross, J., Viscosity of Nitrogen, Helium, Neon, and Argon from 78.5 to 100°C. below 200 Atm, *J. Chem. Phys.,* vol. 38, p. 154, 1963.
13. Rankine, O. A., On a Method of Determining the Viscosity of Gases Especially those Available only in Small Quantities, *Proc. Roy. Soc. (London),* vol. A84, pp. 265 and 516, 1910.
14. Michels, A., and Gibson, R. O., The Measurements of the Viscosity of Gases at High Pressures—the Viscosity of Nitrogen to 1000 Atm, *Proc. Roy. Soc. (London),* vol. 134A, p. 288, 1931.
15. Ubbelohde, L., The Simplest and Most Accurate Viscometer and Other Instruments with Suspended Level, *J. Inst. Petrol. Tech.,* vol. 19, p. 376, 1933.
16. Cannon, M. R., Manning, R. E., and Bell, J. D., Viscosity Measurement, the Kinetic Energy Correction and a New Viscometer, *Anal. Chem.,* vol. 32, p. 355, 1960.
17. Gilchrist, L., An Absolute Determination of the Viscosity of Air, *Phys. Rev.,* vol. 1, p. 124, 1913.
18. Kellström, G., A New Determination of the Viscosity of Air by the Rotating Cylinder Method, *Phil. Mag.,* vol. 23, p. 313, 1937.
19. Bearden, J. A., A Precision Determination of the Viscosity of Air, *Phys. Rev.,* vol. 56, p. 1023, 1939.
20. Maxwell, J. D., On the Viscosity of Internal Friction of Air and Other Gases, *Phil. Trans. Roy. Soc. (London),* vol. 156, p. 246, 1866.
21. Kestin, J., and Leidenfrost, W., An Absolute Determination of the Viscosity of Eleven Gases over a Range of Pressure, *Physica,* vol. 25, p. 1033, 1959.
22. Kestin, J., and Persen, L. N., Small Oscillations of Bodies of Revolution in a Viscous Flow, *Proc. Int. Cong. Appl. Mech., 9th, Brussels,* 1956.
23. Newell, G. F., Theory of Oscillation Type Viscometers V Disc Oscillating between Fixed Plates, *ZAMP,* vol. 10, p. 160, 1959.
24. Kearsley, E. A., An Analysis of an Absolute Torsional Pendulum Viscometer, *Trans. Soc. Rheol.,* vol. 3, p. 69, 1959.
25. Jakob, M., "Heat Transfer," vol. 1, 1949 and vol. 2, 1957, Wiley, New York.
26. Tsederburg, N. V., "Thermal Conductivity of Gases and Liquids," MIT, Cambridge, Mass., 1965.
27. Tye, R. P. (ed.), "Thermal Conductivity," vol. 1 and vol. 2, Academic, New York, 1969.
28. Touloukian, Y. S., Liley, P. E., and Saxena, S. C., TPRC Data Series, vol. 3, Plenum, New York, 1970.
29. Schmidt, E., Uber Trocknungsvorgänge, *Z. Ges. Kälte Ind.,* vol. 43, p. 75, 1936.
30. Michels, A., and Sengers, J. V., The Thermal Conductivity of Carbon Dioxide in the Critical Region, *Prog. Int. Res. Thermodynamic Transport Prop., Symp., 2nd, Princeton, January,* Academic, New York, 1962.
31. Michels, A., and Botzen, A., A Method for the Determination of the Thermal Conductivity of Gases at High Pressures, *Physica,* vol. 18, p. 605, 1952.
32. Michels, A., Sengers, J. V., and Vandeklundert, L. J. M., The Thermal Conductivity of Argon at Elevated Densities, *Physica,* vol. 29, pp. 149–160, 1963.

33. Fritz, W., and Poltz, H., Absolutbestimmung der Wärmeleitfähigkeit, von Flüssigkeiten. 1. Kritische Versuche an einerneuen Platten Apparatur, *Int. J. Heat Mass Transfer*, vol. 5, p. 307, 1962.

34. Poltz, H., and Iugel, R., Thermal Conductivity of Liquids—IV Temperature Dependence of Thermal Conductivity, *Int. J. Heat Mass Transfer*, vol. 10, p. 1075, 1967.

35. Schleiermacher, A., Über die Wärmeleitung der Gase, *Ann. Phys.*, vol. 26, p. 287, 1885.

36. Schmidt, E., and Sellschopp, W., Wärmeleitfähigkeit des Wassers bei Temperaturen bis 270°C, *Forsch. Ing. Wes.*, vol. 2, pp. 165–178 and 213–217, 1931.

37. Keyes, F. G., Measurements of Heat Conductivity of Nitrogen–Carbon–Dioxide Mixtures, *Trans. ASME*, vol. 73, p. 597, 1951.

38. Ziebland, H., The Thermal Conductivity of Toluene—New Determinations and an Appraisal of Recent Experimental Work, *Int. J. Heat Mass Transfer*, vol. 2, p. 273, 1961.

39. Johannin, P., Wilson, M., and Vodar, B., Heat Conductivity of Compressed Helium at Elevated Temperatures, *Symp. Thermophysical Prop., ASME, 2nd, Princeton, January*, Academic, New York, pp. 418–433, 1962.

40. Thodos, G., and Misic, D., The Thermal Conductivity of Hydrocarbon Gases at Normal Pressures, *A. I. Ch. E. J.*, vol. 7, p. 264, 1961.

41. Riedel, L., New Thermal Conductivity Measurements of Organic Liquids, *Chem. Ing. Tech.*, vol. 23, pp. 321–324, 1951.

42. Schmidt, E., and Leidenfrost, W., Der Einfluss Elektrisher Felder auf den Wärme Transport in Flüssigne elektrischen Nichtleitern, *Forsch. Ing. Wes.*, vol. 19, pp. 65–80, 1953.

43. Leidenfrost, W., An Attempt to Measure the Thermal Conductivity of Liquids, Gases, and Vapors with a High Degree of Accuracy over Wide Ranges of Temperature (−180°C. to 500°C.) and Pressure (Vacuum to 500 Atm), *Int. J. Heat Mass Transfer*, vol. 7, pp. 447–478, 1964.

44. Leidenfrost, W., and Tree, D., Prediction of Minor Heat Losses in a Thermal Conductivity Cell and Other Calorimeter Type Cells, *Proc. Conf. Thermal Conductivity, 7th, Washington, D.C.*, NBSSP302, pp. 633–658, 1967.

45. Leidenfrost, W., Theory Design and Use of the Multi-Purpose Instrument for Determining Many Properties and a New Unsteady Temperature-Free Measurement of Heat Transfer and Thermal Properties, in O. G. Martynenko (ed.), "Progress in Heat and Mass Transfer," vol. IV, pp. 333–407, Pergamon, New York, 1972.

46. Viskanta, R., and Grosh, R. J., Heat Transfer by Simultaneous Conduction and Radiation in an Absorbing Media. Trans. ASME, *J. Heat Transfer*, vol. 84C, pp. 63–72, 1963.

47. Poltz, H., Die Wärmeleitfähigkeit von Flüssigkeiten. II. Der Strahlungsanteil der effektiven Wärmekeitfähigkeit, *Int. J. Heat Mass Transfer*, vol. 8, p. 515, 1965.

48. Poltz, H., Die Wärmeleitfähigkeit von Flüssigkeiten. III. Abhangigkeit der Wärmeleitfähigkeit von der Schicht dicke bei Organischen Flüssigkeiten, *Int. J. Heat Mass Transfer*, vol. 8, pp. 609–620, 1965.

49. Kohler, M., Einfluss der Strahlung auf den Wärmetransport durch eine Flüssigkeitsschicht, *Z. Aog. Phys.*, vol. 18, p. 356, 1965.

50. Schödel, G., "Kombinierte Wärmeleitung und Wärmestrahlung in Konvektionsfreien Flüssigkeitsschichten," Ph.D. thesis, Technische Hochschule München, 1969, and Schödel, G., and Grigull, G., Paper R 2.2 *Heat Transfer, Int. Heat Transfer Conf., 4th, Paris*, vol. III, 1970.

51. Leidenfrost, W., Critical Analysis of the Experimental Determination of the Thermal Conductivity of Steam, *Int. Conf. Prop. Steam, 7th, Tokyo*, September, 1968.

52. Leidenfrost, W., and Latko, R. T., Nongrey Radiant Heat Transfer Corrections to Thermal Conductivity Measurements, *Wärme Stoffübertragung*, vol. 3, pp. 114–119, 1970.

53. Dickens, B. G., The Effect of Accommodation on Heat Conduction through Gases, *Proc. Roy. Soc. (London)*, vol. A143, p. 577, 1934.

54. Stalhane, B., and Pyk, S., Metod for Bestamming av Varmelednings-Coefficienter, *Teknisk Tidskrift*, vol. 61, p. 389, 1931.

55. Van der Held, E. F. M., and Van Druven, F. G., A Method of Measuring the Thermal Conductivity of Liquids, *Physica*, vol. 15, p. 866, 1949.

56. Grassmann, P., Straumann, W., Widmer, F., and Jobst, W., Measurements of Thermal Conductivities of Liquids by an Unsteady State Method, *Prog. Ing. Res. Thermodynamic Transport Prop., Symp. Thermophysical Prop., ASME, 2nd, Princeton*, p. 447, Academic, New York, 1962.

57. Briggs, D. G., Goldstein, R. J., and Ibele, W. E., Precision Measurements of the Thermal Conductivity of Gases in a Transient Hot-wire Cell, *Sym. Thermophysical Properties, ASME, 4th*, p. 452, 1968.

58. Allen, P. H. G., Fluid Thermal Conductivity by a Transient Method, *Thermodynamic Transport Prop. Gases, Liquids, Solids, Symp. Thermal Prop., ASME, Purdue University, Lafayette, Indiana, January*, p. 350, McGraw-Hill, New York, 1959.

59. Harrison, W. B., Boteler, W. C., and Spurlock, J. M., Thermal Diffusivity of Nitrogen as Determined by the Cyclic Heat Transfer Method, *Thermodynamic Transport Prop. Gases, Liquids, Solids, Symp. Thermal Prop., ASME*, p. 304, McGraw-Hill, New York, 1959.

60. Tarmy, B. L., and Bonilla, C. F., The Application of Frequency Response Analysis to Thermal Conductivity Measurements, *Progr., Int. Res. Thermodynamic Transport Prop., ASME*, J. T. Masi and D. H. Tsai (eds.), pp. 404–411, Academic, New York, 1962.

61. Westenberg, A. A., and Walker, R. E., Experiments in the Molecular Diffusion of Gases at High Temperatures, *Thermodynamic Transport Prop. Gases, Liquids, Solids, Symp. Thermal Properties, ASME*, p. 314, McGraw-Hill, New York, 1959.

62. Westenberg, A. A., and deHaas, N., High Temperature Gas Thermal Diffusivity Measurements with the Line Source Technique, *Symp. Thermal Properties, ASME, 2nd, Princeton, January*, p. 412, Academic, New York, 1962.

63. Sengers, J. V., Transport Properties of Compressed Gases, in A. C. Eringen (ed.), "Recent Advances in Engineering Sciences," vol. 3, p. 157, Gordon-Breach, New York, 1968.

64. Sogin, H. H., and Thompson, H. A., Apparatus to Measure Transport Properties of Gases by Means of the Rayleigh-Jeffreys Instability, *Symp. Thermophysical Properties, ASME, 4th, College Park, Maryland, April*, p. 416, 1968.

65. Eckert, E. R. G., and Irvine, T. F., Jr., A New Method to Measure Prandtl Number and Thermoconductivity of Fluids, *Appl. Mech.*, vol. 24, no. 1, pp. 25–28, March 1957.

66. Smiley, E. F., "The Measurements of the Thermal Conductivity of Gases at High Temperatures with a Shock Tube," Ph.D. thesis, Catholic University of America, Washington, D.C., 1957.

67. Lauver, M. R., Evaluation of Shock Tube Heat Transfer Experiments to Measure Thermal Conductivity of Argon from 700° to 8600°K, *NASA* D-2117, 1964.

68. Carey, C. A., Carnevale, E. H., and Marshall, T., Heat Transfer and Ultrasonic Methods for the Determination of High Temperature Gas Transport Properties, *Conf. Thermal Conductivity, AFML, 6th, Dayton, Ohio*, p. 177, 1966.

69. Leidenfrost, W., and Kennedy, W. E., Analysis of a Method and a System to Determine Heat Transfer and Properties of Fluids in the Absence of Temperature Measurements and a Modified Fourier Equation, *Wärme und Stoffübertragung*, vol. 2, pp. 114–126, 1973.

70. Ho, C. Y., and Taylor, R. E. (eds.), *Proc. Conf. Thermal Conductivity, 8th*, Plenum, New York, 1969.

71. Shanks, H. R. (ed.), *Conf. Thermal Conductivity, 9th, Ames, Iowa, October, 1969*, AEC-Div. Tech. Int. Rep. CONF-691002, 1970.

72. Flynn, D. R., and Didion, D. A., A Steam Calorimeter Apparatus for Refractories, *Conf. Thermal Conductivity, Battelle, Columbus, Ohio, October*, p. 81, 1961.

73. Flynn, D. R., Thermal Conductivity of Semi-Conductive Solids. Methods for Steady-state Measurements on Small Disk Reference, *NBS Rep.* 7367 and 7323.

74. Black, T. A., and Glaser, P. E., Thermal Conductivity Tests of Cryogenic Insulation, *Conf. Thermal Conductivity, 2nd, Ottawa, Canada, October*, p. 111, 1962.

75. Messmer, J. H., The Thermal Conductivity of Porous Media IV Sandstone: The Effect of Temperature and Saturation, *Conf. Thermal Conductivity, 5th, Denver Univ., Colo.*, 1, II-E-1, 1965.

76. Parker, W. J., Jenkins, R. J., Butler, C. P., and Abbott, C. L., Flash Method of Determining Thermal Diffusivity, Heat Capacity, and Thermal Conductivity, *J. Appl. Phys.*, vol. 32, pp. 1679–1684, 1961.

77. Beck, J. V., Analytical Determination of Optimum Transient Experiments for Measurements of Thermal Properties, *Int. Conf. Heat Transfer Trans., 3rd, Chicago*, 1966.

78. Kasper, J., and Zehms, E. H., Thermal Diffusivity Measurements for Very High Temperatures, *Conf. Thermal Conductivity, 4th, U.S. Naval Rad. Def. Laboratory, San Francisco*, V-E, 1964.

79. Nakata, M. M., A Radial Heat Flow Technique for Measuring Thermal Diffusivity, *Conf. Thermal Conductivity, 2nd, Div. Appl. Physics, NRC, Ottawa*, 1962.

80. Kohlrausch, F., Uber den Stationären Temperatur-zustand eines Elektrisch Geheizten Leiters, *Ann. Phy.*, vol. 4, no. 1, pp. 132–158, 1900.

81. Bode, K. H., Eine neue Methode zur Messung der Wärmeleitfähigkeit von Metallen bei hohen Temperaturen, *Allg. Wärmetech.*, vol. 10, pp. 110–120 and 125–142, 1961.

82. Angell, M. F., Thermal Conductivity at High Temperatures, *Phys. Rev.*, vol. 33, pp. 421–432, 1911.

83. Powell, R. W., and Schofield, F. H., The Thermal and Electrical Conductivities of Carbon and Graphite to High Temperatures, *Proc. Phys. Soc.*, vol. 51, pp. 153–172, 1939.

84. Rudkin, R. L., Parker, W. J., and Jenkins, R. J., Measurements of the Thermal Properties of Metals at Elevated Temperatures, *U.S. Dept. Comm. Publ.* PB 171185hsNRDL-TR-419, pp. 1–24, 1960.

85. Krishnan, K. S., and Jain, S. C., Determination of Thermal Conductivities at High Temperatures, *Brit. J. Appl. Phys.*, vol. 5, pp. 426–430, 1954.

86. Powell, R. W., DeWitt, D. P., and Nalbantyan, M., The Precise Determination of Thermal Conductivity and Electrical Resistivity of Solids at High Temperatures by Direct Electrical Heating Methods, AFML-TR-67-241, Wright-Patterson AFB, Ohio, August 1967.

87. Beckett, C. W., and Cezairliyan, A., High Speed Thermodynamic Measurements and Related Techniques, in John P. McCullough (ed.), "Experimental Thermodynamics, Calorimetry of Non-Reacting Systems," vol. I, chap. 1, Butterworth, London.

88. Powell, R. W., Experiments Using a Simple Thermal Comparator for Measurements of Thermal Conductivity, Surface Roughness and Thickness of Foils or of Surface Deposits, *J. Sci. Instrum.*, vol. 34, p. 485, 1957.

89. Dahl, A. I., and Jones, D. W., Thermal Conductivity Studies with the Powell Method, Paper 60-HT-30, *ASME–AIChE Heat Transfer Conf., Buffalo, N.Y.*, 1960.

90. Ginnings, D. C., Powell Comparator Method for Determining Thermal Conductivities (a discussion), *Conf. Thermal Conductivity Methods, Battelle*, p. 287, 1961.

91. Powell, R. W., Thermal Conductivity Measurements by the Thermal Comparator Method, *Black Hills Summer Conf. Transport Properties, Rapid City, S.D.*, 1962.

92. Kollie, T. G., and McElroy, D. L., Thermal Comparator Apparatus for Thermal Conductivity Measurements from 40 to 400°C, *Conf. Thermal Conductivity Methods, Battelle*, p. 289, 1961.

93. Hoch, M., Nitti, D. A., Gottschlich, C. F., and Blackburn, P. E., New Method for the Determination of Thermal Conductivities between 1000°C and 3000°C, *Progr. Int. Res. Therm. Transport Prop., ASME,* Academic, New York, 1962.

94. Neuer, G., Verfahren zur Messung der Wärmeleitfähigkeit poroser Stoffe bei hohen Temperaturen, *High Temps.-High Press.*, vol. 4, pp. 389–393, 1972.

95. Moser, H., Messung der Wahren Spezifischen Wärme von Silber, Nickel, Quarz Kristall und Quarz Glas Zwischen 50 and 700°C, *Phys. Z.,* vol. 37, p. 512, 1936.

96. Schmidt, E. O., and Leidenfrost, W., Optimierung eines Adiabatischen Kalorimeters zur Genauen Messung von Wahren Spezifschen Wärmen Schlecht Wärmeleitender Substanzen, *Int. J. Heat Mass Transfer,* vol. 5, pp. 267–275, 1962.

97. West, E. D., and Ginnings, D. C., An Adiabatic Calorimeter for the Range 30° to 500°C, *J. Res. Nat. Bur. Stand. (U.S.),* vol. 60, pp. 309–316, 1958.

98. Stansbury, E. E., McElroy, D. L., Picklesimer, M. L., Elder, G. E., and Pawel, R. E., Adiabatic Calorimeter for Metals in the Range 50–100°C, *Rev. Sci. Instrum.,* vol. 30, pp. 121–126, 1959.

99. West, E. D., Heat Exchange in Adiabatic Calorimeter, *J. Res. Nat. Bur. Stand. (U.S.),* vol. 67A, pp. 331–341, 1963.

100. Krischer, O., Über die Bestimmung der Wärmeleitfähigkeit der Wärmekapazität und der Wärmeeindringzahl in einem Kurzzeitverfahren, *Chem. Ing. Tech.,* vol. 26, p. 42, 1954.

101. Douglas, T. B., The Dropping–Type Calorimeter, *Conf. Thermal Conductivity, 3rd, Metals Ceramics Div., Oak Ridge Nat. Lab.,* 1962.

102. Taylor, R. E., Determining Specific Heat and Other Properties by Pulse Heating, *Conf. Thermal Conductivity, 3rd, Metals Ceramics Div., Oak Ridge Nat. Lab.,* 1962.

103. Cezairliyan, A., Measurement of the Heat Capacity of Graphite in the Range 1500 to 3000 K by a Pulse Heating Method, *Proc. Symp. Thermophys. Prop., 6th, Atlanta, Georgia, August,* P. E. Liley, (ed.), ASME, pp. 279–285, 1973.

104. Callendar, H. L., and Barnes, H. T., II–Continuous Calorimetry, *Phil. Trans.,* vol. 199A, p. 149, 1902.

105. Schlesinger, H., *Phys. Z.,* vol. 10, p. 210, 1909.

106. Landis, F., and Nilson, E. N., The Determination of Thermodynamic Properties by Direct Differentiation Techniques, *Prog. Int. Res. Thermodynamic Transport Prop., Symp. Thermophys. Prop., ASME, 2nd,* p. 218, Academic, New York, 1962.

12 Transport property measurements in the Heat Transfer Laboratory, University of Minnesota

W. IBELE
Department of Mechanical Engineering
University of Minnesota

12.1 INTRODUCTION

There are two reasons why it is scientifically profitable to perform transport property measurements in the environment of a heat transfer laboratory. There is need for increasingly accurate knowledge of transport properties to complement the increased sophistication and accuracy of various heat transfer analyses. Secondly, the experimental investigations of heat transfer problems on occasion reveal phenomena that may be employed as a basis for measuring transport properties. These stimuli are responsible for the interest in and contributions to the area of thermophysical properties at the Heat Transfer Laboratory.

12.2 PRANDTL NUMBER OF DETERMINATIONS OF GASES AND GAS MIXTURES BY DIRECT MEASUREMENT OF THE RECOVERY FACTOR

The quantity $\eta C_p/\lambda$, called the Prandtl number Pr, is a dimensionless number formed by the viscosity η, the heat capacity at constant pressure C_p, and the thermal conductivity λ. It enters into all convective heat transfer calculations. Nevertheless, the uncertainty that earlier attended the various properties that comprise the dimensionless quantity caused discrepancies to exist even for familiar gases such as air. For example, Irvine [1] shows differences of over 6% between various estimates of air Prandtl number at 711K.

To remedy this situation a program of measurements was undertaken beginning with the above work in 1956. The analytical model serving as a basis for the measurements is a flat plate, exchanging energy only by convective heat transfer with a high velocity stream at constant pressure. For these conditions the surface of the flat plate, under the competing influences of fluid friction and heat conduction transverse to the flow, assumes a temperature between the static

523

temperature T_S and total temperature T_T of the gas stream called the recovery temperature T_R. This temperature appears in a quantity called the recovery factor, σ, defined as

$$\sigma = \frac{T_R - T_S}{T_T - T_S} = \frac{T_R - T_S}{V^2/2C_p} \tag{12.1}$$

Solution of the boundary layer equation for the conditions specified by Pohlhausen [2] in 1921 established a very simple and accurate relation between the Prandtl number and the recovery factor, i.e., $\sigma = \sqrt{\text{Pr}}$. Rearranging the terms in Eq. (12.1), and using the relationship between the pressure ratio across an isentropic nozzle and the velocity developed V, gives the equation

$$\sigma = \sqrt{\text{Pr}} = 1 - \frac{T_T - T_R}{T_T[1 - (P_s/P_t)^{(\gamma-1)/\gamma}]} \tag{12.2}$$

Three measurements are necessary to determine Prandtl numbers: the difference between total and recovery temperature $T_T - T_R$, the total temperature T_T, and the pressure ratio P_s/P_t. The physical apparatus designed to obtain the necessary measurements prescribed by the analysis is shown in Fig. 12.1. The key measurement is provided by the differential thermocouple which is strung axially along the nozzle axis and serves as the flat plate. The upstream junction senses the total temperature T_T and the downstream junction just beyond the nozzle exit senses the recovery temperature T_R.

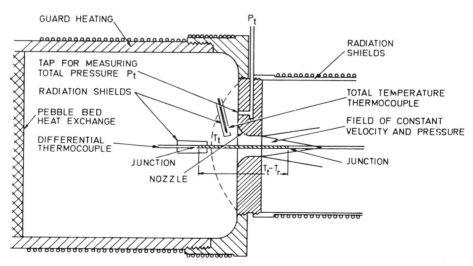

FIG. 12.1 Schematic diagram of Prandtl number measuring device.

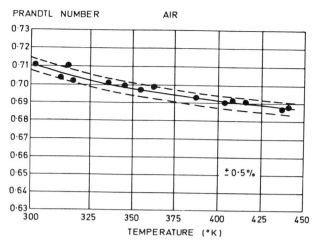

FIG. 12.2 Prandtl number measurements for air at 1 atm pressure.

12.2.1 Pure Gases

Early results for air in the temperature range 293 to 448K were obtained with an 0.46 mm dia. manganin, manganin–constantan thermocouple and are shown in Fig. 12.2 [3]. The Prandtl numbers determined by such measurements are reproducible to ± 0.5%, a precision attributed to the well-defined environment provided by the flat-plate differential thermocouple. This occurs through the use of a blow-down system that provides gas at constant conditions of pressure and temperature to the nozzle for a period of 2 to 3 minutes, an interval adequate for obtaining the required measurements. The test gas may be heated to high temperatures by a pebble-bed heater which provides the apparatus with a temperature capability of room temperature to 670K. The results obtained for carbon dioxide, CO_2, a polyatomic gas, are reproducible with the same precision, ± 0.5%, and are shown in Fig. 12.3.

Helium gas results are interesting since kinetic theory predicts a constant value of the Prandtl number for monatomic gases, yet the values calculated from the "best" values of viscosity, heat capacity, and thermal conductivity available at the time of the National Bureau of Standards Study in 1954 [4] showed departure from this prediction. The departure is attributed to variation in the helium thermal conductivities then available. The Prandtl number determinations from recovery factor measurements agree with theory, as shown in Fig. 12.4. Comparisons are also given with other predictive schemes described in [5, 6].

Argon, a stable monatomic gas of medium molecular weight, also has a predicted constant Prandtl number behavior with temperature and is useful as a calibrating gas. Results obtained for argon appear in Fig. 12.5.

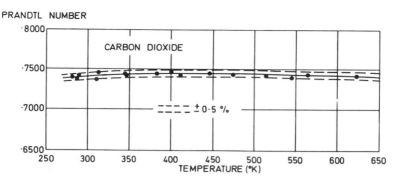

FIG. 12.3 Prandtl number measurements for CO_2 at 1 atm pressure.

12.2.2 Mixtures

The first mixture investigated was the helium–air mixture, due to the interest in helium as a transpiration coolant for surfaces exposed to aerodynamic heating. Measurements were conducted using the same apparatus and techniques employed in the previous measurements given for pure gases. Results are shown in Fig. 12.6 together with predictions made by various schemes described in [8]. The interesting result was the determination of a minimum in the Prandtl

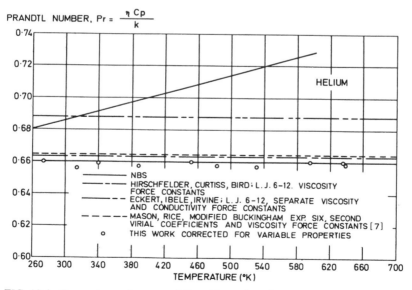

FIG. 12.4 Comparison of measured Prandtl numbers for helium with values calculated from kinetic theory and by combining the "best" experimental values of viscosity, heat capacity, and thermal conductivity.

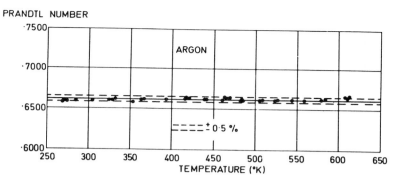

FIG. 12.5 Prandtl number measurements for argon at 1 atm pressure.

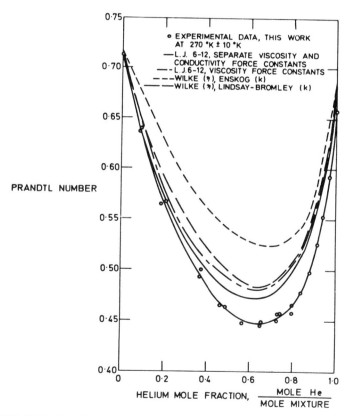

FIG. 12.6 Prandtl number measurements for helium–air mixtures at room temperature.

number (~ 0.45) at 0.65 mole fraction helium appreciably different than the values for the component gases (air 0.71, helium 0.66). The same behavior, though much diminished, is given by Fig. 12.7 for the carbon dioxide–air mixture.

The measurements described above may be used as the basis for extending our knowledge of the Prandtl number and the transport properties, viscosity and thermal conductivity for gas mixtures. The procedure is reported in detail in [8] for helium–air mixtures. Briefly, it recognizes that the superior accuracy of heat capacity C_p data (0.1%), viscosity η data ($\sim 1\%$), and the precision of the Prandtl number determinations ($\sim \pm 1.0\%$) permit the calculation of gas and gas mixture thermal conductivities which generally exceed the accuracy of those determined directly by the various cell methods. The careful selection of an appropriate molecular model to yield the best agreement with measured viscosities and thermal conductivities provides a basis for extrapolating the values of these properties to higher temperatures with reasonable confidence. The properties of the pure components are then combined according to one of a number of mixture rules, the most accurate being given in [5]. The solid curve shown in Fig. 12.6 indicates the result of such a procedure in predicting Prandtl numbers for helium–air mixtures at room temperature. The final step of the process develops working charts of viscosity (Fig. 12.8), thermal conductivity (Fig. 12.9), and Prandtl number (Fig. 12.10). Mixtures treated in such a manner are helium–air [8], carbon dioxide–air [9], nitrogen–carbon dioxide [10], and hydrogen–methane [11]. Subsequent work employs a second Prandtl number apparatus of stainless steel with a temperature capability of 1360K, and 200% excess air [12]. Measurements have been performed on air, argon (calibrating gas) and gas mixtures of N_2, O_2, CO_2 and H_2O in proportions representing the combustion products of the fuel $(CH_2)_n$ for stoichiometric, 100%, and 200% excess air. The results are published as a NASA Technical Note [12].

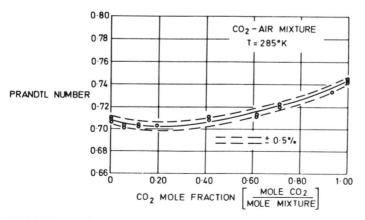

FIG. 12.7 Prandtl number measurements for carbon dioxide–air mixtures at room temperature and 1 atm pressure.

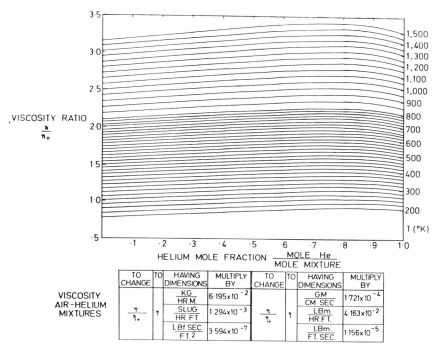

VISCOSITY AIR–HELIUM MIXTURES

TO CHANGE	TO	HAVING DIMENSIONS	MULTIPLY BY	TO CHANGE	TO	HAVING DIMENSIONS	MULTIPLY BY
$\dfrac{\eta}{\eta_o}$	η	$\dfrac{KG}{HR.M.}$	$6 \cdot 195 \times 10^{-2}$	$\dfrac{\eta}{\eta_o}$	η	$\dfrac{GM}{CM.SEC}$	$1 \cdot 721 \times 10^{-4}$
		$\dfrac{SLUG}{HR.FT.}$	$1 \cdot 294 \times 10^{-3}$			$\dfrac{LBm}{HR.FT.}$	$4 \cdot 163 \times 10^{-2}$
		$\dfrac{LBf.SEC}{FT.^2}$	$3 \cdot 594 \times 10^{-7}$			$\dfrac{LBm}{FT.SEC.}$	$1 \cdot 156 \times 10^{-5}$

FIG. 12.8 Air–helium mixture viscosities at various temperatures.

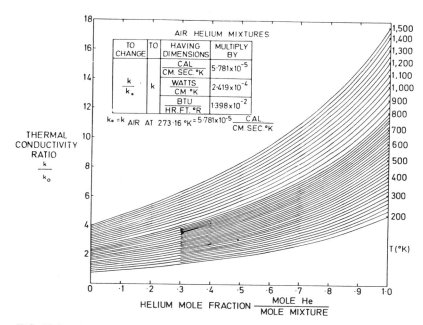

FIG. 12.9 Air–helium mixture thermal conductivities at various temperatures.

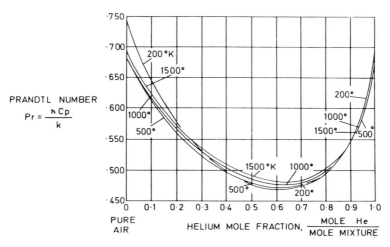

PRANDTL NUMBER

$$Pr = \frac{\hbar\, Cp}{k}$$

FIG. 12.10 Prandtl numbers for air–helium mixtures at various temperatures.

12.3 COEFFICIENT OF ORDINARY DIFFUSION–THERMAL DIFFUSION FACTOR

Earlier heat transfer investigations [13–17] using a low molecular weight gas injected to shield a surface from a high temperature involved at one stage a situation where diffusion of mainstream gas and transpiring gas, though originally at the same temperature, caused large temperature variations on the surface to be observed. This is attributed to the thermal effects associated with the diffusion of unlike gases, i.e., diffusion-thermo effect and the magnitude of the temperature rise (as large as 17K in some instances) suggested its use in measuring the properties connected with the process: the coefficient of ordinary diffusion D_{12} for binary diffusion, and the thermal diffusion factor $(\alpha_T)_{12}$ for the diffusion process arising from the generated temperature field.

Previous measurements of binary diffusion have been concerned primarily with the influence of temperature on the ordinary coefficient of diffusion. The method developed here permits measurements of both the temperature and composition dependence of this property, and the thermal diffusion factor as well. The experiment involves a combination of convective and diffusive flow in a test section where the concentration and temperature at a given location are measured simultaneously and both radial and axial distributions of these quantities are obtained with sufficient accuracy to permit the determination of diffusion coefficients and thermal diffusion factors.

The diffusion chamber (25 mm dia. by 50 mm long) is shown in Fig. 12.11. The gases for which coefficients are to be measured are denoted by 1 and 2, 1 being the heavier gas. The vertical tube is divided into two regions by a porous section made of layers of filter paper. Gas 2 enters at the top of the tube, gas 1 at the same temperature drifts slowly past the tube exit in a direction normal to the tube axis. Both gases flow at velocities so low that the total pressures in both sections of the tube are constant. The presence of gas 1 at the lower end of the test section and gas 2 above the porous section establishes a concentration gradient in the vertical direction and a diffusion mass flow ensues. There follows from this a temperature field due to the diffusion-thermo effect and an additional mass flow occurs in the tube due to the inverse thermal diffusion effect (Fig. 12.12) [18]. Applying the mass conservation law to the mass flux of gas 1 and gas 2 in the vertical tube (y-direction) yields the following expression [19]:

$$\dot{m}_2 = \rho D_{12} \frac{M_2}{M} \frac{1}{x_1} \frac{dx_1}{dy} + x_2 (\alpha_T)_{12} \frac{1}{T} \frac{dT}{dy} \qquad (12.3)$$

where ρ and M are mixture density and molecular weight respectively, M_2 the

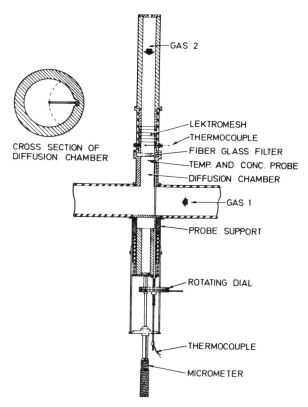

CROSS SECTION OF DIFFUSION CHAMBER

GAS 2

LEKTROMESH
THERMOCOUPLE
FIBER GLASS FILTER
TEMP. AND CONC. PROBE
DIFFUSION CHAMBER

GAS 1

PROBE SUPPORT

ROTATING DIAL

THERMOCOUPLE

MICROMETER

FIG. 12.11 Apparatus for measuring the mass diffusion coefficient and the thermal diffusion factor.

molecular weight of component 2, x the mole fraction, and T the temperature. The contributions of ordinary diffusion and the coupled diffusion thermo effect are apparent. The assumptions of $\partial P/\partial y = 0$ and $\partial x_1/\partial r = 0$ necessary to generate Eq. (12.3) were verified experimentally. The temperature-concentration probe is a 0.89 mm outer diameter, hollow, stainless steel tube containing a 40 gage iron–constantan thermocouple. Concentration is determined from the gas sampled by a thermal conductivity cell.

For determining diffusion coefficients, the second term of Eq. (12.3) is neglected since it is approximately two orders of magnitude less than the first, leaving

$$D_{12} = \frac{RT\dot{m}_2}{M_2 p} \frac{1}{d(\ln x_1)/dy} \tag{12.4}$$

The mass rate of flow of the lighter gas, $\dot{m}_2$, is measured by a carefully calibrated capillary tube submerged in a constant temperature bath.

Results for the gas pair helium–nitrogen at 298K are given in [18] from which Fig. 12.13 is taken. A dependence of D_{12} upon concentration is clear, as well as the consistency of the results obtained when compared to those of earlier investigations. Also shown are the predictions using two molecular models and two calculation schemes. Similar results were obtained at various temperatures over a range of 200 to 400K and are given in [18]. An analysis of the errors attending the measurement of the mass flow, temperature, pressure, and the

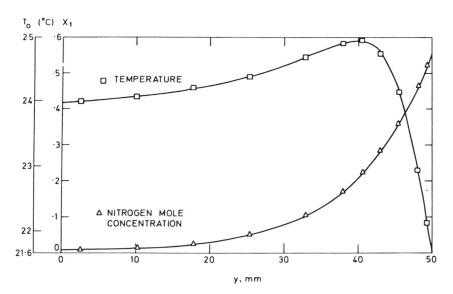

FIG. 12.12 Concentration and temperature profiles along center line for Gas II (helium) mass flow rate of 0.00114 kg/s m².

D_{12} in cm^2/sec

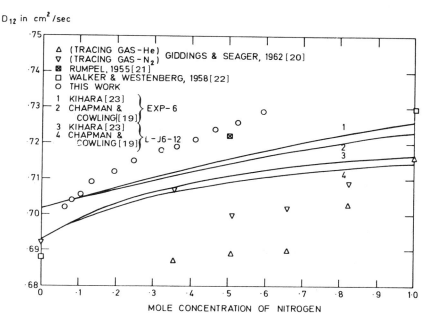

FIG. 12.13 Mass diffusion coefficient for He-N_2 mixture at 298K and 1 atm.

measurement and analytical representation of the concentration yields an estimated accuracy of 2%.

12.4 HOT WIRE CELL DETERMINATION OF GAS THERMAL CONDUCTIVITY

Steady-state means of measuring gas thermal conductivities based on Fourier's law of heat conduction each encounter their own special problems. The parallel plate method encounters errors due to end and convective effects and, at high temperature, radiation errors. The concentric geometries pose construction problems because of the required symmetry, particularly for the sphere, but also for the cylinder, and both encounter radiation errors at high temperature. The line source method introduces a temperature perturbation across a laminar gas stream at known uniform temperature and determines the thermal conductivity from observations of the decay of the perturbation along the flow path [24]. The method yields good results but requires a precise measurement of the gas velocity. The Prandtl number method (described earlier) experiences material problems at higher temperatures due to the large regions and number of parts continuously exposed to such an inhospitable environment, with some parts subject to mechanical stress as well.

The hot wire cell is an extreme form of the concentric cylinder since it uses a very small diameter (0.025 mm) wire as the inner cylinder. When the outer cylinder diameter is 6.35 mm the symmetry problem and radiation error at elevated temperature that usually beset the concentric cylinder method are greatly diminished in importance. Primary concern is directed instead to the temperature-jump effect associated with the small size of the inner wire (its diameter is the order of the mean molecular path length).

The test cell is shown in Fig. 12.14 and consists of a center platinum wire, 0.025 mm diameter, which serves as the line heat source for the test gas, but is also calibrated as a resistance thermometer. The voltage drop along the 40 mm of central wire is measured by means of the 0.0051 mm diameter platinum voltage taps connected to 0.051 mm diameter platinum leads on either side of the central wire. To measure the cell temperature, a very small current (400–600 μA) is permitted to flow through the center wire and the voltage drop measured, the temperature calculated and verified by comparing it to readings of three thermo-couples (Pt − Pt-10% Rh) attached to the outer wall of the outside cylinder. This establishes the cell and wall T_w temperature.

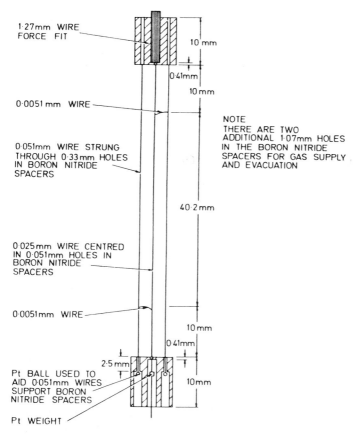

FIG. 12.14 Test section of hot wire gas thermal conductivity cell.

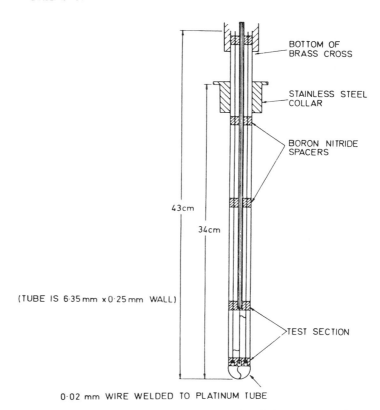

FIG. 12.15 Profile of test section enclosed by platinum outer tube.

The current flow in the central wire is increased to 40–60 mA, enough to raise the temperature of the center wire T_f 60–70K above the wall temperature. The resistance of the central wire (and therefore its temperature) is determined by measuring the voltage drop again and the current. The current is determined by measuring the voltage drop across a standard resistance in series with the central wire. The wall thermocouples are read again to verify that the wall temperature continues to be that indicated by the initial measurement of the central wire used as a resistance thermometer. The cell chamber of 6.35 mm is chosen to eliminate free convection effects by holding the product of Grashof and Prandtl numbers below 1000.

The central wire assembly is installed in the 6.35 mm diameter platinum tube of 0.38 mm wall thickness, Fig. 12.15, and placed in the tubular chamber of the high temperature furnace, a Lindberg Hevi-Duty Laboratory Furnace of 1770K capability. Prior measurements of the furnace internal temperatures located a zone that is isothermal (± 0.5K at 773K) over the 40 mm long test section.

To eliminate the temperature-jump effect and arrive at the true gas thermal conductivity, a succession of measurements are taken at successively lower pressure (1300, 500, 100 mm of Hg-pressure) of test gas in the hot wire cell. Readings are taken at each pressure level and an apparent thermal conductivity; λ_{APP}, calculated from

$$\lambda_{APP} = \frac{Q \ln (r_w/r_f)}{2\pi L (T_f - T_w)} \tag{12.5}$$

where Q is the heat flow through the gas sample, obtained from the measured current flow through the central wire and the known resistance between the two voltage taps, r_f is the radius of the central wire, r_w the inner radius of the outer cylinder, L the length of the test section (40 mm between voltage taps), T_f the central wire temperature, and T_w the outer cylinder wall temperature. Values of $1/\lambda_{APP}$ are plotted against $1/p$ and extrapolated to the zero value of $1/p$. It is known that for the pressure range over which measurements are made, the influence of the pressure on the gas thermal conductivity is insignificant and the departures of λ_{APP} from the true conductivity are due entirely to the temperature-jump effects. The value of $1/\lambda_{APP}$ at $1/p$ equals zero, obtained by the above extrapolation procedure, is the true conductivity value unaffected by temperature-jump effects in accordance with Kennard's [25] representation

$$\lambda = \lambda_{APP} \left(1 + \frac{C}{p} \right) \tag{12.6}$$

where C is a constant.

Results obtained for argon gas are shown in Fig. 12.16, taken from [26, 27]. The references cited are those given in [26] and refer to measurements of other investigators. A careful analysis of the sources of error estimates the most probable error as ± 1.4% at 773K and ± 2.8% at 1273K. Subsequent improvements in measurement circuitry and instrumentation have reduced these errors by a third.

12.5 CONCLUSION

Three measurement schemes for determining quantities important in heat transfer have been presented. The first takes the flat plate boundary layer equation solution for recovery factor as a basis for determining the dimensionless heat transfer quantity Prandtl number, an important factor in all convective heat transfer. The second takes phenomena noted during transpiration heat transfer measurements and designs an optimum test chamber and probe to measure important diffusion properties. In the third instance, the familiar cylindrical geometry is taken to the limit of a very small central wire serving as both heat source and resistance thermometer to measure gas thermal conductivities.

λ (cal /cm sec K) x 10⁻⁵

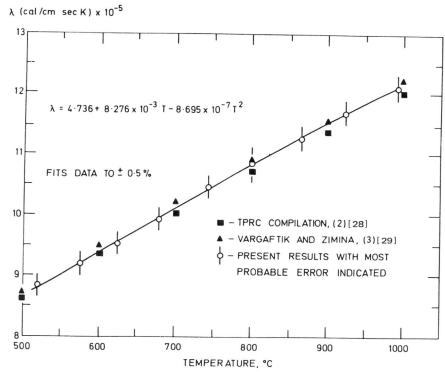

$$\lambda = 4{\cdot}736 + 8{\cdot}276 \times 10^{-3}\ T - 8{\cdot}695 \times 10^{-7}\ T^2$$

FITS DATA TO ± 0·5 %

■ – TPRC COMPILATION, (2) [28]
▲ – VARGAFTIK AND ZIMINA, (3) [29]
⚬ – PRESENT RESULTS WITH MOST
PROBABLE ERROR INDICATED

TEMPERATURE, °C

FIG. 12.16 Thermal conductivity of argon at 1 atm pressure measured by hot wire cell.

NOMENCLATURE

C	constant
C_p	heat capacity at constant pressure
D_{12}	binary diffusion coefficient
k	thermal conductivity coefficient (also λ)
L	length
M	molecular weight
$\dot{m}$	mass flow rate
p	pressure
p_s	static pressure
Pr	Prandtl number
p_t	total pressure
Q	heat flow
r	radius
r_f	wire radius
r_w	wall radius
$\bar{R}$	gas constant

T temperature absolute
T_f wire temperature
T_R recovery temperature
T_T total temperature
T_w wall temperature
V velocity
x mole fraction
y distance
$(\alpha_T)_{12}$ binary thermal diffusion factor
γ adiabatic exponent (C_p/C_v)
η viscosity coefficient
λ thermal conductivity coefficient (also k)
λ_{APP} apparent thermal conductivity
ρ density
σ recovery factor

Subscripts

1   gas component 1
2   gas component 2
n integer

REFERENCES

1. Irvine, T. F., Jr., "A New Method for the Experimental Determination of Prandtl Numbers and Thermal Conductivities of Gases: Results for Air," Ph.D. thesis, University of Minnesota, Minneapolis, 1956.
2. Pohlhausen, E., Der Warmeaustausch Zwischen festen Korpern und Flussigkeiten mit Kleiner Riebung und Kleiner Warmleitung, *Zeit. angew. Math. Mech.*, vol. 1, pp. 115–121, 1921.
3. Eckert, E. R. G., and Irvine, T. F., Jr., A New Method to Measure Prandtl Number and Thermal Conductivity of Fluids, *J. Appl. Mech.*, vol. 24, p. 25, 1957.
4. Hilsenrath, J., and Touloukian, Y. S., The Viscosity, Thermal Conductivity, and Prandtl Number for Air, O_2, N_2, NO, H_2, CO, CO_2, H_2O, He, and A, *Trans. ASME*, vol. 76, pp. 967–983, 1954.
5. Hirschfelder, J. O., Curtiss, C. F., and Bird, R. B., "Molecular Theory of Gases and Liquids," Wiley, New York, 1954.
6. Stroom, P. D., Ibele, W. E., and Irvine, G. F., Jr., Helium Prandtl Number Measurements and Calculated Viscosity and Thermal Conductivity, in "Int. Developments in Heat Transfer," pt. IV, sect. B, no. 105, pp. 870–875, Am. Soc. of Mech. Eng., New York, 1961.
7. Mason, E. A., and Rice, W. E., The Intermolecular Potential of Helium and Hydrogen, *J. Chem. Phys.*, vol. 22, no. 2, p. 522, 1954.
8. Eckert, E. R. G., Ibele, W. E., and Irvine, T. F., Jr., Prandtl Number, Thermal Conductivity and Viscosity of Air-Helium Mixtures, *NASA TN D-533*, Washington, D.C., Sept. 1960.
9. Novotny, J. L., and Irvine, T. F.; Jr., Thermal Conductivity and Prandtl Number of Carbon Dioxide and Carbon Dioxide-Air Mixtures at One Atmosphere, *J. Heat Transfer*, vol. 83, pp. 125–132, 1961.

10. Ibele, W. E., and Briggs, D. G., Prandtl Number Measurements and Transport Property Calculations for N_2-CO_2 Mixtures, *4th Symp. on Thermophysical Properties (ASME), University of Maryland, College Park, Md.*, pp. 392–397, 1968.

11. Ibele, W. E., and Desmond, R. M., Transport Properties of the Hydrogen-Methane System, *Progress in Heat and Mass Transfer*, vol. 2, Pergamon, Oxford, 1969.

12. Anderson, K. M., Pulkrabek, W. W., Ibele, W. E., and Eckert, E. R. G., Measurement of Prandtl Number and Thermal Conductivity, *NASA CR54634, Heat Transfer Laboratory Report No. 88*, Dec. 1968.

12. Tewfik, O. E., Eckert, E. R. G., and Jurewicz, L. S., Diffusion-Thermo Effects on Heat Transfer from a Cylinder in Cross Flow, *AIAA J.*, vol. 1, p. 1537, 1963.

14. Tewfik, O. E., Eckert, E. R. G., and Shirtliffe, C. J., Thermal Diffusion Effects on Energy Transfer in a Turbulent Boundary Layer with Helium Injection, *Proc. 1962 Heat Transfer and Fluid Mechanics Inst.*, Stanford Univ. Press, Stanford, Calif., 1962.

15. Tewfik, O. E., and Yang, J. W., The Thermodynamic Coupling between Heat and Mass Transfer in Free Convection, *Int. J. Heat Mass Transfer*, vol. 6, p. 915, 1963.

16. Sparrow, E. M., Minkowycz, W. J., Eckert, E. R. G., and Ibele, W. E., The Effect of Diffusion Thermo and Thermal Diffusion for Helium Injection into Plane and Axisymmetric Stagnation Flow of Air, *J. Heat Transfer*, vol. 86, p. 311, 1964.

17. Sparrow, E. M., Minkowycz, W. J., and Eckert, E. R. G., Transpiration-Induced Buoyancy and Thermal Diffusion-Diffusion Thermo in a Helium-Air Free Convection Boundary Layer, *J. Heat Transfer*, vol. 86, p. 508, 1964.

18. Yang, J. W., "A New Method of Measuring Mass Diffusion Coefficient and Thermal Diffusion in a Binary Gas System," Ph.D. thesis, University of Minnesota, Minneapolis, 1966.

19. Chapman, S., and Cowling, T. G., "The Mathematical Theory of Non-Uniform Gases," Cambridge Univ. Press, London, 1952.

20. Giddings, J. C., and Seager, S. C., Method for Rapid Determination of Diffusion Coefficients, *I. & E. C. Fundamentals*, vol. 1, p. 277, 1962.

21. Rumpel, W. F., Interferometric Studies of Refractive Indices and Diffusion Coefficients in Selected Gases, *Rept. No. CM 851, Naval Research Laboratory, University of Wisconsin*, 1955.

22. Walker, R. E., and Westenberg, A. A., Molecular Diffusion Studies in Gases at High Temperature. V. Results for H_2-Ar System, *J. Chem. Phys.*, vol. 36, p. 3499, 1962.

23. Mason, E. A., High Temperature Approximations for the Transport Properties of Binary Mixtures. I. General Formulas, *J. Chem. Phys.*, vol. 27, p. 75, 1954; II. Applications, *J. Chem. Phys.*, vol. 27, p. 182, 1954.

24. Westenberg, A. A., and deHaas, N., Gas Thermal Conductivity Studies at High Temperature, Line Source Techniques, and Results in N_2, CO_2, and N_2-CO_2 Mixtures, *Phys. Fluids*, vol. 5, p. 266, 1962.

25. Kennard, E. H., "Kinetic Theory of Gases," McGraw-Hill, New York, 1939.

26. Desmond, R. M., "Measurement of the Thermal Conductivity of Gases at High Temperature with a Hot Wire Cell," Ph.D. thesis, University of Minnesota, Minneapolis, 1968.

27. Desmond, R. M., and Ibele, W. E., The Thermal Conductivity of Helium at Atmospheric Pressure, *Proceedings, Symposium on Thermophysical Properties (ASME), 5th, Boston*, 1970.

28. TPRC Compilation, "Thermophysical Properties Research Center Data Book," vol. 2, June 1966 Supplement, Purdue University, Lafayette, Indiana, 1966.

29. Vargaftik, N. B., and Zimina, N. H., Heat Conductivity of Nitrogen at High Temperature, *Teplofiz Vysokikh. Temp.*, vol. 2, p. 869, 1964. (Translated in *High Temp.*, p. 782, 1964.)

13 The laser-Doppler anemometer

R. J. GOLDSTEIN
University of Minnesota, Minneapolis
D. K. KREID
*Battelle, Pacific Northwest Labs,
Richland, Washington*

If a source of acoustic or electromagnetic waves of constant frequency ν is put in motion the frequency of the waves received by a fixed observer will change. This is the well known Doppler effect, which is sometimes called the Fizeau effect when applied to light waves. A similar phenomenon is observed when the source is fixed but the waves are scattered or reflected from a body in motion before being received by the fixed observer. In either case, the apparent change in frequency, or the Doppler shift, ν_D is related to the velocity in the following way:

$$\frac{\nu_D}{\nu} \sim \frac{\upsilon_r}{a}$$

where υ_r is the speed of the source (or reflector) relative to the observer, and a is the wave propagation speed.

For acoustic waves, a is the speed of sound and υ_r/a is the Mach number. With sound waves, the Doppler shift is often of the same order of magnitude as the source frequency ν. The shift is then usually considerably larger than the source bandwidth and is readily detectable.

For light or electromagnetic radiation, $a = c$, the speed of light, and $\upsilon_r/c \ll 1$, for all but a few exceptional cases. The change in frequency relative to the source frequency is very small and usually difficult to detect. Conventional light sources usually have a bandwidth considerably larger than the Doppler shift (at reasonable velocities) and the Doppler shift cannot easily be measured from the broadened signal. A notable exception is the red shift or apparent recession noted by astronomers in studying the emission spectra of stellar bodies [1]. The red shift is the displacement of the spectra of the source toward the red end of the spectrum. This is apparently due to a Doppler shift caused by the relative recession velocity between the stellar source and the earth. Star velocities of the order of 40,000 Km/sec (or, about 0.13 times the velocity of light) at distances of 250 million light years have been detected [2], while still higher velocities are postulated for quasars. Velocities of this order of magnitude are rare in nonastronomical systems and thus, for light sources with normal bandwidths, the optical Doppler effect is rarely employed for velocity detection.

The introduction of lasers permits measurement of terrestial values of velocity by optical heterodyne techniques. The laser is a source of essentially coherent radiation of extremely narrow bandwidth. A single axial mode of a He-Ne laser operating at 632.8 nm ($\nu \sim 5 \times 10^{14}$ Hz) has a bandwidth of about 10 Hz. In comparison, the green mercury line at 546.1 nm ($\nu \sim 6 \times 10^{14}$ Hz) has a bandwidth of 8×10^8 Hz, which is representative of line widths associated with conventional monochromatic light sources. The Doppler shift of light from a laser source caused by motion of an emitter or reflector moving at moderate speed is still small compared to the source frequency, but it is large compared to the source bandwidth and is detectable by heterodyne techniques. Yeh and Cummins [3–4] first suggested and successfully demonstrated the laser-Doppler technique for measuring fluid velocities in 1964. From the Doppler shift of laser light scattered by particles in flowing water, they measured the laminar velocity profile in a round tube.

The purpose of this chapter is to indicate the factors to be considered in the design and application of a laser-Doppler anemometer, LDA (sometimes called a laser-Doppler velocimeter, LDV), for the measurement of fluid velocity. The advantages, capabilities, and limitations of the system are pointed out and analyzed, and a bibliography of recent work in this area is included. A limited amount of data from various investigations is presented for the purpose of illustrating the type of information available, and specific variations in system design are considered to illustrate design alternatives. The necessary design criteria are specified sufficiently to permit the design of different systems, each best suited for the particular flow to be studied.

13.1 VELOCITY MEASUREMENT FROM DOPPLER SHIFT

The Doppler shift is as familiar as the well-known whistle-tooting trainman. The effect for scattered waves is somewhat different than that for emitted ones. A heuristic, nonrelativistic derivation can be made with reference to Fig. 13.1.

Monochromatic radiation of wavelength λ_i and speed c emanates from a stationary source. In Fig. 13.1, the radiation passes in the direction defined by the unit vector $\hat{n}_i$ and illuminates a particle having a velocity $\mathbf{v}$ (where $|\mathbf{v}| \ll c$).

If the particle were motionless, the number of wave fronts passing—or striking—it per unit time would be c/λ_i (or ν_i). Since, from a nonrelativistic viewpoint, the difference between the velocity of the particle and the illumination is

$$c - \mathbf{v} \cdot \hat{n}_i$$

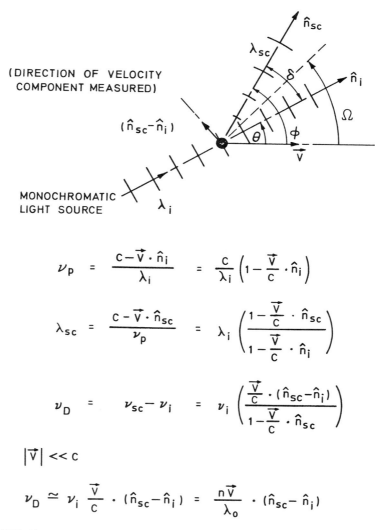

$$\nu_p = \frac{c - \vec{V} \cdot \hat{n}_i}{\lambda_i} = \frac{c}{\lambda_i}\left(1 - \frac{\vec{V}}{c} \cdot \hat{n}_i\right)$$

$$\lambda_{sc} = \frac{c - \vec{V} \cdot \hat{n}_{sc}}{\nu_p} = \lambda_i \left(\frac{1 - \frac{\vec{V}}{c} \cdot \hat{n}_{sc}}{1 - \frac{\vec{V}}{c} \cdot \hat{n}_i}\right)$$

$$\nu_D = \nu_{sc} - \nu_i = \nu_i \left(\frac{\frac{\vec{V}}{c} \cdot (\hat{n}_{sc} - \hat{n}_i)}{1 - \frac{\vec{V}}{c} \cdot \hat{n}_{sc}}\right)$$

$$|\vec{V}| \ll c$$

$$\nu_D \simeq \nu_i \frac{\vec{V}}{c} \cdot (\hat{n}_{sc} - \hat{n}_i) = \frac{n\vec{V}}{\lambda_0} \cdot (\hat{n}_{sc} - \hat{n}_i)$$

FIG. 13.1 Frequency shift for light scattered from a moving particle.

the number of wave fronts incident upon the particle per unit time (or the apparent frequency to the particle) is

$$\nu_p = \frac{c - \mathbf{v} \cdot \hat{\mathbf{n}}_i}{\lambda_i} \tag{13.1}$$

This is also the number of wave fronts scattered per unit time by the moving particle.

Consider now a fixed observer toward which a moving particle is emitting—or scattering—radiation in the direction defined by the unit vector $\hat{n}_{sc}$, the number of wave fronts scattered or emitted being ν_p. After the scattering of one wave front, the particle moves toward that wave front with speed $\mathbf{v} \cdot \hat{n}_{sc}$. Thus, when the next wave front is scattered after a time interval given by $1/\nu_p$, the first wave front is a distance

$$\frac{c - \mathbf{v} \cdot \hat{n}_{sc}}{\nu_p}$$

away from the particle. Thus, to a fixed observer, the apparent wavelength of the scattered radiation is

$$\lambda_{sc} = \frac{c - \mathbf{v} \cdot \hat{n}_{sc}}{\nu_p} = \lambda_i \frac{c - \mathbf{v} \cdot \hat{n}_{sc}}{c - \mathbf{v} \cdot \hat{n}_i} \tag{13.2}$$

The frequency of the scattered radiation to the fixed observer is

$$\nu_{sc} = \frac{c}{\lambda_{sc}} = \frac{c}{\lambda_i} \frac{c - \mathbf{v} \cdot \hat{n}_i}{c - \mathbf{v} \cdot \hat{n}_{sc}} = \nu_i \left[\frac{1 - (\mathbf{v} \cdot \hat{n}_i/c)}{1 - (\mathbf{v} \cdot \hat{n}_{sc}/c)} \right] \tag{13.3}$$

The total Doppler shift ν_D is determined from

$$\nu_D = \nu_{sc} - \nu_i \tag{13.4}$$

or

$$\nu_D = \nu_i \left[\frac{1 - (\mathbf{v} \cdot \hat{n}_i/c)}{1 - (\mathbf{v} \cdot \hat{n}_{sc}/c)} \right] - \nu_i = \frac{\nu_i}{c} \left[\frac{\mathbf{v} \cdot (\hat{n}_{sc} - \hat{n}_i)}{1 - (\mathbf{v} \cdot \hat{n}_{sc}/c)} \right] \tag{13.5}$$

It should be noted that even though the foregoing derivation is simplified, the same equation is obtained in the complete relativistic derivation [5].

Since $|\mathbf{v}| \ll c$,

$$\nu_D = \frac{n\mathbf{v}}{\lambda_0} \cdot (\hat{n}_{sc} - \hat{n}_i) \tag{13.6}$$

where λ_0 is the vacuum wavelength of the incident radiation and n is the index of refraction in the medium surrounding the particle.

Note that, if the directions of the incoming and scattered light beams are fixed (which would be the usual case), the frequency shift gives the component of velocity in the direction given by $(\hat{n}_{sc} - \hat{n}_i)$. This differs from measurements taken with either an impact tube or hot wire anemometer and greatly enhances our techniques for measuring the direction of the velocity.

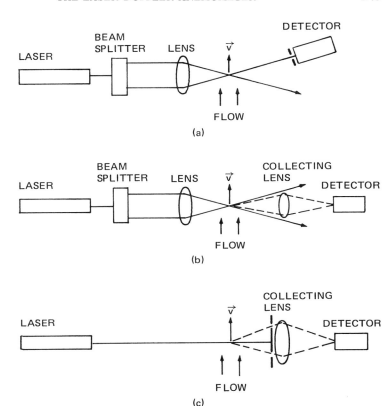

FIG. 13.2 Alternative LDA optical configurations. (*a*) Reference beam. (*b*) Dual scatter (fringe). (*c*) Symmetric scatter.

In practice, the Doppler frequency can be determined by an optical heterodyne technique (Fig. 13.2). An optical heterodyne receiver uses a photocathode to mix the scattered (Doppler shifted) beam with a reference beam from the same laser to generate a current whose ac component has a frequency equal to the difference frequency. The assumption in this technique is that the probability for the emission of an electron from the photocathode—and thus the photoelectric current—is proportional to the intensity of the light which is incident upon it. However, the intensity of the light is proportional to the square of the electric field amplitude. If two coherent sine waves of different frequency are superimposed on a photocathode, the result is an output signal whose amplitude is modulated at the difference frequency, or Doppler frequency, in the present study. Since the photocathode acts as a square law detector, this output signal will vary sinusodially at the difference frequency.

Consider two coherent monochromatic light beams of slightly different frequency combined on the surface of a photocathode. The two light beams may be represented by

$$E_1 = E_{10} \sin 2\pi\nu_0 t$$

$$E_2 = E_{20} \sin 2\pi(\nu_0 + \nu_D)t$$

The output current is proportional to the square of the total electric field incident on it:

$$i \propto (E_1 + E_2)^2 \tag{13.7}$$

Although the photomultiplier is capable of following frequencies up to several hundred megahertz, the frequency of visible light, ν_0, is on the order of 5×10^{14} Hz. Terms in the expansion of i having frequencies of the order of ν_0 will result in a dc current proportional to the time average of those terms. If ν_D is within the frequency response limit of the photomultiplier, there will also be an ac component. For $\nu_D \ll \nu_0$, Eq. (13.7) reduces to

$$i \propto \left[\frac{E_{10}^2 + E_{20}^2}{2}\right] + E_{10}E_{20} \sin 2\pi(\nu_D t + \xi) \tag{13.8}$$

where ξ is a phase angle which is constant if the two beams are coherent. The first term is the dc current and the second term is the ac or Doppler current. To determine the velocity, the spectrum of the photomultiplier current is analyzed and the resulting Doppler frequency is used in Eq. (13.6).

A second type of phenomena has been suggested [6, 7] to explain the detector signal, which leads to the *fringe* theory, or *dual scatter* approach. According to the fringe theory, two coherent light beams that intersect at an angle interfere with one another, giving rise to a three-dimensional fringe pattern in their region of intersection as illustrated in Fig. 13.3. A particle passing through the fringe pattern scatters a burst of light whose amplitude is modulated at a frequency that varies directly with the component of the particle velocity normal to the fringes and inversely with the fringe spacing. By this explanation, the Doppler signal arises due to amplitude modulation in the scattered light as opposed to a shift in its optical frequency.

In recent years the LDA technique has become increasingly popular as a research topic in and of itself and as a basic research tool for the investigation of flow phenomena. The literature on the subject is extensive, and a comprehensive review is beyond the scope of this work. Review articles [8, 9] and bibliographies [10, 11] are available that contain references to most of the major articles and reports published in the decade since the inception of the LDA technique. The brief review which follows is intended to illustrate the wide range of flow phenomena to which the LDA technique is applicable. For further details and more references on selected topics the reader may consult the bibliographies cited above and the current literature.

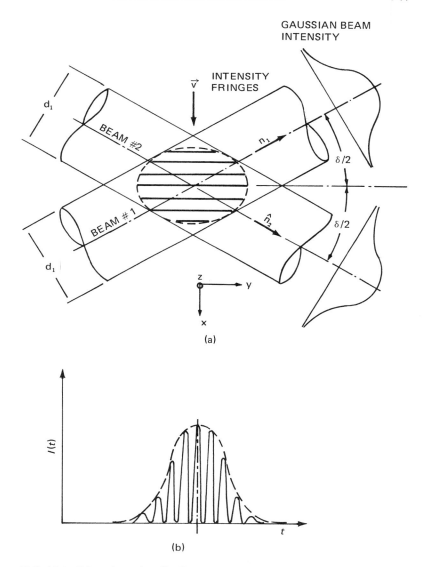

FIG. 13.3 Fringe intensity distribution and burst signal. (*a*) Fringe pattern in the scattering volume. (*b*) Individual Doppler burst from detector.

Yeh and Cummins first demonstrated the LDA technique in 1964 [4] when they measured the fully developed laminar velocity profile for water flowing in a cylindrical duct. Other early measurements in water included the study of steady laminar flow development in square ducts [12] and in round ducts [13], turbulent pipe flow [14, 15] and pipe flow for the full range of Reynolds number from laminar to turbulent flow utilizing dilute water-polymer mixtures

[16]. Air flow measurements [17] were also obtained very early in the development of the LDA. More recently, the technique has been applied under more
severe and demanding conditions, including supersonic flow [18], atmospheric
measurements [19, 20], and studies in flames [21], plasma arcs [22], and
simulated blood [23]. Studies that illustrate the physical scale of phenomena to
which an LDA is applicable range from the blood flow study [23], where
velocity profiles were measured in channels less than half a millimeter in diameter, to the atmospheric studies [19, 20], where velocity measurements may be
obtained at distances of hundreds of meters. Transient and unsteady phenomena
may also be studied using an LDA, as illustrated by measurements in pulsatile flow
[24], in unsteady thermal convection [25], and in separated flow [26].

From the above remarks we may summarize the advantages of an LDA as
compared to other flow measuring systems:

1. An LDA is noninstrusive in that all information is transmitted into and out of
   the flow via light beams. Only minute quantities of very small particles are
   normally required as light scatterers such that flow perturbation can be
   negligible. This is particularly important in measurements near flow boundaries, in recirculating flows or in fluctuating flows where flow reversals
   occur. In the flow of two phase fluids, such as blood, emulsions, water-
   polymer mixtures, and fiber suspensions, problems with probe degradation
   and contamination often arise when using pitot tubes or hot-wire anemometers. These problems do not occur with an LDA. In corrosive fluids or in
   extreme environments, use of an LDA may also be advantageous.

   For completeness, it should be noted that the velocity measured by an
   LDA is the particle velocity which is equal to the fluid velocity only if the
   slip velocity is zero. In liquids this condition is usually easily satisfied;
   however, in gases significant slip can occur, even for very small particles,
   because of the large differences in density. In regions of sharp velocity
   gradients such as near a shock or in highly turbulent flow, extreme care must
   be exercised in evaluating test results from an LDA.

2. An LDA does not require calibration. The relationship between the Doppler
   frequency and the particle velocity, Eq. (13.6), is exact, depending only on
   the frequency of the source, the refractive index of the fluid, and the beam
   geometry. In most cases the laser frequency is known to extreme accuracy
   and measurement of the angle of intersection between the two beams is a
   simple matter.

3. An LDA measures a single component of the velocity, independently of other
   velocity components which might exist. By using multiple beams, detectors,
   and signal processing electronics, two [27] or three [28] components of the
   velocity may be determined simultaneously to establish the full instantaneous
   velocity vector.

4. Equation (13.6) is linear, which considerably enhances the capability for
   studying fluctuating flows. With the frequency shifting apparatus discussed in a

later section, both magnitude and direction of the velocity can be measured, a capability usually lacking in other flow measuring techniques.

5. An LDA is capable of fine spatial resolution. Since all information traverses the system on optical beams which can be accurately focused, the scattering volume can be kept very small, allowing accurate velocity measurements in regions of steep velocity gradient and near flow boundaries.

It would be inappropriate to neglect some of the disadvantages encountered when using a laser-Doppler anemometer. A key one has already been mentioned—not the velocity of the fluid, but the velocity of a scattering center is measured and the scattering center is not a part of the homogeneous fluid in which it is carried. The scattering center will not necessarily follow the flow perfectly, and if the concentration of added scattering centers is sufficiently high, they may affect the basic properties of the flow as well. The absence of a probe with an LDA can also introduce some uncertainty in the location of the measuring region. An optical technique requires a transparent container or some means of introducing a light beam into the working region plus a transparent and relatively homogeneous (optically) working fluid. The cost and complexity of the optical and signal analysis equipment would also appear as disadvantages and should keep us from discarding our impact probes and manometers. As will be discussed below, there are other limitations in certain measurements with an LDA, especially those related to the properties of a turbulent flow.

13.2 OPTIMIZATION OF THE SCATTERING GEOMETRY

13.2.1 Alternative Geometries

Three basic LDA geometries currently in use are shown in Fig. 13.2. There are numerous variations on each of the illustrated geometries.

The physical appearance and the analytical models of the alternative geometries are at first glance considerably different, yet the process by which the signal is generated is essentially the same in all cases. Consequently, the three systems may be treated alike in many important respects. The following discussion on optimization of parameters is valid for all LDA systems except where specific differences are noted.

13.2.2 Intensity of Scattered Light

A number of analyses have been performed concerning optimization of the signal to noise ratio (SNR) with regard to various parameters [29–34]. In summary, it has been found that the SNR varies directly with the scattered beam intensity and inversely with the particle velocity. Generally speaking therefore, it is advantageous to use as much laser intensity as one can, within limits of expense and safety.

The scattered intensity and the SNR are also strongly dependent on the size of the particle and the geometry of the focusing and collecting optics. For Mie scattering, where the particle diameter is on the order of λ, the scattered intensity distribution is very strongly forward directed with minor maxima and minima occurring with increasing angle from the direction of the incident beam. Therefore, for a high scattered intensity (and thus high SNR) the angle between the beams should be small for both the reference beam and dual scatter geometries. In addition, in the dual scatter geometry, the detector should lie on the symmetry axis of the optical system for maximum SNR [32].

For operation in the continuous wave mode (see section on Doppler signal) the scattered intensity and SNR are both generally increased by increasing the particle concentration. There is, however, a maximum particle concentration at which no gain in SNR can be achieved by adding more particles; beyond this signal degradation begins due to multiple scattering effects. Analytical prediction of the optimum concentration is tenuous but the optimum particle concentration can be determined experimentally for each application. If single particle signals are desired, the particle number density should be such that on the average only one particle exists in the scattering volume at any time.

The last parameter affecting the scattered light intensity, and thus the SNR, is the particle size. In general, the scattered intensity increases with increasing particle size; however, it does not follow that larger particles give a larger SNR. The particles should be smaller than the fringe spacing [32], and the following approximate criterion has been established:

$$b \leqslant \frac{\lambda}{4 \sin (\delta/2)} \qquad (13.9)$$

where b is the particle diameter and δ is the angle between the beams. Heuristically, one can see that if the particle is larger than the fringe spacing, the total energy incident upon the particle is nearly constant as it passes through the fringes because it is illuminated by one or more complete fringes at all times. Therefore the amplitude modulation in the scattered beam would be small, and a poor SNR would be obtained.

13.2.3 Magnitude of Frequency Shift

A second parameter which must be considered in optimizing the system geometry is the velocity detection sensitivity, or the magnitude of frequency shift per velocity increment. In many applications, it may be desirable to get the maximum sensitivity to increase accuracy. However, if very large velocities are to be considered, the system sensitivity may have to be decreased to avoid surpassing the frequency range of the detection equipment. The following analysis indicates the optical design to stay within the frequency range of the analyzing equipment for a given velocity range, retaining the highest system sensitivity.

The Doppler shift is given by Eq. (13.6). Referring to Fig. 13.1 and performing the indicated dot product

$$\nu_D = \frac{n\upsilon}{\lambda_0} (\cos \phi - \cos \theta) \tag{13.10}$$

or

$$\nu_D = -2 \frac{n\upsilon}{\lambda_0} \sin \frac{\delta}{2} \sin (\Omega) \tag{13.11}$$

where $\delta = \phi - \theta$ and $\Omega = (\phi + \theta)/2 = \theta + \delta/2$.

The equation is expressed in the latter form to simplify the calculations involved in optimizing the measurement with respect to the scattering angle δ and the incidence angle θ.

It should be noted from Eq. (13.6) that the maximum velocity detection sensitivity is obtained for direct back scatter (i.e., $\theta = 0$, $\phi = \delta = \pi$ or $\phi = 0$, $\theta = \delta = \pi$). In general for a given value of $\delta < \pi$, maximum sensitivity is obtained when the difference vector $(\hat{n}_{sc} - \hat{n}_i)$ is the direction of the velocity vector $\mathbf{v}$ (i.e., $\Omega = \pi/2$). The system is insensitive to the velocity if the difference vector is perpendicular to the velocity vector.

If the frequency limit of the detection apparatus is ν_m and the maximum velocity which is to be measured is υ_m, then the limitation on the angles δ and θ can be specified from Eq. (13.11)

$$\frac{\lambda_0 \nu_m}{2n\upsilon_m} \geq | \sin \frac{\delta}{2} \sin (\Omega)| \tag{13.12}$$

The inequality equation (13.12) must be satisfied to stay within the frequency range of the detection equipment. In many applications, frequency limitations will not be significant and the optimal configuration for maximum sensitivity will correspond to $\Omega = \pm \pi/2$. This arrangement has the added advantages of symmetry, which facilitates system alignment, and minimum bandwidth as indicated below.

13.2.4 Signal Bandwidth

The third condition to be considered is that of minimum signal bandwidth. The analysis will be made first on the basis of geometrical optics, neglecting diffraction effects associated with the wave properties of light. Assuming ν_D to be a function of θ and ϕ and taking differentials of Eq. (13.10), we get:

$$d\nu_D = \frac{n\upsilon}{\lambda_0} (\sin \theta \, d\theta - \sin \phi \, d\phi) \tag{13.13}$$

The angular increments $d\theta$ and $d\phi$ may be interpreted to represent the angular spread in the incident and/or scattered light beams due to the focusing and/or collecting optics. We will deal first with the general form of Eq. (13.13), where specific interpretations for $d\theta$ and $d\phi$ are discussed below for typical LDA systems.

To obtain the relative signal bandwidth, we divide Eq. (13.13) by Eq. (13.11), where application of trigonometric identities gives the result

$$\frac{dv_D}{v_D} = \frac{1}{2}(d\phi - d\theta) \cot\left(\frac{\phi - \theta}{2}\right) + \frac{1}{2}(d\phi + d\theta) \cot\left(\frac{\phi + \theta}{2}\right) \quad (13.14)$$

$$\frac{dv_D}{v_D} = \frac{d\delta}{2} \cot\frac{\delta}{2} + d\,\Omega \cot\Omega \quad (13.15)$$

We note at this point that the optimal geometry for velocity sensitivity $\Omega = \pi/2$ is also optimal for minimum bandwidth (for fixed δ), since $\cot \pi/2 = 0$ eliminating the second term in Eq. (13.15). All systems considered from this point on will be assumed to be symmetric (i.e., $\Omega = \pi/2$).

At this point the *systematic* broadening may be given by

$$\frac{dv_D}{v_D} = \cot\frac{\delta}{2}\frac{d\delta}{2} \quad (13.16)$$

We refer to this as "systematic" broadening because it is entirely dependent on the optical configuration of the particular system employed. As discussed below, this type of broadening is also referred to as *transit time* broadening and as *doppler ambiguity* broadening [15] where different yet essentially equivalent explanations for its existence are proposed; the different arguments all yield essentially the same results. The simple arguments employed here are based on geometrical optics and, though lacking in rigor, the results have been found to be functionally correct and sufficiently accurate for many purposes.

From the viewpoint of geometrical optics the angular increment $d\delta$ in Eq. (13.15) may be interpreted differently for different systems. For example, in a simple dual scatter system such as that illustrated in Fig. (13.2b), a given point on the detector receives light scattered simultaneously from both illuminating beams. Some of the scattered light arises from an illuminating beam at angle ϕ, defined to within an uncertainty of $\pm d\phi/2$ and similarly a second scattered beam arises from a second illuminating beam at angle ϕ, defined to within $\pm d\theta/2$. Since $\delta = \phi - \theta$ and ϕ and θ are statistically uncorrelated, the maximum possible uncertainty in δ may be given approximately by the sum of the magnitudes of the angular uncertainties in the illuminating beams

$$|d\delta| = d|\phi - \theta| \quad (13.17)$$

$$|d\delta| \cong |d\phi| + |d\theta| \tag{13.18}$$

It can be shown that the latter result is also valid for the simple reference beam system in Fig. 13.2(a).

For most systems, $|d\phi| = |d\theta| \simeq B/f$ where B is the unfocused beam diameter and f is the effective focal length of the lens. Equation (13.16) for the signal broadening now becomes

$$\frac{\Delta\nu_D}{\nu_D} \cong \frac{B}{f} \cot \frac{\delta}{2} \tag{13.19}$$

Equation (13.19) is in qualitative agreement with experimental results for the flow of water with a multiparticle signal in reference beam systems [12, 30, 33] and in a dual scatter system [30].

For systems utilizing a number of apertures and compound lenses in the focusing and collecting optics, the identification of $d\delta$ is more complex; however, the functional form of the broadening is still governed by $\cot \delta/2$. In addition, it is often difficult to separate the various broadening effects, so that it is difficult to quantify the agreement of an experimental value with the prediction.

It is interesting to note that Eq. (13.15) can also be derived by considering the wave properties of light [8, 15] as opposed to the strictly geometric approach employed above. In the volume from which scattered light is observed, a particle is present for only a short time, and thus scatters a wave train of finite length. A finite wave train of duration Δt from a monochromatic source has an effective frequency range of the order of the reciprocal of the time Δt. In fact [35]

$$\Delta t \, \Delta\nu \geqslant \frac{1}{4\pi} \tag{13.20}$$

This relation is analogous to the Heisenberg uncertainty principle in quantum mechanics. In many cases of practical application, the inequality may be replaced by an order of magnitude sign.

A laser operated in the TEMoo mode has an approximately Gaussian intensity distribution, and near the focus of an unapertured laser beam, the intensity is also Gaussian and the wave fronts are flat and parallel [36]. For a diffraction limited lens, the diameter of the laser beam at the focus d_1 and the depth of field d_2 are given by [36]

$$d_1 = \frac{2\lambda}{\pi}\left(\frac{f}{B}\right) \tag{13.21}$$

$$d_2 = 4\lambda\left(\frac{f}{B}\right)^2 \tag{13.22}$$

where the spot diameter d_1 and beam diameter B are both measured at the $1/e$ points of the intensity distribution.

Assuming a simple symmetrical system such as that in Fig. 13.2(a) or 13.2(b), the scattering volume is ellipsoidal with a maximum diameter in the direction of the flow of $\sqrt{2}\,d_1/[\cos(\delta/2)]$ [31]. The time required for a particle to pass through the volume is given approximately by

$$\Delta t \cong \frac{\sqrt{2}\,d_1/\cos(\delta/2)}{v} \qquad (13.23)$$

and, from Eq. (13.20), the signal spectral width is given approximately by

$$\Delta \nu \sim \frac{1}{4\pi\,\Delta t} = \frac{v\cos(\delta/2)}{\sqrt{2}\,4\pi\,d_1} \qquad (13.24)$$

Finally, utilizing Eq. (13.21) for d_1 and Eq. (13.11) for ν_D we get, for $\Omega = \pi/2$

$$\frac{\Delta \nu_D}{\nu_D} \sim \frac{\sqrt{2}}{32}\left(\cot\frac{\delta}{2}\right)\frac{B}{f} \qquad (13.25)$$

Equation (13.25) is a lower bound on $\Delta\nu_D/\nu_D$ due to inequality (13.20), however we note that the functional form is essentially the same as in Eq. (13.16).

A result similar to Eqs. (13.16) and (13.25) may also be derived by considering the Fourier spectrum of a segment of sine wave of finite length. From these considerations, the term *transit time* broadening has arisen for the phenomenon described here as systematic broadening.

A third explanation of the systematic broadening arises from the literature on Doppler radar. For a signal generated by simultaneous scattering from a large number of particles, scattered bursts from the various particles arrive at the detector with randomly varying phase. The statistical averaging of these phase variations, which occurs in the detector, gives rise to spectral broadening, often referred to as *ambiguity* broadening [33, 34]. This theory is too complex to treat in any detail in this work. The results are, however, in good agreement with the simple arguments presented above.

Figure 13.4 is a plot of signal bandwidth, measured at the half power point, vs. Reynolds number, for velocity measurements on the centerline of a circular duct. At very low Re (small Doppler frequency, ν_D) the major contribution to signal bandwidth is due to the bandwidth of the spectrum analyzer employed. As the Doppler frequency becomes large compared with the instrument bandwidth, the relative bandwidth ($\Delta\nu_D/\nu_D$) approaches a constant value in laminar flow, as predicted by Eq. (13.16). The data in Fig. 13.4 were taken employing two spectrum analyzers with bandwidths of 6 and 200 Hz respectively. The data

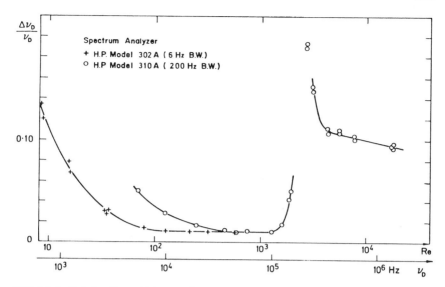

FIG. 13.4 Relative line widths as a function of Reynolds number.

corresponding to the analyzer with the larger bandwidth are considerably broadened at low Doppler frequency but approach the same value of $(\Delta\nu_D/\nu_D)$ as the data from the small bandwidth analyzer as ν_D get large. The sudden increase in bandwidth at Re $\sim$ 2300 is due to the transition to turbulent flow.

The net result of the above is that the signal broadening, $(\Delta\nu_D/\nu_D)$, is proportional to the cotangent of the scattering angle δ. Thus, to minimize broadening, δ should be made as large as the particular application will allow. This is in opposition to the requirement that δ be as small as possible to obtain high signal intensity. The actual design of a laser Doppler system depends on the particular application at hand and no generalization can be made which will satisfy all requirements completely.

Another factor that influences the signal band width is a velocity gradient in the flow. The magnitude of the broadening due to gradients can be estimated from Eq. (13.10), from which we get

$$\frac{\Delta\nu_D}{\nu_D} = \frac{1}{v}\frac{\partial v}{\partial y}\,\Delta y \qquad (13.26)$$

where Δy can be interpreted as the dimension of the scattering volume perpendicular to the direction of the velocity component measured. The broadening predicted by Eq. (13.26) must, in general, be added to the result of Eq. (13.16). If a large velocity gradient is present, Eq. (13.26) will, in fact, represent the major cause of signal broadening. For slowly varying or uniform flows, gradient broadening may be neglected, as is the case for the data shown in Fig. 13.4.

13.2.5 Signal-to-Noise Ratio

Numerous investigators have studied the relationships between various system parameters and the signal to noise ratio (SNR) in the Doppler signal [29-34]. There are so many contributing factors that all analyses are necessarily approximate; however, a number of useful facts have arisen from these studies, some of which are mentioned in an earlier section.

All systems are sensitive to the quality of the optics employed and to their degree of alignment. Crossed beam systems in either reference beam or dual scatter mode require accurate positioning of the focal points of the two beams to ensure that the beams intersect and that their focal points coincide. Crossed beam systems are inherently easier to align than geometries where a second beam is extracted before or after scattering occurs from the illuminating beam. In addition, seeing the beam crossing can be of considerable value when setting up the apparatus.

Analyses [29, 33] and experiments have shown that the SNR is improved by increasing the scattered intensity. In addition, the analyses tend to predict that maximum SNR is obtained when the scattered and reference beams are equal in a reference beam system and when the two illuminating beams are equal in a dual scatter or fringe system. In practice [32], equal beams give maximum SNR for dual scatter systems but, in reference beam systems [12], it is often found advantageous to use a somewhat stronger reference beam.

As mentioned earlier, maximum SNR is obtained when the detector is on the axis of symmetry of the optical system. In addition, some gain in SNR can be achieved by using a collecting lens with large aperture followed by a pin hole spatial filter to remove light which does not come from the scattering volume. In any mode of operation, forward scatter systems will have better SNR than back scatter systems because of the greater light scattering.

For continuous and individual particle signals there are optimum particle concentrations that should be used, as mentioned in a previous section. Using high particle concentrations, reference beam systems will give better SNR, while at low particle concentrations, the dual scatter geometry is better [7].

Finally, the SNR is greatly affected by the instrumentation employed. No specific guidelines can be given here because of the wide variety of detectors, signal processors and other apparatus employed. Use of Bragg cells or other frequency shifters is often advantageous for better SNR and for other reasons which are discussed below.

13.2.6 Effect of Path Length

Another requirement in mixing the two light beams is that the optical path lengths of the scattered and reference light beams should be the same or differ by integral multiples of $2L$, where L is the laser resonator cavity length. However, the allowable difference in path length is much larger for lasers than

RELATIVE MODE INTENSITY $(I/\bar{I})$

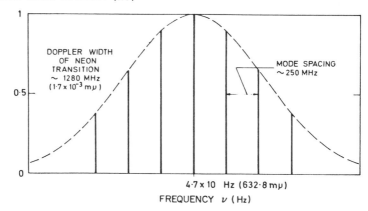

FREQUENCY ν (Hz)

FIG. 13.5 HeNe laser output spectrum, 60 cm length seven excited axial modes.

for noncoherent light sources owing to the high spatial coherence of the laser radiation. For a given laser cavity, the allowable difference in path length varies inversely with the number of excited modes in the laser output. For example, employing a sixty cm laser with seven excited axial modes (Fig. 13.5), a path length difference of ten cm results in signal intensity loss of about 20% whereas at forty cm the signal is no longer detectable. Curves for relative signal amplitude as a function of path length difference, for lasers with one to twenty-one axial modes are available [37].

13.2.7 Optical Geometry

Figure 13.2 illustrates three fundamentally different LDA geometries, as briefly discussed above. Various investigators have employed numerous variations on the three basic systems, where the number of alternative geometries available is essentially equal to the number of investigators who have utilized LDA techniques. We shall discuss some typical geometries in somewhat more detail than is given in Fig. 13.2, whereby a few commonly used devices and techniques can be illustrated.

Figure 13.6 illustrates three reference beam LDA systems which employ basically different means for accomplishing the beam superposition. The figure is drawn to show only the basic geometry of the alternative systems. Placement of mirrors, filters, lenses, etc. in the figure are representative of particular systems which have been successfully employed, however, the geometry can be varied considerably from that shown.

The design in Fig. 13.6(a) accomplishes the beam superposition quite simply with the added benefit that one can readily detect visually the point at which

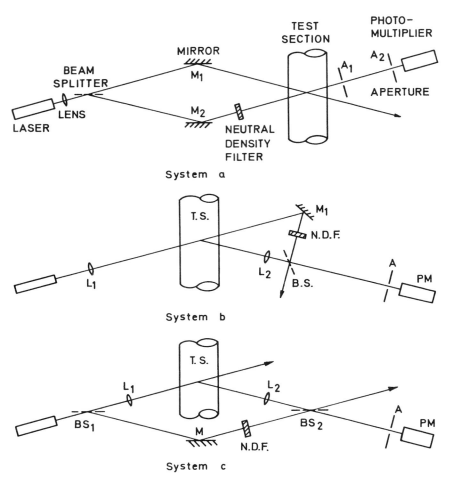

FIG. 13.6 Typical reference beam LDA systems.

the velocity measurement is being made. The laser beam is split into two beams which are focused at the same point in the flow. The photomultiplier is placed in the reference beam behind two apertures which are positioned so that the reference beam just passes through them. Only that portion of the scattered light that comes from a small region within the common focal point will fulfill the coherence requirements so that interference will occur. Aperture A_2 acts as a shield for noncoherent scattered light and background light. Actually, this design can be used with only one aperture. The intersection of the two beams is the region at which the velocity is measured and can easily be observed for alignment. To obtain the optimal signal, fine adjustment of the beams can be made while monitoring the photomultiplier signal on an oscilloscope.

The system illustrated in Fig. 13.6(b) focuses the full laser beam at a point in the flow. Light scattered from the focal region is then recombined with the reference beam which is the transmitted portion of the original beam. The system in Fig. 13.6(c) does essentially the same thing except that the reference beam is split off before scattering and follows a path outside the test section to be recombined with the scattered light. Both of these systems, 13.5(b) and 13.6(c), accomplish the superposition by directing both the scattered and reference beams at the splitter plate in such a way that the reflected portion of the reference beam is coincident with and parallel to the transmitted portion of the scattered beam. This can be very difficult to accomplish because of the severe alignment criteria which must be met in combining the two beams. It should also be noted that systems 13.6(a) and 13.6(c) allow for easy adjustment of beam path length difference for optimal signal as previously discussed. System 13.6(b) must be set up so that the paths differ by $2jL$, where j is an integer, if one wishes to get the maximum possible signal.

Figure 13.7 illustrates three dual scatter geometries, utilizing a single focusing lens in each case, but with different beam splitting techniques to obtain parallel beams. Figure 13.7 illustrates forward scatter systems that can easily be converted to reference beam systems if the detector is placed in one beam as shown. Figure 13.7(c) illustrates both forward scatter and back scatter options in one system. In addition, the geometry in Fig. 13.7(c) may be used to obtain two components of $\mathbf{v}$ by rotating the beam splitter. Flow scanning may be accomplished in the systems illustrated in Fig. 13.7 by translating the test section or by translating the focusing lens.

Numerous other systems have been employed including various types of two- [27] and three-component [28] systems. Compact integrated optical packages are commercially available from a number of sources and numerous companies and individuals are involved in custom design and construction of LDA systems. However, only a minimum amount of optical equipment and expertise are required, and individually assembled systems are often used. Signal analysis is, however, a much more difficult problem, as discussed in detail below.

13.2.8 Scattering Volume

An important advantage of the LDA technique is the ability to obtain highly localized velocity measurements. The size of the scattering volume, defined as the volume from which scattering is observed, can be estimated from the geometry of the focusing and collecting optics. For most crossed beam systems the scattering volume is the region of intersection of the laser beams. In particular, the scattering volume is an ellipsoid [31], as illustrated in Fig. 13.3, with major axes given by

$$d_x = \frac{\sqrt{2}\, d_1}{\cos\left(\delta/2\right)} \tag{13.27}$$

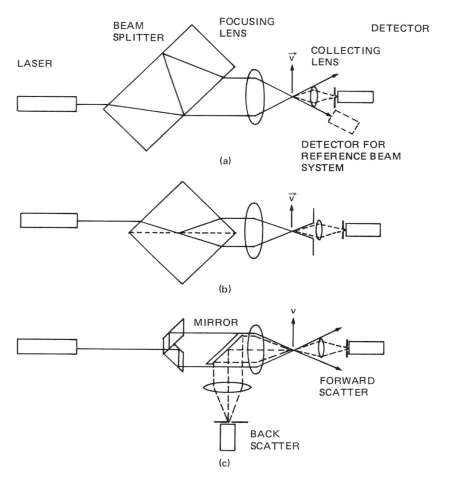

FIG. 13.7 Typical dual-scatter LDA systems.

$$d_y = \frac{\sqrt{2}\, d_1}{\sin{(\delta/2)}} \qquad\qquad (13.28)$$

$$d_z = d_1 \qquad\qquad (13.29)$$

where d_1 is the diameter of the individual beams at the focus. One can arrive at the above result by considering coherence conditions between the scattered and reference light by the classical Doppler approach or, equivalently, by determining the region in which fringes exist in the fringe theory explanation.

It is generally desirable to make the scattering volume as small as one can to obtain the best possible spatial resolution. Since the velocity measured from the Doppler signal is a weighted average of the velocity distribution in the scattering volume, the signal broadening due to gradients in the flow is proportional to the size of the volume—cf. Eq. (13.36). Therefore, to accurately measure very fine features of the flow such as turbulence scale, to measure flow very near to a boundary, and to minimize gradient broadening, we wish to minimize the scattering volume dimensions. Sometimes this can be accomplished by proper orientation of the plane of the two crossed beams relative to the flow, but in general the proper focal length f and beam spread δ must be optimized.

For illustration, assume a flow situation such as that illustrated in Fig. 13.3 where the velocity $\mathbf{v}$ is in the x direction with variation of $\mathbf{v}$ in the y direction only. With the laser beams in the x-y plane symmetric to $\mathbf{v}$, the scattering volume diameter in the direction of flow variation is d_y, as given by Eq. (13.28). For this geometry, minimizing the gradient broadening requires a minimum for d_y which in turn requires a minimum for d and/or a maximum for sin $\delta/2$. From Eq. 13.21 a minimum for d requires a short focal length f in the focusing optics. Therefore minimizing the scattering volume requires a large beam spread angle δ and a short focal length lens.

On the other hand, the ambiguity broadening from Eq. (13.19) goes as $B/f \cot \delta/2$. Since it is also generally desirable to minimize this broadening, a large value of f and a small value of $\cot \delta/2$ are required. The requirement on δ is the same as that for maximum spatial resolution, but the requirement on f is exactly the opposite. Therefore, a trade-off must be made in the choice of f (and the consequent scattering volume dimensions) based on the requirements of the application in question. In general, the problem of selecting the optimum scattering volume reduces to balancing gradient broadening against ambiguity broadening.

It is possible to reduce the effective size of the scattering volume from that described in Eqs. (13.27), (13.28), and (13.29) by using collecting optics that focus on a small region inside the beam intersection volume. A pin-hole spatial filter in the collecting optics helps to reduce the effective scattering volume and eliminates off axis light, thereby improving the SNR. In general, however, anything done to decrease the scattering volume will increase the ambiguity broadening.

13.2.9 The Doppler Signal

One can consider two different mechanisms by which Doppler signals arise. These have been called *coherent* and *noncoherent modes* [7]. Coherent signals arise from optical beating of light scattered through different angles which usually implies light scattering from different particles in the measurement volume. In order for a beat signal to result, the two beams of light must be coherent, which means that their phase difference must not vary appreciably over the exposed area of the detector. The consequence of the coherence requirement is to restrict the size of the detector area that can effectively be used, thereby limiting the available signal intensity. Increasing the detector area beyond that imposed by the coherence condition merely adds to the noise and D.C. level of the signal reducing the SNR.

Noncoherent signals arise from light scattered from a single particle. One might consider the modulation of intensity of light scattered by a particle passing through a sequence of real fringes in the dual scatter system. The useful detector aperture is not limited for such signals, as long as a very small number of particles contribute to the signal at any time [7]. Since large collection apertures can be employed in noncoherent systems, higher intensity and greater SNR can be achieved than for coherent beating under similar conditions [7, 32]. There are, in fact, practical limitations on the collection angle that should be employed for noncoherent systems [32], due primarily to lens aberations and the extreme variation in scattered intensity with angle associated with small particles. The essential point still holds that noncoherent systems are inherently capable of larger collection angles and better SNR for low particle concentration.

Virtually all LDA systems can operate coherently, providing small apertures are used. In a given case, coherent and noncoherent signals may exist simultaneously, where the dominant mode would depend on the specific system employed. For high particle concentration, the coherent mode of operation is best, especially if a reference beam system is used. If, however, low particle concentrations are employed, the noncoherent mode, with a fringe system, is preferable. The choice of which mode to employ may be dictated by specific requirements for high or low particle concentration or by the ultimate choice of signal processing techniques as discussed below.

The nature of the Doppler signal is highly dependent on the particle concentration, both in how it arises and in the resultant output from the detector. At low particle concentrations, where particles traverse the scattering volume one at a time, individual bursts of Doppler signal will appear at the detector output as illustrated in Fig. 13.3(b).

When scattering occurs simultaneously from a large number of particles, numerous bursts of scattered light with randomly varying phase and amplitude reach the detector. The detector output is proportional to the square of the instantaneous intensity incident upon it, generating a Doppler signal that is modulated in both frequency and amplitude. The multiple-particle signal is often referred to as a *continuous wave* (CW) signal, which is a misnomer, because there are periods when the sum of the randomly varying individual bursts cancel. Regions of zero signal amplitude are called signal *fall-out* and tend to complicate signal analysis procedures.

13.3 VELOCITY AND FREQUENCY LIMITATIONS

The range of velocities that lies within the capabilities of the laser-Doppler technique is limited by the inherent properties of the laser. The cavity that constitutes the active element of the laser is an optical resonator. Only those frequencies that satisfy the resonance condition, that the resonant cavity length is an integral multiple of one half the wave length, are amplified and thus constitute the output of the laser. The resonance condition is

$$L = \frac{j\lambda_j}{2}$$

$$\nu_j = \frac{jc}{2L} \tag{13.30}$$

where j is an integer, L is the resonator length, and λ_j and ν_j are corresponding possible resonant wave lengths and frequencies, respectively. In the case of a He-Ne laser, those frequencies which lie within the Doppler width of the Neon transition at 632.8 nm and which satisfy the resonance condition, Eq. (13.23), are amplified and emitted. The resulting spectrum consists of several discrete peaks or modes centered at 632.8 nm and (Fig. 13.5) separated by

$$\Delta\lambda = \frac{\lambda_0^2}{2L} \tag{13.31}$$

Each mode is extremely narrow; the bandwidth can be of the order of 10 Hz. Any velocity resulting in a Doppler shift that is less than this would not be easily resolvable. With this as a criterion, the minimum velocity measurable by Doppler velocity technique may be estimated:

$$\nu_D = \frac{n\mathbf{v}}{\lambda_0} \cdot (\hat{\mathbf{n}}_{\mathrm{sc}} - \hat{\mathbf{n}}_i) \sim \frac{\upsilon}{\lambda_0}$$

$$\upsilon_{\min} \sim 10 \ (\mathrm{Hz}) \ 632.8 \times 10^{-7} (\mathrm{cm})$$

$$\upsilon_{\min} \sim 10^{-3} \ \mathrm{cm/sec}$$

The mode separation given by Eq. (13.24) gives an effective upper limit for velocities, in that if the velocity is high enough to produce a Doppler shift of the order of the mode separation, intermodal beats will occur. For a multi-mode He-Ne laser with cavity length of one meter, the upper velocity limit from this criterion is:

$$\Delta\nu_{\mathrm{ms}} \sim \frac{c}{2L} \sim 150 \ \mathrm{MHz}$$

$$\upsilon_{\max} \sim \Delta\nu_{\mathrm{ms}} \cdot \lambda_0 \sim 100 \ \mathrm{m/sec}$$

The introduction of a laser etalon would greatly reduce the problem of intermodal beats. In general, the maximum usable regional frequency would be limited by the frequency limits of the detector and the signal analysis equipment.

13.3.1 Detector

Both photomultipliers and photodiodes can be used as detectors in LDA applications. Although each has certain advantages, the photomultiplier is probably the best all around choice if the higher costs are acceptable. The type of photomultiplier that should be used depends on the type and power of the laser used and the sensitivity required. Photomultipliers are available that cover the entire visible spectrum from ultraviolet into the near infrared. The He–Ne gas laser is most widely used for LDA applications. Use of Argon ion lasers permits a more intense input beam which is also at a wavelength nearer to the peak sensitivity range of available photocathodes. For either of the above lasers, a type S-20 tube (or extended S-20 with a He–Ne laser) works well, where 10 to 14 dynode stages are normally used to obtain high gain. Recently, considerable progress has been made in the design of photomultipliers with substantial gains in noise reduction and sensitivity.

Photodiodes are often employed in LDA applications because of their lower cost and simplicity of operation. A good photodiode can be obtained for a fraction of the cost of a photomultiplier and in addition the power supply needed to drive it is simpler, safer, and less costly. Recently, integrated diode-amplifier packages have become commercially available that are approaching the sensitivity and noise characteristics of a photomultiplier, but with considerably lower frequency capabilities. The simplicity of operation and lower costs make the integrated diode-amplifier units competitive and possibly preferable in some cases, especially for systems that do not require the high sensitivity and frequency response capability of a photomultiplier. An avalanche photodiode, however, does provide some gain and can be used at high frequencies.

Whether a photomultiplier or diode is used as a detector, considerable benefits are achieved by using a narrow band interference filter as the detector window. The pass-band of the filter should be centered at the frequency of the laser employed to minimize the effects of background radiation. When a filter is used, measurements may be obtained under room lights without interference and the SNR is vastly improved. It is also important to supply a shielded, light tight housing for the detector, especially for a photomultiplier. Cooling is not normally required for either type of detector.

13.3.2 Directional Ambiguity

A weakness common to simple LDA systems is a 180 deg measurement ambiguity in the flow direction. By this we mean that whereas the Doppler frequency is proportional to the magnitude of the velocity along a well defined line in space, it is insensitive to its direction or sign. If the flow direction is known, the ambiguity is of little consequence, but in complex or fluctuating flows where flow reversals occur, the directional ambiguity can be a severe limitation. A recent study [38] reviews alternative means by which the directional ambiguity may be removed, the principal points of which are summarized here.

The most commonly used technique for removing the directional ambiguity involves frequency shifting of one of the laser beams by a known amount ν_f (or possibly by shifting both beams with a net difference of ν_f). As a consequence of this modification, a velocity of $\pm v$ will give rise to a Doppler signal at $\nu_f \pm \nu_D$, where $v \propto \nu_D$, and where ν_f is chosen larger than the maximum expected Doppler shift so that both the magnitude and the direction of the flow may be deduced. An alternate explanation of this phenomenon can be deduced from the fringe theory, whereby superposition of a frequency shifted beam and an unshifted beam results in a fringe pattern whose fringes move at a constant velocity through the scattering volume. Thus, the frequency of the resultant Doppler signal is seen to be a direct consequence of whether the particles are moving in the same or opposite direction as the fringes.

Frequency shifting may be accomplished with three different devices: acousto-optic modulators (Bragg cells) [39, 40], rotating diffraction gratings [41, 42, 43], and other electro-optical devices [44, 45]. Each of these is best suited to a particular frequency range which should be considered in selecting a modulator for a given application. Other factors that weigh upon such a choice are cost, ease of operation, and availability.

An acousto-optic modulator or Bragg cell [39, 40] consists of a liquid or solid medium in which a traveling ultrasonic wave is maintained by excitation with a piezoelectric crystal driven at a frequency ν_f. When a laser beam is passed through the ultra-sonic wave, diffraction occurs due to the periodic density variation in the cell. Several diffracted beams may be generated, depending on the strength of the ultrasonic wave and the angle of incidence of the beam. If the incident angle is equal to the Bragg angle, as much as 90–95% of the incident energy can be diffracted into the plus or minus first order beams. The frequency shift introduced in the diffracted beam is $\pm \nu_f$, the driving frequency, where the sign of the shift depends on whether the $+1$ or -1 order beam is utilized.

In order to obtain sufficient beam separation in a reasonable distance and for efficient operation, Bragg cells should operate at frequencies greater than about 15 to 20 MHz. Most commercially available units are driven at 30 or 40 MHz. These frequencies are generally too high for most signal analyzers, which requires electronic downshifting to the range of the signal analyzer. Alternatively, a second Bragg cell, driven at a slightly different frequency, may be inserted in either of the beams using, for example, one cell at 40 MHz and another at 35 MHz, in which case the Doppler signal would appear centered at the difference frequency, 5 MHz. In this way one can optically downshift the signal to match the frequency capabilities of a given signal analyzer.

A second, commonly used technique for frequency shifting employs a moving diffraction grating [41, 42, 43]. The result is similar to that obtained with the Bragg cell, except that only 10–15% of the incident power is diffracted into the first order beams. The system is less costly than an acousto-optic modulator and simpler to operate. A rotating grating system can operate at very low frequencies which can easily be varied by changing rotational speed. Difficulties encountered include sensitivity to mechanical vibrations and speed fluctuation, beam distortion, and low efficiency. Despite these problems, grating systems have proven useful for some applications.

Frequency shifting can also be accomplished with electro-optic modulators [44, 45]. An electro-optic modulator consists of one or more crystals in which a rotating electric field is generated whereby the incident circularly polarized light is either accelerated or decelerated depending on the direction of the relative rotation of the two fields. The transmitted light can be upshifted or downshifted by proper choice of rotation. Frequencies from a few KHz to over a hundred MHz have been employed. Problems encountered are the presence of harmonics and difficulty in maintaining high conversion efficiency. In addition electro-optic modulators require high voltage power supplies and are generally more difficult to use than Bragg cells or diffraction gratings.

It should be pointed out that frequency shifting is used for several reasons, only one of which is removal of the directional ambiguity. An equally important reason is to shift the signal away from zero frequency and the overpowering noise found there. In addition, the bandwidth requirements of the signal processing apparatus can be considerably relieved. For example, a Doppler signal that varies between 0 and 1 MHz could be made to vary between 9 and 11 MHz (for positive or negative velocities) if an offset of 10 MHz is used. For tracking devices in particular, the latter may pose no difficulty whereas continuous tracking from 0 to 1 MHz has not been accomplished. Consequently, for measurements of turbulence parameters, frequency off-setting is often required [32, 34].

13.3.3 Signal Analysis

The frequency spectrum of the photomultiplier current can be analyzed in several ways depending upon the information sought and the equipment available. Perhaps the simplest technique is to display the voltage across the photomultiplier load resistor directly on an oscilloscope. The Doppler frequency can be determined directly from the oscilloscope screen if the velocity is steady or slowly varying, but the accuracies attainable are limited, and only the time average velocity at a point can be determined in this way. Even for turbulent flows, the signal can be monitored with an oscilloscope while measurements are made with a more suitable instrument. The oscilloscope is also useful when making fine adjustments in the optics since changes in signal quality are readily observed.

Many oscilloscopes accept plug-in spectrum analyzers that allow visual display of the signal spectrum directly on the screen. The time base is converted to read directly in frequency and the ordinate represents signal intensity at the corresponding frequency. The accuracy of this technique is usually limited to about 2–5%.

A standard wave analyzer is a narrow band audio and radio frequency volt meter. It performs basically the same function as the plug-in spectrum analyzer described above except that the tuning mechanism is hand or motor driven and the voltage-frequency display is usually recorded on a strip chart or *x-y* recorder.

Figure 13.8 shows, in simplified form, the type of Doppler frequency spectra obtained for laminar [12] and turbulent [15] flow. The instrument bandwidth is usually fixed although several bandwidths may be available in some models. Wave analyzers are available in a range of frequencies well into the megaHertz range. Due to the hand tuning feature of wave analyzers, they are slow and applicable only for steady and slowly varying signals. They are however quite accurate and readily calibrated. Turbulence measurements can be made from the bandwidth of the recorder Doppler signal [15]; however, the results are not very precise, especially at low turbulence intensities. In spite of its limitations, spectrum analysis is quite useful in many applications and, since such equipment is generally available and simple to use, spectrum analyzers will probably continue to play a significant role in LDA signal analysis.

In recent years, various types of phase-locked loops and frequency-locked loops [48, 49, 50] have come into general use for signal analysis. Although these vary somewhat in design and operation, their basic function is the same. In both systems the Doppler signal is compared with a reference signal from a voltage controlled oscillator (VCO) built into the instrument. If the signal frequency differs from that of the VCO, a feedback control adjusts the control voltage on the VCO to force its frequency to match that of the Doppler signal.

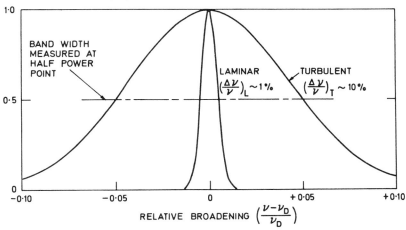

FIG. 13.8 Turbulent broadening of the Doppler signal.

When the loop is in lock the VCO frequency is the same as that of the Doppler signal and the VCO control voltage is linearly proportional to the input signal frequency. The output from a phase- or frequency-locked loop is, therefore, an analog voltage that is proportional to the instantaneous fluid velocity. The output may be further analyzed on-line or from tape recordings from which the mean velocity and various turbulence parameters may be determined as illustrated in Fig. 13.9.

In practice, certain characteristics of LDA signals complicate the design and application of phase and frequency locked loops. Perhaps the biggest problem is caused by the signal fall-out phenomenon wherein the signal amplitude falls to zero or below the minimum discrimination level required by the loop. A clipper amplifier ahead of the loop removes most amplitude fluctuations but some periods of fall-out remain where a simple loop will lose "lock." When this happens, the VCO will suddenly return to its "free running" frequency and

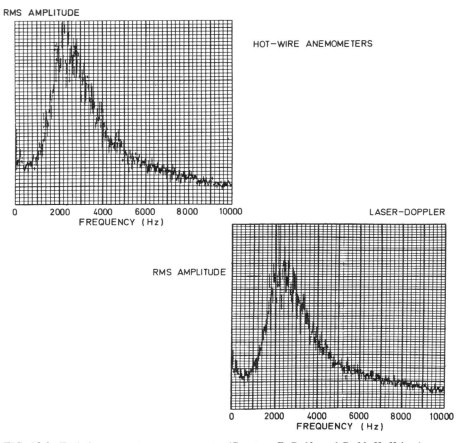

FIG. 13.9 Turbulence spectra measurements. (Courtesy E. Rolfe and R. M. Huffaker.)

either manual or automatic search procedures must be initiated to allow the loop to recapture the signal and return to the locked mode. The result is a highly discontinuous output signal.

The fall-out problem is dealt with by various means. In some cases it is possible to damp the response of the loop such that it responds slowly enough to coast through a fall-out period, providing the frequency and duration of the fall-out periods are not excessive. This procedure is adequate for measuring the mean velocity in a steady or slowly varying flow; however, most turbulent flows could not be tracked by such an instrument. Other more sophisticated means have been developed to ameliorate the fall-out problem. For example, systems have been designed and built [48] that prevent the loop from losing lock during fall-out by artificially forcing the VCO to maintain the frequency it had when the signal vanished. When the signal returns, the loop is again freed and adjusts itself to compensate for any frequency change that may have occurred during the fall-out period. The output signal thus follows the input except for brief flat spots and small step readjustments which occur each time fall-out occurs. Since the fall-out phenomenon is random, the mean velocity and turbulence intensity measurements are unaffected if sufficient averaging times are used. Filtering can also smooth the output; however, proper adjustment of the filter pass-band is critical.

A second problem encountered in utilizing phase- and frequency-locked loops is caused by ambiguity broadening. Most loop systems require continuous (multi-particle) Doppler signals for efficient operation which means ambiguity broadening is encountered. This broadening causes a flattening of the turbulence power spectrum at the high frequency end, thus masking measurements in a region of the frequency spectrum of considerable interest. It has recently been reported [51] that the problem of ambiguity broadening may be eliminated in the measurement of turbulent power spectra obtained with phase and frequency locked loops by using two filters in such a way that the ambiguity broadening is subtracted out. Though detailed experimental verification of this technique is not yet available, this approach is potentially of great significance.

Phase-locked loops are available from numerous manufacturers in integrated circuit form. A simple phase- or frequency-locked loop tracker circuit can be assembled at very little cost, but incorporation of fall-out protection is much more complex. Complete LDA signal trackers are commercially available that have built-in fall-out protection and signal conditioners. In addition, some trackers also utilize a frequency verification circuit that uses a comparison scheme whereby the frequency obtained from the first few cycles in a Doppler burst is compared with that obtained from a larger number of cycles from which a decision is made as to whether the measurement is valid. Similar circuitry is employed in the "counter" type LDA analyzers described below. Although most trackers work best on C.W. signals, trackers that utilize the verification circuit also work on single-particle, burst-type signals.

Various types of frequency counters have been employed [52, 53, 54] for measurement of the Doppler frequency. For CW signals, the same problems of fall-out and ambiguity broadening are encountered when using a counter technique as are encountered in the application of the tracking systems described above. A clipper amplifier and an amplitude detector which triggers the counter only when the signal is above a minimum level can effectively eliminate the fall-out problem, but ambiguity broadening still is a serious problem.

In recent years, considerable effort has gone into the development of frequency counter systems for use in analyzing single particle LDA signals. Individual Doppler bursts are analyzed by frequency counting and various schemes are employed for verifying the measurement, such as the comparison technique described above. It is generally believed that ambiguity broadening does not occur [52, 55] in single particle LDV signals; however, a number of other problems are encountered [55] that appear to be unique to the counter analyzer systems. The nature, the seriousness, and the remedy of these problems are still matters of debate and developmental effort.

The frequency counter systems vary considerably in design, but most current units tend to be electronically complex and require peripheral apparatus (such as a direct computer hook-up). As a consequence, the counter systems can generate considerable data which can be manipulated in many ways. Because many bursts must be analyzed and averaged for a given measurement, storage is generally required. Even so, the system response is usually sufficient to follow rapidly fluctuating flows.

The photon correlation technique for analysis of LDA signals is of recent orign. It has been demonstrated [56, 57] that LDA signals may be analyzed by photon correlation spectroscopy under very low light level conditions where other signal processing devices fail. This technique derives Doppler information from observations of individual photons scattered by particles passing through the scattering volume in a real fringe system. The output from the analyzer is a correlation function that contains information on the mean frequency, the ambiguity broadening, and the turbulence intensity. Because of its recent origin and because complicated and expensive equipment is required, photon correlation spectroscopy has not received the attention it may deserve, particularly in light of its apparent potential for low light level applications.

Finally, we mention the Fabry-Perot interferometer technique [46, 47] which is an optical spectrum analyzer as opposed to the electronic wave analyzer described above. In this system the beat signal is not directly produced on a photodetector, but rather the difference in wavelength between the original and Doppler shifted light is measured. This is quite suitable for high velocity, steady flows. The Fabry-Perot interferometer consists of two partially reflecting mirrors a given distance apart. Light that is transmitted is governed by the plate spacing and the frequency of the light. By moving one mirror, a variable frequency filter is obtained. By oscillating the mirror, the frequency spectrum of the incident light is scanned and the Doppler shift can be measured.

13.3.4 Scattering Centers

With an LDA the velocity of a flowing fluid is measured indirectly by measuring the velocity of particles (scattering centers) carried in the fluid. The validity of the assumption that the particle velocity is the same as the velocity of the fluid that surrounds it is of course dependent upon the nature of the particle, the fluid, and the flow. In a steady uniform flow a small particle that is neutrally bouyant will be carried along with the same velocity as the fluid. For turbulent flows however, or in other situations where the fluid velocity changes or when there are large velocity gradients, velocity differences can exist.

To minimize the drag force the fluid must exert on the particle, the particles should be as small as possible. If, however, the particles are too small, Brownian motions will occur and will cause broadening of the frequency spectrum of the scattered light.

To prevent particles from settling, their density should be as near to the fluid density as possible. An estimate of settling velocity can be made by applying Stokes equation for drag on a sphere at low Reynolds number [58]. Equating the drag force to the bouyancy force, one obtains.

$$v_s = \frac{(\gamma - 1)b^2 g}{18\eta} \tag{13.32}$$

where γ is the particle specific gravity (relative to the solvent fluid), and v_s is the settling velocity. With 0.5 μm polystyrene spheres (specific gravity 1.04) in water, Brownian motion gives rise to Doppler broadening of only 20 Hz [3] and the settling velocity, calculated from Eq. (13.32) is of the order of 0.2 mm/hr. Particles of this type have been satisfactorily employed by several investigators [3, 4, 12, 13] at concentrations of 10–50 parts per million.

It is more difficult to estimate the effect of turbulence in regard to velocity lag between fluid and particle. The micro-scale of isotropic turbulence is given [59] by:

$$s^2 = 15 \frac{D l_e}{\text{Re Tu}} \tag{13.33}$$

where D is the duct diameter, l_e the turbulence macroscale (approximately one half of the duct diameter D), Re is the Reynolds number based on duct diameter and average velocity, and Tu is the turbulence intensity; for $D = 1$ cm, Re $= 10^5$ and Tu $= 5\%$; $s = 0.04$ mm. Thus, the scale of the turbulence is much larger than the particles described above. A study [50] on the behavior of particles in turbulent flow indicates that the small particles usually employed follow satisfactorily the turbulent fluctuations in water. For each particle application similar calculations would have to be done to verify whether the particles involved follow fluid motions closely enough to give information about the turbulent motion of the fluid employed.

Although it is often possible to obtain Doppler signals from scattering produced by naturally occurring contaminants in the fluid, some type of "seed" particles are generally required. Many different types of particles have been used, including polystyrene spheres [3, 4, 12, 13, 62], milk [15], and blood cells [23] in liquids; and smoke [17, 21], oil drops [57, 61], water drops [61], and D.O.P. [61] in air. Although any type of particle can generate Doppler signals, optimum results require particles that are smaller than the fringe spacing but large enough to avoid Brownian motions, as discussed above. Seeding is seldom a problem in liquids. However, seeding of gas flows, especially in high velocity applications, is often difficult.

13.4 CONCLUSION

In a little more than a decade the laser-Doppler anemometer has evolved into a sophisticated and versatile tool for the study of fluid mechanics. Velocity measurements have been obtained with LDA systems in a broad range of fluid flows which firmly establish its accuracy, versatility, and reliability. Through experimentation and analysis, the theory of the LDA has been well established and its capabilities and limitations are now relatively well known. Although some unanswered questions still exist, primarily with regard to interpretation of turbulent flow data, the LDA has apparently reached a point in its evolution where exploitation rather than development is coming to the fore.

NOMENCLATURE

A	aperture diameter
a	wave propagation speed
B	unfocused laser beam diameter
b	particle diameter
c	speed of light
c_0	vacuum speed of light
D	diameter of circular duct and length of side of square duct
d	distance between apertures A_1 and A_2
d_1	diffraction limited spot size of a focused beam
d_2	depth of field of focus
d_a	distance between apertures A_1 and A_2
d_x, d_y, d_z	axes of scattering volume ellipsoid
$E_{1,2}$	electric field amplitude
$E_{10,20}$	maximum of electric field amplitudes
f	focal length of lens
g	gravitational acceleration

$I(t)$	intensity of Doppler burst
I_i	incident beam intensity
I_{sc}	scattered beam intensity
I_r	reference beam intensity
i	photomultiplier current
j	an integer
L	length of laser cavity
l_e	turbulence macroscale
l	distance between photocathode and scattering volume
N	particle number density
n	index of refraction
$\hat{n}_{1,2}$	unit vectors
$\hat{n}_i$	unit vector in direction of incident light
$\hat{n}_{sc}$	unit vector in direction of scattered light
Re	Reynolds number based on bulk velocity $= \bar{U}D/\eta$
Re_c	Reynolds number based on centerline velocity $= U_c D/\eta$
s	turbulence microscale
t	time
Tu	longitudinal turbulence intensity $= u'/U_c$
$\bar{U}$	average or bulk velocity
U_c	longitudinal centerline velocity
u'	longitudinal turbulent velocity component
V_{sc}	scattering volume
$\mathbf{v}$	velocity
$\mathbf{v}_m$	maximum velocity detectable
v_s	particle setting velocity
v	magnitude of velocity $= |\mathbf{v}|$
x, y, z	coordinate directions
Δy	scattering volume dimension perpendicular to the direction of the velocity component measured
γ	specific gravity
δ	scattering angle $= \phi - \theta$
η	kinematic viscosity $= \mu/\rho$
θ	angle between velocity vector and incident light beam
λ	wavelength
λ_0	vacuum wavelength of laser
λ_i	wavelength of incident light
λ_j	wavelength for which resonance is attained
λ_p	wavelength of incident light as seen by the particle
λ_{sc}	wavelength of scattered light
μ	dynamic viscosity
ν_D	Doppler frequency
$\bar{\nu}_D$	average value of the Doppler frequency
$\Delta\nu_D$	bandwidth of the Doppler spectrum at the half power points

ν_f	frequency shift of single side band modulated system
ν_i	frequency of incident light
ν_j	frequency at which resonance is attained
ν_m	maximum frequency detectable
ν_0	frequency of laser
ν_p	frequency of incident light as seen by the particle
ν_{sc}	frequency of scattered light
ρ	density of fluid
ξ	phase angle
ϕ	angle between velocity vector and scattered light beam
Ω	angle between velocity vector and the bisector of δ, $\Omega = \phi + \theta/2$

Subscripts

i	refers to incident beam
sc	refers to scattered beam

REFERENCES

1. Hubble, E., A Relation between Distance and Radial Velocity among Extra-Galactic Nebulae, *Proc. Nat. Sci., Wash.*, vol. 15, p. 168, 1929.
2. Skilling, W. T., and Richardson, R. S., "Astronomy," rev. ed., Henry Holt, N. Y., 1951.
3. Cummins, H. A., Knable, N., and Yeh, Y., Observation of Diffusion Broadening of Rayleigh Scattered Light, *Phys. Rev. Letters*, vol. 12, p. 150, 1964.
4. Yeh, Y., and Cummins, H., Localized Fluid Flow Measurements with a He Ne Laser Spectrometer, *Appl. Phys. Letters*, vol. 4, p. 176, 1964.
5. Temes, C. L., Relativistic Consideration of Doppler Shift, *IRE Trans. Aero & Navig. Elect.*, vol. 6, p. 37, 1959.
6. Rudd, M. J., A New Theoretical Model for the Laser Dopplermeter, *J. Phy. E. Sci. Inst.*, vol. 2, p. 55, 1969.
7. Drain, L. E., Coherent and Noncoherent Methods in Doppler Optical Beat Velocity Measurement, *J. Phys. D. Appl. Phys.*, vol. 5, p. 481, 1972.
8. Goldstein, R. J., Measurement of Fluid Velocity by Laser-Doppler Techniques, *Appl. Mech. Rev.*, vol. 27, p. 753, 1974.
9. Durst, F., Melling, A., and Whitelaw, J. H., Laser Anemometry: A Report on Euromech 36, *J. Fluid Mech.*, vol. 56, p. 143, 1972.
10. Durst, F., and Zare, M., Bibliography of Laser-Doppler Anemometry Literature, Dec. 1974, available upon request from the *Technical Information Dept., DISA Electronics*, 779 Susquehanna Ave., Franklin Lakes, N.J. 07417.
11. Goldstein, R. J., General Bibliography—Measurement of Fluid Velocity By Laser Doppler Techniques, *Appl. Mech. Rev.*, extension of AMR Feature Article, Ref [8] above, 1974.
12. Goldstein, R. J., and Kreid, D. K., Measurement of Laminar Flow Development in a Square Duct Using a Laser Doppler Flowmeter, *J. Apl. Mech.*, vol. 34-E, p. 813, 1967.
13. Foreman, J. W., Lewis, R. D., Thorton, J. R., and Watson, H. J., Laser Doppler Velocimeter for Measurement of Localized Flow Velocities in Liquids, *Proc. I.E.E.E.*, vol. 54, p. 424, 1966.
14. Goldstein, R. J., and Hagen, W. F., Turbulent Flow Measurements Utilizing the Doppler Shift of Scattered Laser Radiation, *Phys. Fluids*, vol. 10, p. 1349, 1967.
15. Pike, E. R., Jackson, D. A., Bourke, P. J., and Page, D. I., Measurement of Turbulent

Velocities from the Doppler Shift in Scattered Laser Radiation, *J. Sci. Inst.*, vol. 1, p. 727, 1968.

16. Goldstein, R. J., Adrian, R. J., and Kreid, D. K., Turbulent and Transition Pipe Flow of Dilute Polymer Solutions, *I. & E. C. Fund.*, vol. 8, p. 498, 1969.

17. Foreman, J. W., George, E. W., and Lewis, R. D., Measurement of Localized Flow Velocities in Gases with a Laser Doppler Flowmeter, *Appl. Phys. Letters*, vol. 7, p. 77, 1965.

18. Jackson, D. A., and Paul, D. M., Measurement of Supersonic Velocity and Turbulence by Laser Anemometry, *J. Phys. E.*, vol. 4, p. 173, 1971.

19. Farmer, W. M., Hornkohl, J. O., and Brayton, D. B., A Relative Performance Analysis of Atmospheric Laser-Doppler Velocimeter Methods, *Opt. Eng.*, vol. 11, p. 24, 1972.

20. Huffaker, R. M., Jelalian, A. V., and Thomson, J. A., Laser-Doppler System for Detection of Aircraft Trailing Vortices, *I.E.E.E. Proc.*, vol. 58, p. 322, 1970.

21. Durao, D. F. G., and Whitelaw, J. H., Instantaneous Velocity and Temperature Measurements in Oscillating Diffusion Flames, *Proc. Roy. Soc. London-A*, vol. 338, p. 479, 1974.

22. Barrault, M. R., Jones, G. R., and Blackburn, T. R., A Laser Doppler Technique for Measuring Flow Velocities in High Current Arc Discharges, *J. Phys. E.*, vol. 7, p. 663, 1974.

23. Kreid, D. K., and Goldstein, R. J., Measurement of Velocity Profiles in Simulated Blood by the Laser-Doppler Technique, *Paper No. 4-2-95, presented at the ISA Joint Symp. on Flow*, Pittsburgh, 1971.

24. Denison, E. B., Stevenson, W. H., and Fox, R. W., Pulsating Laminar Flow Measurements with a Directionally Sensitive Laser Velocimeter, *AIChE J.*, vol. 17, p. 701, 1971.

25. Amenitskii, A. N., Rindevichyus, B. S., and Solovev, G. M., Measurement Using the Doppler Effect of Small Velocities in Flows Occurring in the Free Convection of Fluids, *Sov. Phys. Doklady*, vol. 17, p. 1078, 1973.

26. Durst, F., Melling, A., and Whitelaw, J. H., Low Reynolds Number Flow over a Plane Symmetric Expansion, *J. Fluid Mech.*, vol. 64, p. 111, 1974.

27. Orloff, K. L., Grant, G. R., and Gunter, W. D., Laser Velocimeter for Simultaneous Two-Dimensional Velocity Measurements, *NASA Tech. Brief 73-10267*, 1973.

28. Farmer, W. M., Determination of a Third Orthogonal Velocity Component Using Two Rotationally Displaced Laser Doppler Velocimeter Systems, *Appl. Opt.*, vol. 11, p. 770, 1972.

29. Mayo, W. T., "Laser-Doppler Flowmeters—A Spectral Analysis," Ph.D. thesis, Georgia Institute of Technology, 1970.

30. Edwards, R. V., Angus, J. C., French, M. J., and Dunning, J. W., Spectral Analysis of the Signal from the Laser-Doppler Flowmeter: Time Independent Systems, *J. Phys. D.*, vol. 42, p. 837, 1971.

31. Adrian, R. J., and Goldstein, R. J., Analysis of a Laser-Doppler Anemometer, *J. Phys. E.*, vol. 4, p. 505, 1971.

32. Durst, F. and Whitelaw, J. H., Optimization of Optical Anemometers, *Proc. Roy. Soc. London*, vol. 324, p. 175, 1971.

33. George, W. K., "An Analysis of the Laser-Doppler Velocimeter and Its Application to the Measurement of Turbulence," Ph.D. thesis, Johns Hopkins University, 1970.

34. George, W. K., and Lumley, J. L., The Laser-Doppler Velocimeter and Its Application to the Measurement of Turbulence, *J. Fluid Mech.*, vol. 60, p. 321, 1973.

35. Born, M., and Wolf, E., "Principles of Optics," 2nd ed., Permagon Press, Oxford, 1964.

36. Innes, O. J., and Bloom, A. L., Design of Optical Systems for Use with Laser Beams, *Spectra-Physics Tech. Bul. No. 5*, 1966.

37. Foreman, J. W., Optical Path Length Difference Effects in Photomixing with Multimode Gas Laser Radiation, *Appl. Optics*, vol. 6, p. 821, 1967.

38. Durst, F., and Zare, M., Removal of Pedestals and Directional Ambiguity from Optical

Anemometer Signals, I: A Survey of Available Methods, *Paper No. SFB 80/M/2*, University of Karlsruhe, Germany, 1973.

39. Cummins, H., Knable, N., Gampel, G., and Yeh, Y., Frequency Shifts in Light Diffracted by Ultrasonic Waves in Liquid Media, *Appl. Phys. Letters*, vol. 2, p. 62, 1963.

40. Lanz, O., Johnson, C. C., and Morikawa, S., Directional Laser-Doppler Velocimeter, *Appl. Opt.*, vol. 10, p. 884, 1971.

41. Mazumder, M. K., Laser-Doppler Velocity Measurement without Directional Ambiguity by Using Frequency Shifted Incident Beams, *Appl. Phys. Letters*, vol. 16, p. 462, 1970.

42. Denison, E. B., Stevenson, W. H., and Fox, R. W., Pulsating Laminar Flow Measurements with a Directionally Sensitive Laser Velocimeter, *AIChE J.*, vol. 17, p. 701, 1971. (Earlier paper *Rev. Sci. Inst.*, vol. 41, p. 1475, 1970.)

43. Stevenson, W. H. Optical Frequency Shifting by Means of a Rotating Diffraction Granting, *Appl. Opt.*, vol. 9, p. 649, 1970.

44. Buhrer, C. F., Baird, D., and Conwell, E. M., Optical Frequency Shifting by Electro-Optical Effect, *Appl. Phys. Letters*, vol. 1, p. 46, 1962.

45. Drain, L. E., and Moss, B. C., The Frequency Shifting of Laser Light by Electro-Optical Techniques, *Opto-Electronics*, p. 429, 1972.

46. James, R. N., Babcock, W. R., and Seifert, H. W. S., Application of a Laser-Doppler Technique to the Measurement of Particle Velocity in Gas-Particle Two Phase Flow, *Stanford Univ. Dept. of Aeronautics & Astronautics Rep. No. 265*, 1966.

47. James, R. N., Babcock, W. R., and Seifert, H. W. S., A Laser-Doppler Technique for the Measurement of Particle Velocity, *AIAA J.*, vol. 6, p. 160, 1968.

48. Deighton, M. O., and Sayle, E. A., An Electronic Tracker for the Continuous Measurement of Doppler Frequency from a Laser Anemometer, in *DISA Information Rpt.* #12, 1971.

49. Mattis, J. A., and Camenzind, H. R., A New Phase Locked Loop with High Stability, *Signetics Corporation Tech. Application Note*, Sunnyvale, Calif., 1971.

50. Fridman, J. D., Young, R., and Meister, K., Wide Band Frequency Tracker in Laser-Doppler Velocimeter System, *ISA Proc. of 17th Aerospace Instrumentation Symposium*, 1971.

51. George, W. K., Turbulence Intensity Measurement Using Real-Time Laser-Doppler Velocimetry, *presented at 2nd Annual International LDV Workshop, Purdue University*, Lafayette, Ind., 1974.

52. Asher, J. A., Laser-Doppler Velocimeter System Development and Testing, *General Electric Rpt. No. 72CRD295*, 1972.

53. Brayton, D. B., Kalb, H. T., and Crosswy, F. L., A Two Component, Dual Scatter Laser-Doppler Velocimeter with Frequency Burst Signal Readout, *Appl. Opt.*, vol. 12, p. 1145, 1973.

54. Whiffen, M. C., and Meadows, D. M., Two Axis Single Particle Laser Velocimeter System for Turbulence Spectral Analysis, *presented at 2nd International Workshop on Laser Velocimetry, Purdue University, Lafayette, Ind.*, 1974.

55. Mayo, W. T., A Discussion of Limitations and Extensions of Power Spectrum Analysis of Burst-Counter LDV Data, *presented at 2nd International Workshop on Laser Velocimetry, Purdue University, Lafayette, Ind.*, 1974.

56. Pike, E. R., The Application of Photon Correlation Spectroscopy to Laser-Doppler Measurements, *J. Phys. D.*, vol. 5, p. 51, 1972.

57. Abiss, J. B., Airflow Measurements Using the Doppler Difference Technique, *E. O. Sys. Des.*, vol. 6, p. 29, 1974.

58. Schlichting, H., "Boundary Layer Theory," 4th ed., McGraw-Hill, New York, 1955.

59. Hinze, J. O., "Turbulence," McGraw-Hill, New York, 1959.

60. Hjelmfelt, A. T., and Mockros, L. F., Motion of Discrete Particles in a Turbulent Fluid," *Apl. Sci.*, vol. 16, p. 149, 1966.

61. Durst, F., and Whitelaw, J. H., Local Velocity Measurements in Atomized Spray, *1971 Jahrbuch der DGFLR*, 1971.
62. Kreid, D. K., and Rowe, D. S., Light Scattering Particles for Laser Velocimeter Measurements, *Appl. Opt.*, vol. 15, p. 321, 1976.

14 Operation and application of cooled film sensors for measurements in high-temperature gases

L. M. FINGERSON
Thermo-Systems Inc., St. Paul, Minnesota
A. M. AHMED
McGill University, Montreal, Canada

The cooled film sensor is a device for making hot-wire-anemometry-type measurements in a high temperature environment. In the normal hot-wire or hot-film anemometer the temperature of the sensor must be above the environment temperature under conditions where convection dominates the heat transfer. The essential feature of the cooled film is the addition of a heat sink to permit the operation of the sensor below the environment temperature.

Since the introduction of the cooled-film sensors [1], additional data has been collected on both operational details and applications. This chapter is intended as a review of the work that has been done with cooled probes to date. From this information, a reasonable estimate can be made of the applicability of the cooled film probes to a particular measurement.

The potential of the cooled-film sensors for fluid measurements would appear promising. They have many characteristics in common with hot-wire anemometry systems including small sensor size and high-frequency response. This permits, ideally, an instantaneous measurement at a point in the fluid stream. In addition, it has the capability of making measurements in temperatures of several thousand degrees where few immersion instruments can survive.

14.1 CONSTANT TEMPERATURE ANEMOMETRY WITH WIRES, FILMS, AND COOLED PROBES

The electronic control and data reduction technique for cooled-film sensors are almost identical to that of the standard constant temperature hot-wire or hot-film anemometer. Figure 14.1 shows the basic components of a constant temperature anemometer system.

The first requirement in the system of Fig. 14.1 is that the sensor resistance changes with temperature. For maximum sensitivity, this change of resistance with temperature (temperature coefficient of resistance) should be high. The resistance of the bridge arms are set so that the bridge is in balance with the

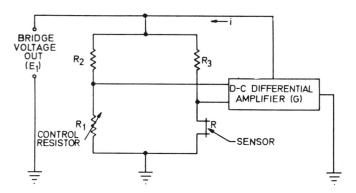

FIG. 14.1 Schematic diagram of constant temperature system.

sensor transferring heat to the environment. The high gain feedback amplifier maintains this condition (a balanced bridge) by adjusting the current through the sensor. For example, as the environment velocity increases, more heat would be transferred between the sensor and its environment. To maintain the sensor temperature, the bridge system increases the current through the sensor. Therefore, the current required by the sensor is a direct measure of the heat transferred between the sensor and its environment. This heat transfer from the sensor (set and maintained at a constant average temperature by the bridge) is, therefore, the basic measurement.

Heat is transferred from the sensor by convection, radiation, and conduction. Similarly, the rate of heat transfer is affected by any property of the environment that influences heat transfer including temperature, velocity, pressure, composition, etc. In the normal application of anemometry, convection is the dominant mode of heat transfer and velocity is the only environment variable. Hence the term *anemometer*.

Limiting the heat transfer to convection, the heat transfer between the sensor and environment where velocity is the only variable, can be expressed approximately by what is commonly referred to as King's law [2].

$$P = I^2 R_s = (A + B\sqrt{\bar{V}})(T_s - T) \tag{14.1}$$

where A, B = constants
V = environment velocity
T_s = sensor temperature
T_∞ = environment temperature

Figure 14.2 shows a calibration curve of a fine hot wire and the calculated curve using two calibration points (end points) and King's Law. Many heat transfer relations have been derived which are a significant improvement on King's law. Still, when discussing hot-wire anemometry the simplicity of the relation in Eq. (14.1) makes it a useful reference.

Figure 14.3 shows two types of sensors commonly used for work in hot-wire anemometry. The fine hot wire is the original type of sensor and still is widely used. The hot film [3, 4] is more recent and has advantages in many applications. It consists of a glass substrate with a thin metallic film on the surface. The glass substrate dominates the physical and thermal characteristics of the sensor while the metal film dominates the electrical characteristics.

Figure 14.4 shows the physical characteristics of a typical cooled film sensor and support and a diagram showing its operation. The entire right-hand side of Eq. (14.1) is simply $-Q_E$ (negative since Q_E is shown as heat transfer to the sensor from the environment) in the figure. Therefore, for an idealized hot-film sensor without cooling

$$P = -Q_E \tag{14.2}$$

and for an idealized hot-film sensor with cooling

$$P = Q_C - Q_E \tag{14.3}$$

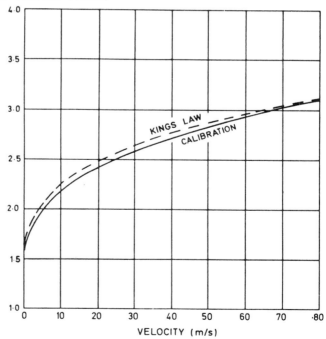

FIG. 14.2   Calibration curve for 0.0038-mm tungsten hot wire and comparison with King's law.

where P = electrical power input to the sensor

Q_C = heat transferred from the sensor surface to the cooling fluid

Q_E = heat transferred from the environment to the sensor surface

Equation (14.3) gives the basic information on maximum environment conditions. Since P must be greater than zero, for proper operation

$$Q_C > Q_E \tag{14.4}$$

The maximum heat transfer from the environment to the sensor is then equal to the cooling rate. On present sensors the practical upper limit for short-term tests

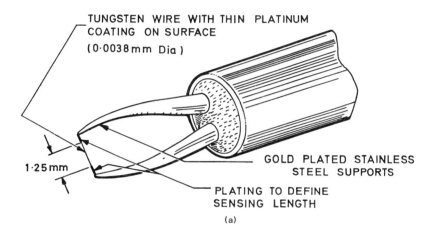

TUNGSTEN WIRE WITH THIN PLATINUM COATING ON SURFACE (0·0038mm Dia)

1·25mm

GOLD PLATED STAINLESS STEEL SUPPORTS

PLATING TO DEFINE SENSING LENGTH

(a)

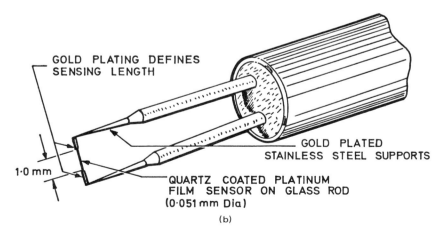

GOLD PLATING DEFINES SENSING LENGTH

1·0 mm

GOLD PLATED STAINLESS STEEL SUPPORTS

QUARTZ COATED PLATINUM FILM SENSOR ON GLASS ROD (0·051mm Dia)

(b)

FIG. 14.3 Sensors commonly used for hot-wire anemometry. (a) Tungsten hot-wire sensor and support needles (diameter: 0.00015 in. or 0.0038 mm). (b) Cylindrical hot-film sensor and support needles (diameter: 0.002 in. or 0.051 mm).

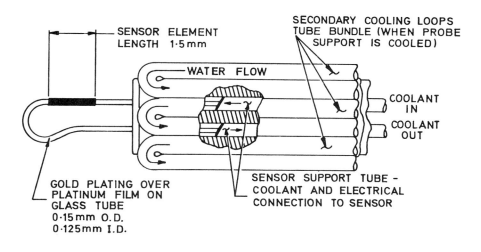

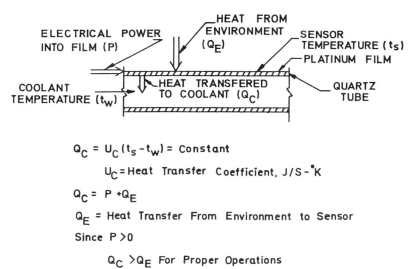

$$Q_C = U_C (t_s - t_w) = \text{Constant}$$

$U_C = $ Heat Transfer Coefficient, $J/S-°K$

$$Q_C = P + Q_E$$

$Q_E = $ Heat Transfer From Environment to Sensor

Since $P > 0$

$$Q_C > Q_E \text{ For Proper Operations}$$

FIG. 14.4 Cooled probe and fundamental heat transfer relation.

is 20 W with 10 W being realistic for continuous use in most types of environments. These figures are for the heat transfer rate in watts from the environment to the sensor surface area (0.15 mm dia by 1.5 mm long). Frequency responses of up to 50 kHz can be attained (-3 db point) using cooled-film sensors.

An important requirement of the cooled-film sensor is that the term Q_c remains constant, independent of external environment conditions. The circuit operates to maintain the average sensor surface temperature constant. If the entering cooling fluid temperature and flow rate is also constant, then ideally the term Q_c will remain constant. A number of factors affect this idealized situation but in actual operation of the cooled probe this assumption must usually be made. Some potential errors in the assumption are discussed later.

In a constant temperature, constant composition environment a calibration curve similar to that shown in Fig. 14.2 could be plotted for a cooled probe. The operation and data reduction technique under these conditions are then essentially identical to those for a hot wire of similar diameter operated at constant temperature. The term Q_c can be handled like the free convection term is for a hot wire.

Perhaps the most important difficulty with the cooled probes is that a high temperature environment seldom, if ever, satisfies the constant temperature condition. In addition composition changes are common due to different constituents, chemical reactions, or even ionization. In this sense measurements with cooled probes resemble more closely measurements in supersonic flows with hot wires. An important difference is that in supersonic flows the hot wire can usually be operated very close to the environment temperature. This permits quite effective separation of velocity and temperature when two probes are used operating at different surface temperature. In high temperature gases this separation is much more difficult since the sensor cannot be operated close to the environment temperature. Other complexities in cooled-probe systems are: The need to water-cool the probe, increasing both cost and size; the need for watertight connections with no condensation that can cause electrical shorting; difficulties in inserting the probe in many high temperature environments; and higher power required from control circuitry. These are practical problems that are largely eliminated by proper design and operation of the system.

At high heat fluxes, the cooling fluid in the sensor will be turbulent to maintain the desired cooling rate. This lowers the signal-to-noise ratio of the system when compared with a normal hot wire, since the low frequency signals get transmitted through the tube to the sensitive film. Finally, the cooling also limits over heat ratios because for a given coolant temperature, the heat flux from the surface to the coolant is determined by the sensor operating temperature. Arbitrary selection of a sensor temperature can result in (a) this cooling rate being excessively high, so the cooling water boils or the sensor burns out, or (b) the cooling rate Q_c being less than maximum heat transfer from the environment to the sensor Q_E. Under these conditions the circuit shuts off and no data is obtained until the inequality in Eq. (14.4) is again satisfied. The result is that it is the maximum expected heat flux Q_E that determines sensor operating temperature rather than a selected overheat ratio as in hot-wire anemometer operation.

14.2 HEAT TRANSFER CORRELATION
FOR COOLED PROBES

For a cooled-film sensor placed normal to a high temperature fluid stream of low Mach number, the forced convective heat transfer to the film is a function of stream temperature, stream velocity, and the fluid transport properties. For a stream composed of a binary mixture, the transport properties of the mixture are functions of the mixture ratio. Thus, a cooled-film anemometer may be calibrated to measure temperature, velocity, and percentage composition of a binary mixture flow. Direct calibration is straightforward for measurements in flows where only one of the independent parameters is a variable, as discussed earlier for the common measurement of velocity in low-speed aerodynamics. For measurements in fluid streams where more than one of the independent parameters is a variable, such as in hypersonic wakes, mixing regions of jets of dissimilar fluids and/or of dissimilar temperatures, and diffusion flames, a direct calibration over the entire range of variables is often tedious and time consuming. For such cases interpretation of measured heat flux data by means of a more general, nevertheless accurate, correlation of the appropriate dimensionless groups involved is more appropriate.

In considering the nature of the cooled-film operation and its application for measurements in diffusion flames and hypersonic wakes, one may stipulate that the required heat transfer correlation must be valid for the following cases:

When the transfer of heat is from the environment to the sensor (i.e., $T_\infty > T_s$)

When the temperature difference between the sensor and the environment is large (i.e., $T_\infty/T_s > 2$)

When the environment is composed of flows of different gases and gas mixtures

When the Reynolds number (based on sensor diameter) of the flow is low, (i.e., Re < 100)

It may be noted that conditions of very low Reynolds numbers, where free and forced convection may interact, also conditions of high Knudsen numbers where free molecular effects may be important, have been left out of the stipulated conditions.

Forced convective heat transfer involving cylinders has been extensively investigated. However none of the investigations, individually or all of them collectively, cover the entire range of conditions mentioned above [5]. It must be emphasized that such correlations as that obtained by careful experimentation by Collis and Williams [6] for hot-wire work are not applicable to precise cooled-film work (when $T_s < T_\infty$). This is because of the different nature of dynamical dissimilarity with temperature loading between cases of heating and cooling [5]. On the other hand, data obtained for the heating of cooled cylinders such as those by Churchill and Brier [7] are for Reynolds numbers above 300. A detailed discussion of the existing correlations has been presented in [5].

In the present investigation the flow conditions were simulated in a plasma jet (jet orifice diameter 12 cm). The jet conditions were maintained such that at the points of heat transfer measurement (i.e., at the potential core of the jet, where the distribution of velocity, temperature, and concentration are uniform) ionization was negligible and recombination was complete. By means of a cooled film the maximum relative intensity of heat flux fluctuations was found to be less than 3%. If it is considered that the heat flux fluctuations are due only to velocity fluctuations, the error in heat transfer measurement will be less than 2% due to turbulence.

Constant-temperature, quartz-coated, cooled films were used as the heat transfer surface. The film is obtained by a deposition of platinum of thickness 1,000 Å–2,000 Å on a loop made on Vycor tube, 0.0152 cm OD and 0.0102 cm ID. The sensitive section of the film is isolated by a heavy gold plating (0.013–0.025 mm thick) on the rest of the loop. The length of the sensing film is 0.103 cm and the thickness of its quartz coating is around 5,000 Å. Along with the heat transfer measured by the cooled film, velocity and temperature were obtained by means of a carbon-tipped pitot probe of orifice diameter 0.103 cm and a Pt-Pt 10% Rh thermocouple of bead diameter 0.127 cm. Further details of experimental setup and procedure may be found in [5]. The variables and the variable ranges considered in the investigation are as follows:

Temperature loading. The range of jet temperature considered was $800°K$ to $1600°K$, and the film temperature was varied in three steps between $350°K$ and $525°K$. From this a temperature loading (T_∞/T_s) range of 1.5 to 4.5 was obtained.

Reynolds number. The range of Re based on cylinder diameter was 4 to 80.

Flow composition. The plasma jet was composed of He, N_2 and mixtures of He-N_2, N_2-CO_2. Two mixture ratios were considered for each mixture.

The true jet temperature was obtained from the thermocouple temperature by applying correction for radiation and conduction errors. Transport property values of the gas species and specie mixtures for data analysis were calculated from the expressions collected in [8].

In the course of data analysis the following points were noted:

Consideration of the film temperature $[T_f = (T_\infty + T_s)/2]$ for the evaluation of fluid properties in the dimensionless parameters Nusselt number and Reynolds number, was not sufficient to eliminate the temperature loading effect. The residual effect caused a decrease in Nusselt number with increased temperature loading at a particular Re for the present case of heating of cylinders.

Replacement of the usual temperature ratio in the temperature loading factor by a kinematic viscosity ratio enabled a unique correlation to be derived for flows composed of gas species whose transport property value variations with temperature are different. This conclusion is supported by arguments presented in [9] for hot wires.

The expected slight influence of the small variation of Prandtl number (because of the consideration of different flow species and specie mixtures and also because of variation of temperature), could not be discerned.

The Re_f dependency of Nu_f was found to be different from that according to King's law.

A discontinuity in the heat transfer curve was noted between $Re_f = 40$ and $Re_f = 55$. Unfortunately, no data were collected between these values of Re_f and no specific investigation was carried out to determine a change in the flow features, (onset of eddy shedding) in this range. For the present, data collected in the Re_f range below $Re_f = 40$ was considered for correlation purpose.

In view of the above discussion, the following form of correlation was considered:

$$Nu_f \left(\frac{\nu_\infty}{\nu_f} \right)^n = C + D \, Re_f^m$$

The constants n, m, C, and D evaluated by the least-square method were found to be as follows:

Re_f	n	m	C	D
$4-40$	0.15	0.45	0.2068	0.4966

Figure 14.5 shows the effectiveness of this relation in correlating the entire body of data below $Re_f = 40$. The rms deviation considering this range of data was 0.0924.

Finally, Fig. 14.6 compares values of Nu_f calculated from some of the correlations obtained for heat transfer from heated cylinders with values calculated from the present correlation. The evaluation is for an identical condition of heat transfer to a cooled cylinder with $T_s = 400K$, $T_\infty = 1200K$ and the flow composed of N_2. Even considering the wide discrepancies in values calculated from correlations obtained for heated cylinders only, the error involved in using them to analyze heat transfer to cooled cylinders is quite apparent.

14.3 TESTS IN SEVERE ENVIRONMENTS

The maximum environment capabilities of the cooled-film probe have been expressed in terms of heat transfer rates from the environment to the sensor. To

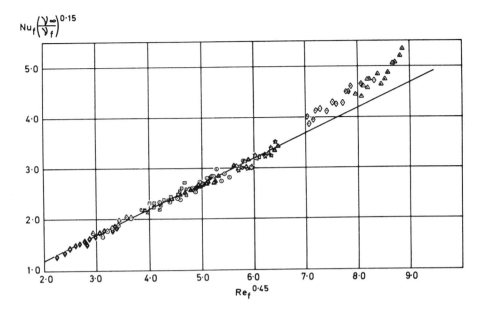

Flow	T K	T_S K	Symbol
Nitrogen	852–1660	361–496	▲
Helium	803–1360	387–532	◆
He 40% by vol. N_2 60% by vol.	900–1269	361–496	○
He 77.5% by vol. N_2 22.5% by vol.	900–1270	387–532	□
He 42% by vol. CO_2 58% by vol.	701–1088	342–465	◇
He 93.5% by vol. CO_2 6.5% by vol.	702–1090	342–465	◐
N_2 50% by vol. CO_2 50% by vol.	703–1278	367–512	◆
N_2 30% by vol. CO_2 70% by vol.	703–1277	367–512	▲

FIG. 14.5 $Nu_m (v_\infty/v_m)$ 0.15 vs. Re_m 0.45 showing all heat transfer data for various species and specie mixtures and for various temperature loadings uniquely correlated by the relation $Nu_m (v_\infty/v_m)$ 0.15 = 0.2068 + 4966 Re_m 0.45 in the Re_m range of 5 to 40.

convert this to temperature and velocity requires an accurate heat transfer relation or tests where the conditions are well known. Although the following data does not completely satisfy either criteria, it does give some indication of the capability of the cooled-film sensors.

Figure 14.7 shows a traverse across the tip of an acetylene torch for two distances from the tip. The inside diameter of the tip is 3.8 mm and maximum heat flux to the sensor is 16 W. Referring to Fig. 14.7, the traverse was made from left to right. The sensor did shift in resistance during each traverse, as shown by the failure of the points on the right to approach closer to the abscissa. A single traverse took approximately 5 minutes. The 16 W represents a heat transfer rate to the 0.15-mm-dia sensor of 2.66 kW/cm^2.

Reference [10] is another application where the cooled-film sensors were exposed to a severe environment. The cooled sensors were used in a combustor that burned ethanol and liquid oxygen at an average chamber pressure of 178 psia. Pressure oscillations of ±15% were sustained at 1190 cps with a siren mounted directly downstream of the exhaust nozzle. Under test conditions the

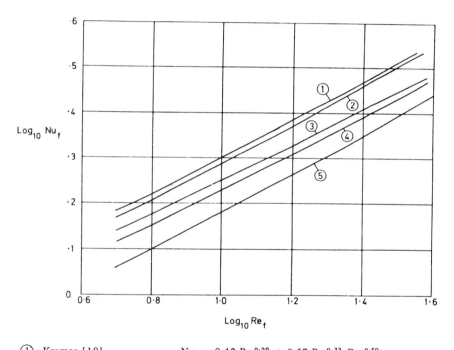

① Kramer [18] $Nu_f = 0.42\ Pr_f^{0.20} + 0.57\ Pr_f^{0.33}\ Re_f^{0.50}$

② Van der Hegge Zijnen [19] $Nu_f = 0.35 + 0.5\ Re_f^{0.5} + 0.001\ Re_f$

③ Collis and Williams [6] $Nu_f = (0.24 + 0.56\ Re_f^{0.45})\ (T_f/T)^{0.17}$

④ Hilpert [20] $Nu_f = 0.821\ [Re_f(T_s/T)^{0.25}]^{0.385}$

⑤ Present result $Nu_f = (0.2068 + 0.496\ Re_f^{0.45})\ (\nu_\infty/\nu_f)^{-0.15}$

$$T_s = 400K \qquad T = 1200K \qquad Flow = nitrogen$$

FIG. 14.6 Comparison of present correlation for forced convective heat transfer to cooled cylinders in heated crossflow with previous correlations for heat transfer from heated cylinders in ambient or near-ambient crossflow.

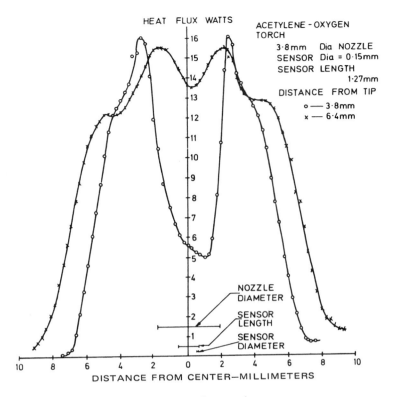

FIG. 14.7 Heat flux traverse of acetylene torch.

average environment temperature was calculated to be 1900K and the maximum velocity (measured from streak photographs) about 110 m/sec. Velocity (possibly temperature and/or composition included) fluctuations gave heat flux variations to the sensor of about 6 to 17 W [10]. In this case the standard sensor diameter (0.15 mm) was used but it was shorter than standard length 1.0 mm.

The value of Q_c set for the experiments of [10] was 20 W, which went down to 18.8 W after the 1-sec run even for "successful" runs. Initially, sensor breakage was a serious problem which was corrected by shielding the sensor during engine startup. Even then, as reported in [10], sensor stability and longevity caused a serious problem for the experiments. The velocities calculated from cooled-sensor data did not agree with the streak photographs. As pointed out in the reference, no detailed calibration was deemed practical so the heat transfer relation used could be suspect, in addition to other sources of error. The maximum Reynolds number of 580 is well beyond the calibration data of Fig. 14.5.

Figure 14.8 shows a typical set of data from the cooled sensor when exposed to the test chamber of [10]. The lower curve is for the sensor when shielded on the inlet side. These data were taken to identify the reverse flow point, since a cylindrical sensor cannot differentiate flow direction.

The environment of [10] would seem to be at the upper heat flux limit for useful data from cooled sensors. Since many environments exceed these conditions (e.g., hydrogen/oxygen combustors, plasmas, etc.) there is a need to extend the range of cooled sensors to higher temperatures. Some efforts have been made in this direction [11] which led to the present sensor design using Vycor rather than the original Pyrex [1]. Although further improvement is always possible, the difficulty of cooling a small tube adequately for survival seems to preclude a significant improvement. Going to larger tubes is not generally desirable since characteristics such as frequency response and spatial resolution would be compromised.

14.4 ACCURACIES WITH COOLED PROBES

Experimental data on the accuracy of the cooled probes is limited. Reference [11] discusses several potential sources of error and gives calculated estimates of the effects while [12] gives details on the error in the two-sensor technique. The consistency of the calibration data in this paper indicates the kind of reproducibility that can be expected for mean measurements.

The primary source of error is in the assumption that the heat transfer from the sensor surface to the cooling fluid Q_c is constant during external environment changes. One source of error is the exposed part of the sensor tube

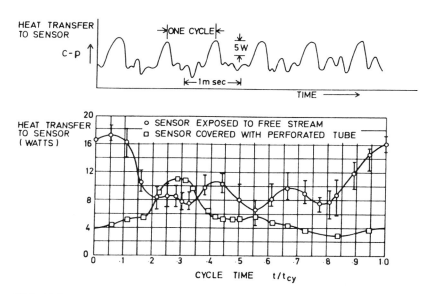

FIG. 14.8 Typical heat-flux sensor output and average heat flux for one cycle of oscillation. (Reprinted from Povinelli and Ingebo, 1967, Ref. [10].)

between the protective cooling jacket and the "sensitive" portion of the sensor. In a high temperature environment the water temperature will rise in this passage, while during the tare reading of Q_c there would be no temperature rise. Another cause of error is the redistribution of sensor surface temperature between the tare reading and the reading in the environment. Some measurements of sensor surface temperature distribution are given in [12]. For the calculated conditions in [11], the error estimate for all the above factors was 5.2% on the heat flux reading.

In measurements with hot-wire and hot-film probes, an assumption is made that the steady state calibration can be used directly to interpret unsteady state data. This seems valid for the very fine hot wires used near atmospheric temperature and pressure conditions. For the larger cooled films at high Reynolds numbers, flow separation effects around the sensor could invalidate the assumption. In addition, the complex nature of the cooled film can cause transient errors due to changes in surface temperature distribution both longitudinally and radially under varying flow conditions.

14.5 APPLICATIONS OF COOLED–FILM SENSORS

Much of the data given above has been concerned with the problems associated with cooled-film sensors. It is important to recognize these problems before undertaking measurements. At the same time, the unique capabilities of the cooled sensors make them not only a valuable tool, but sometimes the only tool that can give the required data in a given situation.

The measurements in a rocket chamber have been discussed as an example of the upper limit of conditions where the cooled films are applicable. Other applications that data are available on are: Measurements of a hypersonic boundary layer and measurements in the wake of a hypersonic projectile. Other applications are certainly feasible and in fact have been made, but no published data are available.

Hypersonic boundary layer measurements were made by McCroskey, Bogdonoff, and McDougall [13], and by McCroskey [14]. Nominal test conditions in the nitrogen tunnel were: $T_{01} = 2000K$, $P_{01} = 135-340$ atm, $M_\infty = 23-26$, and Re/cm = 2,950-5,900. Tests in helium were run at $T_{01} = 297K$, $P_{01} = 15.3$ atm, $M_\infty = 16.5$, and Re/cm = 47,200 [15] with hot wires. Although there is some scatter, most of it is between sensors and the slope for all sensors agrees with Dewey's results (Figure 14.9).

In this work a pitot probe, a cooled-film probe, and the recovery temperature of a hot wire were used to calculate the variables of interest such as velocity, density, and pressure near the leading edge of a sharp flat plate. In the measurements with cooled probes, one of the problems that came up was inadequate resolution. Even though the total temperature of the environment was high, the total heat flux to the probe was 0.1 to 0.5 W. Therefore, to increase sensitivity and lower dependence on entering coolant temperature, nitrogen

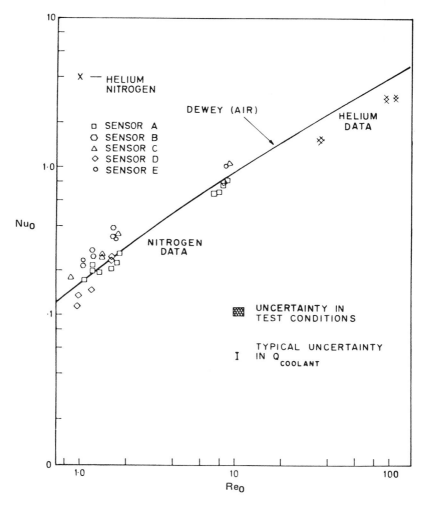

FIG. 14.9 Calibration data for cooled-film sensors. (Reprinted from McCroskey, 1965, Ref. [14].)

rather than water was used for cooling. This increased resolution by permitting a much higher temperature difference between sensor surface and coolant, while keeping the total power dissipation to the coolant low.

An extensive investigation of turbulence characteristics of hypersonic wakes by means of cooled-film anemometers is being carried out at the Canadian Armament Research and Development Establishment (CARDE) at Val Cartier, Quebec [12, 16, 17]. Owing to the high temperature encountered in the near-wake region of hypersonic projectiles, the application of the cooled-film technique appears appropriate. Further, cooled films have proved to be sufficiently robust to survive the hypersonic range environment at least for a sufficient length of time to record a signal of several thousand body diameters in duration [15].

The above experiments are performed in the CARDE Hypersonic Range No. 5 which consists of a light gas gun with a 102-mm barrel capable of launching projectiles into a depressurized tank of 122 m length at velocities in excess of 4150 m/sec [17]. Two constant-temperature cooled-film sensors with different sensor surface temperatures, positioned several thousandths of a centimeter apart are located near the flight axis of the projectile [16]. The cooled-film anemometer bridge voltage is recorded by means of oscilloscopes viewed by Wollensak Fastax cameras.

The two cooled-film sensors are operated at two different surface temperatures to attempt the separation of environment temperature and velocity. It is important that the temperature difference between sensors be large to optimize the accuracy using this technique [11, 12]. At the same time, it is best to have both anemometer circuits operating at about the same power level for nearly equivalent frequency response and sensitivity. One way to satisfy these criteria is to use water as the coolant for the sensor with low surface temperature and an oil (such as Fluorolube FS or Silicon Oil 704) for the sensor with high surface temperature [12]. The method of reduction of recorded voltage data in terms of velocity and temperature distribution in the wake, to determine their power spectral density functions has been shown in [12].

14.6 CONCLUSIONS

From the data presented, some tentative conclusions can be drawn for cooled-film sensors presently available:

The maximum heat fluxes from the environment to the sensor are: About 10 W for good sensor stability and longevity and up to 20 W maximum with decreasing stability at increasing heat fluxes.

The use of a coolant other than water is often desirable for a given measurement situation.

Accuracies of better than ± 5% on heat flux are probably not possible unless the calibration covers a range that includes the test conditions exactly.

The heat transfer correlation presented indicates that the cooled sensors can be calibrated in high temperature environments. Of particular interest is the successful use of a transport property (kinematic viscosity) to correlate different compositions.

The cooled-film sensor greatly extends the temperature range of the hot-wire or hot-film anemometer. It retains many of the important features such as small size, high-frequency response, and high resolution which make the hot-wire anemometer a valuable tool in fluid mechanics research. Also, like the hot wire, it has definite limitations in both accuracy and environment conditions which must be recognized before measurements are attempted.

NOMENCLATURE

A, B, C, D	constants
d	sensor diameter
I	current in sensor
M	Mach number
m, n	constants
P	electrical power input to sensor
Q_c	heat transferred from sensor surface to cooling fluid
Q_E	heat transferred from environment to sensor surface
Re	Reynolds number (Vd/ν)
R_s	sensor operating resistance
t	temperature
T	absolute temperature
V	environment velocity
ν	kinematic viscosity

Subscripts

s	sensor surface
f	arithmetic mean —when referring to fluid properties, signifies they are evaluated at the arithmetic mean temperature, e.g., $T_f = (T_\infty + T_s)/2$.
∞	free stream

REFERENCES

1. Fingerson, L. M., "A Heat Flux Probe for Measurements in High Temperature Gases," Ph.D. thesis, University of Minnesota, Minneapolis, 1961.
2. King, L. V., On the Convection of Heat from Small Cylinders in a Stream of Fluid: Determination of the Convective Constants of Small Platinum Wires with Applications to Hot-Wire Anemometry, *Proc. Roy. Soc. (London)*, vol. 214A, no. 14, p. 373, 1914.
3. Ling, S. L., and Hubbard, P. G., The Hot-Film Anemometer: A New Device for Fluid Mechanics Research, *J. Aeronaut. Sci.*, vol. 23, p. 890, 1956.
4. Lowell, H. H., Response of Two-Material Laminated Cylinders to Simple Harmonic Environment Temperature Change, *J. Appl. Phys.*, vol. 24, no. 12, p. 1473, 1953.
5. Ahmed, A. M., Forced Convective Heat Transfer to Cooled Cylinders at Low Reynolds Numbers and with Large Temperature Difference, *McGill MERL. T.N.* 67-5, September 1967.
6. Collis, D. C., and Williams, M. J., Two Dimensional Convection from Heated Wires at Low Reynolds Numbers, *J. Fluid Mech.*, vol. 6, p. 357, 1959.
7. Churchill, S. W., and Brier, J. C., Convective Heat Transfer from a Gas Stream at High Temperature to a Circular Cylinder Normal to the Flow, *Chem. Eng. Progr. Symp. Series*, no. 17, pp. 57-66, 1955.
8. Brokaw, R. S., Alignment Charts for Transport Properties Viscosity, Thermal Conductivity and Diffusion Coefficients for Nonpolar Gases and Gas Mixtures at Low Density, *NASA* TR-R81, 1961.

9. Davies, P. O. A. L., and Fisher, M. J., Heat Transfer from Electrically Heated Cylinders, *Proc. Roy. Soc. (London),* vol. 280A, pp. 486–527, 1964.

10. Povinelli, F. P., and Ingebo, R. D., Evaluation of a Thin-Film, Heat-Flux Probe for Measuring Gas Velocities in an Unstable Rocket Combustor, *NASA Tech. Mem.* TM X-1333, February 1967.

11. Fingerson, L. M., Research on the Development and Evaluation of a Two-Sensor Enthalpy Probe, *Aerosp. Res. Lab. Rep.* ARL 64-161, October 1964.

12. Ellington, D., and Trottier, G., Some Observations on the Application of Cooled-Film Anemometry to the Study of the Turbulent Characteristics of Hypersonic Wakes, *Cana. Arm. Res. Devel. Estab. Rep. CARDE* TN 1773/67, September 1967.

13. McCroskey, W. J., Bogdonoff, S. M., and McDougall, J. G., An Experimental Model for the Leading Edge of a Sharp Flat Plate in Rarefied Hypersonic Flow, *AIAA Paper* 66-31, 1966.

14. McCroskey, W. J., A New Probe for Hot-Wire Anemometry at High Temperature, *Princeton Gas Dynamics Lab. Int. Mem.* 7, August 1965.

15. Dewey, C. F., Hot Wire Measurements in Low Reynolds Number Hypersonic Flows, *J. ARS*, vol. 31, no. 12, p. 1709, December 1961.

16. Trottier, G., Ahmed, A. M., and Ellington, D., Cooled-Film Anemometer Measurements in the Hypersonic Wake, *CARDE* TN 1720/66, May 1966.

17. Staff of Aerophysics Wing (Compiled by D. Heckman), Re-Entry Physics Research Program on Turbulent Wakes, *CARDE* TN 1741/67, January 1967.

18. Krmaer, H., *Physics,* vol. 12, no. 2-3, p. 61, 1946.

19. Van der Hegge Zijnen, B. G., *Appl. Sci. Res. A,* vol. 6, p. 129, 1956.

20. Hilpert, R., *Forsch. Geb. Ingenieur.,* vol. 4, p. 215, 1933.

Appendix

The International Practical
Temperature Scale of 1968

Adopted by the Comité International des Poids et Mesures

FOREWORD

The Comité International des Poids et Mesures (CIPM) at its meeting in
October 1968 agreed to adopt the International Practical Temperature
Scale of 1968 (IPTS-68) in accordance with the decision of the 13^e
Conférence Générale des Poids et Mesures. Resolution 8. of October 1967.
This resolution gave the CIPM permission to introduce a new Scale to
replace the existing IPTS-48 as soon as the Comité Consultatif de Ther-
mométrie (CCT) made its recommendations.

There were two important reasons for revising the IPTS-48. The first
of these was the need to extend the Scale to lower temperatures to achieve
a unification of the existing national scales in the region 10K-90K. Secondly the Scale had not been revised significantly over the range from
$-183°$C to $1063°$C since its inception in 1927, and modern gas thermometer measurements showed that the Scale gave values considerably different
from thermodynamic temperatures.

The CCT agreed on the basic data and methods of interpolation for a
new definition of the IPTS at its 8th meeting held in Washington and
Ottawa in September 1967. Provision was made to extend the Scale down
to 13.81K, the triple point of equilibrium hydrogen, and for the use of the
best known values of thermodynamic temperatures. The value of c_2 used in
the Planck equation was also revised as a result of a better knowledge of
the values of fundamental physical constants. The CCT appointed a small
sub-committee to prepare a text of the definition of the IPTS-68 in
accordance with its recommendations, with instructions to resolve some
points of detail. The final draft of the English text, which is reproduced
below, received the approval of all members of the CCT before it was

Offprint from *Metrologia, International Journal of Scientific Metrology*, vol. 5, no. 2,
pp. 35–44, Springer-Verlag, Berlin, April 1969.

The text in French of this Scale is published in Comptes rendus de la Treizième
Conférence Générale des Poids et Mesures, 1967–1968, Annexe 2. and Comité Consultatif de
Thermométrie, 8e session, 1967, Annexe 18.

submitted to the CIPM. The official text of the IPTS-68 is the French text which was prepared by the Bureau International des Poids et Mesures from the English text, and is published as an annexe to Comptes rendus des séances de la Treiziéme Conférence Générale des Poids et Mesures.

C. R. Barber

I. INTRODUCTION

The basic temperature is the thermodynamic temperature, symbol T, the unit of which is the kelvin, symbol K. The kelvin is the fraction 1/273.16 of the thermodynamic temperature of the triple point of water.[1]

The Celsius temperature, symbol t, is defined by

$$t = T - T_0 \tag{1}$$

where $T_0 = 273.15K$. The unit employed to express a Celsius temperature is the degree Celsius, symbol °C, which is equal to the kelvin. A difference of temperature is expressed in kelvins; it may also be expressed in degrees Celsius.

The International Practical Temperature Scale of 1968 (IPTS-68) has been chosen in such a way that the temperature measured on it closely approximates the thermodynamic temperature; the difference is within the limits of the present accuracy of measurement.

The International Practical Temperature Scale of 1968 distinguishes between the International Practical Kelvin Temperature with the symbol T_{68} and the International Practical Celsius Temperature with the symbol t_{68}; the relation between T_{68} and t_{68} is

$$t_{68} = T_{68} - 273.15K \tag{2}$$

The units of T_{68} and t_{68} are the kelvin, symbol K, and degree Celsius, symbol °C, as in the case of the thermodynamic temperature T and the Celsius temperature t.

The International Practical Temperature Scale of 1968 was adopted by the International Committee of Weights and Measures at its meeting in 1968 according to the power given to it by Resolution 8 of the 13th General Conference of Weights and Measures. It replaces the International Practical Temperature Scale of 1948 (amended edition of 1960).

II. DEFINITION OF THE INTERNATIONAL PRACTICAL TEMPERATURE SCALE OF 1968[2] (IPTS-68)

1. Principle of the IPTS-68 and Defining Fixed Points

The IPTS-68 is based on the assigned values of the temperatures of a number of reproducible equilibrium states (defining fixed points) and on standard instru-

---

[1] 13th General Conference of Weights and Measures (1967), Resolutions 3 and 4.

[2] In this document Kelvin temperatures are used, in general, below 0°C and Celsius temperatures are used above 0°C. This avoids the use of negative values and conforms with general usage.

ments calibrated at those temperatures. Interpolation between the fixed point temperatures is provided by formulae used to establish the relation between indications of the standard instruments and values of International Practical Temperature.

The defining fixed points are established by realizing specified equilibrium states between phases of pure substances. These equilibrium states and the values of the International Practical Temperature assigned to them are given in Table 1.

The standard instrument used from 13.81K to 630.74°C is the platinum resistance thermometer. The thermometer resistor must be strain-free, annealed pure platinum. The resistance ratio $W(T_{68})$, defined by

$$W(T_{68}) = \frac{R(T_{68})}{R(273.15\text{K})} \tag{3}$$

where R is the resistance, must not be less than 1.392 50 at $T_{68} = 373.15\text{K}$. Below 0°C the resistance-temperature relation of the thermometer is found from a reference function and specified deviation equations. From 0°C to 630.74°C two polynomial equations provide the resistance-temperature relation.

The standard instrument used from 630.74°C to 1064.43°C is the platinum-10% rhodium/platinum thermocouple, the electromotive force-temperature relation of which is represented by a quadratic equation.

Above 1337.58K (1064.43°C) the International Practical Temperature of 1968 is defined by the Planck law of radiation with 1337.58K as the reference temperature and a value of 0.014 388 metre kelvin for c_2.

2. Definition of the International Practical Temperature of 1968 in Different Temperature Ranges

a) The Range from 13.81K to 273.15K

From 13.81K to 273.15K the temperature T_{68} is defined by the relation

$$W(T_{68}) = W_{\text{CCT-68}}(T_{68}) + \Delta W(T_{68}) \tag{4}$$

where $W(T_{68})$ is the resistance ratio of the platinum resistance thermometer and $W_{\text{CCT-68}}(T_{68})$ is the resistance ratio as given by the reference function set out in Table 2.[3] The deviations $\Delta W(T_{68})$ at the temperatures of the defining fixed points are obtained from the measured values of $W(T_{68})$ and the corresponding values of $W_{\text{CCT-68}}(T_{68})$, see Table 4. To find $\Delta W(T_{68})$ at intermediate temperatures interpolation formulae are used. The range between 13.81K and 273.15K is divided into four parts in each of which $\Delta W(T_{68})$ is defined by a polynomial in T_{68}. The constants in the polynomials are determined from the values of $\Delta W(T_{68})$ at the fixed points and the condition that there should be no discontinuity in $d\Delta W(T_{68})/dT_{68}$ at the junctions of the temperature ranges.

---

[3] For the relation between the IPTS-68 and the "national scales" from which it was in part derived see Appendix I.

From 13.81K to 20.28K the deviation function is

$$\Delta W(T_{68}) = A_1 + B_1 T_{68} + C_1 T_{68}^2 + D_1 T_{68}^3 \tag{5}$$

where the constants are determined by the measured deviations at the triple point of equilibrium hydrogen, the temperature of 17.042K and the boiling point

TABLE 1 Defining fixed points of the IPTS-68[a]

Equilibrium state	Assigned value of International Practical Temperature	
	T_{68} (K)	t_{68} ($^{\circ}$C)
Equilibrium between the solid, liquid and vapour phases of equilibrium hydrogen (triple point of equilibrium hydrogen)	13.81	-259.34
Equilibrium between the liquid and vapour phases of equilibrium hydrogen at a pressure of 33 330.6 N/m^2 (25/76 standard atmosphere)	17.042	-256.108
Equilibrium between the liquid and vapour phases of equilibrium hydrogen (boiling point of equilibrium hydrogen)	20.28	-252.87
Equilibrium between the liquid and vapour phases of neon (boiling point of neon)	27.102	-246.048
Equilibrium between the solid, liquid and vapour phases of oxygen (triple point of oxygen)	54.361	-218.789
Equilibrium between the liquid and vapour phases of oxygen (boiling point of oxygen)	90.188	-182.962
Equilibrium between the solid, liquid and vapour phases of water (triple point of water)[c]	273.16	0.01
Equilibrium between the liquid and vapour phases of water (boiling point of water)[b, c]	373.15	100
Equilibrium between the solid and liquid phases of zinc (freezing point of zinc)	692.73	419.58
Equilibrium between the solid and liquid phases of silver (freezing point of silver)	1235.08	961.93
Equilibrium between the solid and liquid phases of gold (freezing point of gold)	1337.58	1064.43

[a]Except for the triple points and one equilibrium hydrogen point (17.042K) the assigned values of temperature are for equilibrium states at a pressure $p_0 = 1$ standard atmosphere (101 325 N/m^2). In the realization of the fixed points small departures from the assigned temperatures will occur as a result of the differing immersion depths of thermometers or the failure to realize the required pressure exactly. If due allowance is made for these small temperature differences, they will not affect the accuracy of realization of the Scale. The magnitudes of these differences are given in section III.

[b]The equilibrium state between the solid and liquid phases of tin (freezing point of tin) has the assigned value of $t_{68} = 231.9681^{\circ}$C and may be used as an alternative to the boiling point of water.

[c]The water used should have the isotopic composition of ocean water, see Section III.4.

TABLE 2 The reference function W_{CCT-68} (T_{68}) for platinum resistance thermometers for the range from 13.81K to 273.15K[a]

$$T_{68} = \left\{ A_0 + \sum_{i=1}^{20} A_i \left[\ln W_{CCT-68}(T_{68})\right]^i \right\} K \tag{22}$$

Coefficients A_i:

i	A_i	i	A_i
0	$0.273\ 15 \times 10^3$	11	$0.767\ 976\ 358\ 170\ 845\ 8 \times 10$
1	$0.250\ 846\ 209\ 678\ 803\ 3 \times 10^3$	12	$0.213\ 689\ 459\ 382\ 850\ 0 \times 10$
2	$0.135\ 099\ 869\ 964\ 999\ 7 \times 10^3$	13	$0.459\ 843\ 348\ 928\ 069\ 3$
3	$0.527\ 856\ 759\ 008\ 517\ 2 \times 10^2$	14	$0.763\ 614\ 629\ 231\ 648\ 0 \times 10^{-1}$
4	$0.276\ 768\ 548\ 854\ 105\ 2 \times 10^2$	15	$0.969\ 328\ 620\ 373\ 121\ 3 \times 10^{-2}$
5	$0.391\ 053\ 205\ 376\ 683\ 7 \times 10^2$	16	$0.923\ 069\ 154\ 007\ 007\ 5 \times 10^{-3}$
6	$0.655\ 613\ 230\ 578\ 069\ 3 \times 10^2$	17	$0.638\ 116\ 590\ 952\ 653\ 8 \times 10^{-4}$
7	$0.808\ 035\ 868\ 559\ 866\ 7 \times 10^2$	18	$0.302\ 293\ 237\ 874\ 619\ 2 \times 10^{-5}$
8	$0.705\ 242\ 118\ 234\ 052\ 0 \times 10^2$	19	$0.877\ 551\ 391\ 303\ 760\ 2 \times 10^{-7}$
9	$0.447\ 847\ 589\ 638\ 965\ 7 \times 10^2$	20	$0.117\ 702\ 613\ 125\ 477\ 4 \times 10^{-8}$
10	$0.212\ 525\ 653\ 556\ 057\ 8 \times 10^2$		

The reference function $W_{CCT-68}(T_{68})$ is continuous at $T_{68} = 273.15$K in its first and second derivatives with the function $W(t_{68})$ given by Eqs. (9) and (10) for $\alpha = 3.925\ 966\ 8 \times 10^{-3}\,°C^{-1}$ and $\delta = 1.496\ 334°C$.

[a] A tabulation of this reference function, sufficiently detailed to allow interpolation to an accuracy of 0.0001K, is available from the Bureau International des Poids et Mesures, 92-Sèvres, France. A skeleton tabulation appears in this text as Table 3.

of equilibrium hydrogen and by the derivative of the deviation function at the boiling point of equilibrium hydrogen as derived from Eq. (6).

From 20.28K to 54.361K the deviation function is

$$\Delta W(T_{68}) = A_2 + B_2 T_{68} + C_2 T_{68}^2 + D_2 T_{68}^3 \tag{6}$$

where the constants are determined by the measured deviations at the boiling point of equilibrium hydrogen, the boiling point of neon and the triple point of oxygen and by the derivative of the deviation function at the triple point of oxygen as derived from Eq. (7).

From 54.361K to 90.188K the deviation function is

$$\Delta W(T_{68}) = A_3 + B_3 T_{68} + C_3 T_{68}^2 \tag{7}$$

where the constants are determined by the measured deviations at the triple point and the boiling point of oxygen and by the derivative of the deviation function at the boiling point of oxygen as derived from Eq. (8).

TABLE 3 Values of W_{CCT-68} (T_{68}), according to the equation given in Table 2, at integral values of T_{68}

T_{68} K	W_{CCT-68} (T_{68})	T_{68} K	W_{CCT-68} (T_{68})	T_{68} K	W_{CCT-68} (T_{68})
		50	0.075 377 56	100	0.286 302 01
		51	0.079 101 23	101	0.290 621 54
		52	0.082 875 95	102	0.294 938 41
		53	0.086 698 59	103	0.299 252 45
		54	0.090 566 00	104	0.303 563 59
		55	0.094 475 15	105	0.307 871 83
		56	0.098 423 36	106	0.312 177 10
		57	0.102 407 74	107	0.316 479 39
		58	0.106 425 83	108	0.320 778 56
		59	0.110 475 06	109	0.325 074 67
		60	0.114 553 12	110	0.329 367 65
		61	0.118 657 89	111	0.333 657 51
		62	0.122 787 22	112	0.337 944 16
13	0.001 230 61	63	0.126 939 14	113	0.342 227 68
14	0.001 459 73	64	0.131 111 89	114	0.346 508 00
15	0.001 745 41	65	0.135 303 63	115	0.350 785 19
16	0.002 094 74	66	0.139 512 84	116	0.355 059 10
17	0.002 515 12	67	0.143 738 00	117	0.359 329 89
18	0.003 014 28	68	0.147 977 73	118	0.363 597 54
19	0.003 599 62	69	0.452 230 58	119	0.367 861 99
20	0.004 277 80	70	0.156 495 41	120	0.372 123 31
21	0.005 054 95	71	0.160 771 08	121	0.376 381 51
22	0.005 936 68	72	0.165 056 43	122	0.380 636 57
23	0.006 928 04	73	0.169 350 49	123	0.384 888 51
24	0.008 033 16	74	0.173 652 40	124	0.389 137 32
25	0.009 255 04	75	0.177 961 17	125	0.393 383 16
26	0.010 595 85	76	0.182 276 05	126	0.397 625 94
27	0.012 056 90	77	0.186 596 28	127	0.401 865 67
28	0.013 639 01	78	0.190 921 07	128	0.406 102 42
29	0.015 342 61	79	0.195 249 92	129	0.410 336 28
30	0.017 167 68	80	0.199 582 12	130	0.414 567 09
31	0.019 113 63	81	0.203 917 14	131	0.418 795 07
32	0.021 179 44	82	0.208 254 45	132	0.423 020 15
33	0.023 363 43	83	0.212 593 44	133	0.427 242 33
34	0.025 663 35	84	0.216 933 88	134	0.431 461 69
35	0.028 076 45	85	0.221 275 23	135	0.435 678 31
36	0.030 599 53	86	0.225 617 12	136	0.439 892 10
37	0.033 229 16	87	0.229 959 16	137	0.444 103 22
38	0.035 961 55	88	0.234 301 05	138	0.448 311 59
39	0.038 793 05	89	0.238 642 48	139	0.452 517 30
40	0.041 719 68	90	0.242 983 15	140	0.456 720 33
41	0.044 737 60	91	0.247 322 90	141	0.460 920 77
42	0.047 842 92	92	0.251 661 28	142	0.465 118 61
43	0.051 031 78	93	0.255 998 36	143	0.469 313 87
44	0.054 300 36	94	0.260 333 69	144	0.473 506 60
45	0.057 644 86	95	0.264 667 18	145	0.477 696 82
46	0.061 061 61	96	0.268 998 70	146	0.481 884 59
47	0.064 546 79	97	0.237 328 07	147	0.486 069 85
48	0.068 096 90	98	0.277 655 16	148	0.490 252 74
49	0.071 708 35	99	0.281 979 88	149	0.494 433 19
50	0.075 377 56	100	0.286 302 01	150	0.498 611 35

TABLE 3 (*continued*)   Values of W_{CCT-68} (T_{68}), according to the equation given in Table 2, at integral values of T_{68}

T_{68} K	W_{CCT-68} (T_{68})	T_{68} K	W_{CCT-68} (T_{68})	T_{68} K	W_{CCT-68} (T_{68})
150	0.498 611 35	200	0.704 966 94	250	0.907 383 09
151	0.502 787 07	201	0.709 050 04	251	0.911 397 53
152	0.506 960 58	202	0.713 131 61	252	0.915 410 74
153	0.511 131 72	203	0.717 211 74	253	0.919 422 74
154	0.515 300 65	204	0.721 290 26	254	0.923 433 43
155	0.519 467 37	205	0.725 367 33	255	0.927 442 83
156	0.523 631 80	206	0.729 442 88	256	0.931 451 01
157	0.527 794 09	207	0.733 516 90	257	0.935 458 05
158	0.531 954 17	208	0.737 589 47	258	0.939 463 71
159	0.536 112 11	209	0741 660 59	259	0.943 468 22
160	0.540 267 92	210	0.745 730 26	260	0.947 471 52
161	0.544 421 67	211	0.749 798 41	261	0.951 473 52
162	0.548 573 36	212	0.753 865 18	262	0.955 474 30
163	0.552 722 91	213	0.757 930 43	263	0.959 473 85
164	0.556 870 48	214	0.761 994 30	264	0.963 472 19
165	0.561 016 06	215	0.766 056 72	265	0.967 469 31
166	0.565 159 58	216	0.770 117 70	266	0.971 465 13
167	0.569 301 12	217	0.774 177 30	267	0.975 459 80
168	0.573 440 76	218	0.778 235 45	268	0.979 453 25
169	0.577 578 48	219	0.782 292 23	269	0.983 445 41
170	0.581 714 23	220	0.786 347 56	270	0.987 436 42
171	0.585 848 06	221	0.790 401 51	271	0.991 426 14
172	0.589 979 99	222	0.794 454 09	272	0.995 414 71
173	0.594 110 08	223	0.798 505 23	273	0.999 401 99
174	0.598 238 35	224	0.802 555 06		
175	0.602 364 78	225	0.806 603 52		
176	0.606 489 31	226	0.810 650 54		
177	0.610 612 08	227	0.814 696 25		
178	0.614 733 10	228	0.818 740 59		
179	0.618 852 29	229	0.822 783 64		
180	0.622 969 72	230	0.826 825 31		
181	0.627 085 40	231	0.830 865 61		
182	0.631 199 39	232	0.834 904 61		
183	0.635 311 64	233	0.838 942 24		
184	0.639 422 13	234	0.842 978 57		
185	0.643 530 94	235	0.847 013 53		
186	0.647 638 07	236	0.851 047 26		
187	0.651 743 52	237	0.855 079 63		
188	0.655 847 30	238	0.859 110 69		
189	0.659 949 47	239	0.863 140 46		
190	0.664 049 96	240	0.867 168 94		
191	0.668 148 86	241	0.871 196 11		
192	0.672 246 07	242	0.875 221 99		
193	0.676 341 76	243	0.879 246 57		
194	0.680 435 77	244	0.883 269 94		
195	0.684 528 25	245	0.887 292 00		
196	0.688 619 13	246	0.891 312 69		
197	0.692 708 41	247	0.895 332 24		
198	0.696 796 17	248	0.899 350 49		
199	0.700 882 32	249	0.903 367 44		
200	0.704 966 94	250	0.907 383 09		

TABLE 4 Values of W_{CCT-68} (T_{68}), according to the data given in Table 2, at the fixed-point temperatures

Fixed point	T_{68} (K)	t_{68} (°C)	W_{CCT-68}
e-H$_2$ triple	13.81	− 259.34	0.001 412 06
e-H$_2$ 17.042	17.042	− 256.108	0.002 534 44
e-H$_2$ boiling	20.28	− 252.87	0.004 485 17
Ne boiling	27.102	− 246.048	0.012 212 72
O$_2$ triple	54.361	− 218.789	0.091 972 52
O$_2$ boiling	90.188	− 182.962	0.243 799 09
	273.15	0	1
H$_2$O boiling	373.15	100	1.392 596 68
Sn freezing	505.1181	231.9681	1.892 570 86

From 90.188K to 273.15K the deviation function is

$$\Delta W(T_{68}) = A_4 t_{68} + C_4 t_{68}^3 (t_{68} - 100°C) \tag{8}$$

where $t_{68} = T_{68} - 273.15$K and the constants are determined by the measured deviations at the boiling point of oxygen and the boiling point of water.[4]

b) The Range from 0°C (273.15K) to 630.74°C

From 0°C to 630.74°C t_{68} is defined by

$$t_{68} = t' + 0.045 \left(\frac{t'}{100°C}\right)\left(\frac{t'}{100°C} - 1\right)\left(\frac{t'}{419.58°C} - 1\right)\left(\frac{t'}{630.74°C} - 1\right)°C \tag{9}$$

where t' is defined by the equation:

$$t' = \frac{1}{\alpha} [W(t') - 1] + \delta \left(\frac{t'}{100°C}\right)\left(\frac{t'}{100°C} - 1\right) \tag{10a}$$

where $W(t') = R(t')/R(0°C)$. The constants $R(0°C)$, α and δ are determined by measurement of the resistance at the triple point of water, the boiling point of water (or the freezing point of tin, see Note[b], Table 1) and the freezing point of zinc.

Equation (10a) is equivalent to the equation

$$W(t') = 1 + At' + Bt'^2 \tag{10b}$$

where $A = \alpha(1 + \delta/100°C)$ and $B = - 10^{-4} \alpha \delta°C^{-2}$.

___

[4] If the freezing point of tin (see Note[b], Table 1) is used as a fixed point instead of the boiling point of water, $W(100°C)$ for the platinum thermometer should be calculated from Eqs. (9) and (10).

c) The Range from 630.74°C to 1064.43°C

From 630.74°C to 1064.43°C t_{68} is defined by the equation

$$E(t_{68}) = a + bt_{68} + ct_{68}^2 \qquad (11)$$

where $E(t_{68})$ is the electromotive force of a standard thermocouple of rhodium-platinum alloy and platinum, when one junction is at the temperature $t_{68} = 0°C$ and the other is at temperature t_{68}. The constants a, b and c are calculated from the values of E at 630.74°C ± 0.2°C, as determined by a platinum resistance thermometer, and at the freezing points of silver and gold.

The wires of the standard thermocouple shall be annealed and the purity of the platinum wire shall be such that the ratio $W(100°C)$ is not less than 1.3920. The rhodium-platinum wire shall contain nominally 10% rhodium and 90% platinum by weight. The thermocouple shall be such that the electromotive forces $E(630.74°C)$, $E(Ag)$ and $E(Au)$ shall satisfy the following relations:

$$E(Au) = 10\ 300\ \mu V \pm 50\ \mu V \qquad (12)$$

$$E(Au) - E(Ag) = 1183\ \mu V + 0.158\ [E(Au) - 10\ 300\ \mu V] \pm 4\ \mu V \qquad (13)$$

$$E(Au) - E(630.74°C) = 4766\ \mu V + 0.631\ [E(Au) - 10\ 300\ \mu V] \pm 8\ \mu V \qquad (14)$$

d) The Range above 1337.58K (1046.43° C)

Above 1337.58K (1064.43°C) the temperature T_{68} is defined by the equation

$$\frac{L_\lambda(T_{68})}{L_\lambda[T_{68}(Au)]} = \frac{\exp\ [c_2/\lambda T_{68}(Au)] - 1}{\exp\ (c_2/\lambda T_{68}) - 1} \qquad (15)$$

in which $L_\lambda(T_{68})$ and $L_\lambda[T_{68}(Au)]$ are the spectral concentrations at temperature T_{68} and at the freezing point of gold, $T_{68}(Au)$, of the radiance of a blackbody at the wavelength λ^5; $c_2 = 0.014\ 388$ metre kelvin.

III. SUPPLEMENTARY INFORMATION

The apparatus, methods and procedures described in this section represent good practice at the present time.

1. Standard Resistance Thermometer

A standard platinum resistance thermometer should be so designed and constructed that the four-terminal resistance element is as free as possible from

---

[5] Since $T_{68}(Au)$ is close to the thermodynamic temperature of the freezing point of gold and c_2 is close to the second radiation constant of the Planck equation, it is not necessary to specify the value of the wavelength to be employed in the measurements [see Metrologia 3, 28 (1967)].

strain and will remain so in use. Satisfactory resistors have been made with platinum wires of uniform diameter between 0.05 and 0.5 mm and with at least a short portion of each lead adjacent to the resistor also made of platinum. A commonly used value of $R(0°C)$ is ≈ 25 ohms and the measuring current for such a thermometer is normally 1 or 2 milliamperes. All thermometer components in close proximity to the resistor must be clean and non-reactive with platinum. During fabrication it is recommended that the thermometer be evacuated while at about 450°C and then filled with dry gas and hermetically sealed. It is desirable to have oxygen present in the gas filling to ensure that trace impurities in the platinum will remain in an oxidised state. After completion, the resistance element should be stabilized by heating at a temperature higher than its intended maximum operating temperature and in any case not lower than 450°C.

The insulation resistance of the components supporting the resistance element and leads must be high enough to avoid significant shunting of the element. For example, care must be taken to avoid condensation of water vapour between the leads at low temperatures, and intrinsic leakage in the insulators themselves at high temperatures. The insulators are usually fabricated from mica, silica or alumina, and these materials normally give adequate intrinsic insulation up to 500°C. However, as the temperature approaches 630°C, the problem becomes more critical and errors of 1 mK or greater may easily occur. In the case of mica, there is the additional difficulty that significant amounts of water may be released during its exposure to temperatures above 450°C, and unless this moisture is removed by periodic pumping or by a desiccant the insulation resistance will deteriorate rapidly.

To ensure adequate stability in the resistance and the temperature coefficients of resistivity, the resistor of a standard platinum resistance thermometer should be maintained, as far as possible, in an annealed state. Added resistivity may arise both from the accidental cold working that results from normal thermometer handling and also as a result of rapid cooling when a thermometer is transferred rapidly from an environment above 500°C to room temperature. This latter increase in resistance is due to quenched-in, non-equilibrium concentrations of vacancy defects and is retained as long as the thermometer remains below 200°C. Much of the cold work and all of the quenched-in resistance may be removed by annealing at 500°C for 30 min.

Significant errors can be caused by radiation loss from the thermometer by total reflection in the constructional components, particularly if these are of silica. Such loss in the sheath, but not in the internal components, can be suppressed by blackening the outer surface of the sheath (e.g., with a colloidal graphite suspension) or by sand-blasting the surface to produce a matt finish.

The completed thermometer should be tested to establish that the depth of immersion is sufficient to avoid heat conduction errors. An effective way of doing this is to confirm that the apparent temperature gradient in a metal freezing point is in agreement with that to be expected from hydrostatic effects (see Table 5).

For temperatures below 90K it is usual to use a small platinum resistance thermometer, generally not larger than 5 mm in diameter and 60 mm in length, that can be totally immersed in a uniform temperature zone with heat conduction down the leads being suppressed by attaching them to a suitable guard ring. In order to achieve good thermal contact between the resistor and its surroundings the resistor is contained in a thin sheath, commonly of platinum about 0.25 mm thick, which is filled with helium.

A useful criterion by which the efficiency of the annealing and the reliability of the thermometer may be judged is the constancy of its resistance at a reference temperature. The temperatures of the triple point of water (273.16K) and the boiling point of helium (4.215K) are commonly used for this purpose. The first of these is convenient for most high temperature thermometers, while the second is not only often conveniently attained for thermometers built into cryogenic apparatus but has the additional advantage that the resistance is relatively insensitive to temperature variations. In practice it is found that variations of resistance at the triple point of water for commercially produced high temperature thermometers should not exceed $4 \times 10^{-6} R(0°C)$ (equivalent to ≈ 1 mK above 40K), and will not exceed $5 \times 10^{-7} R(0°C)$ over a reasonable period of use for the very best thermometers when these are handled with extreme care. For resistance thermometers used only at temperatures of 100°C or less, variations should not exceed $5 \times 10^{-7} R(0°C)$.

The small temperature rise of the thermometers caused by the measuring current may be determined by measurements at two currents.

2. Standard Thermocouple

Satisfactory standard thermocouples have been made of wires of a uniform diameter between 0.35 and 0.65 mm. The thermocouple wires must be thoroughly annealed in order to ensure constancy of e.m.f. in use. For this purpose it

TABLE 5   Effect of pressure on the freezing-point temperatures of metals

Metal	Freezing point at 1 standard atmosphere (°C)	Pressure coefficient	
		Kelvins per atmosphere	Kelvins per centimetre of liquid
Mercury	− 38.862	+ 0.005 4	+ 0.000 071
Indium	156.634	+ 0.004 9	+ 0.000 033
Tin	231.9681	+ 0.003 3	+ 0.000 022
Bismuth	271.442	− 0.003 5	− 0.000 034
Cadmium	321.108	+ 0.006 2	+ 0.000 048
Lead	327.502	+ 0.008 0	+ 0.000 082
Zinc	419.58	+ 0.004 3	+ 0.000 027
Antimony	630.74	+ 0.000 85	+ 0.000 005

is necessary to heat the platinum wire to a temperature of at least 1100°C and the platinum-rhodium wire to 1450°C. If the annealing is done before the wires have been mounted in their insulators the completed thermocouple must be heated again to a temperature of at least 1100°C until its electromotive force has been stabilized and local inhomogeneities caused by strain have been removed. When this has been satisfactorily accomplished the thermocouple e.m.f. should not be changed by changes in the temperature gradients along the wires; it should not change, for example, with increase of depth of immersion in an enclosure at a uniform temperature.

3. Pressure

In practice pressures are usually determined by means of a mercury column. The mean density of pure mercury at the temperature t_{68} in a barometric column supported by the pressure p being measured is given, with sufficient accuracy over the temperature range from 0°C to 40°C and for the pressures relevant to these measurements, by the relation

$$\varrho\left(t_{68}\ \frac{p}{2}\right) = \frac{\varrho(20°C_{p_0})}{[1 + A(t_{68} - 20°C) + B(t_{68} - 20°C)^2] \times [1 - \chi(p/2)p_0)]} \tag{16}$$

where
$$A = 18\ 115 \times 10^{-8}°C^{-1}$$
$$B = 0.8 \times 10^{-8}°C^{-2}$$
$$\chi = 4 \times 10^{-11}\ N^{-1}\ m^2$$

$\varrho(20°C, p_0) = 13\ 545.87$ kg/m³ is the density of pure mercury at $t_{68} = 20°C$ under a pressure $p_0 = 1$ standard atmosphere (101 325 N/m²)

A sufficiently accurate value of the local gravity may be obtained by using the Potsdam system and applying a correction of -14×10^{-5} m/s² $(-14$ milligals).[6]

Hydrostatic head pressures within the fixed points cells cause small but significant temperature effects; these are summarized in Table 5.

4. Triple Point of Water

The temperature of the triple point of water can be realized in sealed glass cells containing only water of high purity and of substantially the isotopic composition of ocean water. The cells have an axial well for the thermometers and the

---

[6] By Resolution 1 (1968), the International Committee of Weights and Measures decided that for metrological purposes, the value of the acceleration due to gravity at Potsdam, the reference point of this System, should be taken as 9.812 60 m/s² instead of 9.812 74 m/s², the value adopted initially.

triple-point temperature is obtained wherever the ice is in equilibrium with a liquid-vapour surface. At a depth h below the liquid-vapour surface, the equilibrium temperature t_{68} between ice and liquid water is given by

$$t_{68} = A + Bh \qquad (17)$$

where $A = 0.01°C$ and $B = -7 \times 10^{-4}$ m$^{-1}|°C$. The method recommended for preparing a triple-point cell consists of forming a thick layer of ice around the thermometer well by cooling from within, then melting enough of this sheath, also from within, to produce a new water-ice interface adjacent to the well. During the first hours following the preparation of the cell the temperature measured in the thermometer well rises fairly rapidly by a few ten thousandths of a kelvin becoming stable after from 1 to 3 days. This initial change of temperature is probably caused by the growth of the ice crystals or by the slow disappearance of strain in the crystals. A cell prepared in this way and kept in an ice bath is capable of maintaining a temperature constant to about 0.0001K for several months. Even with cells from various sources, when used in this way, the differences in the temperatures obtained should in any case not exceed 0.0002K. A significant rise in temperature of the thermometer above the triple-point temperature may be caused by artificial light or sunlight falling on the ice-covered cell and it is therefore recommended that measurements should be made with the cell suitably shielded from radiation.

Variations in the isotopic content of naturally occurring water are such that they will result in detectable differences in the triple-point temperature. Ocean water contains about 0.016 moles of deuterium, ^{2}H, per 100 moles of hydrogen, ^{1}H, and 0.04 moles of ^{17}O and 0.2 moles of ^{18}O per 100 moles of ^{16}O. This proportion of heavy isotopes is substantially the highest to be found in naturally occurring water. Continental surface water normally contains about 0.015 moles of ^{2}H per 100 moles of ^{1}H; water coming from polar snow may occasionally contain as little as 0.01 moles of ^{2}H per 100 moles of ^{1}H.

The operation of purifying the water may slightly modify its isotopic composition and the isotopic composition at an ice-water interface is slightly dependent on the freezing technique.

An increase of 0.001 moles of ^{2}H per 100 moles of ^{1}H corresponds to an increase of temperature of the triple point of 0.000 04K; this is the difference between the triple points for ocean water and the normally occurring continental surface water. The extreme difference in the triple-point temperatures of naturally occurring water is 0.000 25K.

5. Triple Point, 17.042K Point and Boiling Point of Equilibrium Hydrogen

Hydrogen has two molecular modifications, designated by the prefixes *ortho* and *para*, caused by different relative orientations of the two nuclear spins in the diatomic molecules. The equilibrium *ortho para* composition is temperature dependent and at room temperature is about 75% orthohydrogen and 25% parahydrogen (so-called "normal hydrogen"). On liquefaction the composition changes slowly with time and there are corresponding changes in the physical properties. At the boiling point the equilibrium composition is 0.21% ortho- and 99.79% parahydrogen and the temperature is lower than that of normal hydrogen by about 0.12K. The name equilibrium hydrogen means in this document that the hydrogen has its equilibrium *ortho-para* composition at the relevant temperature. In order to avoid errors in the realization of these fixed points caused by indeterminate composition it is advisable to use equilibrium hydrogen converted by the use of a catalyst such as ferric hydroxide. Hydrogen of high chemical purity should be used as may be obtained by diffusion through palladium.

The temperature of equilibrium between solid, liquid and vapour phases of hydrogen can be realized by using a sufficient quantity of liquid hydrogen with some catalyst in a cavity in a copper block in which platinum resistance thermometers are immersed and which is surrounded by a vacuum space. The temperature of the block is reduced until the hydrogen is solidified and then the temperature is allowed to rise slowly and the transition at the triple point is observed. The flat portion of the time-temperature curve can be constant to 0.0001K for 30 min or more.

The temperature of equilibrium between liquid and gaseous hydrogen is normally realized by the static method. In this method a cavity in a block of metal of high thermal conductivity is maintained at a temperature close to the boiling point by immersing it in liquid hydrogen. In order to avoid temperature gradients due to hydrostatic pressure the liquid hydrogen makes contact with the top of the block only, the lower part being shielded by a vacuum jacket. The cavity contains a small quantity of very pure liquid hydrogen together with some catalyst. The vapour pressure of this hydrogen is transmitted by a thin tube of low heat conduction connected to a manometer outside the enclosure. Precautions must be taken to avoid direct radiation down this tube into the cavity and to ensure that the tube at every point is at a higher temperature than the temperature at the surface of the liquid hydrogen in the cavity. Comparisons are made between the vapour-pressure thermometer thus formed and platinum resistance thermometers immersed in closely fitting holes in the metal block adjacent to the cavity.

The validity of the measurements may be checked by showing that the values obtained are independent of the ratio of the volume of the liquid hydrogen to the volume of the vapour in the cavity.

The temperature T_{68} as a function of the vapour pressure of equilibrium hydrogen is given to an accuracy of a few millikelvins for the range from 13.81K to 23K by the equation

$$\lg \frac{p}{p_0} = A + \frac{B}{T_{68}} + CT_{68} + DT_{68}^2 \tag{18}$$

where $A = 1.711\ 466$
$\qquad B = -44.010\ 46K$
$\qquad C = 0.023\ 590\ 9K^{-1}$
$\qquad D = -0.000\ 048\ 017K^{-2}$
$\qquad p_0 = 101\ 325\ N/m^2$

6. Boiling Point of Neon

The boiling point of neon can be realized in a manner similar to that described for hydrogen. The normal isotopic composition of neon is 0.0026 moles of ^{21}Ne and 0.088 moles of ^{22}Ne per 0.909 moles of ^{20}Ne.

The temperature T_{68} as a function of the vapour pressure of neon is given to an accuracy of $\pm$ 0.0002K for the range from 27K to 27.2K by the equation

$$T_{68} = \left[27.102 + 3.3144\left(\frac{p}{p_0} - 1\right) - 1.24\left(\frac{p}{p_0} - 1\right)^2 + 0.74\left(\frac{p}{p_0} - 1\right)^3\right]K \tag{19}$$

7. Triple Point and Boiling Point of Oxygen

The triple point and the boiling point of oxygen can be realized in a manner similar to that described for hydrogen. Particular care must be given to the purity of the oxygen in the vapour-pressure thermometer. The oxygen is sufficiently pure when the normal boiling point remains constant with the removal of successive fractions of vapor. The temperature T_{68} as a function of the vapour pressure of oxygen is given to an accuracy of $\pm$ 0.0001K for the range from 90.1K to 90.3K by the equation

$$T_{68} = \left[90.188 + 9.5648\left(\frac{p}{p_0} - 1\right) - 3.69\left(\frac{p}{p_0} - 1\right)^2\right] + 2.22\left(\frac{p}{p_0} - 1\right)^3\right]K \tag{20}$$

8. Boiling Point of Water

The temperature of equilibrium between liquid water and its vapour is usually realized by the dynamic method with the thermometer in the saturated vapour. For precise calibration it is preferable to use closed systems in which the boiler and manometer are connected to a manostat filled with air or preferably helium.

The boiler must be constructed in such a way that contamination of the water is avoided. The thermometer must be protected from radiation emitted by bodies which are at temperatures different from the boiling-point temperature. If the equilibrium temperature has been obtained the observed temperature (reduced to a constant pressure) will be independent of the time elapsed, variations in the rate of heat supplied to the liquid, and the depth of immersion of the thermometer.

A change in the proportion of deuterium in the water produces a change in the boiling point of water in the same direction as in the triple point but to about one third the extent.

The temperature t_{68} as a function of the vapour pressure of water is given, to an accuracy of ± 0.0001K for the range from 99.9°C to 100.1°C by the equation

$$t_{68} = \left[100 + 28.0216\left(\frac{p}{p_0} - 1\right) - 11.624\left(\frac{p}{p_0} - 1\right)^2 + 7.1\left(\frac{p}{p_0} - 1\right)^3 \right]°C$$

(21)

9. Freezing Points of Tin and Zinc

Very reproducible temperatures can be realized by observing the flat part of the temperature versus time curve obtained during the slow freezing of very pure metals.

The melting and freezing of tin and zinc may be carried out in a crucible of very pure artificial graphite (99.999% by weight) about 5 cm in diameter provided with an axial thermometer well. The depth of immersion of the thermometer in the metal must be sufficient to eliminate the influence of thermal conduction along the thermometer leads on the temperature of the sensitive element. It is convenient to contain the crucible and ingot of metal in a pyrex or silica tube under an inert atmosphere and to heat it in a metal-block furnace.

The aim of the cooling technique in the determination of the freezing point is to ensure that the thermometer sensor is as nearly as possible enclosed by, and is in thermal equilibrium with, a solid-liquid interface: shortly after nucleation there should be present either a completed solid shell nucleated on and thickening from the crucible wall or a completed solid mantle induced on the thermometer well.

The equilibrium temperature between solid and liquid metal varies slightly with pressure. The magnitudes of these variations are given in Table 5.

High purity tin (99.9999% by weight) supercools by 20K to 30K before solidification when cooled from the liquid state. The following technique has been successful for nucleating an ingot of tin while avoiding excessive undercooling of the freezing-point furnace. Starting with the temperature a few kelvins above the freezing point, the furnace is slowly cooled at about 0.1 kelvin per minute until the melt reaches its liquidus-point temperature; the sample holder containing the melt and a monitoring resistance thermometer is then either withdrawn into the throat of the furnace or removed from the furnace entirely. Under either of these conditions the sample cools rapidly; when the rapid rise in temperature that indicates general nucleation is detected the sample holder is quickly replaced in the furnace which is still cooling slowly. Thereafter, as solidification proceeds slowly, a characteristic cooling curve for a high purity metal is realized having a temperature plateau that is reproducible for a particular sample to ±0.0001K for durations dependent on the rate of furnace cooling.

High purity zinc (99.9999% by weight) is treated in a somewhat different manner since it does not supercool excessively. A thin layer of solid metal is produced on the central thermometer well by removing the thermometer when the melt reaches the liquidus temperature, and either cooling the thermometer to room temperature and replacing it, or inserting a silica rod in its place for about 30 s before replacing the thermometer.

A criterion for checking that the purity of a sample of zinc or of tin is satisfactory is that its melting range is not more than 0.001K.

10. Freezing Points of Silver and Gold

The equilibrium temperatures between the solid and liquid phases of silver and of gold may be realized in covered crucibles of either very pure artificial graphite, or ceramic material or vitreous silica. If graphite is used it is advisable to prevent the access of air to the crucible so that the graphite is not oxidised. Molten silver should be protected so as to prevent the solution of oxygen into it and the consequent depression of the freezing point.

The ingot of metal should be heated to a uniform temperature a few kelvins above the melting point of the metal and then cooled slowly. The thermocouple to be calibrated, mounted in a protecting tube of suitable refractory material, with refractory insulators separating the two wires, is immersed in the molten metal which is then allowed to freeze. The depth of immersion of the thermo-couple in the metal must be sufficient to eliminate the influence of thermal conduction along the wires of the thermocouple.

That the equilibrium temperature is obtained can be checked by the follow-ing criteria: the electromotive force of the thermocouple should be independent of small variations in the depth of immersion in the ingot of metal during successive freezes, and should remain constant for at least 5 min during one freeze.

A blackbody at the freezing point of gold is required for the establishment of the reference temperature for radiation temperature measurements. For the realization of such a blackbody the crucible containing the gold must be modi-fied to provide a uniform temperature enclosure immersed in the gold. A blackbody enclosure is more readily achieved if the material employed for its construction has already a high emissivity and therefore graphite is very suitable for this purpose.

11. Secondary Reference Points

As well as the defining fixed points of the IPTS-68 given in Table 1 other reference points are available. Some of these points and their temperatures on the IPTS-68 are given in Table 6. Except for the triple points and the vapour pressure-temperature equations each temperature is that for a system in equilib-rium under the pressure of one standard atmosphere.

TABLE 6 Secondary reference points

Equilibrium state	International Practical Temperature	
	T_{68} (K)	t_{68} (°C)
Equilibrium between the solid, liquid and vapour phases of normal hydrogen (triple point of normal hydrogen)	13.956	259.194
Equilibrium between the liquid and vapour phases of normal hydrogen (boiling point of normal hydrogen)	20.397	− 252.753

$$\lg \frac{p}{p_0} = A + \frac{B}{T_{68}} + CT_{68} + DT_{68}^2 \qquad (23)$$

$A = 1.734\ 791, \quad B = -\ 44.623\ 68\text{K}, \quad C = 0.023\ 186\ 9\text{K}^{-1},$
$D = -\ 0.000\ 048\ 017\text{K}^{-2}$

for the temperature range from 13.956K to 30K.

Equilibrium between the solid, liquid and vapour phases of neon (triple point of neon)	24.555	− 248.595
Equilibrium between the liquid and vapour phases of neon		

$$\lg \frac{p}{p_0} = A + \frac{B}{T_{68}} + CT_{68} + DT_{68}^2 \qquad (24)$$

$A = 4.611\ 52, \quad B = -\ 106.3851\text{K}, \quad C = -\ 0.036\ 833\ 1\text{K}^{-1},$
$D = 4.248\ 92 \times 10^{-4}\,\text{K}^{-2}$

for the temperature range from 24.555K to 40K.

Equilibrium between the solid, liquid and vapour phases of nitrogen (triple point of nitrogen)	63.148	− 210.002
Equilibrium between the liquid and vapour phases of nitrogen (boiling point of nitrogen)	77.348	− 195.802

$$\lg \frac{p}{p_0} = A + \frac{B}{T_{68}} + C \lg \frac{T_{68}}{T_0} + DT_{68} + ET_{68}^2 \qquad (25)$$

$A = 5.893\ 139, \quad B = -\ 404.131\ 05\text{K}, \quad C = -\ 2.3749,$
$D = -\ 0.014\ 250\ 5\text{K}^{-1}, \quad E = 72.5342 \times 10^{-6}\,\text{K}^{-2}$

for the temperature range from 63.148K to 84K.

Equilibrium between the liquid and vapour phases of oxygen

$$\lg \frac{p}{p_0} = A + \frac{B}{T_{68}} + C \lg - \frac{T_{68}}{T_0} + DT_{68} + ET_{68}^2 \qquad (26)$$

$A = 5.961\ 546, \quad B = -\ 467.455\ 76\text{K}, \quad C = -\ 1.664\ 512,$
$D = -\ 0.013\ 213\ 01\text{K}^{-1}, \quad E = 50.8041 \times 10^{-6}\,\text{K}^{-2}$

for the temperature range from 54.361K to 94K.

Equilibrium between the solid and vapour phases of carbon dioxide (sublimation point of carbon dioxide)	194.674	− 78.476

$$T_{68} = \left[194.674 + 12.264\left(\frac{p}{p_0} - 1\right) - 9.15\left(\frac{p}{p_0} - 1\right)^2 \right]\text{K} \qquad (27)$$

for the temperature range from 194K to 195K.

TABLE 6 (*continued*)   Secondary reference points

Equilibrium state	International Practical Temperature	
	T_{68} (K)	t_{68} (°C)
Equilibrium between the solid and liquid phases of mercury (freezing point of mercury)[a]	234.288	−38.862
Equilibrium between ice and air-saturated water (ice point)	273.15	0
Equilibrium between the solid, liquid and vapour phases of phenoxybenzene (diphenyl ether) (triple point of phenoxybenzene)	300.02	26.87
Equilibrium between the solid, liquid and vapour phases of benzoic acid (triple point of benzoic acid)	395.52	122.37
Equilibrium between the solid and liquid phases of indium (freezing point of indium)[a]	429.784	156.634
Equilibrium between the solid and liquid phases of bismuth (freezing point of bismuth)[a]	544.592	271.442
Equilibrium between the solid and liquid phases of cadmium (freezing point of cadmium)[a]	594.258	321.108
Equilibrium between the solid and liquid phases of lead (freezing point of lead)[a]	600.652	327.502
Equilibrium between the liquid and vapour phases of mercury (boiling point of mercury)	629.81	356.66

$$t_{68} = \left[356.66 + 55.552\left(\frac{p}{p_0} - 1\right) - 23.03\left(\frac{p}{p_0} - 1\right)^2 \right.$$
$$\left. + 14.0\left(\frac{p}{p_0} - 1\right)^3\right]°C \quad (28)$$

for $p = 90 \times 10^3$ N/m² to 104×10^3 N/m².

Equilibrium state		
Equilibrium between the liquid and vapour phases of sulphur (boiling point of sulphur)	717.824	444.674

$$t_{68} = 444.674 + 69.010\left(\frac{p}{p_0} - 1\right) - 27.48\left(\frac{p}{p_0} - 1\right)^2$$
$$+ 19.14\left(\frac{p}{p_0} - 1\right)^3\Big]°C \quad (29)$$

for $p = 90 \times 10^3$ N/m² to 104×10^3 N/m².

Equilibrium state		
Equilibrium between the solid and liquid phases of the copper-aluminum eutectic	821.38	548.23
Equilibrium between the solid and liquid phases of antimony (freezing point of antimony)[a]	903.89	630.74
Equilibrium between the solid and liquid phases of aluminium (freezing point of aluminium)	933.52	660.37
Equilibrium between the solid and liquid phases of copper (freezing point of copper)	1357.6	1084.5
Equilibrium between the solid and liquid phases of nickel (freezing point of nickel)	1728	1455

[a]See Table 5 for the effect of pressure variations on these freezing points.

TABLE 6 (*continued*)   Secondary reference points

	International Practical Temperature	
Equilibrium state	T_{68} (K)	t_{68} (°C)
Equilibrium between the solid and liquid phases of cobalt (freezing point of cobalt)	1767	1494
Equilibrium between the solid and liquid phases of palladium (freezing point of palladium)	1827	1554
Equilibrium between the solid and liquid phases of platinum (freezing point of platinum)	2045	1772
Equilibrium between the solid and liquid phases of rhodium (freezing point of rhodium)	2236	1963
Equilibrium between the solid and liquid phases of iridium (freezing point of iridium)	2720	2447
Equilibrium between the solid and liquid phases of tungsten (temperature of melting tungsten)	3660	3387

APPENDIX I. HISTORY OF THE DEVELOPMENT OF INTERNATIONAL TEMPERATURE SCALES; DIFFERENCES BETWEEN THE IPTS-68 AND THE IPTS-48

The International Temperature Scale was adopted in 1927 to overcome the practical difficulties of the direct realization of thermodynamic temperatures by gas thermometry and to unify the existing national temperature scales. It was introduced by the 7th General Conference of Weights and Measures with the intention of providing a practical scale of temperature which was easily and accurately reproducible and which gave as nearly as possible thermodynamic temperatures.

The International Temperature Scale was revised in 1948. The experimental procedures by which the Scale was realized remained substantially unchanged, but two amendments were made to the definition of the Scale resulting in appreciable changes in the numerical values assigned to measured temperatures. The change in the value of the temperature of the freezing point of silver from 960.5°C to 960.8°C changed temperatures measured with the standard thermocouple (range 630°C to 1063°C); the maximum difference was about 0.4K near 800°C. The adoption of the value of 0.014 38 metre kelvin instead of 0.014 32 metre kelvin for the radiation constant c_2 changed all temperatures above the freezing point of gold, while the use of the Planck radiation formula instead of the Wien formula affected the very high temperatures. Temperatures above the freezing point of gold were decreased, for example, by 2.2K at 1500°C and by 6K at 2000°C. Also at this revision, in order to secure international uniformity of nomenclature, the 9th General Conference of Weights and Measures decided to abandon the word "Centigrade" and its French equivalent "Centésimal" in favour of the name "Celsius." That is "°C" was now regarded as the abbreviation of "degree Celsius."

An amended edition of the 1948 Scale was adopted by the 11th General Conference of Weights and Measures under the new title "International Practical Temperature Scale of 1948 (amended edition of 1960)," the numerical values of temperature remaining the same as in 1948. The new edition incorporated the new definition of the kelvin given by defining the thermodynamic temperature of the triple point of water as exactly 273.16K (10th General Conference of Weights and Measures 1954, Resolution 3). It was also recognized at this time that the IPTS no longer represented the thermodynamic temperature as closely as possible and the text included a section on the differences between them. The IPTS-68 has been devised to bring these differences within the limits of accuracy to which the thermodynamic temperatures are known (see Table 7) and to extend the Scale to lower temperatures.

The IPTS-68 differs from the IPTS-48 in the following ways. The lower limit of the Scale is now 13.81K instead of 90.18K. The values assigned to the defining fixed points are modified where necessary to conform as nearly as possible to the thermodynamic temperatures, the only points remaining unchanged being the triple point of water, which is permanently fixed by definition, and the boiling point of water. The interpolation instruments remain the same as before but the standard platinum resistance thermometer is now required to have $W(100°C)$ at least equal to 1.3925 instead of 1.3920. Over the temperature range from 90.188K to 273.15K the Callendar-Van Dusen equation is no longer used for interpolation, instead the reference function W_{CCT-68} (T_{68}) is employed. Above 0°C, the Callendar equation is modified so that interpolated values of temperature conform more nearly with thermodynamic-temperature

TABLE 7   Estimated uncertainties of the assigned values of the defining fixed points in terms of thermodynamic temperatures

Defining fixed point	Assigned value	Estimated uncertainty (K)
Triple point of equilibrium hydrogen	13.81K	0.01
17.042K point	17.042K	0.01
Boiling point of equilibrium hydrogen	20.28K	0.01
Boiling point of neon	27.102K	0.01
Triple point of oxygen	54.361K	0.01
Boiling point of oxygen	90.188K	0.01
Triple point of water	273.16K	Exact by definition
Boiling point of water	100°C	0.005
Freezing point of tin	231.9681°C	0.015
Freezing point of zinc	419.58°C	0.03
Freezing point of silver	961.93°C	0.2
Freezing point of gold	1064.43°C	0.2

values. Finally the latest value of c_2, namely 0.014 388 metre kelvin is introduced in the Planck equation for determining temperatures above the freezing point of gold. The effect of all these changes is summarised in Table 8 which gives the differences between the values of temperature derived from the IPTS-68 and the IPTS-48.

In the range from 13.81K to 90.188K the IPTS-68 is based on the average of four "national scales" and on chosen "best" temperatures for the defining fixed points. These national scales are each defined in terms of platinum resistance thermometers calibrated against a gas thermometer and are highly reproducible.

The differences between the IPTS-68 and the national scales are published in Metrologia, Vol. 5, p. 47, (1969). This allows the use of the national scales and these differences to give a close approximation to the IPTS-68.

The text of this definition of the IPTS-68 conforms with the decision of the 13th General Conference of Weights and Measures to denote the unit of thermodynamic temperature by the name "kelvin," symbol K, and to denote a temperature interval by the same unit and symbol or by "degree Celsius" or "°C."

APPENDIX II. PRACTICAL SCALES OF TEMPERATURE FOR USE OVER THE RANGE FROM 0.2K TO 5.2K

Temperatures can be derived from measured vapour pressures of ^{4}He and ^{3}He. The upper limits for use are set by the critical points of the gases (5.2K for ^{4}He and 3.3K for ^{3}He) and the lower limit by the vapour pressures becoming too low for practical measurement. The "1958 ^{4}He scale" and the "1962 ^{3}He scale" are the recommended scales[7] in which temperatures are denoted by T_{68} and T_{62} respectively.

The "1958 ^{4}He scale," recommended in 1958 by the International Committee of Weights and Measures, is defined by a table of ^{4}He vapour pressures versus temperature (Comité Consultatif de Thermométrie, 5e session 1958, p. T 192 and Procès-Verbaux CIPM, 26-A, 1958, p. T 192).[8]

The "1962 ^{3}He scale," recommended by the International Committee of Weights and Measures in 1962, is defined by an equation giving the vapour pressure of ^{3}He as a function of temperature (Comité Consultatif de Thermométrie, 6e session 1962, p. 184).[9]

In the temperature range between 0.9K and the critical temperature of ^{3}He the temperatures T_{68} and T_{62} are believed to be in agreement to within 0.3 mK.

[7] Recent measurements by the acoustic thermometer give temperatures higher than those of the He vapour pressure scales; the difference at the boiling point of ^{4}He is about 0.008K.

[8] An expanded form of this table together with auxiliary information is given in the J. Res. Nat. Bur. Standards 64A.1 (1960).

[9] A table of values and information on measuring vapour pressures are given in the J. Res. Nat. Bur. Standards 68 A, 547, 559, 567, 579 (1964).

TABLE 8 Approximate differences $(t_{68} - t_{48})$, in kelvins, between the values of temperature given by the IPTS of 1968 and the IPTS of 1948

t_{68} °C	0	−10	−20	−30	−40	−50	−60	−70	−80	−90	−100
− 100	0.022	0.013	0.003	−0.006	−0.013	−0.013	−0.005	0.007	0.012	0.029	0.022
− 0	0.000	0.006	0.012	0.018	0.024	0.029	0.032	0.034	0.033		

t_{68} °C	0	10	20	30	40	50	60	70	80	90	100
0	0.000	−0.004	−0.007	−0.009	−0.10	−0.010	−0.010	−0.008	−0.006	−0.003	0.000
100	0.043	0.004	0.007	0.012	0.016	0.020	0.025	0.029	0.034	0.038	0.043
200	0.073	0.047	0.051	0.054	0.058	0.061	0.064	0.067	0.069	0.071	0.073
300	0.076	0.074	0.075	0.076	0.077	0.077	0.077	0.077	0.077	0.076	0.076
400	0.079	0.075	0.075	0.075	0.074	0.074	0.074	0.075	0.076	0.077	0.079
500	0.150	0.082	0.085	0.089	0.094	0.100	0.108	0.116	0.126	0.137	0.150
600	0.39	0.165	0.182	0.200	0.23	0.25	0.28	0.31	0.34	0.36	0.39
700	0.67	0.42	0.45	0.47	0.50	0.53	0.56	0.58	0.61	0.64	0.67
800	0.95	0.70	0.72	0.75	0.78	0.81	0.84	0.87	0.89	0.92	0.95
900	1.24	0.98	1.01	1.04	1.07	1.10	1.12	1.15	1.18	1.21	1.24
1000		1.27	1.30	1.33	1.36	1.39	1.42	1.44			

t_{68} °C	0	100	200	300	400	500	600	700	800	900	1000
1000	0	1.5	1.7	1.8	2.0	2.2	2.4	2.6	2.8	3.0	3.2
2000	3.2	3.5	3.7	4.0	4.2	4.5	4.8	5.0	5.3	5.6	5.9
3000	5.9	6.2	6.5	6.9	7.2	7.5	7.9	8.2	8.6	9.0	9.3

AUTHOR INDEX

SUBJECT INDEX

IN THE DOG AND MUFF, TONYPANDY.

HERE'S EMILY, LIAM, MICHAEL AND KURT, REGULARS AT THE DOG...

LIKE I WAS SAYING, I'VE ALWAYS FANCIED A FULL ARM TATTOO...

OH YEAH?

I FOUND THIS LOT OUTSIDE BECOMING LUNCH TO A GIANT BABY. THAT'S WHAT COMES OF BEING TOO EMOTIONAL.

IT'S ALL A QUESTION OF SCALE. IT WOULD COVER MY WHOLE ARM, SO IT WOULD BE A FULL ARM TATTOO, BUT IT WOULD BE CHEAP BECAUSE...

...I'VE GOT REALLY SKINNY ARMS.

THAT'S RIGHT. THAT'S RIGHT. IT'S ALL A QUESTION OF WHERE YOU LOOK...

LITTLE DOES KURT KNOW, BUT HIS CANNY REALISATION LEADS US RIGHT INTO ART IDEA NUMBER TWO...

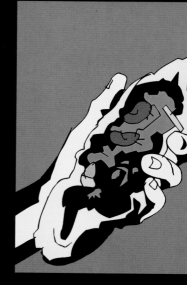

IF LIFE IS JUST A DRAG, THEN WHO'S TRYING TO BE WHO?

WOMEN AND MEN

CREATIVE TYPES OF ALL SORTS TAKE A LOT OF NOTICE OF THINGS THAT MAKE THE WORLD WHAT IT IS - PERHAPS NOTHING MORE SO THAN THE DIFFERENCES BETWEEN WOMEN AND MEN.

THESE DIFFERENCES AREN'T JUST A QUESTION OF BIOLOGY OR THE DIFFERENT PLACES THAT WOMEN AND MEN SEEM TO TAKE IN LIFE.

PERHAPS MORE IMPORTANT THAN THESE DIFFERENCES FOR ARTISTS IS THE IDEA THAT WOMEN AND MEN EXPERIENCE THINGS COMPLETELY DIFFERENTLY - THAT THE EXPERIENCE OF BEING A WOMAN IS COMPLETELY DIFFERENT TO THE EXPERIENCE OF BEING A MAN.

...SO THERE ARE TWO UNDERLYING TYPES OF DRAG IN THE WORLD - THE DRAG OF BEING A MAN AND THE DRAG OF BEING A WOMAN. ONWARD, BOYS AND GIRLS...*

* FOR MORE FUNKY STUFF ABOUT PEOPLE WHO'VE THOUGHT ABOUT WOMEN AND MEN, LOOK AT: www.wikipedia.org/wiki/feminism AND www.wikipedia.org/wiki/Simone_De_Beauvoir AND www.wikipedia.org/wiki/Andrea_Dworkin

WHILE DAI'S MATES ARE PONDERING INDIVIDUALITY, LET'S TAKE A LOOK AT WHAT DISTINGUISHES PEOPLE FROM EACH OTHER...

ART iDEA NUMBER FOUR

AUDIENCE AND EXPERIENCE

THINKING BACK TO THE IDEAL THAT ARTISTS, MUSICIANS AND THAT HAVE OF 'THE GOOD READER', THIS IDEA IS SORT OF THE FLIP-SIDE.

IF ARTISTS WANT THEIR WORK TO BE COMPLETELY UNDERSTOOD, THEY ACTUALLY KNOW THAT IT DOESN'T HAPPEN VERY OFTEN. THE REASON FOR THIS IS THAT PEOPLE'S EXPERIENCES DIFFER WIDELY.

YOU CAN'T EXPECT SOMEONE AGED 18 TO HAVE EXPERIENCED GROWING UP IN THE 1960S, OR SOMEONE WHO ONLY SPEAKS ENGLISH TO KNOW MUCH JAPANESE LITERATURE. FAIR ENOUGH. SO ARTISTS OFTEN HAVE TO THINK ABOUT WHO IS GOING TO EXPERIENCE THEIR ART, IF IT'S NOT GOING TO BE 'THE GOOD READER'.★

NOW LET'S SEE WHAT HAPPENS NEXT...

AS THE TEXTS, MAILS AND IMAGES FLOOD IN, IT MIGHT BE A GOOD MOMENT TO GET STUCK INTO ART IDEA NUMBER FIVE;

ART IDEA NUMBER FIVE

AUTHENTICITY

ARTISTS FREQUENTLY TRY TO TEST OUT IN THEIR WORK WHAT THEY THINK IS TRUE, PARTICULARLY WHEN IT COMES TO PHOTOGRAPHS, FILMS AND THE LIKE THAT ARE INTENDED TO SHOW THE TRUTH.

FOR INSTANCE, IS AN ELEPHANT THAT HAS ONLY BEEN SEEN ON TELEVISION AS MUCH OF AN ELEPHANT AS ONE THAT YOU'VE SEEN WITH YOUR OWN EYES? IS IT AN ELEPHANT AT ALL?

INCREASINGLY, PEOPLE EXPERIENCE A GROWING WORLD SECOND HAND AS MUCH AS THEY EXPERIENCE IT FIRST HAND. IN SOME WAYS, THIS IS PART OF THE WAY IN WHICH ANY PICTURE OF SOMETHING WORKS.

TAKE A LOOK AT THIS PICTURE OF A PIPE BY BELGIAN ARTIST RENE MAGRITTE. UNDERNEATH HE HAS WRITTEN IN FRENCH 'THIS IS NOT A PIPE'. NO CONTRADICTION THERE - THIS IS A PICTURE OF A PIPE, NOTHING MORE.*

Ceci n'est pas une pipe.

IS THERE AN ELEPHANT IN THE CREW'S FRONT ROOM?..